AF902712

RAILWAY NOISE AND VIBRATION:

MECHANISMS, MODELLING AND MEANS OF CONTROL

RAILWAY NOISE AND VIBRATION: MECHANISMS, MODELLING AND MEANS OF CONTROL

DAVID THOMPSON

Institute of Sound and Vibration Research
University of Southampton
Southampton, UK

with contributions from
CHRIS JONES and PIERRE-ETIENNE GAUTIER

ELSEVIER

AMSTERDAM · BOSTON · HEIDELBERG · LONDON · NEW YORK · OXFORD
PARIS · SAN DIEGO · SAN FRANCISCO · SYDNEY · TOKYO

Elsevier

The Boulevard, Langford Lane, Kidlington,Oxford, OX5 1GB
Radarweg 29, PO Box 211, 1000 AE Amsterdam, The Netherlands

First edition 2009
Reprinted 2010
Copyright © 2009 Elsevier Ltd. All rights reserved

No part of this publication may be reproduced, stored in a retrieval system or transmitted in any form or by any means electronic, mechanical, photocopying, recording or otherwise without the prior written permission of the publisher

Permissions may be sought directly from Elsevier's Science & Technology Rights Department in Oxford, UK: phone (+44) (0) 1865 843830; fax (+44) (0) 1865 853333; email: permissions@elsevier.com. Alternatively you can submit your request online by visiting the Elsevier web site at http://elsevier.com/locate/permissions, and selecting Obtaining permission to use Elsevier material

Notice
No responsibility is assumed by the publisher for any injury and/or damage to persons or property as a matter of products liability, negligence or otherwise or from any use or operation of any methods, products, instructions or ideas contained in the material herein

British Library Cataloguing in Publication Data
Thompson, David
Railway noise and vibration: mechanisms, modelling and means
1. Railroad tracks - Noise 2. Railroad tracks - Vibration
3. Railroad trains - Noise 4. Railroad trains - Vibration
5. Noise control 6. Damping (Mechanics)
I. Title
625.1′4

Library of Congress Cataloguing in Publication Data
A catalogue record for this book is available from the Library of Congress

Library of Congress Control Number: 2008934415

ISBN-13: 978-0-08-045147-3

Printed and bound in Great Britain

10 11 12 13 14 8 7 6 5 4 3 2

Working together to grow
libraries in developing countries

www.elsevier.com | www.bookaid.org | www.sabre.org

ELSEVIER BOOK AID International Sabre Foundation

Contents

Preface

Speak through the earthquake, wind and fire,
O still small voice of calm!

J.G. Whittier (1807–1892)

When our children were young we took them to the railway station to see off some visitors. They were excited at the prospect of seeing the trains. But, while we were waiting, an express train thundered through the station and it was all too much for their sensitive ears. 'We don't like trains now' they sobbed. 'Don't worry', I said, 'it's my job to make trains quieter' and that seemed to reassure them. But then they wanted to know: 'How *do* you make trains quieter?' Well, perhaps this book gives the answer; it is in any case the result of over 25 years of trying to 'make trains quieter'.

On graduating in 1980 I was privileged to join British Rail Research. After a 'training period' working on various projects it was suggested to me by Alistair Gilchrist, then Head of Civil Engineering Research that I should join the Acoustics Unit. I have to confess that until then I didn't really know what acoustics was!

I was to work on rolling noise; Alistair suggested that I should be able to solve the rolling noise problem in six months or a year and then could get on with the 'really interesting topic' of ground vibration. It wasn't until nearly 20 years later that this could be fulfilled, although even now rolling noise is not completely 'solved'.

While at BR I was fortunate to be able to register as an external student at the Institute of Sound and Vibration Research (ISVR) at the University of Southampton where I studied for a PhD on the topic of rolling noise modelling. When this was completed in 1990 I joined the Low Noise Design group at TNO in Delft where I continued to work mainly on railway noise problems. Then in 1996 I moved 'back' to ISVR as a lecturer and latterly as professor. The challenge of teaching courses at Masters level on topics such as noise control and structural vibration has helped to put the things I was already doing in railway noise into a more structured academic context. Railway noise spans a wide range of disciplines within acoustics and vibration, such as multiple degree of freedom systems, analytical modelling of beam and plate vibration, finite element and boundary element analysis, signal processing, modal analysis, vibroacoustics and aeroacoustics.

This book brings together research in the area of railway noise and vibration, much of which has been published in various papers. The intention is to present it in a way that provides a coherent introduction to the field. While inevitably many of the references are our own, the book is not just a record of the authors' work and also owes much to the many colleagues we have worked with over the years.

It is not practical in a book of this sort to provide an introduction to sound and vibration for those completely unfamiliar with them. Some basic background to acoustics is therefore assumed. For those seeking a good introduction to acoustics, a number of books could be recommended, including *Fundamentals of Noise and*

Vibration, by F. Fahy and J. Walker, published by Spon Press and *Sound and Structural Vibration*, by F. Fahy and P. Gardonio, published by Academic Press.

A few words about notation are perhaps in order. For frequency, both f in cycles/s (Hz) and ω, the circular frequency in radians/s, are used, depending on the context. These are, of course, related by $\omega = 2\pi f$. Complex notation is used with an implicit time dependence of $e^{i\omega t}$. Consequently, some published results (based on the alternative $e^{-i\omega t}$ form) have been modified to be consistent with this. The imaginary unit is denoted by i, rather than j, out of personal preference.

An attempt has been made to use a consistent coordinate system throughout (explained in Chapter 3). This has necessitated modifying the notation from some published work that is quoted. Symbols have also been standardised where possible into a reasonably consistent notation. Many of the example calculations are based on sets of parameters that have been chosen for illustration purposes. They are not intended to represent any particular case but are in some sense 'typical'. By following the various calculations using this example set of parameters the progression in modelling is hopefully not distracted by the choice of parameter values.

The frequency range of interest is determined essentially by the audible frequency range, nominally 20 Hz to 20 kHz (apart from the section on ground vibration where a lower range is appropriate). As the ear responds to sound logarithmically (a doubling of frequency is a constant musical interval, an octave), logarithmic scales are widely used for the frequency axis. Also for amplitudes, the decibel scale is widely used. Where it is not stated, standard reference values have been assumed for decibel values: 2×10^{-5} Pa for sound pressure level, 10^{-12} W for sound power level. The decibel scale has its limitations, but it can also be very forgiving – a measurement accuracy of better than ±1 dB is hard to achieve, but this corresponds to $\pm12\%$ in amplitude which in other fields of engineering would be considered an excessive margin of error.

Frequency response functions (FRF's) are mostly presented in terms of mobility (velocity/force) although in places receptance or accelerance are used where published results are in this form. In common with much current practice in vibration these are preferred here over mechanical impedance (or dynamic stiffness or apparent mass).

Damping is almost entirely represented by using a hysteretic (constant loss factor) damping model rather than a viscous model. Justification of this in relation to the track is given in Chapter 3. For the wheel, the damping is so light that it makes little difference which damping model is used.

The text deals with mechanisms of sound generation, modelling techniques for representing them and means of control, particularly those applied at the source. Hopefully it will also find a readership beyond the railway community. By looking at *noise control principles* applied to a real problem it provides an extended worked example of how to combine various techniques, theoretical and experimental, in first understanding the problem deeply enough before proposing solutions and testing them. It attempts to strike a balance between mathematical treatment and practical examples, between exploring fundamentals and discussing application, between text and pictures, between equations and physical explanations.

David Thompson
Southampton, October 2008

Acknowledgements

There are many people I wish to thank for their help in this project. First of all my two co-authors, Chris Jones and Pierre-Etienne Gautier. While the book has essentially been my project, in many ways it couldn't have been written without them. Their direct contributions are found in Chapters 8, 12, 13 and 14, but their influence can also be seen throughout the book. Chris was a colleague in my BR days and again since 1997 at ISVR. It has been a very fruitful and enjoyable collaboration and friendship. Pierre-Etienne I have known since 1991 when TWINS was first 'born'. More recently he has generously invited me to spend several periods of time at SNCF Direction de l'Innovation et de la Recherche in Paris as 'visiting professor' from 2005 onwards and considerable parts of the book have been written during these times.

I am grateful to the ISVR and the University of Southampton for the freedom to write this. It is a real privilege to have a job that one can enjoy. I wish also to thank Trinity College, Cambridge for a visiting scholarship in 2005 during which the book was started, the Department of Engineering at the University of Cambridge for hosting me during that period and in particular Hugh Hunt and Robin Langley.

The book has also gained much inspiration from the 'Savoir' *International Course on Noise and Vibration from Rail Transport Systems* which began in 1991 as a collaborative venture between ISVR, TNO, SNCF and STUVA and has been run ten times since then. Initially my fellow-lecturers were Tjeert ten Wolde, John Walker, Friedrich Krüger and Eric Tassilly. Over the years other colleagues took over lecturing and responsibility for the organisation, notably Michael Dittrich as well as Laurent Guccia and Chris Jones. Each has taught *me* a lot as well as the delegates on the course.

Thanks are due to the following people who have supplied photographs, diagrams and information or who have proofread parts of the text: Olly Bewes, Estelle Bongini, Steve Cox, Virginie Delavaud, Pieter Dings, Don Eadie, Dieter Hoffmann, Marcel Janssens, Rick Jones, Toshiki Kitagawa, Jan Lub, Florence Margiocchi, Kerri Parsley, Franck Poisson and Edwin Verheijen. I am also grateful to the staff at Elsevier, especially Melanie Benson and Susan Li for all their help and patience.

Looking back, there are many people I wish to thank, especially the people who have 'believed' in me and who have created the environment that has allowed this work to come to fruition. It was Alistair Gilchrist who first interviewed me for a job at BR Research and who later pointed me in the direction of acoustics. Many others at BR Research were influential, including Charles Frederick, Richard Gostling, Colin Stanworth and Brian Hemsworth. Bob White was a great inspiration as my PhD supervisor at ISVR. Tjeert ten Wolde encouraged me to move to TNO; it was a pleasure to work with many colleagues there, especially Jan Verheij, Michael Dittrich and Marcel Janssens. Through ERRI it was a privilege also to work with Paul Remington, Nicolas Vincent, Maria Heckl and the late Manfred Heckl. At ISVR, Joe Hammond was brave enough to take me on; he, Phil Nelson, Steve Elliott and Mike Brennan have all encouraged me as my 'line managers'.

Funding for the research over the years has come from many sources including British Rail, ORE (later ERRI) through its Committee C163, the Dutch Ministries of Environment and Transport, the EU, EPSRC and various companies including Corus, Pandrol, Bombardier, DB, ProRail, SNCF and RFF. All these are gratefully acknowledged.

The ISVR is a great place to work where I have many supportive colleagues. I want to thank especially the many research students and post-docs I have had the privilege of supervising, including: Tianxing Wu, Xiaozhen Sheng, Woo Sun Park, Anand Thite, Tristan Armstrong, Gang Xie, Ji Woo Yoo, Olly Bewes, Andrew Monk-Steel, Angela Müller, Jungsoo Ryue, Toshiki Kitagawa, Zhenyu Huang, Azma Putra, Nazirah Ahmad, Briony Croft, David Herron, Becky Broadbent, Nuthnapa Triepaischajonsak. Many of you will see some of your work reflected in the enclosed. I gladly acknowledge that I couldn't have done it without you. I also want to mention Robin Ford who spent a productive sabbatical with us at ISVR working on contact filtering.

Finally, I want to thank my family and friends for all their support, especially during the times of stress involved in writing this. Thank you to the many friends at St John's church Rownhams. Thank you to Mum and Dad for all you gave me. Thank you to Alison and Fiona (the real twins) and to Sandra for all your love and for all the fun we have together. And thank you Claire, simply for everything over the last 25 years. There is so much I could not have done without you.

Copyright Acknowledgements

Figs 1.1, 2.14, 7.7, 7.8, 7.12, 7.15, 7.19, 7.22, 7.25, 7.26, and 7.28 Reprinted from *Proceedings of the Institution of Mechanical Engineers, Part F, Journal of Rail and Rapid Transit* 220(4) D.J. Thompson and P.E. Gautier. A review of research into wheel/rail rolling noise reduction. 385-408, ISSN: 0954-4097, DOI: 10.1243/0954409JRRT79, Copyright (2006), with permission from Professional Engineering Publishing.

Fig. 1.2 Reprinted from *Journal of Sound and Vibration* 231, D.J. Thompson and C.J.C. Jones, A review of the modelling of wheel/rail noise generation, 519-536, Copyright (2000), with permission from Elsevier.

Figs 2.5, 7.11, 7.18 and 7.27 Reprinted from *Noise and Vibration from High-Speed Trains*, ed. V.V. Krylov, published by Thomas Telford, Copyright (2001), with permission from Thomas Telford Ltd.

Fig. 2.7 Reprinted from *Journal of Sound and Vibration* 193, D.J. Thompson. On the relationship between wheel and rail surface roughness and rolling noise, 149-160, Copyright (1996), with permission from Elsevier.

Figs 2.11 and 2.12 Reprinted from *Journal of Sound and Vibration* 120, D.J. Thompson. Predictions of acoustic radiation from vibrating wheels and rails, 275-280, Copyright (1988), with permission from Elsevier.

Fig. 2.13 Reprinted from *Proceedings of the Institution of Mechanical Engineers, Part F, Journal of Rail and Rapid Transit* 205(F2) D.J. Thompson. Theoretical modelling of wheel-rail noise generation. 137-149, ISSN: 0954-4097, DOI: 10.1243/PIME_PROC_1991_205_227_02, Copyright (1991), with permission from Professional Engineering Publishing.

Figs 2.14, 2.16a, 9.6 and 14.5 Reproduced from *Handbook of Railway Vehicle Dynamics*, ed. S.D. Iwnicki. Copyright (2006) by Taylor & Francis Group LLC. Reproduced with permission of Taylor & Francis Group LLC via Copyright Clearance Center.

Fig. 3.10 Reprinted from *Proceedings of the Institution of Mechanical Engineers, Part F, Journal of Rail and Rapid Transit* 213(4), D.J. Thompson, C.J.C. Jones, T.X. Wu and G. de France, The influence of the non-linear stiffness behaviour of railpads on the track component of rolling noise. 233-241, ISSN: 0954-4097, DOI: 10.1243/0954409991531173, Copyright (1999), with permission from Professional Engineering Publishing.

Figs 3.42, 3.43 and 3.44 Reprinted from Acustica united with Acta Acustica 86, T.X. Wu and D.J. Thompson, The influence of random sleeper spacing and ballast stiffness on the vibration behaviour of railway track, 313-321, Copyright (2000), with permission from S. Hirzel Verlag.

Fig. 3.45, 3.46 and 3.47 Reprinted from *Journal of Sound and Vibration* 203, D.J. Thompson, Experimental analysis of wave propagation in railway tracks, 867-888, Copyright (1997), with permission from Elsevier.

Fig. 5.26 Reprinted from *Journal of Sound and Vibration* 267, D.J. Thompson, The influence of the contact zone on the excitation of wheel/rail noise, 523-535, Copyright (2003), with permission from Elsevier.

Figs 6.22, 6.23, 6.24 and 6.25 Reprinted with permission from *Journal of the Acoustical Society of America*, 113, D.J. Thompson, C.J.C. Jones, N. Turner, Investigation into the validity of two-dimensional models for sound radiation from waves in rails, 1965-1974, 2003. Copyright (2003) American Institute of Physics.

Figs 6.27 and 6.28 Reprinted from *Journal of Sound and Vibration* 293, T. Kitagawa and D.J. Thompson, Comparison of wheel/rail noise radiation on Japanese railways using the TWINS model and microphone array measurements, 496-509, Copyright (2006), with permission from Elsevier.

Figs 6.41, 6.42a, 6.43 and 6.44 Reprinted from *Journal of Sound and Vibration* 267, C.J.C. Jones, D.J. Thompson, Extended validation of a theoretical model for railway rolling noise using novel wheel and track designs, 509-522, 2003, Copyright (2003), with permission from Elsevier.

Fig. 6.42b Reprinted from *Journal of Sound and Vibration* 193, D.J. Thompson, P. Fodiman and H. Mahé. Experimental validation of the TWINS prediction program, part 2: results, 137-147, Copyright (1996), with permission from Elsevier.

Fig. 7.4 Reprinted from *Journal of Sound and Vibration* 231, D.J. Thompson and P.J. Remington. The effects of transverse profile on the excitation of wheel/rail noise, 537-548, Copyright (2000), with permission from Elsevier.

Fig. 7.10 Reprinted from *Journal of Sound and Vibration* 231, D.J. Thompson and P.J. Remington. The effects of transverse profile on the excitation of wheel/rail noise, 537-548, Copyright (2000), with permission from Elsevier.

Fig. 7.23 Reprinted from *Journal of Sound and Vibration* 231, C.J.C. Jones and D.J. Thompson. Rolling noise generated by wheels with visco-elastic layers, 779-790, Copyright (2000), with permission from Elsevier.

Fig. 7.30 Reprinted from *Applied Acoustics* 68, D.J. Thompson, C.J.C. Jones, T.P. Waters and D. Farrington. A tuned damping device for reducing noise from railway track, 43-57, 2007, Copyright (2007), with permission from Elsevier.

Figs 8.13 and 8.22a Reprinted from *Journal of Sound and Vibration* 267, C. Talotte, P.E. Gautier, D.J. Thompson, C. Hanson, Identification, modelling and reduction potential of railway noise sources: a critical survey, 447-468, 2003, Copyright (2003), with permission from Elsevier.

Fig. 8.15 Reprinted from *Journal of Sound and Vibration* 293, K. Nagakura, Localization of aerodynamic noise sources of Shinkansen trains, 547-565, Copyright (2006), with permission from Elsevier.

Fig. 8.17 Reprinted from *Journal of Sound and Vibration* 231, T. Kitagawa and K. Nagakura, Aerodynamic noise generated by Shinkansen cars, 913-924, Copyright (2000), with permission from Elsevier.

Fig. 9.9 Reprinted from *Journal of Sound and Vibration* 293, A.D. Monk-Steel, D.J. Thompson, F.G. de Beer and M.H.A. Janssens. An investigation into the influence of longitudinal creepage an railway squeal noise due to lateral creepage. 766-776, Copyright (2006), with permission from Elsevier.

Figs 10.1 and 10.2 Reprinted from *Vehicle System Dynamics* 34, T.X. Wu and D.J. Thompson. Theoretical investigation of wheel/rail non-linear interaction due to roughness excitation. 261-282, 2000, Taylor & Francis Ltd, http://www.informaworld.com.

Figs 10.6, 10.7, 10.8, 10.9, 10.10 and 10.11 Reprinted from *Journal of Sound and Vibration* 251, T.X. Wu and D.J. Thompson. A hybrid model for the noise generation due to railway wheel flats, 115-139, Copyright (2002), with permission from Elsevier.

Figs 11.7 and 11.20 Reprinted from *Journal of Sound and Vibration* 193, M.H.A. Janssens and D.J. Thompson, A calculation model for noise from steel railway bridges, 295-305, Copyright (1996), with permission from Elsevier.

Figs 11.10, 11.11 and 11.21 Reprinted from *Journal of Sound and Vibration* 293, O.G. Bewes, D.J. Thompson, C.J.C. Jones and A. Wang. Calculation of noise from railway bridges and viaducts: experimental validation of a rapid calculation model, 933-943, Copyright (2006), with permission from Elsevier.

Fig. 11.13 Reprinted from *Proceedings of the Institution of Mechanical Engineers, Part F, Journal of Rail and Rapid Transit* 214 M.F. Harrison, D.J. Thompson and C.J.C. Jones, The calculation of noise from railway viaducts and bridges. 214(3), 125-134, ISSN: 0954-4097, DOI: 10.1243/0954409001531252, Copyright (2000), with permission from Professional Engineering Publishing.

Figs 12.12, 12.14, 12.15 and 12.17 Reprinted from *Journal of Sound and Vibration* 272, X. Sheng, C.J.C. Jones, and D.J. Thompson, A theoretical study on the influence of the track on train-induced ground vibration, 909-936, Copyright (2004), with permission from Elsevier.

Figs 12.18 and 12.19 Reprinted from *Journal of Sound and Vibration* 228, X. Sheng, C.J.C. Jones and M. Petyt, Ground vibration generated by a load moving along a railway track, 129-156, Copyright (1999), with permission from Elsevier.

Figs 12.24, 12.25 and 12.26 Reprinted from *Journal of Sound and Vibration* 267, X. Sheng, C.J.C. Jones, and D.J. Thompson, A comparison of a theoretical model for quasi-statically and dynamically induced environmental vibration from trains with measurements, 621-636, Copyright (2003), with permission from Elsevier.

Figs 12.32 and 12.33 Reprinted from *Journal of Sound and Vibration* 293, X. Sheng, C.J.C. Jones, and D.J. Thompson, Prediction of ground vibration from trains using the wavenumber finite and boundary element methods, 575-586, Copyright (2006), with permission from Elsevier.

Fig. 13.3 Reprinted from *Proceedings of the Institution of Civil Engineers, Transport Journal* 153, C.J.C. Jones, D.J. Thompson and M. Petyt, A model for ground vibration from railway tunnels. 121-129, Copyright (2002), with permission from Thomas Telford Ltd.

CHAPTER 1

Introduction

1.1 THE NEED FOR NOISE AND VIBRATION CONTROL IN RAILWAYS

To some of us the sound of a passing train is music to the ears. Audio recordings of trains are sold; the sound of a steam engine labouring up a gradient or passing at speed may evoke a strong impression of its power or the nostalgia of a lost age. But to many people the noise from passing trains is unwanted and can be considered a disturbance.

It has always been so. The early railways were often subject to considerable opposition. The following was written in 1825, in a letter to the *Leeds Intelligencer* (quoted in [1.1]): 'Now judge, my friend, of my mortification, whilst I am sitting comfortably at breakfast with my family, enjoying the purity of the summer air, in a moment my dwelling … is filled with dense smoke, … . Nothing is heard but the clanking iron, the blasphemous song, or the appalling curses of the directors of these infernal machines.' Nevertheless, although some objections such as this were attributed to environmental reasons including noise, most were based on economic or aesthetic arguments [1.1]. An interesting example occurred as early as 1863, when the Manchester, Buxton, Matlock and Midlands Junction Railway (later to become part of the Midland Railway) in England was forced to build its line in a shallow cut-and-cover tunnel almost 1 km long so that it should not be visible from the Duke of Rutland's home at Haddon Hall [1.2]. Today such schemes and changes in alignment are not uncommon to mitigate noise, but the idea is clearly not new.

Particularly since the 1960s, environmental noise has become an increasingly important issue. Noise is often identified as a source of dissatisfaction with the living environment by residents. As environmental noise levels have increased, the population has become increasingly aware of noise as a potential issue. This applies to railway noise in common with many other forms of environmental noise. Opposition to new railway lines is therefore now often focused on their potential noise impact. This may be because noise is quantifiable in a way that aesthetics are not, so that complaints about the railway as such become focused on the issue of noise. However, as noted by the Wilson Report of 1963 [1.3]: 'There is a considerable amount of evidence that, as living standards rise, people are less likely to tolerate noise.'

It was estimated in 1996 that 20% of the population of Western Europe lived in areas where the ambient noise level[1] was over 65 dB and as many as 60% in areas where the noise level was over 55 dB [1.4]. The major source of this noise is road traffic, which accounts for around 90% of the population exposed to levels of noise over 65 dB (i.e. 18 of the 20%). However, railways and aircraft are also important sources of noise in the community. Rail traffic accounts for noise levels over 65 dB for 1.7% of the population.

While not everyone reacts to noise in the same way, it is no surprise that in terms of annoyance, between 20 and 25% of the population are annoyed by road traffic noise and between 2 and 4% by railway noise [1.4]. Nevertheless it has been found that for the same level of noise, railways are less annoying than road traffic [1.5], leading to a 'railway bonus' in some national standards, recommendations and guidelines, notably a 'bonus' of 5 dB in Germany [1.6].

The prevalence of high noise levels and increased public awareness of noise has led to the introduction of legislation to limit sound levels, both at receiver locations and at the source. European legislation has existed since the 1970s to control the sound emitted by road vehicles and aircraft. For road vehicles, reductions in the levels obtained during the drive-by test of 8–11 dB have been achieved between 1973 and 1996. However, it is widely recognized that this does not translate into equivalent reductions of noise in traffic, due to a mismatch between the test conditions and typical traffic conditions. In the former (low speed acceleration under high engine speed) engine noise dominates whereas in traffic usually tyre noise dominates for speeds of 50 km/h and above [1.7]. Changes in the test procedure are proposed to overcome this. Rubber tyres are clearly not quiet, being responsible for much of the noise exposure due to transport (see also box on page 4).

By contrast aircraft noise has been reduced by stricter noise certification and by night-time flying bans and other operational measures. The introduction of high bypass ratio turbo-fan engines has reduced noise by 20–30 dB since the early 1970s, although the rapid increase in the number of flights means that the noise exposure has in many situations close to airports continued to rise or at least remained steady.

For railway noise, the difficulty of separating the influence of track and vehicle and the consequent difficulty of defining a unique source value for a particular vehicle have contributed to the long delay in the introduction of source limits. Legal limits on the noise emitted by individual rail vehicles have only been introduced in Europe since 2002. These have been achieved through the means of 'Technical Specifications for Interoperability' (TSI) [1.8, 1.9], which are intended primarily to *allow* interoperability of vehicles between different countries in Europe. Such limits have the potential to reduce railway noise in the long term.

Noise limits at receiver locations apply in many countries. These were mostly introduced initially to apply to new lines or altered situations, providing for mitigation such as noise barriers or secondary glazing where limits were exceeded. However, the recent introduction of the Environmental Noise Directive (END) has led to the requirement to produce noise maps of existing sources and to develop Action Plans to reduce noise in identified 'hot spots' [1.10]. These, too, will mean that railway operators and infrastructure companies will have to consider how to minimize noise.

[1] Expressed as a long-term average A-weighted sound level, L_{Aeq}.

As well as noise, vibration from railways can cause annoyance. This may be due to feelable vibration, usually in the range 2 to 80 Hz, or due to the radiation of low frequency sound transmitted through the ground, usually in the range 30 to 250 Hz. Vibration may also cause objects to rattle, adding to the sensation.

Noise at a particular receiver location can be reduced by secondary measures, either in the transmission path such as noise barriers or at the receiver such as by installing windows with improved acoustic insulation. To a lesser extent vibration can also be dealt with by secondary measures, such as mounting sensitive buildings on isolation springs. Nevertheless, reduction of noise and vibration *at the source* is generally more cost effective. On the other hand, it is also generally true that noise control at source is more complex, as it requires a good knowledge of the mechanisms operating within the source. It is important that safe and economic operation of the equipment, in this case the railway, is not impeded by changes aimed at reducing noise. The railway is often seen as a conservative industry where there is reluctance to change the way things are done, particularly because of potential implications for safety or operational efficiency. Nevertheless, it is the author's belief and experience that significant noise reductions are possible by careful study of the sources, appropriate modelling, and use of those models for optimization, while taking into account the many non-acoustic factors.

Railways are generally acknowledged to be an environmentally-friendly means of transport with the potential to operate with considerably less pollution, energy use and CO_2 emissions per passenger-km than road or air. High speed trains have been found to compete effectively with air transport on routes up to 3 hours or more, achieving large market shares on routes such as between London, Paris and Brussels. Mass transit systems hold the key to urban mobility. Rail freight is growing across Europe. In order to improve the market share of rail transport, and thereby improve sustainability, it is imperative that noise is reduced.

1.2 THE NEED FOR A SYSTEMATIC APPROACH TO NOISE CONTROL

The problem of reducing railway noise can be used to illustrate the classical approach to noise control. (Note that it is beyond the scope of this book to introduce the reader to the fundamentals of acoustics and vibration. For this, there are many good text books, e.g. [1.11, 1.12, 1.13]. Familiarity is assumed, for example, with the decibel scale, frequency analysis and complex notation for harmonic motion.)

The first step in noise control is to *identify the dominant source*. There are many different sources of noise from a railway, and in different situations the dominant source may vary. Notably on North American freight railways, a major issue for environmental noise is related to locomotive warning signals. It is obligatory to sound the horn in an extended sequence on the approach to road crossings [1.14]. There are many such crossings, especially in populated areas and so it is a major source of annoyance, particularly from operations at night. In other situations, such as stations in urban areas, the public address system may be the major source of noise in the immediate neighbourhood. However, the most important source of noise from railway operations is usually rolling noise caused by the interaction of wheel and rail during running on straight track. Other sources include curve squeal, bridge

'Have you ever thought of using rubber wheels?'

In discussing railway noise control, people often ask: 'Have you ever thought of using rubber wheels?' The simple answer is: 'Yes, of course.' The introduction of flexibility at the wheel/rail contact is known to be beneficial in reducing the excitation of wheel and rail [1.17] (see also Chapter 7). Yet, behind the question is usually the naïve assumption that 'rubber wheels' must be quieter because rubber is somehow a 'quiet' material. This is not necessarily the case.

To pose a slightly different question: Are trains actually louder than lorries? In both cases the dependence on speed is similar so noise levels are compared here at a common speed of 80 km/h. In [1.7], for example, it was shown that the A-weighted SEL[2] for a single heavy road vehicle travelling at 80 km/h and measured at a distance of 7.5 m has remained around 87 dB between 1972 and 1998. These levels are dominated by tyre noise. A-weighted sound pressure levels at 25 m from a freight train consisting of four-axle 100 tonne tank wagons travelling at 80 km/h are given in [1.18] as about 84 dB. To compare these two figures, corrections for distance and to sound exposure level (SEL) are required. As shown in the table this gives an SEL of 88 dB for the rail vehicle which is quite close to the result for the lorry.

A-weighted sound pressure levels in dB at 80 km/h from lorries and rail vehicles

	Correction	Rail vehicle	Lorry
Average level at 25 m		84	
Average level at 7.5 m	$10 \log_{10}(25/7.5) = 5$ dB	89	
SEL at 7.5 m for single vehicle	$10 \log_{10}(T) = -1$ dB ($T = 0.8$ s for vehicle length 20 m)	88	87
Normalize by load factor	$10 \log_{10}(2.5) = 4$ dB	88	91
Update to current practice	Rail vehicle 10 dB quieter due to braking system; lorry on porous road surface	78	86

However, each 100 tonne tank wagon can carry about two to three times as much as a 40 tonne lorry. Or put another way, a rubber-tyred rail vehicle would require two to three times as many wheels as the lorry. Taking this loading factor as 2.5, this gives the rail vehicle an additional advantage of 4 dB.

Finally, it may be noted that the levels given for tank wagons in [1.18] are for vehicles fitted with cast-iron brake blocks; more modern vehicles with either disc brakes or composite brake blocks can be expected to be about 10 dB quieter (see Chapter 7). Similarly, quiet road surfaces, such as porous asphalt, can reduce the tyre noise by around 5 dB; on a new surface the reduction may be somewhat more than this. This gives levels of 78 and 86 dB, respectively. Therefore this rough comparison shows that the carriage of freight by rail has a potential noise advantage of around 8 dB over carriage of the same load at the same speed by road. Clearly, the supposition that rubber

Continued

[2] The SEL is the sound pressure level of an event of duration 1 s containing the same sound energy as the original event, in this case the passage of a single rail vehicle.

'Have you ever thought of using rubber wheels?'—Cont'd

wheels would be advantageous is misplaced in this context. A similar comparison can be made for passenger traffic and would, if anything, give even more advantage to rail (depending on the load factors assumed).

The various features of the two systems are compared in the table below, from which it is clear that the high damping and low radiation efficiency of a rubber tyre are offset by the high amplitudes of excitation at the tyre/road interface and by the 'horn effect' which amplifies sound radiation from the tread region near the contact zone.

Features of rubber tyres and steel wheels

	Rubber tyre	**Steel wheel**
Damping loss factor	~ 0.1	$\sim 3 \times 10^{-4}$
Radiation efficiency	Very low (also influenced by 'horn effect')	~unity
Excitation mechanisms	Road/tread roughness Block impact/snap-out Air pumping	Wheel/rail roughness
Amplitude of roughness excitation	~mm	~μm

Other comparisons could be made: if the tread pattern could be eliminated and the road surface replaced by a smoother track, tyre noise levels may be reduced considerably, perhaps by as much as 20 dB. Such systems exist in mass transit systems where rubber-tyred vehicles are used, but clearly they cannot operate on conventional tracks. However, the final argument in favour of the system of steel wheel on steel rail is that the rolling resistance, and hence energy use, is much less than for a rubber tyre.

noise, traction noise and aerodynamic noise. Noise inside the vehicle also includes all of these sources, as well as others such as air-conditioning fan noise.

Having identified the dominant source, the next step is to *quantify the various paths or contributions*. Focusing on exterior rolling noise, the vibration of the wheel and the rail can be identified as potential sources. Early attempts to understand the problem tended to be polarized into attributing the noise solely to one or the other [1.15]. More recently, however, it has become widely recognized that both wheels and rails usually form important sources which make similar contributions to the overall sound level [1.16]. Prediction models allow their relative contributions to be quantified (measurement methods can also be used). Clearly, effective noise control requires both sources to be tackled. For example, in a situation where wheel and rail contribute equally to the overall level, a reduction of 10 dB in one of them, while the other is unchanged, will produce a reduction of only 2.5 dB in the total (see also box on p. 224).

The next step is to *understand how each source can be influenced*. Here, the theoretical models allow the sensitivity of the noise to various design parameters to be investigated (measurements alone do not). Noise control principles can be considered in terms of reduced excitation, increased damping, vibration isolation, acoustic shielding or absorption.

From these principles, *actual designs can be developed and tested*, first using the prediction model, then in laboratory tests and ultimately in practical tests on the operational railway. Tests should be carried out in a controlled situation; where possible not just the noise but intermediate parameters such as vibration should be measured.

It would, of course, be risky to proceed straight to this last step. The source or path that is treated may not be the dominant one, or the modification introduced may not influence the source as intended. Yet there are many examples in railway noise control where this has been done, often leading to the conclusion 'we've tried that and it doesn't work'.

Before noise control measures can be applied in normal operation, *practical constraints have to be taken into account*. The measures that have been developed in principle have to satisfy many other requirements of the operating environment. In the case of the railway these are particularly related to safety. At this point compromise is often required in the acoustic design. Ideally such constraints should be considered as early as possible in the design process, provided that they don't stifle innovation altogether.

The approach described in this book is based on the above principles, particularly the development of theoretical models with an appropriate level of detail to understand the source mechanisms and then the use of these models to develop and understand mitigation measures.

1.3 SOURCES OF RAILWAY NOISE AND VIBRATION

There are many sources of noise and vibration in the railway system. The main ones are summarized in this section. The dominant source at most speeds is rolling noise which increases with train speed V at a rate of about $30 \log_{10} V$, i.e. a 9 dB increase in sound level for a doubling of speed. Traction noise is much less dependent on train speed so that it is often dominant only at low speeds. Conversely, aerodynamic noise has a much greater speed dependence than other sources and so becomes dominant at high speeds.

The audible frequency range extends from 20 to 20,000 Hz and this broadly defines the range of interest for acoustic analysis. Within this range some sources, such as bridge noise or ground-borne noise are concentrated at low frequencies, while squeal nose can occur at very high frequencies.

1.3.1 Rolling noise

The most important source of noise from railways is rolling noise caused by wheel and rail vibrations induced at the wheel/rail contact. Roughness on the wheel and rail running surfaces induces vertical vibration of the wheel and rail systems according to their dynamic properties. Figure 1.1 indicates this in the form of a flowchart, while Figure 1.2 shows the mechanism visually. The main wavelengths of roughness that are relevant to rolling noise are between about 5 and 500 mm. This vibration is transmitted into the wheel and track structures leading to sound radiation. Often,

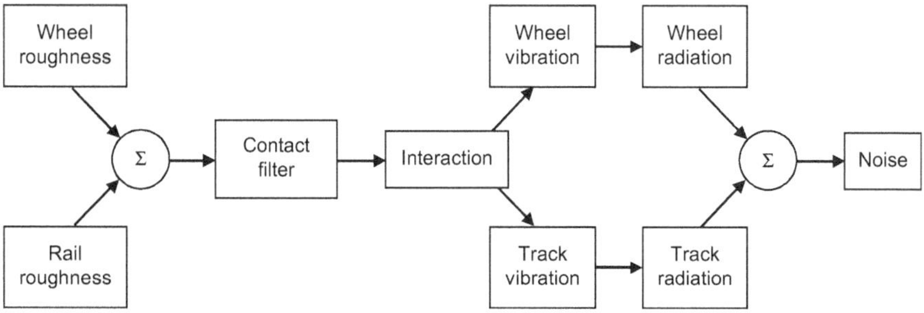

FIGURE 1-1 Model for rolling noise generation

both wheel and track vibration are important to the overall noise level and both must be taken into account. Rolling noise is fairly broad-band in nature, the relative importance of higher frequency components increasing as the train speed increases.

Another complication is that the vibrations of both wheel and rail are induced by the combination of their roughnesses. A typical situation is that *wheels* fitted with cast-iron brake blocks have large roughness with wavelengths around 40–80 mm. For a train speed of 100 km/h this roughness excites frequencies where the *track* vibration radiates most sound. In such a situation, is the vehicle or the track responsible for the noise? Clearly, both are.

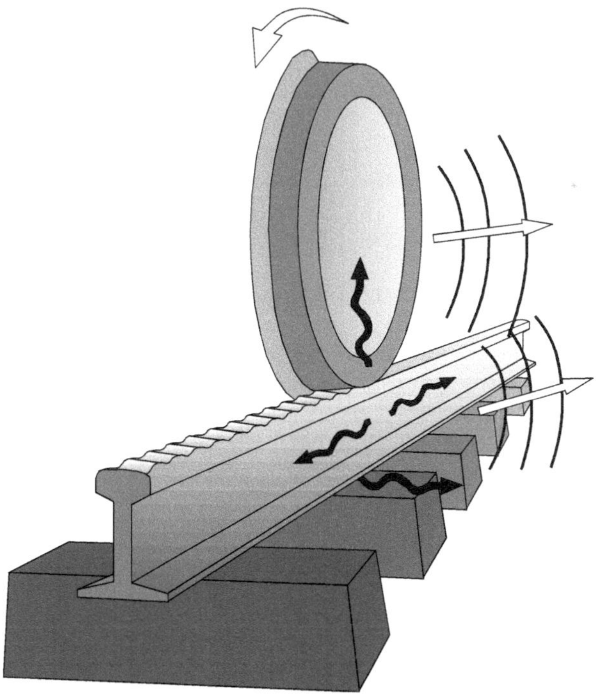

FIGURE 1-2 Illustration of the mechanism of generation of rolling noise

Compared with the other sources, rolling noise has been the subject of the greatest amount of research over the years, and will be treated in most detail in the first half of the book.

Impact noise is excited by irregularities such as wheel flats, rail joints, dipped welds or switches and crossings. This has similarities to rolling noise as the excitation is primarily vertical, but it differs in that non-linearity in the contact zone is important and cannot be neglected.

1.3.2 Curve squeal

Curve squeal noise is also caused by the interaction between wheel and rail but has a quite different character. It is strongly tonal, being associated with the vibration of the wheel in one of its resonances. This is excited by unsteady transverse forces at the contact occurring during curving. It is also necessary to distinguish between squeal caused by lateral creepage, or 'top-of-rail' squeal, and 'flange squeal' or 'flanging noise'. The latter has a more intermittent nature and generally has a higher frequency content, consisting also of many more harmonics rather than a single dominant tone.

A similar phenomenon is brake squeal in which tonal or multi-tonal noise is emitted during braking.

1.3.3 Bridge noise

When a train runs over a bridge the noise emitted can increase considerably, depending on the type of bridge. The bridge is excited by dynamic forces acting on it from the track. Bridges vary greatly in construction. Steel bridges with direct fastenings are usually the noisiest and can be more than 10 dB noisier than plain ballasted track, in some cases up to 20 dB. Moreover, the increase in noise on a bridge is usually greatest at low frequency, so that the A-weighted sound level does not fully take account of this effect.

1.3.4 Aerodynamic noise

In contrast to most other railway noise sources, which are caused by the radiation of sound by the vibration of solid structures, aerodynamic noise is caused by unsteady air flow over the train. Aerodynamic sources of noise generally increase much more rapidly with speed than mechanical sources, typically between $60 \log_{10} V$ and $80 \log_{10} V$. Both broad-band and tonal noise can be generated by air flow over various parts of the train, but much of the sound energy is concentrated in the lower part of the frequency region. Considerable understanding of the sources of aerodynamic noise has been obtained in recent years but modelling is much more involved than for vibroacoustic problems and the models are still at a relatively early stage.

1.3.5 Ground vibration and noise

As noted already, vibration transmitted through the ground can be experienced in two ways. Low frequency vibration between about 2 and 80 Hz is perceived as feelable 'whole body' vibration. This tends to be associated most with heavy freight trains at particular sites. Higher frequency ground-borne vibration from about 30 to 250 Hz causes the walls, floors and ceilings of rooms to vibrate and radiate low frequency noise. This is associated more with trains in tunnels in urban areas such as metro operations but can also be significant for surface railways, particularly where noise barriers block out the direct airborne sound.

Whereas air is essentially a uniform medium, the ground can have a layered structure and its properties vary considerably from one site to another and even within a single site. Thus prediction of absolute levels of vibration and ground-borne noise is extremely difficult, requiring complex models of the ground and in turn detailed information about ground properties, as well as models of the train and track and possibly also the buildings at the receiver location.

1.3.6 Other sources of railway noise

Other sources of railway noise, not discussed further in this book, include:

- Traction noise from diesel engines, exhaust and intake, from traction motors and fans, from gearboxes, turbochargers, etc.
- Warning signals from trains (horns, etc.) and fixed installations (e.g. level crossings).
- Track maintenance equipment.
- Shunting noise, in particular the noise from impacts between vehicles.

1.3.7 Internal noise and vibration

Many of the sources that are relevant to environmental noise also lead to noise inside the railway vehicle. However, their frequency content is modified by the transmission paths into the vehicle, which may be both structure-borne and airborne. Consequently noise spectra inside vehicles tend to be dominated by low frequency sound, as is also the case for automobiles. In addition, low frequency vibration inside the vehicle affects ride comfort.

1.4 STRUCTURE OF THE BOOK

The first part of the book describes models for rolling noise generation. This commences in Chapter 2 with an overview of rolling noise, followed by a detailed treatment of the vibration properties of track and wheels in Chapters 3 and 4, respectively. Wheel/rail interaction due to roughness excitation is dealt with in Chapter 5 which includes a discussion of the measurement of roughness. Sound radiation from the vibrations of wheel and track is described in Chapter 6. The validation of the complete model using experimental data is also addressed. The model for rolling noise is used as a basis to explain the operation and efficiency of many potential noise mitigation measures in Chapter 7.

Other sources of noise are discussed in Chapters 8 to 11: aerodynamic noise, curve squeal, impact noise due to rail joints and wheel flats, and bridge noise. Apart from aerodynamic noise, these draw on parts of the model for rolling noise while also introducing new elements.

Chapters 12 and 13 describe low frequency ground-borne vibration and ground-borne noise. The final chapter discusses vehicle interior noise.

While the book is primarily concerned with models of noise sources and their use to develop noise control solutions, standard measurement methods are discussed briefly in Appendix A. A brief glossary of railway terminology is included in Appendix B.

REFERENCES

1.1 W.R. Siddall. No nook secure: transportation and environmental quality, *Comparative Studies in Society and History*, **16**, 2–23, 1974.

1.2 Anon, Haddon Tunnel, *Wikipedia*, http://en.wikipedia.org/wiki/Haddon_Tunnel (accessed 22/3/07).

1.3 A.H. Wilson *et al. Committee on the problem of noise – final report, Cmnd 2056*. HMSO, London, 1963.

1.4 Future Noise Policy, European Commission Green Paper, COM(96) 540 final, Brussels, 4 November 1996.

1.5 U. Moehler, M. Liepert, R. Schuemer, and B. Griefahn. Differences between railway and road traffic noise. *Journal of Sound and Vibration*, **231**, 853–864, 2000.

1.6 F. Krüger *et al. Schall- und Erschütterungsschutz im Schienenverkehr*. Expert Verlag, Renningen, 2001.

1.7 U. Sandberg. Noise emission of road vehicles – effect of regulations. Final report, I-INCE, 2001.

1.8 Commission Decision 2002/735/EC concerning the Technical Specification for Interoperability (TSI) relating to the rolling stock subsystem of the trans-European high-speed rail system. *Official Journal of the European Communities*, **L245**, 402–506, 2002.

1.9 Directive 2001/16/EC of the European Parliament and of the Council of 19 March 2001 on the interoperability of the trans-European conventional rail system. *Official Journal of the European Communities*, **L110**, 1–27, 2001.

1.10 Directive 2002/49/EC of the European Parliament and of the Council of 25 June 2002 relating to the assessment and management of environmental noise. *Official Journal of the European Communities*, **L189**, 12–25, 2002.

1.11 F.J. Fahy and J.G. Walker (eds.). *Fundamentals of Noise and Vibration*. E&FN Spon, London, 1998.

1.12 L.E. Kinsler, A.R. Frey, A.B. Coppens and J.V. Sanders. *Fundamentals of Acoustics*, 4th edition. John Wiley & Sons, New York, 1999.

1.13 S.S. Rao. *Mechanical Vibrations*, 4th edition. Prentice Hall, Upper Saddle River, NJ, 2003.

1.14 L. Meister and H. Saurenman. Noise impacts from train whistles at highway/rail at-grade crossings. *Proceedings of Internoise 2000*, Nice, France, 1038–1043, 2000.

1.15 P.J. Remington. Wheel/rail rolling noise: what do we know? What don't we know? Where do we go from here? *Journal of Sound and Vibration*, **120**, 203–226, 1988.

1.16 D.J. Thompson. Predictions of the acoustic radiation from vibrating wheels and rails. *Journal of Sound and Vibration*, **120**, 275–280, 1988.

1.17 D.J. Thompson and P.J. Remington. The effects of transverse profile on the excitation of wheel/rail noise. *Journal of Sound and Vibration*, **231**, 537–548, 2000.

1.18 B. Hemsworth. Railway noise prediction – data base requirements. *Journal of Sound and Vibration*, **87**, 275–283, 1983.

CHAPTER 2

Introduction to Rolling Noise

2.1 THE SOURCE OF ROLLING NOISE

Before turning to a detailed consideration of the models for rolling noise in the next four chapters, a brief introduction will be given in this chapter to the character of rolling noise, mainly on the basis of measured data.

Figure 2.1 shows typical time histories of the sound pressure level during a train pass-by. The upper figure shows the sound measured from a four-coach electric multiple unit (EMU) with cast-iron brake blocks. The lower figure shows a measurement of a diesel high speed train (HST or Intercity 125) with two diesel power cars at the ends (which have cast-iron brake blocks) and seven disc-braked trailer vehicles between them. The measurements were made at a distance of 4 m from the near rail. Before the train arrives the sound level gradually increases; similarly it reduces as the train recedes. This is associated with the 'singing' of the rail. The sound level reaches a maximum as each bogie passes the microphone, suggesting that the wheels are an important source. At this distance of 4 m only each bogie or bogie pair can be heard distinctly but closer to the track individual wheels can be detected.

In fact, as will be seen, rolling noise is generated by the *interaction* of the wheel and rail at their contact area and both of them radiate significant proportions of the noise. The relative contributions of the wheel and track to the overall noise will be discussed in Section 2.4.

Neither the wheel nor the rail running surfaces are entirely smooth. Their unevenness causes the wheel and rail to move relative to one another. If a wave with a wavelength λ is present on the rail (or wheel), and this is traversed at a speed V, sinusoidal vibration will be generated with a frequency

$$f = \frac{V}{\lambda} \tag{2.1}$$

NB if λ is given in m and f is in Hz, V must be given in m/s, not km/h.

The resulting high frequency vibrations are transmitted into both structures (irrespective of which surface contains the largest undulations) and from the vibrating structures sound is radiated into the air (see Figure 1.2). The wavelengths of importance for rolling noise can be determined from equation (2.1) and cover the range from tens of centimetres down to the size of the contact area (about 1 cm).

This surface unevenness is commonly referred to as 'roughness', a term which will also be used throughout this book. However, this is perhaps a rather unfortunate

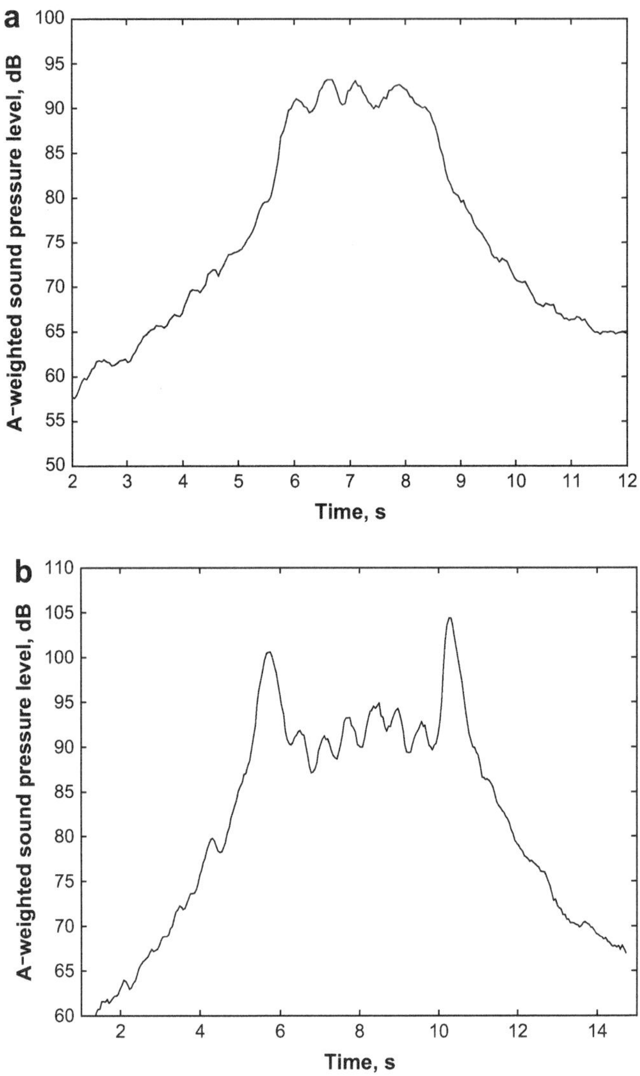

FIGURE 2-1 Examples of time histories of sound pressure levels from passing trains measured at 4 m from the near rail (up track near Beaulieu Road station, March 2001). A-weighted sound pressure level is shown at 0.05 s intervals with overlapping averages of length 0.2 s. (a) Four-coach Mk I EMU travelling at 120 km/h; (b) HST (two power cars and seven trailers) travelling at 135 km/h

term as roughness more usually refers to a wavelength scale of the order of a millimetre or less (micro-roughness). While micro-roughness is essential for adhesion (traction or braking), longer wavelength evenness (macro-roughness) is undesirable.

Measurement and analysis of roughness will be discussed in Chapter 5, but the effect of roughness will be considered qualitatively in the next section.

2.2 SPEED AND ROUGHNESS DEPENDENCE

Rolling noise levels increase with increasing train speed. The A-weighted sound pressure level is usually taken to be proportional to the logarithm of the speed, according to an equation such as:

$$L_p = L_{p0} + N.\log_{10}\left(\frac{V}{V_0}\right) \tag{2.2}$$

where L_{p0} is the sound level at a reference speed V_0. Values of the speed 'exponent' N, determined from measurements on the basis of linear regression, are usually found to be between about 25 and 35, with a typical value of 30. This means that a doubling of speed corresponds to an increase in A-weighted level of 8 to 10 dB.

Figure 2.2 shows a series of results measured at 7.5 m from the track during the passage of one type of train plotted against speed on a logarithmic scale [2.1]. Also shown is a regression line of the form of equation (2.2), with $N = 33.3$. Omitting the obvious outliers the measured data has a standard deviation of 1.5 dB relative to the regression line. In each case the track was good quality continuously welded rail; measurements of rail roughness were made and were found to be similar to the ISO 3095 limit (see Appendix A). Small corrections for differences in roughness between these tracks of up to 1.5 dB have been applied.

Examples of the average sound levels for various types of train measured at 25 m from the track during a train pass-by are shown in Figure 2.3 plotted against speed. These results are taken from various sources [2.2–2.5]. It can be seen that the results

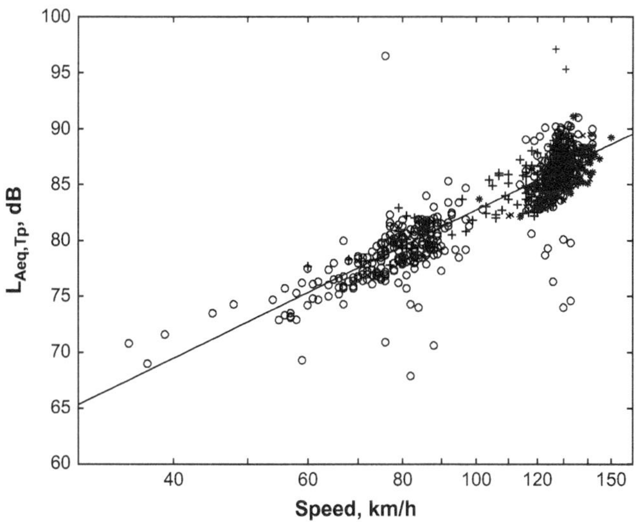

FIGURE 2-2 Sound levels at 7.5 m from the track, measured data and regression line for NS double deck multiple units (DDM) (from [2.1]). Different symbols represent different measurement sites (two tracks at each of two sites). Courtesy of ProRail

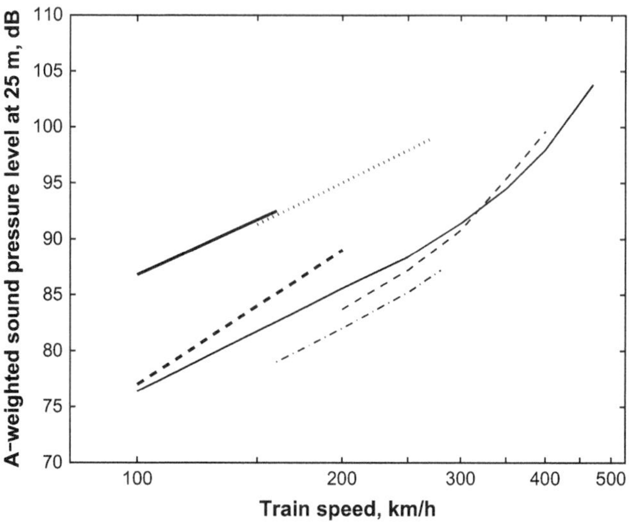

FIGURE 2-3 Rolling noise from various types of train. Tread braked vehicles: **—**, BR Mk II coaches; ····, TGV-PSE. Disc-braked vehicles: **- - -**, BR Mk III coaches; —, TGV-A, Duplex and Thalys (from Chapter 8); – – –, ICE/V; – · – ·, Talgo (drum brakes) [2.2–2.5]

for different types of trains tend to fall basically into two groups. On further inspection it is found that these groups correspond to trains with cast-iron block brakes and those without (e.g. disc braked). These two groups are 8 to 10 dB apart for a given speed. The Talgo vehicles, as well as being drum braked, have fewer wheels per unit length as they are articulated with only a single wheel on each side between each pair of vehicles.

The first generation French high speed trains (TGV-PSE) had cast-iron block brakes in addition to disc brakes, whereas the later TGV-A, Duplex and Thalys are purely disc-braked trains apart from the power cars, which in some cases have supplementary composite block tread brakes. The German ICE has disc brakes; the wheels are also fitted with 'absorbers' (dampers). For both TGV and ICE the noise levels increase at a rate of roughly $30 \log_{10} V$ up to around 300 km/h, above which aerodynamic noise becomes more important leading to an increased slope. Although high speed trains run at twice the speeds of much traditional stock, their lack of block brakes means that the noise levels of both TGV and ICE are similar at 280–300 km/h to traditional cast-iron block-braked stock at 140–160 km/h.

Figure 2.4 shows examples of roughness profiles measured on a wheel which is braked using cast-iron brake blocks and one which is unbraked [2.6]. The roughness of the block-braked wheel is characterized by a regular pattern of roughness (or corrugation) with a wavelength in this case of about 50 mm. The unbraked wheel can be seen to be mostly much smoother. As is often the case, sharp pits (holes) are found in the running surface, particularly of the braked wheel here. These have no effect on the noise and should therefore be removed mathematically from the roughness signal before determining spectra (see Chapter 5).

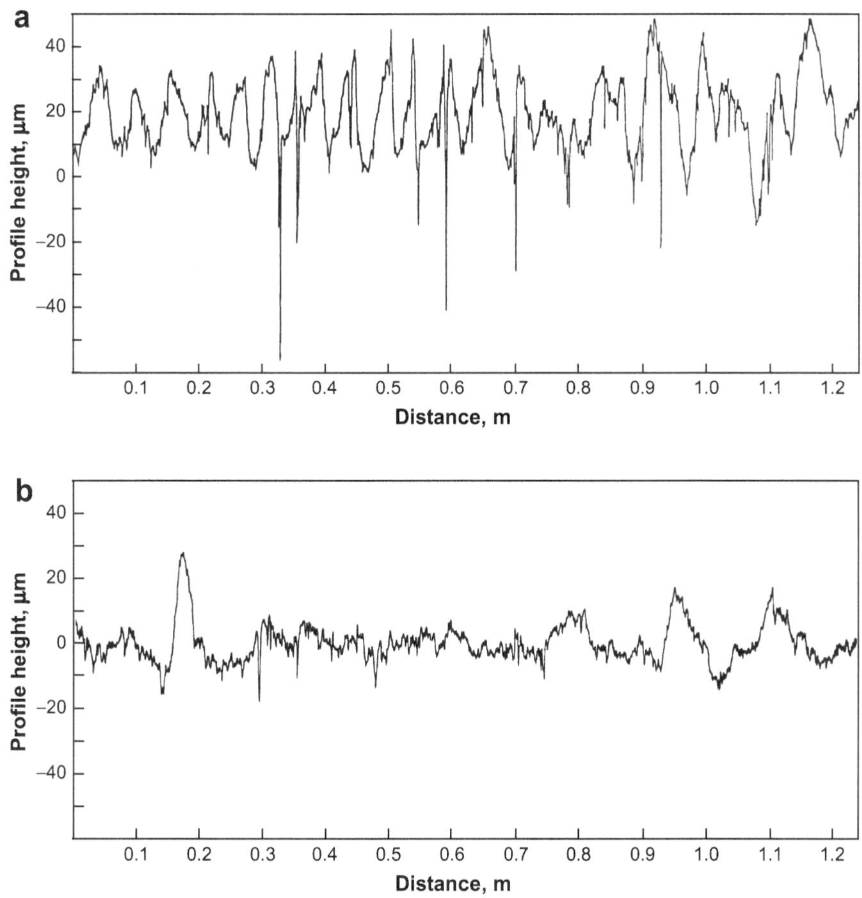

FIGURE 2-4 Wheel roughness profiles measured on (a) a wheel with cast-iron block brakes and (b) an unbraked wheel [2.6]

Although the amplitudes of around 50 µm peak-to-trough may appear small, when traversed at high speed this becomes a significant high frequency vibration. A wavelength of 50 mm traversed at 160 km/h (44 m/s) corresponds to 900 Hz, see equation (2.1). A vibrational displacement of 50 µm peak-to-trough (i.e. 17 µm r.m.s.) at this frequency is equivalent to an r.m.s. vibration velocity of 0.1 m/s, and an r.m.s. acceleration of 560 m/s^2 (e.g. to convert from displacement to velocity amplitude it is necessary to multiply by $\omega = 2\pi f$).

The difference in roughness seen in Figure 2.4 is now known to be the reason why cast-iron block-braked stock is consistently noisier than disc-braked stock (as seen in Figure 2.3). It should be pointed out that the use of composite brake blocks, instead of cast iron, can avoid the formation of wheel corrugation and thus achieve levels of roughness similar to that found for disc brakes. This will be discussed further in Chapter 7.

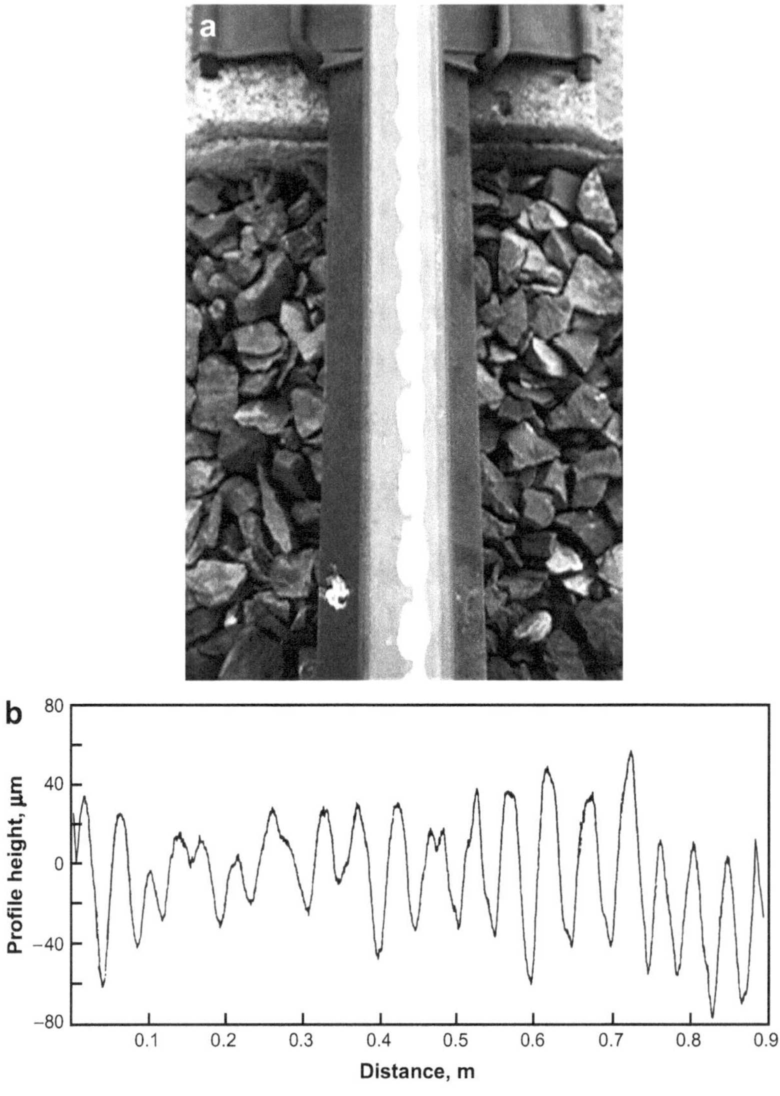

FIGURE 2-5 An example of rail corrugation (BR Derby–Burton line, Willington up track, 1986) and a measured profile at the same site (from [2.7])

Not only the wheels, but also the rails can suffer from periodic wear or corrugation problems (also known as 'roaring rail'). Figure 2.5 shows a photograph of a corrugated rail, in which bright areas can be seen at the typical 50 mm pitch. Also shown is a measurement of the roughness profile from the same site.

Rail corrugation can lead to an increase in the sound emission of cast-iron block-braked rolling stock of up to 10 dB; for disc-braked stock, which is quieter on smooth rails, the increase can be up to 20 dB. On corrugated track disc-braked and block-braked stock thus have similar noise levels to one another, emphasizing that it is the sum of the wheel and rail roughness which determines the excitation. The reasons for the occurrence of corrugation are not discussed further in this book. The interested reader should refer to, e.g., [2.8–2.11].

2.3 FREQUENCY CONTENT

Figure 2.6 shows some examples of rolling noise spectra for BR coaches with cast-iron block brakes and with disc brakes (from [2.2]). The effect of corrugated track can be clearly seen (although note the difference in measurement position means that the levels are not directly comparable between the two figures; the axes have been adjusted to allow approximately for this effect). The noise is essentially broad-band, but with its maximum level around 1–2 kHz at a train speed of 160 km/h. On corrugated track a broad peak is clearly seen at the frequency corresponding to the corrugation wavelength, here 800 Hz.

In Figure 2.7 spectra are shown for disc-braked stock (BR Mk III coaches) on six tracks ranging from very smooth to corrugated [2.7, 2.12]. The highest levels (—) were measured at the corrugated site shown in Figure 2.5. Also shown are spectra from corresponding rail roughness measurements, which indicate the same trends, although no account has been taken of wheel roughness.

Figure 2.8 shows the effect of severe wheel corrugation on the spectra of Dutch Intercity coaches with disc brakes and supplementary cast-iron block brakes. This gives a noise increase of about 7 dB in this case [2.13]. Again, similar differences are found in the wheel roughness spectra. At this site the rail roughness was also measured and found to be lower than that of either wheel type.

As speed increases, the sound in each frequency band does not increase at the same rate. At lower speeds the spectrum peaks at lower frequencies, whereas at higher speeds the peak moves towards higher frequencies. This is illustrated in

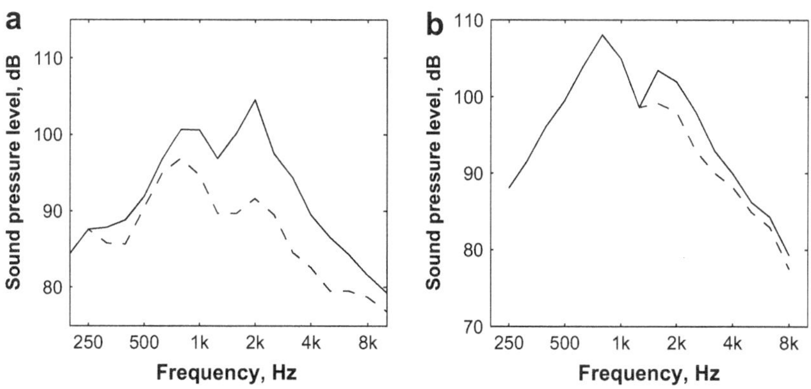

FIGURE 2-6 Noise spectra for BR Mk II coaches with cast-iron block brakes (—) and BR Mk III disc-braked coaches (– – –) at 160 km/h [2.2]. (a) On good track (measured at 2 m from track); (b) on corrugated track (measured at 5 m from track)

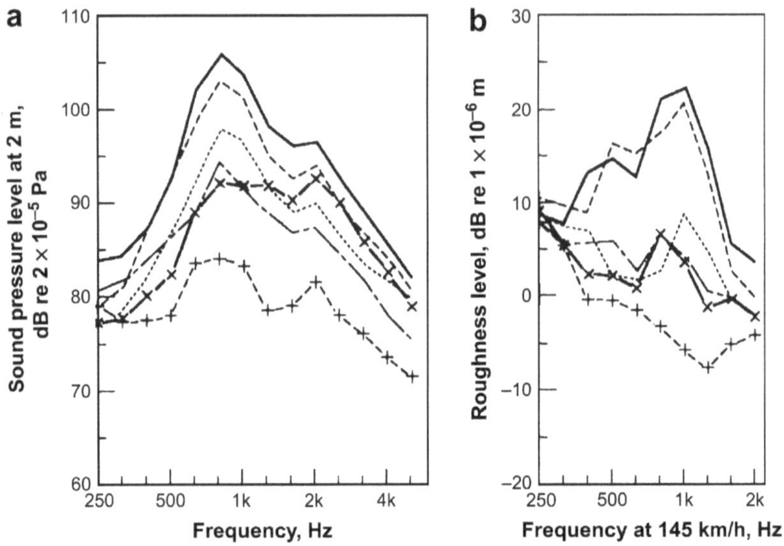

FIGURE 2-7 (a) Noise spectra for BR Mk III disc-braked stock at 145 km/h for track ranging from very smooth to corrugated (measured at 2 m from track); (b) rail roughness spectra at the same sites [2.7, 2.12]

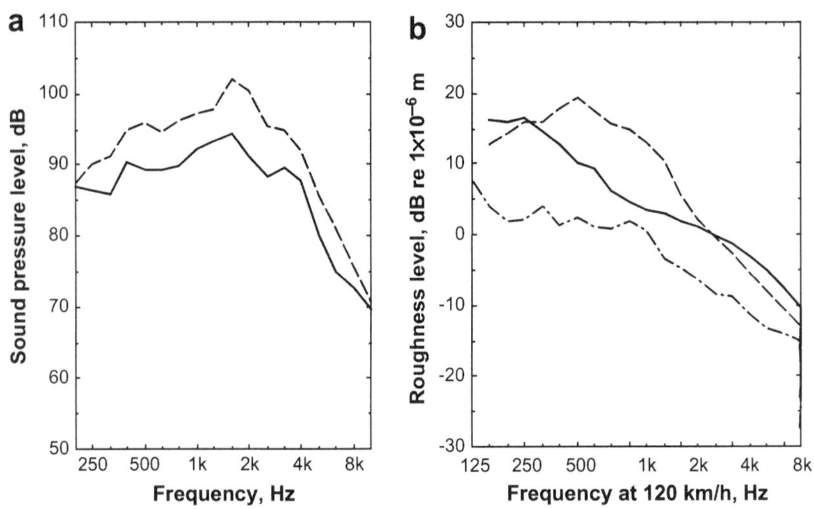

FIGURE 2-8 (a) Sound pressure level measured at 1.0 m from near rail for NS Intercity coaches travelling at 120 km/h: —, disc brakes with supplementary cast-iron block brakes, – – –, disc brakes only; (b) corresponding wheel roughness spectra, – · – ·, rail roughness [2.13]

Figure 2.9 which shows example measurements of BR trains on smooth and corrugated track at three different speeds. Especially for the corrugated track, a clear peak can be seen at 500 Hz at 80 km/h and at 800–1000 Hz at 160 km/h, corresponding to the corrugation wavelength of 50 mm, see equation (2.1).

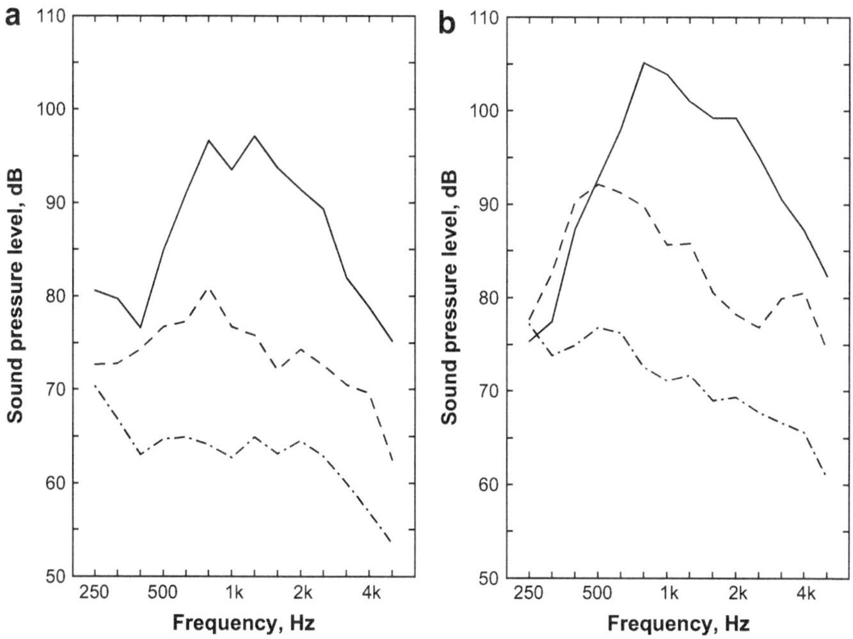

FIGURE 2-9 Effect of speed on sound pressure level from trains measured at 7.5 m from track, Commonwealth wheels with cast-iron brake blocks (a) on smooth track, (b) on corrugated track. —, 160 km/h; – – –, 80 km/h; – · – ·, 40 km/h (data from [2.6])

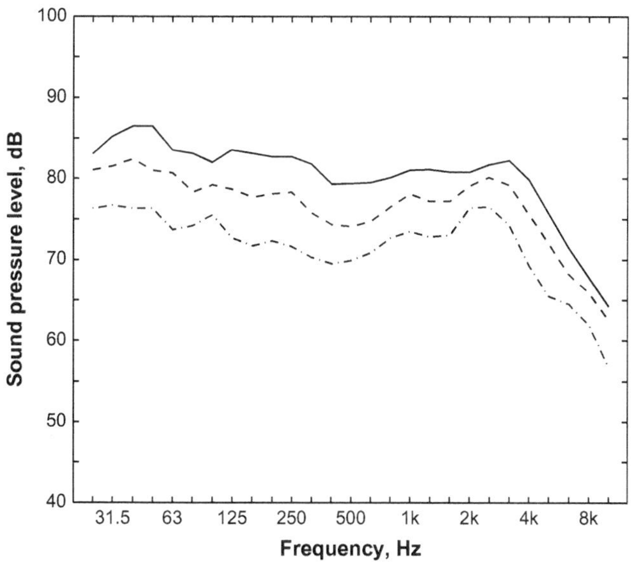

FIGURE 2-10 Sound pressure spectra from TGV-Duplex at 25 m from track (data from [2.14]). – · – ·, 200 km/h; – – –, 250 km/h; —, 300 km/h

Figure 2.10 compares frequency spectra of TGV-Duplex trains at various speeds. These trains have disc brakes on the trailer axles and composite brake blocks on the powered axles. Note that the peak in the spectrum in Figure 2.10 is at about 2–3 kHz, which is higher in frequency than the peak in Figure 2.9 for trains at conventional speeds. The high broad-band levels at low frequency are caused by aerodynamic noise (see Chapter 8).

2.4 IS IT THE WHEEL OR IS IT THE RAIL?

As discussed in Section 1.2, to achieve noise reduction of *any system* it is of the utmost importance to know what the source of the noise is, or if there is more than one source, what their relative importance is. Otherwise, a great deal of effort may be spent in reducing a component of noise which turns out to give little reduction to the overall noise level.

Therefore, the question of whether rolling noise is caused by wheel or rail has long been a subject of investigation, often with apparently conflicting results. It is now generally recognized that both the wheel and the rail contribute significant proportions of the noise. The exact result will, however, vary from one situation to another.

Early attempts at answering the question consisted of vibration measurements on the rail in combination with a simple radiation model for the rail [2.15, 2.16]. However, without the check of wheel vibration measurements it is easy to attribute radiation to the wheel which is actually due to discrepancies in the rail vibration or radiation estimate and conflicting results were found.

It is clearly much more desirable to have available the wheel vibration, measured using train-board instrumentation, simultaneously with the trackside measurements of rail vibration and noise. Such measured data was obtained, for example, in 1984 by tests in the UK sponsored by ORE [2.6], with six measuring positions on the wheel, and ten on the rail. This was carried out for four speeds (from 40 to 160 km/h) and a variety of roughness conditions, giving A-weighted noise levels spanning a range of more than 40 dB. A single type of wheel and track was used.

Along with such data, reliable models of the noise radiation of both wheel and rail are required. Development of such models allowed results to be determined which were published in [2.17]. These are reproduced in Figure 2.11. This shows the noise spectra predicted from the measured vibration minus the measured noise spectra. A level of less than 0 dB, therefore, corresponds to an underprediction of the noise; greater than 0 dB to an overprediction. The results have then been averaged over all speeds and roughness conditions. It can be seen that the prediction falls within one standard deviation of the measured result for the frequency range 500 to 5000 Hz, and that the standard deviation itself is less than 3 dB for frequencies between 500 and 2500 Hz, despite the wide range of input data.

The separate components of wheel and rail noise (again relative to the *total* measured noise) are shown in Figure 2.12. It can be seen that the respective components are in fixed proportions for each frequency band (despite the large variation in conditions covered by the data), at least for a single design of wheel and track. As the actual frequency content of the radiated noise varies with speed, as seen in Figures 2.9 and 2.10, the relative contributions of the wheel and rail to the total sound level will also be speed dependent. For the test results discussed here, this is

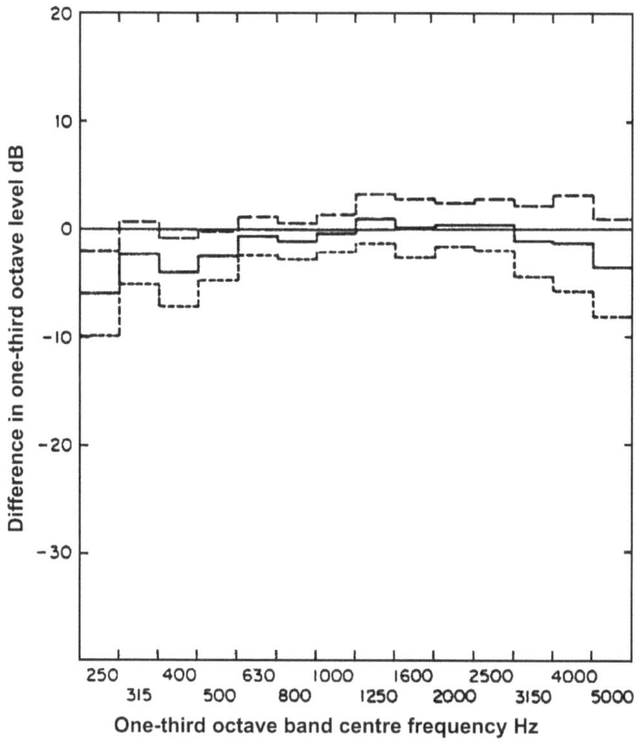

FIGURE 2-11 Estimated total noise from wheel and rail minus measured total noise: —, mean; – – –, mean + standard deviation; - - -, mean – standard deviation [2.17]

shown in Figure 2.13, taken from [2.13], which indicates that the rail has a higher contribution at low speed while the wheel contribution becomes more important at high speed.

It can be seen from Figure 2.12 that the wheel only contributes significantly to the total noise at higher frequencies (above 1 kHz), whereas the rail is a significant source for all frequencies. The low frequency limit of the wheel contribution will vary somewhat according to wheel design but is generally between 1 and 2 kHz. The particular wheel considered in these tests had a diameter of 1.06 m, which is larger than average; smaller wheels will radiate effectively from a higher frequency, as discussed in Chapters 4 and 7.

At low frequencies, the predictions appear inadequate, and it is thought that the sleepers were also an important source of noise here, not taken into account in [2.6] and [2.17].

Remington [2.19, 2.20] used a similar procedure based on measured vibration data and theoretical models of radiation to identify the relative importance of the various sources (as well as predicting the vibration levels) with similar conclusions. These tests were carried out for relatively low speeds (30–130 km/h). The total prediction was mostly within about 2 dB of the measured spectrum, but differences of up to 5 dB were found in the mid frequencies due to an overprediction of the rail noise component. The wheel appeared less significant than the rail, but in these tests

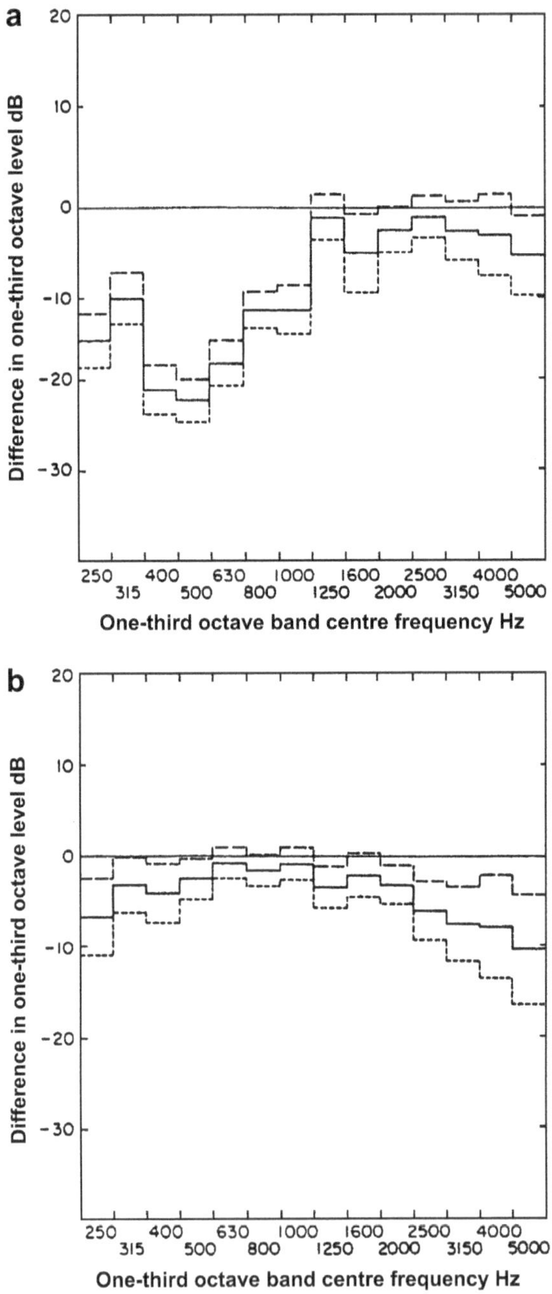

FIGURE 2-12 Estimated noise components from (a) wheel and (b) rail minus measured total noise: —, mean; – – –, mean + standard deviation; - - -, mean – standard deviation [2.17]

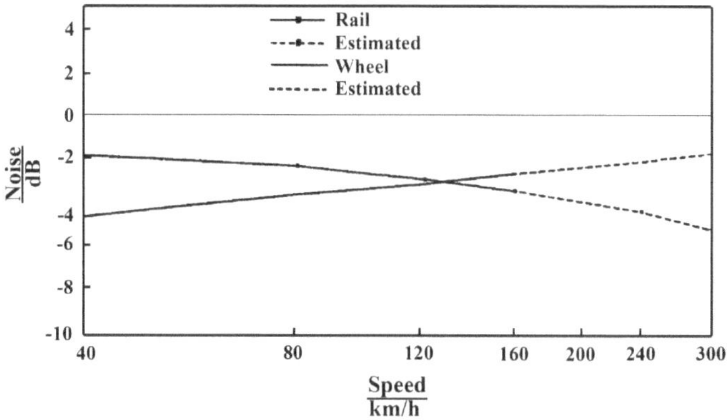

FIGURE 2-13 Estimated components of wheel and rail noise relative to total A-weighted level as a function of train speed [2.18]

it was rather small with a diameter of only 0.76 m and therefore contributed most strongly to the overall noise only above 2.5 kHz. These predictions therefore tended to give most prominence to the rail.

Further results are found in [2.21] using both measured and predicted vibration levels on wheels, rails and sleepers in combination with radiation models. A typical result is shown in Figure 2.14, based on predicted vibration levels. These results are for a freight vehicle travelling at 100 km/h on a track with a relatively soft rail pad (200 MN/m). Here, the noise radiated by the rail is the dominant component, the wheel component being about 4 dB lower. Although the sleepers are seen to be important at low frequencies, their contribution to the overall A-weighted level is small in this case. It would be greater for stiffer rail pads, see Chapter 7.

A number of researchers have also used experimental techniques in an attempt to 'localize' the source of rolling noise. The most important such technique is a microphone array placed at the trackside [2.22]. Arrays have also been widely used to

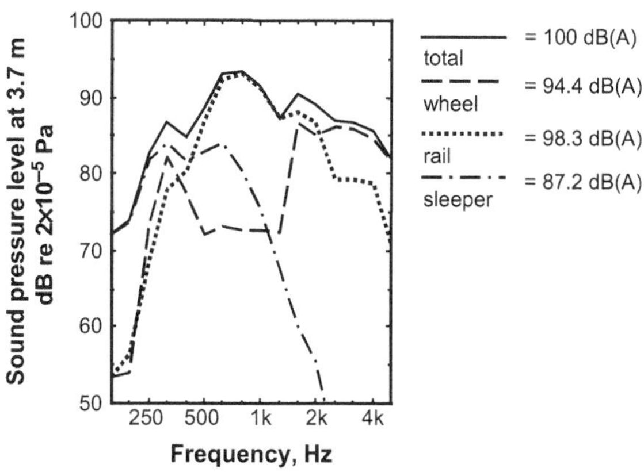

FIGURE 2-14 Results from typical prediction using TWINS model showing contributions of wheel, rail and sleeper components to total noise [2.21]

locate aerodynamic sources on high speed trains (see Chapter 8). When used in separating wheel and rail contributions, however, they often give most emphasis to the wheels. Recent work by Kitagawa [2.23] has helped to explain this. It has been found that a microphone array orientated perpendicular to the track will not 'see' the rail source and may therefore underestimate the rail contribution to the noise by as much as 10 dB. This is discussed further in Chapter 6.

2.5 OVERVIEW OF THE GENERATION MECHANISM

Figure 2.15 shows a more detailed version of the flowchart shown in Figure 1.1. This framework is the basis of a model of rolling noise generation, developed in [2.24] (see also [2.25–30] and [2.21]). This was derived from a model originally presented by Remington [2.19].

It is usually convenient to express this model in the frequency domain, that is to calculate the vibration and sound radiation at a series of frequencies, assuming in this case a steady state random process. Time domain approaches are also possible and will be discussed briefly in Chapter 10. They are also used for curve squeal, see Chapter 9. The frequency domain approach is limited to situations in which the model is linear, but its main advantage is that it usually provides much greater insight into the physical mechanisms.

The input is the combination of the wheel and rail roughness. This is modified by the contact filter, discussed in Chapter 5, so that the excitation due to roughness with wavelengths shorter than the contact patch length (typically 10–15 mm) is greatly attenuated; at 160 km/h this is equivalent to a cut-off frequency of about 4 kHz.

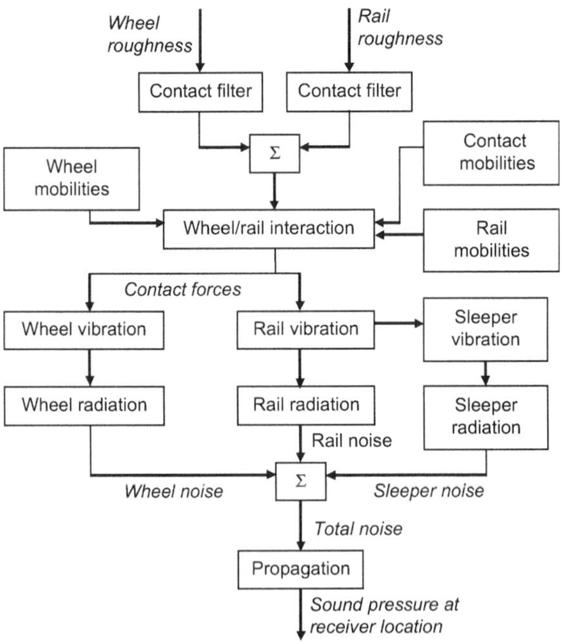

FIGURE 2-15 Schematic diagram of wheel/rail rolling noise generation mechanism

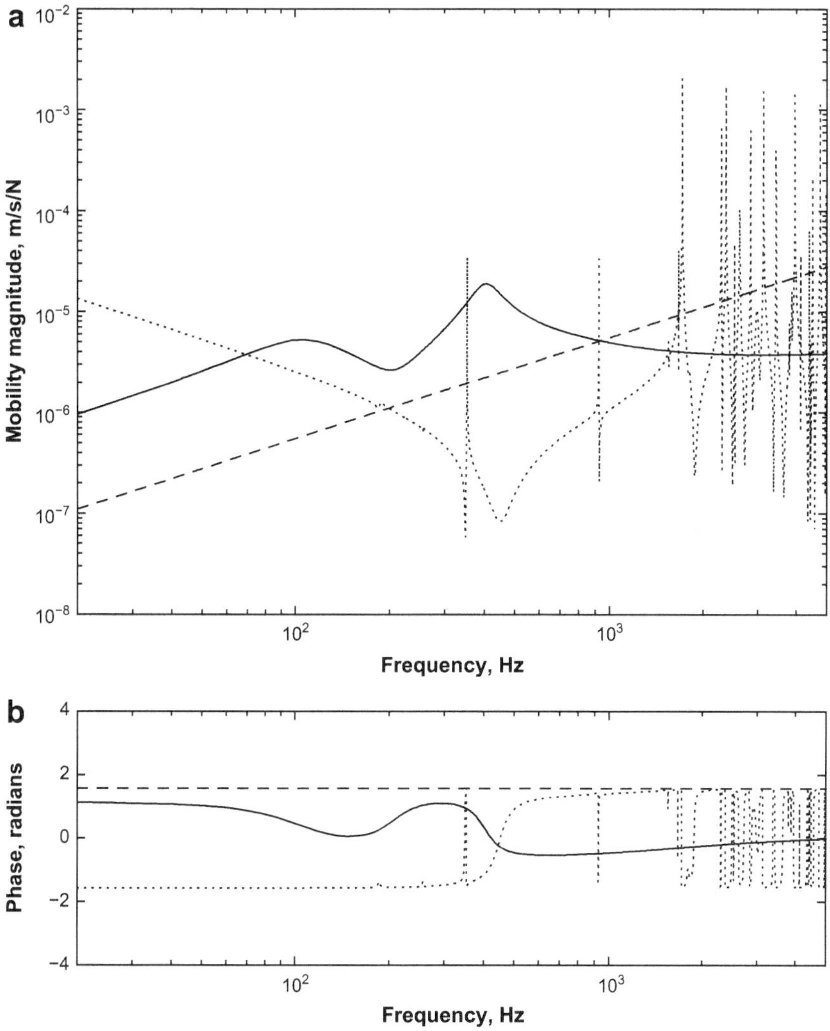

FIGURE 2-16 Vertical mobilities of the wheel/rail system. ····, Radial mobility of UIC 920 mm freight wheel; —, vertical mobility of track with moderately soft pads (parameters in Table 3.3); – – –, contact spring mobility

High frequency dynamic interaction forces are generated by this imposed relative displacement. The dynamic properties of the wheel and rail, as expressed by their frequency response functions (FRFs),[1] then determine the extent to which the displacement induced by the roughness is taken up as wheel vibration or as

[1] In [2.25–29] these were expressed as receptances (displacement per unit force). Mostly in this book they will be given in terms of mobility (velocity for a unit force). Sometimes accelerance (acceleration per unit force) is used.

rail vibration. These vibrating surfaces then radiate noise into the air, see also Figure 1.2.

Figure 2.16 compares typical mobilities of a wheel, a track and the contact spring. For frequencies where the track has the highest mobility (here between 70 and 1000 Hz), it is found that the rail vibrates with the highest amplitude. Conversely, the wheel vibration (and noise) are greater in the high frequency region where a number of lightly damped modes are present. These characteristics are reflected in the frequency regions in which track and wheel are dominant, as seen in Figure 2.14.

The framework of Figure 2.15 will be considered in more detail in subsequent chapters, beginning with the dynamic behaviour of the track and the wheels, and followed in Chapter 5 by a discussion of how they together influence their interaction. Noise radiation is discussed in Chapter 6.

REFERENCES

2.1 E. Verheijen. Geluidemissie van de spoormaterieelcategorieën, dBvision report PRO020-01-16 produced for ProRail, February 2007.

2.2 B. Hemsworth. Recent developments in wheel/rail noise research. *Journal of Sound and Vibration*, **66**, 297–310, 1979.

2.3 R. Wettschureck and G. Hauck. Geräusche und Erschütterungen aus dem Schienenverkehr, Chapter 16. In: M. Heckl and H.A. Müller (eds.). *Taschenbuch der Technischen Akustik*, 2nd edition. Springer Verlag, Berlin, 1994.

2.4 W.F. King, III. The components of wayside noise generated by high-speed tracked vehicles. *Proceedings of Inter Noise 90*, Gothenburg, Sweden, 375–378, 1990.

2.5 B. Mauclaire. Noise generated by high speed trains. New information acquired by SNCF in the field of acoustics owing to the high speed test programme. *Proceedings Inter Noise 90*, Gothenburg, Sweden, 371–374, 1990.

2.6 O.R.E. Wheel/rail contact noise – comparative study between theoretical model and measured results. Question C163, Railway Noise, Report RP7, Utrecht, 1986.

2.7 O.R.E. Wheel/rail contact noise – an experimental comparison of various systems for measuring the rail roughness associated with train rolling noise. Question C163, Railway Noise, Report RP9, Utrecht, 1988.

2.8 S.L. Grassie and J. Kalousek. Rail corrugation: characteristics, causes and treatments. *Proceedings of the Institution of Mechanical Engineers, Part F: Journal of Rail and Rapid Transit*, **207F**, 57–68, 1993.

2.9 S.L. Grassie. Rail corrugation: advances in measurement, understanding and treatment. *Wear*, **258**, 1224–1234, 2005.

2.10 J.C.O. Nielsen, R. Lundén, A. Johansson, and T. Vernersson. Train-track interaction and mechanisms of irregular wear on wheel and rail surfaces. *Vehicle System Dynamics*, **40**, 3–54, 2003.

2.11 B. Jacobson and J.J. Kalker. *Rolling Contact Phenomena*. Springer Verlag, Berlin, 2000.

2.12 D.J. Thompson. On the relationship between wheel and rail surface roughness and rolling noise. *Journal of Sound and Vibration*, **193**, 149–160, 1996.

2.13 D.J. Thompson and P.J. Remington. The effects of transverse profile on the excitation of wheel/rail noise. *Journal of Sound and Vibration*, **231**, 537–548, 2000.

2.14 P. Vinson and P. Pinconnat. TGV-Duplex à 350 km/h, bruit rayonné dans l'environnement au passage en champ libre, derrière Merlon et Écran. SNCF report AEF-D R03030/02D-127, 2003.

2.15 E.K. Bender and P.J. Remington. The influence of rails on train noise. *Journal of Sound and Vibration*, **37**, 321–334, 1974.

2.16 S. Peters, B. Hemsworth, and B. Woodward. Noise radiation by a railway rail. *Journal of Sound and Vibration*, **35**, 146–150, 1974.

2.17 D.J. Thompson. Predictions of acoustic radiation from vibrating wheels and rails. *Journal of Sound and Vibration*, **120**, 275–380, 1988.

2.18 D.J. Thompson. Theoretical modelling of wheel-rail noise generation. *Proceedings of the Institution of Mechanical Engineers. Journal of Rail and Rapid Transit*, **205F**, 137–149, 1991.
2.19 P.J. Remington. Wheel/rail rolling noise, I: Theoretical analysis. *Journal of the Acoustical Society of America*, **81**, 1805–1823, 1987.
2.20 P.J. Remington. Wheel/rail rolling noise, II: Validation of the theory. *Journal of the Acoustical Society of America*, **81**, 1824–1832, 1987.
2.21 D.J. Thompson, P. Fodiman, and H. Mahé. Experimental validation of the TWINS prediction program, Part 2: results. *Journal of Sound and Vibration*, **193**, 137–147, 1996.
2.22 B. Bariskow, W.F. King, III, and E. Pfizenmaier. Wheel/rail noise generated by a high speed train investigated by a line array of microphones. *Journal of Sound and Vibration*, **118**, 99–122, 1987.
2.23 T. Kitagawa and D.J. Thompson. Comparison of wheel/rail noise radiation on Japanese railways using the TWINS model and microphone array measurements. *Journal of Sound and Vibration*, **293**, 496–509, 2006.
2.24 D.J. Thompson. Wheel-rail noise: theoretical modelling of the generation of vibrations. PhD thesis, University of Southampton, 1990.
2.25 D.J. Thompson. Wheel-rail noise generation, Part I: Introduction and interaction model. *Journal of Sound and Vibration*, **161**, 387–400, 1993.
2.26 D.J. Thompson. Wheel-rail noise generation, Part II: Wheel vibration. *Journal of Sound and Vibration*, **161**, 401–419, 1993.
2.27 D.J. Thompson. Wheel-rail noise generation, Part III: Rail vibration. *Journal of Sound and Vibration*, **161**, 421–446, 1993.
2.28 D.J. Thompson. Wheel-rail noise generation, Part IV: Contact zone and results. *Journal of Sound and Vibration*, **161**, 447–466, 1993.
2.29 D.J. Thompson. Wheel-rail noise generation, Part V: Inclusion of wheel rotation. *Journal of Sound and Vibration*, **161**, 467–482, 1993.
2.30 D.J. Thompson, B. Hemsworth, and N. Vincent. Experimental validation of the TWINS prediction program, Part 1: Method. *Journal of Sound and Vibration*, **193**, 123–135, 1996.

CHAPTER 3

Track Vibration

3.1 INTRODUCTION

3.1.1 Track structure

A typical railway track structure is shown in cross-section in Figure 3.1. The rails are held by clips to transverse sleepers, which may be made of concrete, wood or steel. These in turn are supported in a layer of stones known as ballast, the depth of which is typically up to 30 cm beneath the sleepers. Details of the subgrade, etc. below the ballast are not of interest at this stage.

Figure 3.2 shows a typical rail fastening system. An important component of the track, which has a considerable influence on the rolling noise behaviour, is the rail pad. This is made of rubber or synthetic rubber and is located between the rail foot and the sleepers, as can be seen in Figure 3.2. These pads have the primary function of protecting the sleepers, particularly concrete ones, from high impact loads which may cause cracking but, as will be seen, their stiffness also affects the noise produced by the track.

Ballastless, or slab, tracks have been used increasingly in recent years but are still much less common than conventional ballasted track. They are generally more expensive to install but have lower maintenance costs [3.1]. In some forms of slab track, conventional concrete sleepers are embedded in concrete to form the slab; in other cases the rails are directly attached to the concrete slab. In each case, importantly, the rail fasteners used on slab tracks involve more flexible pads or baseplates than on ballasted track as they are required to replace the resilience of the ballast.

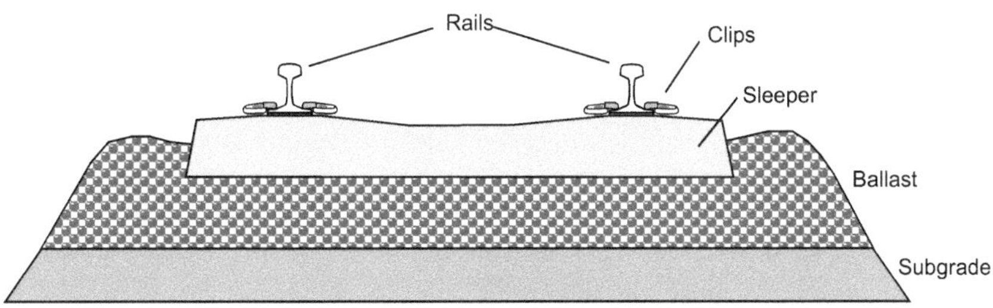

FIGURE 3-1 Typical track construction showing rails, sleepers and ballast

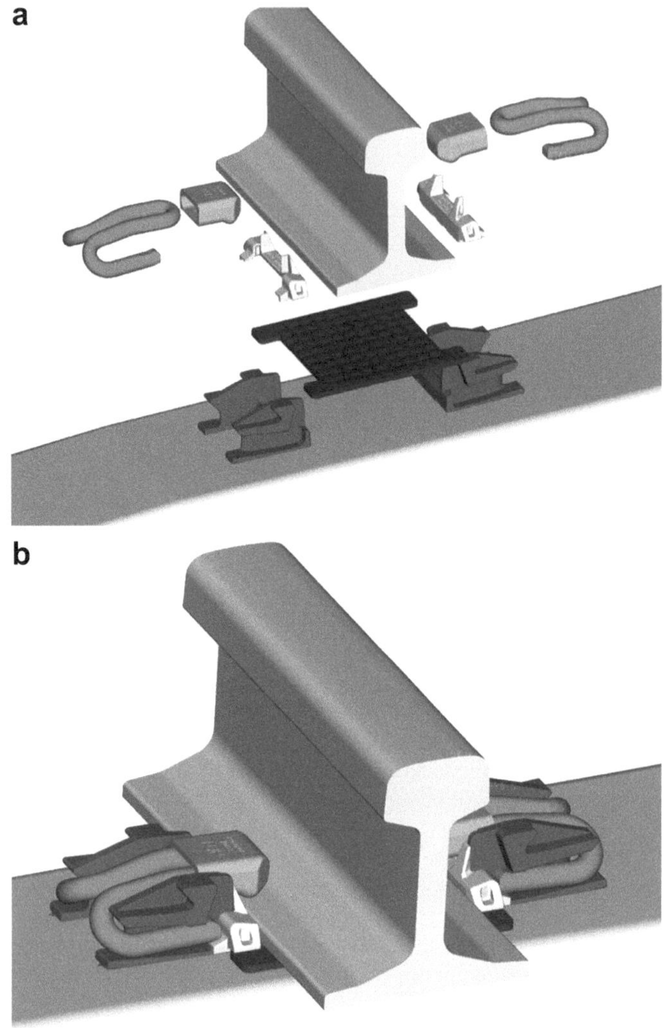

FIGURE 3-2 Pandrol Fastclip rail fastening system. (a) Exploded view showing the rail, rail pad, clips, isolators and brackets for attaching to the sleeper. (b) Assembled fastening (drawing courtesy of Pandrol, used by permission)

The wheel, being a finite and lightly damped structure, has a clear set of modes of vibration which are important in characterizing its vibration, as seen in Figure 2.16 and discussed further in Chapter 4. In contrast, the track is essentially an infinite structure. It therefore acts as a waveguide, permitting one or more structural waves to propagate along it. Moreover, the intrinsic damping of the track structure is much higher than for the wheel. Although the support structure introduces behaviour that can be understood in terms of modes of vibration at low frequency, the rail does not have resonances in the same way as the wheel. Instead, at a given frequency a number of different structural waves can exist in the rail. Initially, at low frequencies, these are vertical and lateral bending, torsional and longitudinal waves but at higher frequencies additional waves occur involving deformation of the rail cross-section.

In this section an introduction is given to track vibration on the basis of some examples of measured data. These results will then be explained in more detail in subsequent sections through the development of models for track vibration of increasing complexity.

3.1.2 Frequency responses

The dynamic behaviour of the track can be understood primarily from its frequency response functions (FRFs). Measured vertical FRFs of a track with concrete bi-bloc sleepers and quite stiff rail pads are shown in Figure 3.3, replotted from [3.2] in the form of mobilities. They were measured by exciting the rail using an instrumented impact hammer, which includes an integral force transducer; the response was measured using an accelerometer mounted on the railhead close to the point of excitation.

Some differences can be seen between the results obtained when the measurement is taken above a sleeper and between sleepers. These differences are associated with the 'pinned–pinned mode'. For the vertical direction this occurs at about 1 kHz, depending on the sleeper spacing. At this frequency half the bending wavelength in the rail corresponds to the sleeper spacing. A peak is found in the FRF between sleepers and a sharp dip above a sleeper.

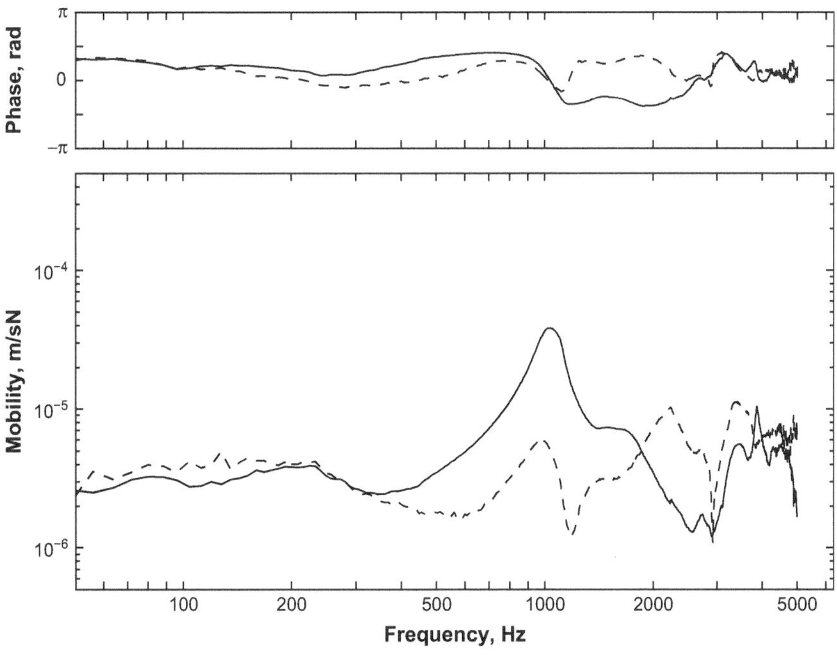

FIGURE 3-3 Vertical mobility for bi-bloc track with stiff pad. —, measured at mid-span; – – –, measured above a sleeper (adapted from [3.2])

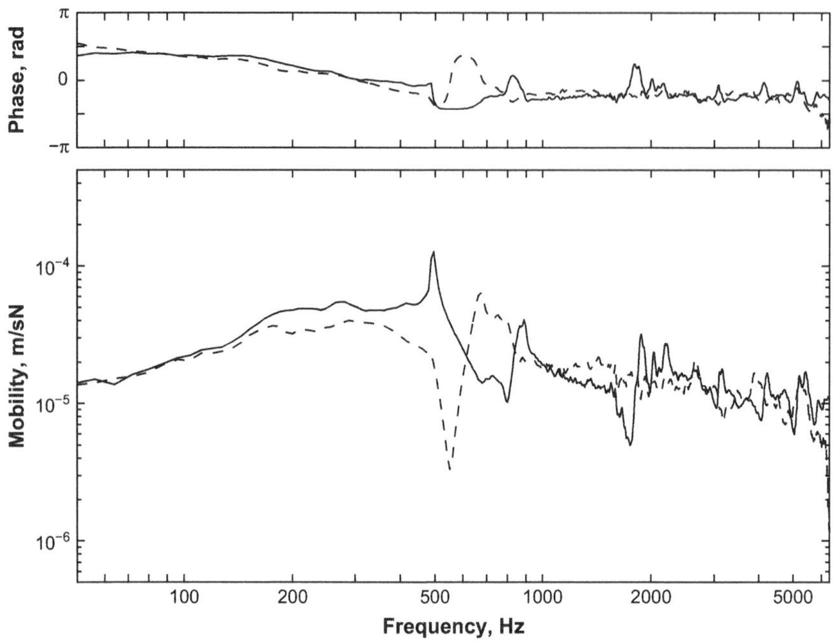

FIGURE 3-4 Lateral mobility for bi-bloc track with stiff pad. —, measured at mid-span; – – –, measured above a sleeper (adapted from [3.2])

It can be seen that the rail frequency response functions are much flatter than those of the wheel. For the wheel its many resonance peaks and anti-resonance dips mean that the response has a dynamic range of at least 80 dB (see Figure 2.16), whereas for the rail the range is much more limited. This is a consequence of the infinite nature of the track system.

Figure 3.4 shows corresponding measurements for the lateral direction. Here, the pinned–pinned mode occurs at about 500 Hz and is much less pronounced. The general level of the mobility is higher than for the vertical direction.

The coupling between vertical and lateral motions is very sensitive to the lateral position on the railhead. For the ideal case of a completely symmetrical rail and sleeper there should, theoretically, be no coupling at all. For positions further from the centreline the coupling increases due to rotation of the rail section. Figure 3.5 shows an example measurement of the cross mobility (lateral response to a vertical force, or vice versa) for a position 25 mm from the rail centreline.

In order to illustrate the effect of rail pad stiffness, Figure 3.6 shows the vertical mobility for a track with a much lower pad stiffness than the above results. From [3.2] the vertical pad stiffness is 1.3×10^9 N/m for the track in Figure 3.3 and about 2×10^8 N/m for the track in Figure 3.6. The pinned–pinned mode is much less pronounced but still occurs at 1 kHz since the frequency is determined by the sleeper spacing, not the pad stiffness. A broad peak occurs at about 450 Hz which corresponds to a resonance of the rail mass on the pad stiffness. This actually

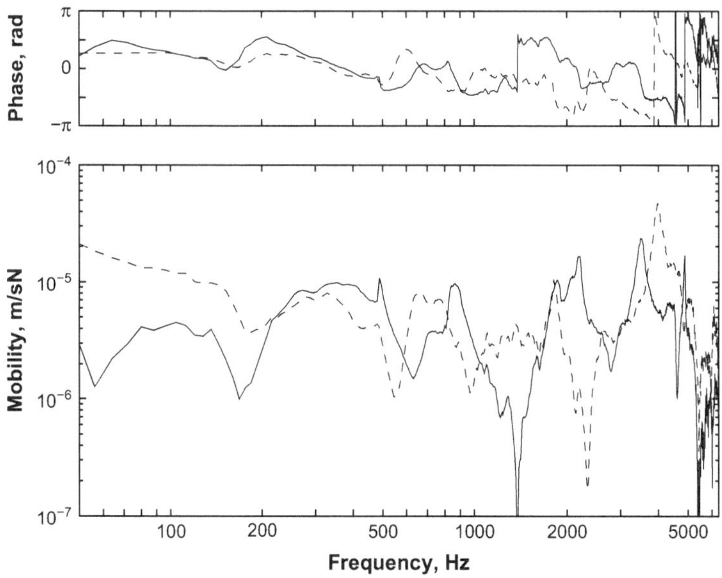

FIGURE 3-5 Vertical-lateral cross mobility for bi-bloc track with stiff pad at 25 mm from rail centreline. —, measured at mid-span; – – –, measured above a sleeper (adapted from [3.2])

occurs at about 1 kHz for the stiffer track in Figure 3.3, making it difficult to identify separately from the pinned–pinned mode. For both tracks, at around 100 Hz there is a highly damped peak at which the whole track mass bounces on the ballast stiffness.

3.1.3 Track decay rates

Although the rail experiences much greater damping than the wheel, this damping does not significantly affect the point frequency responses of the track, at least for frequencies above 500–1000 Hz where free wave propagation occurs in the rail. The phase of the mobility, especially in Figures 3.4 and 3.6, indicates the dominance of 'damping' terms (phase around 0°) rather than mass (~−90°) or stiffness (~90°), but this 'damping' is principally a result of the transmission of energy away from the driving point along the infinite rail. It would also be present for an undamped infinite beam, as will be seen in Section 3.2.

Damping introduced through the support system is, nevertheless, an important parameter for the track because it influences the decay of vibration along the track, and hence determines the effective length which is vibrating. The longer the section of rail which vibrates for each wheel, the more noise is radiated. This damping has two sources: (i) losses occurring in the resilient fastening systems (rail pads, etc.), and (ii) energy transmitted into the sleepers and the ground which also appears as damping to the rail. In addition, the support system blocks the transmission of vibration along the rail at low frequencies. Although this gives high decay rates these are not associated with damping.

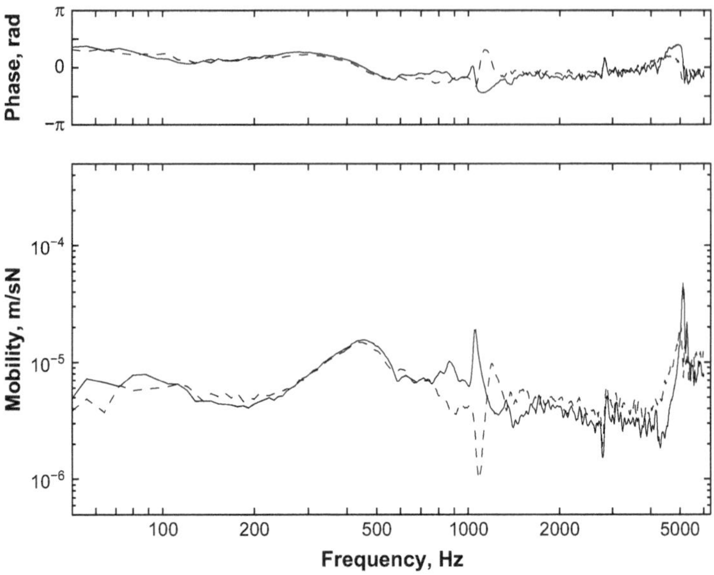

FIGURE 3-6 Vertical mobility for bi-bloc track with softer rail pad. —, measured at mid-span; – – –, measured above a sleeper (adapted from [3.2])

The rail pad stiffness affects the damping of the rail and the degree of coupling between the sleepers and the rail, as indicated in Figure 3.7. For soft rail pads the sleepers are well isolated from the rail vibration, but the vibration can propagate relatively freely along the rail. Conversely, for stiff pads the rail vibration is restricted by the coupling to the sleepers and damping of the pads but the sleeper vibration is greater. This affects the balance of noise produced by the rail and the sleeper, as discussed further in Chapter 7.

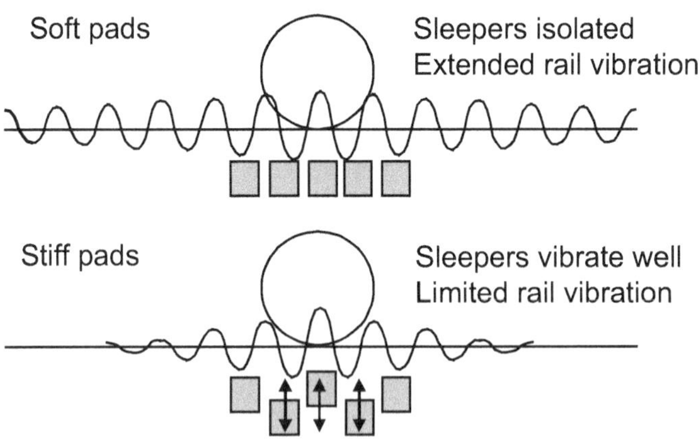

FIGURE 3-7 Illustration of the effect of the rail pad stiffness on the coupling between rail and sleeper and on the damping of waves in the rail

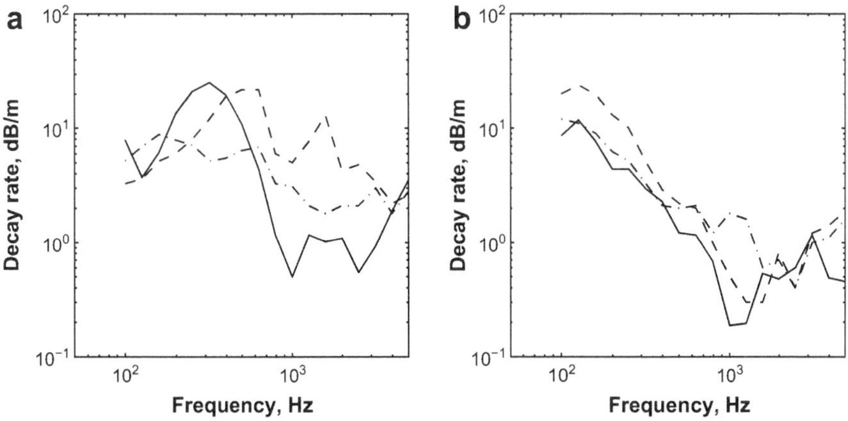

FIGURE 3-8 Measured attenuation with distance in track. —, track with bi-bloc sleepers and softer rail pad; – – –, track with bi-bloc sleepers and stiff rail pad; – · – ·, track with wooden sleepers; (a) vertical motion, (b) lateral motion

The rail vibration amplitude decays approximately exponentially with distance along the track. The greater the damping of the track, the greater this decay. The parameter used to describe this is the track decay rate, usually expressed in dB/m. This can be determined experimentally by measuring the transfer frequency responses of the track using an impact hammer and accelerometer for a number of distances between force and response locations. It is usually easiest to keep the accelerometer fixed and move the force position. The FRFs are then averaged into frequency bands (e.g. one-third octaves) and the rate of decay with distance is determined either by curve fitting or, better, by numerical procedures [3.3]. It is also possible to measure the track decay rates using the rail vibration during a train pass-by [3.4].

Examples of measured track decay rates are shown in Figure 3.8 for three types of track, replotted from [3.2]. Two are from the same tracks as the results in Figures 3.3 and 3.6, both with concrete sleepers but with different rail pad stiffnesses. The third has wooden sleepers with very stiff fasteners.

For the concrete sleepered track with resilient pads, a broad peak is seen in the decay rate for the vertical motion around 300 Hz. This occurs due to the sleeper and pad acting as a 'dynamic absorber'.[1] Above about 500 Hz the decay rate drops to less than 1 dB/m as waves propagate freely along the track. For the track with the stiffer pad the vertical decay rate remains high for much of the frequency range. The main peak is shifted to 400–600 Hz. For both these tracks the lateral decay rate is much lower, however, especially in the region 1–2 kHz. This means that significant noise generation may occur due to the lateral motion, even though, at the contact point,

[1] A dynamic absorber or neutralizer is a mass–spring system added to a host structure, which effectively pins the host structure at its resonance frequency [3.5]. Depending on the application, they can be used to neutralize a particular forcing frequency, to add damping or to alter the natural frequencies of the host structure. The application of these to a railway track is discussed further in Chapter 7.

the vertical motion is usually higher due to the roughness excitation acting in the vertical direction.

For the wooden sleepered track, the vertical decay rate is moderately high across the frequency range but the strong peak found for the concrete sleepers does not occur. This is because the sleeper mass is much lower, so the 'dynamic absorber' effect is smaller. Because of the much stiffer connection between the rail and sleeper on this track, much more vibration is transmitted to the sleepers, which causes the additional attenuation. Noise from the rail will therefore be less. On the other hand, the sleepers will also radiate noise, and since their surface area is greater than that of the rail and the amplitudes are similar, the sleepers will actually be the dominant component of track radiation up to 1 kHz.

The ratio of the vibration measured on the sleeper to that on the rail is shown in Figure 3.9 for these same three tracks, replotted from [3.2]. These were measured during a train pass-by, although similar results are obtained using hammer excitation [3.2]. The sleeper vibration is similar in amplitude to the rail vibration up to about 250 Hz for the concrete sleepered track with resilient pads, above which its vibration reduces with increasing frequency. For the stiffer pad, the sleeper is only isolated from the rail above about 600 Hz. The wooden sleeper vibrates with a similar amplitude to the rail for frequencies up to 1 kHz.

To illustrate more directly the effect of pad stiffness on the track decay rates, Figure 3.10 shows measurements of the track decay rate for a single track fitted with rail pads of three different stiffnesses [3.6]. The decay rate drops sharply from its low frequency value of around 10 dB/m. The frequency at which this drop occurs increases as the pad stiffness increases.

To explain these phenomena in more detail, and for use in predicting noise radiation, track models of various degrees of complexity will be introduced in the

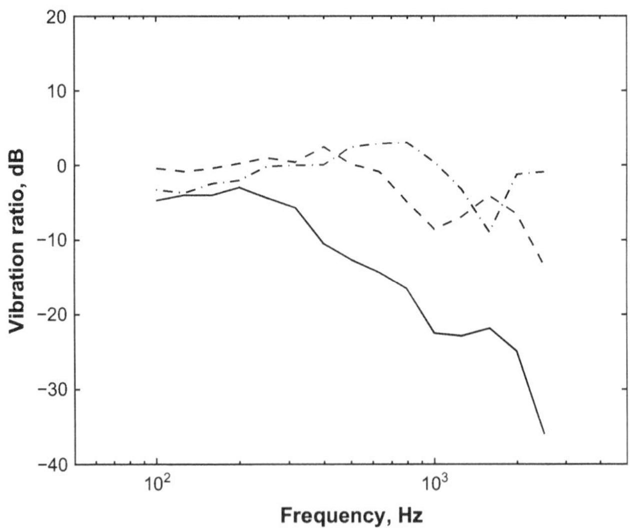

FIGURE 3-9 Ratio of sleeper to rail vibration, measured during a train pass-by. —, track with bi-bloc sleepers and softer rail pad; – – –, track with bi-bloc sleepers and stiff rail pad; – · – ·, track with wooden sleepers

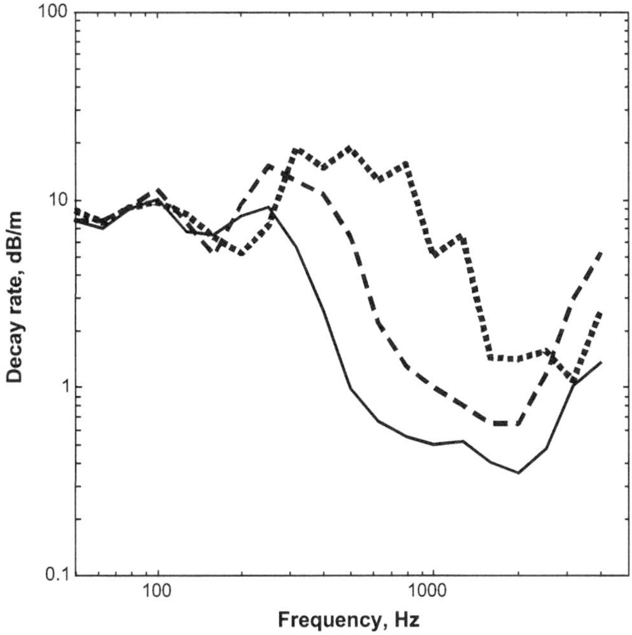

FIGURE 3-10 Measured effect of rail pad stiffness on track decay rates: same track fitted with three rail pads of different stiffnesses (approximately 125, 270 and 950 MN/m) [3.6]

following sections. Results will focus on the vertical vibration of the rail as this is often the more important for noise radiation, but the same models can also be applied to lateral vibration. In Sections 3.2 to 3.4 beam models of the track will be developed in some detail as these give physical understanding of the phenomena seen in the track vibration. More complex models involving discrete supports and cross-sectional deformation of the rail will be described briefly in Sections 3.5 and 3.6. In Section 3.7 the vibration of the sleeper will be discussed briefly along with the effect of the ballast stiffness. Finally, in Section 3.8 the dynamic properties of rail pads will be addressed.

3.2 SIMPLE BEAM MODELS

A useful model to consider first is that of a simple infinite beam, representing the rail, on an elastic foundation, as shown in Figure 3.11. This is often referred to as a Winkler foundation (see, e.g., [3.7]). While this does not contain all the features of a track, it demonstrates a number of important aspects of its dynamic behaviour.

3.2.1 Free wave propagation

The equation of free motion of an Euler–Bernoulli beam on an elastic foundation in the absence of damping is, e.g., [3.8]

$$EI\frac{\partial^4 u}{\partial x^4} + su + m'_r\frac{\partial^2 u}{\partial t^2} = 0 \qquad (3.1)$$

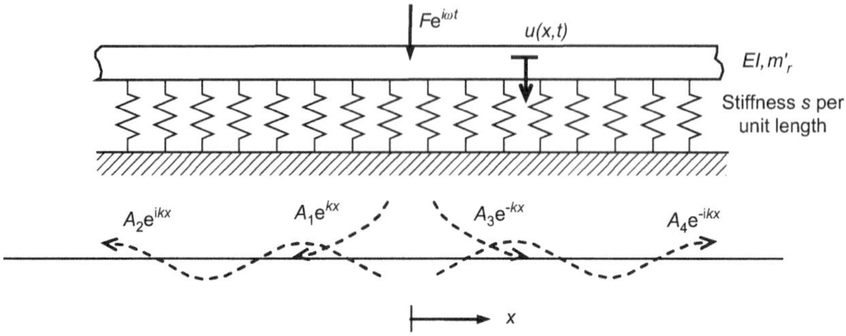

FIGURE 3-11 A beam on an elastic foundation showing the waves generated by a point force at $x = 0$

where E is Young's modulus, I is the second moment of area of the cross-section (EI is the bending stiffness), m'_r is the mass per unit length of the beam and s is the stiffness per unit length of the foundation. The bending vibration of the beam is denoted $u(x, t)$ where x is the coordinate along the beam and t is time (see box on page 39 for the coordinate system adopted). Considering harmonic motion at frequency ω, free wave solutions can exist of the form

$$u(x,t) = Ue^{i\omega t}\, e^{-ikx} \tag{3.2}$$

where U is the complex amplitude.[2] Real positive values of the wavenumber, k, correspond to waves travelling in the positive x direction and negative values to negative travelling waves. However, as will be seen, the wavenumber k can be complex. The real part of k is the phase change per unit distance, equal to $2\pi/\lambda$, with λ the wavelength of vibration. It can be likened to the spatial equivalent of frequency and has units rad/m. The imaginary part corresponds to decay with distance (see Section 3.2.3 below). Substituting this form of solution into equation (3.1) gives the 'dispersion relation' between k and ω:

$$EI\, k^4 + s - m'_r\omega^2 = 0 \tag{3.3}$$

This can be rearranged into the form

$$k^2 = \pm\sqrt{\frac{m'_r\omega^2 - s}{EI}} = \pm\sqrt{\frac{m'_r(\omega^2 - \omega_0^2)}{EI}} \tag{3.4}$$

where it is helpful to define the frequency ω_0 as

$$\omega_0 = \sqrt{\frac{s}{m'_r}} \tag{3.5}$$

[2] The usual convention of using complex amplitudes is followed in which the actual solution is understood to be the real part of equation (3.2).

Axis systems

In order to minimize confusion a common axis system is used throughout this book. The choice of a universally applicable axis system necessarily involves a compromise. The axes adopted are shown below. They are chosen as a right-handed axis system in which

- x is in the longitudinal direction, positive in the direction of travel;
- y is in the transverse direction, positive away from the centre of the track;
- z is vertically downwards;
- the origin is at the centre of the contact between wheel and rail.

For a left-hand wheel, the y coordinate is positive towards the centre of the track. Local coordinates can be defined at the contact (Chapter 5) in which z is normal to the contact plane.

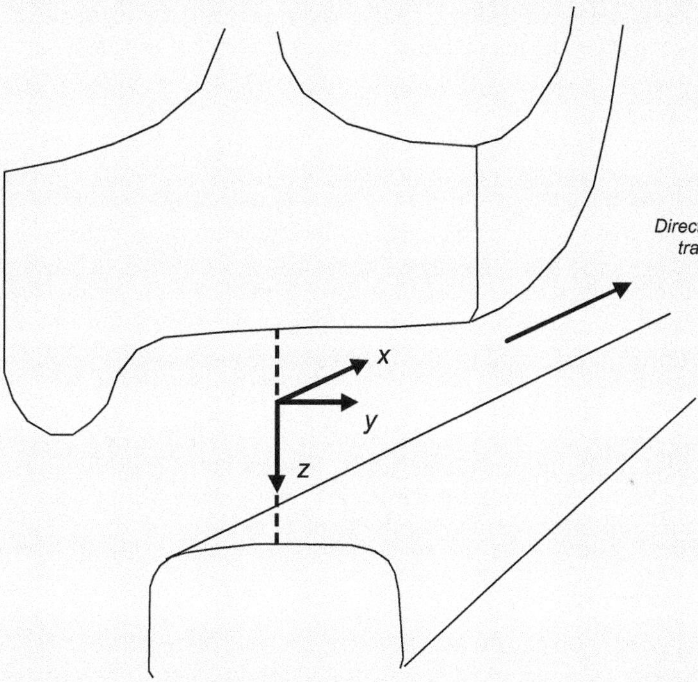

Axis system with origin at centre of wheel/rail contact zone

For the wheel, in Chapter 4, cylindrical polar coordinates are used, which for consistency with the above are denoted (r, θ, y).

For sound propagation to the wayside, in Chapter 6, it is less intuitive to use a vertical coordinate which is positive downwards, so heights h $(=-z)$ will be introduced.

The frequency ω_0 corresponds to the 'resonance' of the mass of an elementary length of the beam, $m'_r dx$ on the corresponding support stiffness $s\,dx$. For frequencies above ω_0, the solutions for k are either real or imaginary, depending on the sign of the square root taken in equation (3.4). The real wavenumbers correspond to propagating waves, the imaginary ones to near-field or evanescent waves, in the same way as for an unsupported beam [3.8]. These wavenumbers are always less than the corresponding ones for the unsupported rail, which are given by

$$k_B = \left(\frac{m'_r \omega^2}{EI}\right)^{1/4} \tag{3.6}$$

but they tend towards k_B at high frequency. However, for frequencies below ω_0, k^4 in equation (3.4) is negative. The wavenumber k cannot then be real and free wave propagation does not occur. Instead, all four waves have a wavenumber with a non-zero imaginary part, equal in magnitude to the real part. For $\omega \ll \omega_0$:

$$k \approx \frac{\pm 1 \pm i}{\sqrt{2}} \left(\frac{s}{EI}\right)^{1/4} \tag{3.7}$$

Here the real and imaginary parts of k are equal in magnitude and independent of frequency. Note that $(s/EI)^{1/4}$ is the value of the free beam wavenumber, k_B, at ω_0.

The frequency ω_0 is called a 'cut-on' frequency; as frequency increases the free waves begin to propagate above this frequency (sometimes the terminology 'cut-off' is preferred, e.g. [3.8]). At the cut-on frequency, $k \to 0$ and the wavelength of free wave propagation becomes infinite, meaning that the whole beam moves in phase along its length. The phase speed, $c = \omega/k$, becomes infinite at this frequency.

The *group velocity*, the speed at which energy is transmitted along the beam, is given by the slope of the dispersion relation, $c_g = \partial\omega/\partial k$ for real positive k. As k is not proportional to ω the waves are dispersive and $c_g \neq c$. Differentiating equation (3.3) with respect to k allows the group velocity to be found as

$$c_g = \frac{\partial \omega}{\partial k} = \frac{2EIk^3}{\omega m'_r} \tag{3.8}$$

From this it can be seen that when $k \to 0$ the group velocity tends to zero indicating that no energy is transmitted along the beam.

3.2.2 Effect of damping

The model described up to this point contains no damping. In practice, damping is present in the support and this can be included most simply by replacing the support stiffness by a complex stiffness, making use of the damping loss factor η [3.5, 3.9]. Thus s is replaced by $s(1 + i\eta)$. Similarly damping can be added to the beam itself by making E complex with the form $E(1 + i\eta_r)$ where η_r is the loss factor of the rail. This results in a modified form of equation (3.3):

$$EI(1 + i\,\eta_r)k^4 + s(1 + i\,\eta) - m'_r\,\omega^2 = 0 \tag{3.9}$$

The implications of this damping model will be discussed further in Section 3.2.7 below. Substituting k_B from equation (3.6) for the free wavenumber in the *unsupported* (and undamped) beam, the wavenumber in the supported beam can be expressed as

$$k = k_B(1+i\eta_r)^{-1/4}\left(1-\frac{s(1+i\eta)}{m_r'\omega^2}\right)^{1/4} \approx k_B\left(1-\frac{i\eta_r}{4}\right)\left(1-\frac{(1+i\eta)}{(\omega/\omega_0)^2}\right)^{1/4} \tag{3.10}$$

where it is understood that the solution with positive real part and (small) negative imaginary part is taken. Three particular cases can be considered:

(i) At high frequency, $\omega \gg \omega_0$, the real part $\mathrm{Re}(k) \approx k_B$. The imaginary part is given by

$$\mathrm{Im}(k) \approx -k_B\left(\frac{\eta}{4}\left(\frac{\omega_0}{\omega}\right)^2 + \frac{\eta_r}{4}\right) \tag{3.11}$$

Since $k_B \propto \omega^{1/2}$, the first term is proportional to $\eta\omega^{-3/2}$, whereas the second term is proportional to $\eta_r\omega^{1/2}$. Thus, although generally $\eta \gg \eta_r$, at high enough frequency the rail damping will predominate over that in the support.

(ii) At the cut-on frequency, $\omega = \omega_0$, $k \approx (-i\eta s/EI)^{1/4}$, the damping in the rail having negligible effect. Of the various roots, the one with the smallest imaginary part (and hence the lowest decay) has $k = e^{-i\pi/8}(\eta s/EI)^{1/4}$. Hence:

$$\mathrm{Re}(k) = 0.924\left(\frac{\eta s}{EI}\right)^{1/4}, \quad \mathrm{Im}(k) = -0.383\left(\frac{\eta s}{EI}\right)^{1/4} \tag{3.12}$$

(iii) For $\omega \ll \omega_0$, the damping has negligible effect and from equation (3.7):

$$\mathrm{Re}(k) = -\mathrm{Im}(k) \approx \frac{1}{\sqrt{2}}\left(\frac{s}{EI}\right)^{1/4} \tag{3.13}$$

To illustrate these results, use will be made of the set of typical parameters given in Table 3.1. The beam parameters apply to a UIC60 rail section in vertical bending. Figure 3.12 shows the real and imaginary parts of the wavenumbers plotted against frequency f. The cut-on frequency in this case is $f_0 = \omega_0/2\pi = 205$ Hz, shown by the vertical dotted line.

Below ω_0 the real and imaginary parts both tend to a constant value of 1.41 rad/m as predicted by equation (3.13). As well as these curves, Figure 3.12 shows the wavenumber of the unsupported beam, k_B, which forms the asymptote for the real part of k at high frequencies. The imaginary part follows the trend given in equation (3.11), falling according to $\omega^{-3/2}$ in the absence of rail damping. Also shown in the figure is the imaginary part for the case when damping is added to the rail with $\eta_r = 0.02$. As shown in equation (3.11), the beam damping dominates the imaginary part at frequencies well above ω_0. The real part of k is unaffected by damping for small values of η and η_r. As will be seen, this value of η_r, although too large to represent losses in the rail itself, provides a good fit to experimental results (e.g. Figures 3.8 and 3.10).

TABLE 3-1 TYPICAL PARAMETERS USED FOR RAILWAY TRACK IN SECTION 3.2

Vertical bending stiffness	EI	6.42 MN/m^2
Mass per unit length	m_r'	60 kg/m
Damping loss factor of rail	η_r	0.02
Stiffness per unit length of foundation	s	100 MN/m^2
Damping loss factor of foundation	η	0.1

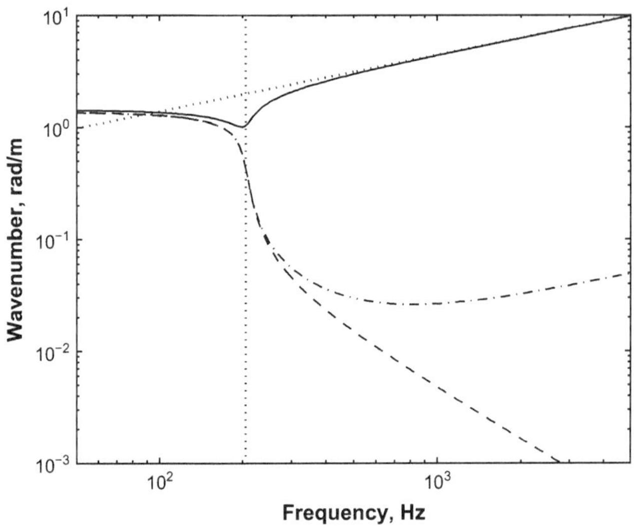

FIGURE 3-12 Wavenumbers of beam on damped elastic support. —, real part; – – –, imaginary part (negative); · · · ·, free beam; – · – ·, imaginary part (negative) for damped beam. The cut-on frequency is marked by the vertical line

3.2.3 Decay rate

The rate of attenuation of vibration along the rail is important for the noise radiated by the rail, as will be discussed in Section 6.4.5. For a single wave with a complex wavenumber $k = k_r + ik_i$ (k_i is negative) the amplitude reduces by a factor $\exp(k_i)$ over a distance of 1 m. Therefore, the decay rate in dB/m is given by

$$\Delta = -20\log_{10}(\exp(k_i)) = -8.686\,k_i \tag{3.14}$$

This is plotted in Figure 3.13 for the above parameters. At the cut-on frequency, from equation (3.12), $\mathrm{Im}(k) = -0.428$. Hence, $\Delta = 3.72$ dB/m at 205 Hz for the above parameters. Note that the decay rate well below the cut-on frequency is around 10 dB/m. This value is independent of the damping of the support but rather is given from equation (3.13) by

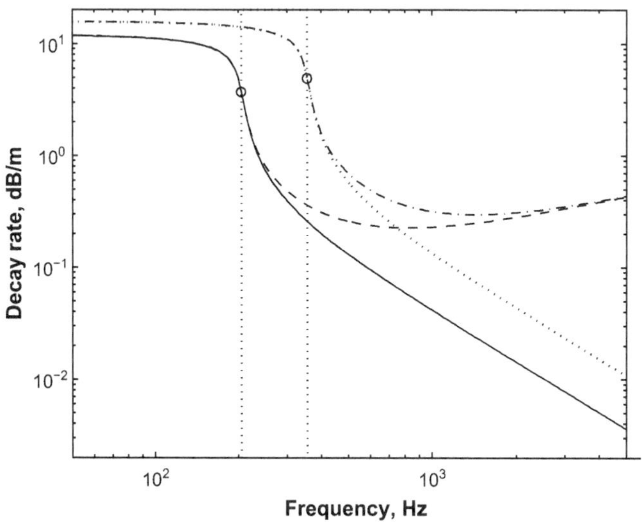

FIGURE 3-13 Wave decay rates of an undamped beam on a damped elastic support: —, $s = 100$ MN/m; ···, $s = 300$ MN/m, and a damped beam on a damped elastic support: – – –, $s = 100$ MN/m; – · – ·, $s = 300$ MN/m. The cut-on frequencies are marked by the vertical lines. The values of decay rate calculated for the cut-on frequency are shown by ○

$$\Delta = -8.686\ \mathrm{Im}(k) \approx \frac{8.686}{\sqrt{2}}\left(\frac{s}{EI}\right)^{1/4} \quad \text{for } \omega \ll \omega_0 \tag{3.15}$$

Clearly, this depends on the support stiffness s. However, even for a change in s by a factor of 3, this low frequency value of decay rate Δ is changed only by a factor of 1.3 as shown in Figure 3.13. In practice, on a conventional track the low frequency rail support stiffness is dominated by the ballast, which does not usually vary by more than a factor of about 3 in stiffness from one location to another. Therefore a low frequency decay rate of around 10 dB/m is commonly found in practice, see also Figures 3.8 and 3.10.

The effect of changing the damping loss factors is shown in Figure 3.14, where the result is shown of increasing either the loss factor of the support or that of the beam (rail) by a factor of 5. Increasing the damping of the rail is effective over a much wider frequency range than increasing the damping of the support. This has implications for noise control, which will be discussed in terms of practical systems for rail damping in Chapter 7.

3.2.4 Frequency response

Turning now to the forced response, consider a harmonic point force, $Fe^{i\omega t}$, acting on the beam at $x = 0$, as shown in Figure 3.11. The response consists of two free waves on each side of the forcing point. Above ω_0 these are a propagating wave and an evanescent wave; below ω_0 they are both highly attenuated as k is complex, as seen above.

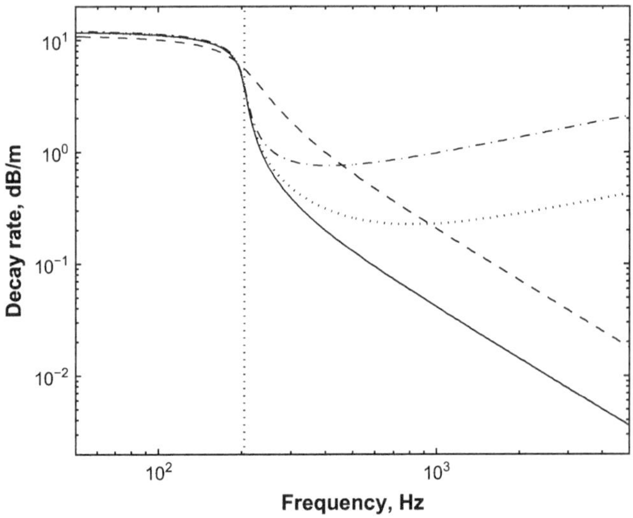

FIGURE 3-14 Wave decay rates undamped beam on damped support: —, $\eta = 0.1$; – – –, $\eta = 0.5$; damped beam (loss factor η_r) on damped support ($\eta = 0.1$): · · · ·, $\eta_r = 0.02$; – · – ·, $\eta_r = 0.1$

So that the solution is finite at $\pm\infty$ the evanescent waves must decay away from $x = 0$. Power can only flow away from the excitation point, hence also only outward travelling waves can exist. In the same way as for an unsupported beam [3.8], the solution is

$$u_-(x) = A_1 e^{k_p x} + A_2 e^{ik_p x} \qquad \text{for} \quad x \leq 0$$

$$u_+(x) = A_3 e^{-k_p x} + A_4 e^{-ik_p x} \quad \text{for} \quad x \geq 0 \tag{3.16}$$

where the time dependence $e^{i\omega t}$ is implicit, A_i are complex wave amplitudes and u is now the complex amplitude of vibration at location x. Here, the wavenumber k_p corresponds to the propagating wave. It is the solution to equation (3.10) with positive real part and (small) negative imaginary part. Four boundary conditions are available at $x = 0$. These are: continuity of displacement, rotation and bending moment and equilibrium of forces, i.e. the discontinuity of shear force is equal to the external force. (Use could also be made of symmetry yielding directly $A_1 = A_3$ and $A_2 = A_4$). Applying these conditions the solution can be found as

$$u(x) = \frac{-iF}{4EIk_p^3}\left(e^{-ik_p|x|} - ie^{-k_p|x|}\right) \tag{3.17}$$

Setting $x = 0$, the point mobility at frequency ω is given by

$$Y(\omega) = \frac{i\omega u(0)}{F} = \frac{(1-i)\omega}{4EIk_p^3} \tag{3.18}$$

This has the same form as the corresponding expressions for an unsupported beam [3.8] except that the wavenumber k_p is different in the present case. Substituting for k_p from equation (3.10):

$$Y(\omega) = \frac{(1-i)\omega}{4EIk_B^3}\left(1 - \frac{(1+i\eta)}{(\omega/\omega_0)^2}\right)^{-3/4} = \frac{(1-i)\omega^{1/2}}{4(EI)^{1/4}m_r'^{3/4}}\left(1 - \frac{(1+i\eta)}{(\omega/\omega_0)^2}\right)^{-3/4} \tag{3.19}$$

Letting $\omega \to 0$, the static stiffness of the track can be found, given by $K_T = \mathrm{i}\omega/Y(0)$. From equation (3.18), substituting for k from equation (3.7) and neglecting damping, it is found that

$$K_T = 2\sqrt{2}(EI)^{1/4}s^{3/4} \qquad \text{for} \quad \omega \ll \omega_0 \tag{3.20}$$

Figure 3.15 shows the point mobility for the example parameters in Table 3.1. A peak can be seen at the cut-on frequency, ω_0, the sharpness of which is governed by the support damping, η. Below this frequency, the mobility corresponds to a stiffness, $Y \approx \mathrm{i}\omega/K_T$, while at high frequency the mobility tends to that of the infinite unsupported beam, which has a slope of $\omega^{-1/2}$ and a phase of $-\pi/4$. This can be expected as, from equation (3.10), $k \to k_B$ at high frequency, which gives the mobility of the unsupported beam when substituted in equation (3.18).

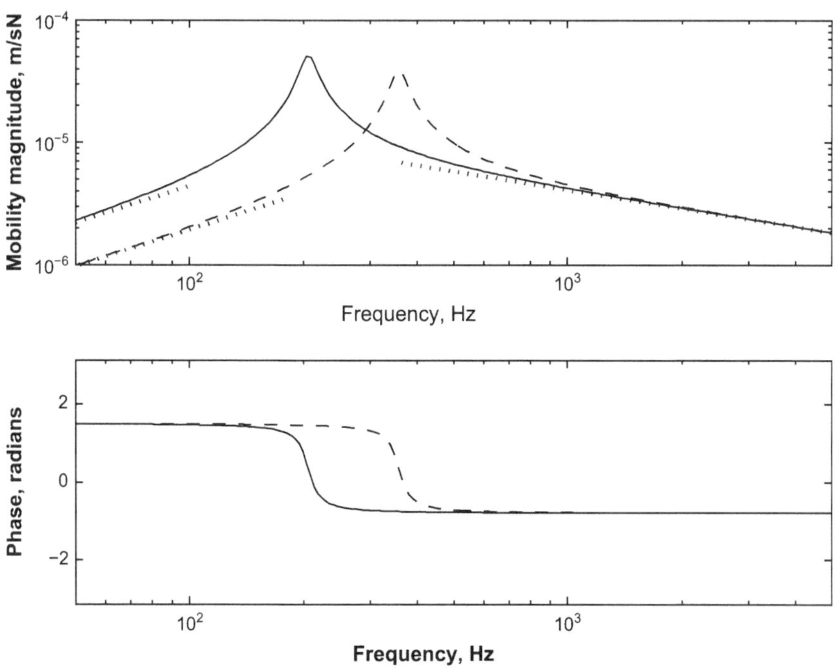

FIGURE 3-15 Point mobility of beam on damped elastic support. —, $s = 100$ MN/m; – – –, $s = 300$ MN/m. The results corresponding to the static stiffness and the free beam are shown as dotted lines

The effect of changing the stiffness s from 100 MN/m^2 to 300 MN/m^2 is also shown in Figure 3.15. This shifts the cut-on frequency by a factor of $3^{1/2} = 1.73$ (see equation (3.5)) and increases the static stiffness by a factor $3^{3/4} = 2.28$ (see equation (3.20)). At high frequency both curves tend to the mobility of the free beam. The effect of this change on the decay rate was already shown in Figure 3.13.

3.2.5 Solution by Fourier transforms

The point mobility can also be determined using a Fourier transform approach. Although it is more complicated mathematically, it will be introduced here as it will be useful later. The equation of motion, equation (3.1), can be modified to include the forcing term at frequency ω:

$$EI\frac{d^4u}{dx^4} + su - m'_r\omega^2 u = F\delta(x-0) \tag{3.21}$$

where the time dependence $e^{i\omega t}$ is implicit and u is now the complex amplitude. Damping has been omitted for clarity, but could be included as before by making E and s complex. A Fourier transform pair can be defined for the displacement

$$U(k) = \int_{-\infty}^{\infty} u(x)e^{ikx}\,dx \tag{3.22}$$

$$u(x) = \frac{1}{2\pi}\int_{-\infty}^{\infty} U(k)e^{-ikx}\,dk \tag{3.23}$$

and equivalently for the force, which has a constant Fourier transform F. Thus the Fourier transform of equation (3.21) is

$$EIk^4U(k) + sU(k) - m'_r\omega^2 U(k) = F \tag{3.24}$$

Note that the wavenumber k here can take any (complex) value. The mobility at wavenumber k (and frequency ω) is

$$Y(k) = \frac{i\omega U(k)}{F} = \frac{i\omega}{EIk^4 + s - m'_r\omega^2} \tag{3.25}$$

and by taking the inverse Fourier transform with respect to k the response at position x is given by

$$Y(x) = \frac{i\omega u(x)}{F} = \frac{i\omega}{2\pi}\int_{-\infty}^{\infty}\frac{e^{-ikx}}{EIk^4 + s - m'_r\omega^2}dk \tag{3.26}$$

To evaluate the integral, use can be made of contour integration, see for example [3.10]. For $x \geq 0$, the appropriate contour consists of the real axis plus a semicircle at infinity in the lower half plane, since $|e^{-ikx}| \to 0$ for $\text{Im}(k) < 0$. This is shown in Figure 3.16. (Conversely, for $x \leq 0$, a semicircle at infinity in the upper half plane is required.) The integrand in equation (3.26) has four poles, which are the free wavenumbers found in equation (3.4) or, with damping, in equation (3.10). The two poles in the upper half plane, with $\text{Im}(k) > 0$ can be denoted $k_2 = -k_p$ and $k_1 = ik_p$

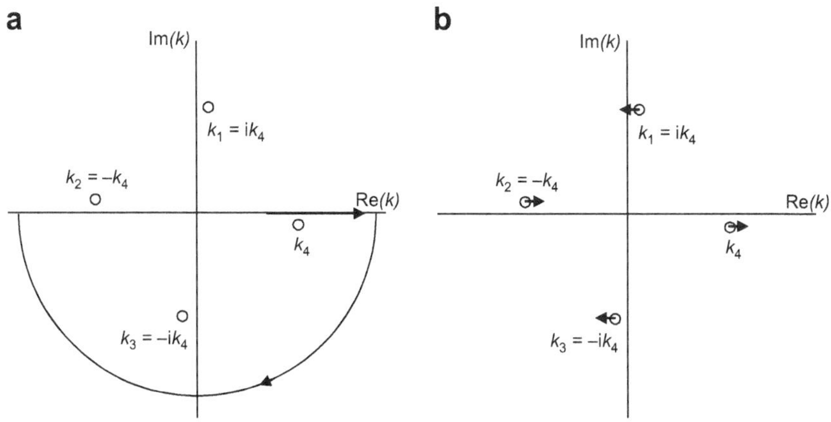

FIGURE 3-16 Poles of equation (3.26), o. (a) Contour for integration of beam response. (b) Effect of a moving excitation

and those in the lower half plane as $k_4 = k_p$ and $k_3 = -ik_p$ where k_p is the root with the largest positive real part, as shown in Figure 3.16(a).

The integral is equal to -2πi times the sum of the residues of the poles enclosed by the contour (with the minus sign due to the fact that the contour encloses them in a clockwise direction). Thus the response is given by

$$Y(x) = -\frac{i\omega}{2\pi} 2\pi i \sum_{n=3}^{4} \text{Res}(k_n) \qquad \text{for} \quad x \geq 0 \tag{3.27}$$

where the residues at the poles $k_n = i^n k_p$ are given by

$$\text{Res}(k_n) = \frac{e^{-ik_n x}}{4EIk_n^3} \tag{3.28}$$

Hence the mobility is given by

$$Y(x) = \frac{\omega}{4EIk_p^3}(e^{-ik_p x} - ie^{-k_p x}) \qquad \text{for} \quad x \geq 0 \tag{3.29}$$

which is equivalent to equation (3.17). Similarly, the equivalent result can be obtained for $x \leq 0$. For $x = 0$ the two expressions are equal. Note that, for $\omega < \omega_0$, the poles move to positions in the middle of each quadrant (equal real and imaginary parts), but the same solution applies.

3.2.6 Moving excitation

For a moving harmonic load of excitation frequency ω and speed V, the equation of motion of the supported beam becomes[3]

---

[3] Caution is required when using hysteretic damping with a constant loss factor. In practice $\omega \gg Vk$ so no significant error is introduced.

$$EI\frac{\partial^4 u}{\partial x^4} + su + m'_r\frac{\partial^2 u}{\partial t^2} = F\delta(x - Vt)e^{i\omega t} \tag{3.30}$$

Changing variables to $\xi = x - Vt$, representing a coordinate moving with the load, this becomes

$$EI\frac{\partial^4 u}{\partial \xi^4} + su + m'_r\left(V^2\frac{\partial^2 u}{\partial \xi^2} - 2V\frac{\partial^2 u}{\partial \xi \partial t} + \frac{\partial^2 u}{\partial t^2}\right) = F\delta(\xi)e^{i\omega t} \tag{3.31}$$

where $\partial/\partial t$ now corresponds to a fixed value of ξ. Solving by Fourier transforms as above, the Fourier transform of equation (3.31) is

$$EIk^4 U + sU - m'_r(Vk + \omega)^2\, U = F \tag{3.32}$$

Thus, by taking the inverse Fourier transform, the response at ξ is given by

$$u(\xi) = \frac{F}{2\pi}\int_{-\infty}^{\infty} \frac{e^{-ik\xi}}{EIk^4 + s - m'_r(Vk + \omega)^2}dk \tag{3.33}$$

For load velocities much lower than the wave speed, the poles of the integrand are close to those for the non-moving load. They can be found approximately by writing $k = k_n(1 + \varepsilon)$ where k_n is the pole for $V = 0$ satisfying equation (3.3) or (3.9). Then

$$EIk_n^4((1 + \varepsilon)^4 - 1) - m'_r((Vk_n\,(1 + \varepsilon) + \omega)^2 - \omega^2) = 0 \tag{3.34}$$

Ignoring second-order terms in V and ε:

$$4\varepsilon EIk_n^4 - 2\omega Vk_n m'_r \approx 0 \tag{3.35}$$

so that, from equation (3.8):

$$\varepsilon \approx \frac{V\omega m'_r}{2EIk_n^3} = i^n\frac{V}{c_g} \tag{3.36}$$

where strictly, in the presence of damping, c_g is complex and is no longer the group velocity. Figure 3.16(b) shows schematically the effect of load motion on the poles. The wavenumbers of the two propagating waves are modified such that the wavenumber with positive real part, k_4, which corresponds to a wave travelling in the positive direction, is increased (its wavelength is reduced) while that travelling in the negative direction, k_2, is reduced (its wavelength is increased). This is the basis of the Doppler effect, although here the waves are also dispersive. The near-field waves are modified by a corresponding *increase* in their real parts. These changes do not affect the contour of integration as all the poles remain in their respective half planes.

The point mobility can be evaluated in the same way as before, using the residues of the poles in either the lower or upper half plane. For modest load speeds, $V \ll c_g$, the effect on the poles is small and it is found that the effect on the mobility is negligible. For example, for the track parameters used in the rest of this section ($s = 100$ MN/m^2) and $V = 50$ m/s, $|\varepsilon|$ is at most 0.2 (at the cut-on frequency of

200 Hz) and by 500 Hz is less than 0.03. This is a consequence of the relatively high wave speeds in a rail; the group velocity of bending waves in a free rail is 2000 m/s at 500 Hz. Thus for a continuously supported rail, the effect of load motion on the point mobility can be ignored (see also [3.19]).

Figure 3.17 illustrates the effect of the load motion on the wave propagation along the rail. These results are for a load speed of 50 m/s and a frequency of 500 Hz. In front of the load ($x > 0$) the wavelength can be seen to be shortened relative to the static case (when viewed in a frame moving with the load), whereas behind the load ($x < 0$) it is lengthened.

3.2.7 Damping models

The hysteretic, or structural, damping model based on a constant loss factor [3.9] has been used throughout this section. This is less common in vibration textbooks than the classical viscous damping model and can only be used in a frequency-domain approach due to problems of causality if it is used in a time-domain model. Nevertheless, it gives a better description of the frequency dependence of the damping of many engineering structures and materials, at least within the frequency range of interest for acoustic problems, and for this reason it is preferred here.

In order to demonstrate the difference between these two models, the beam of Figure 3.11 is considered here with a support including viscous damping C per unit length (for simplicity the rail itself is left with no damping – it is possible to include viscous damping in the rail but this is less intuitive). The equation of free motion becomes

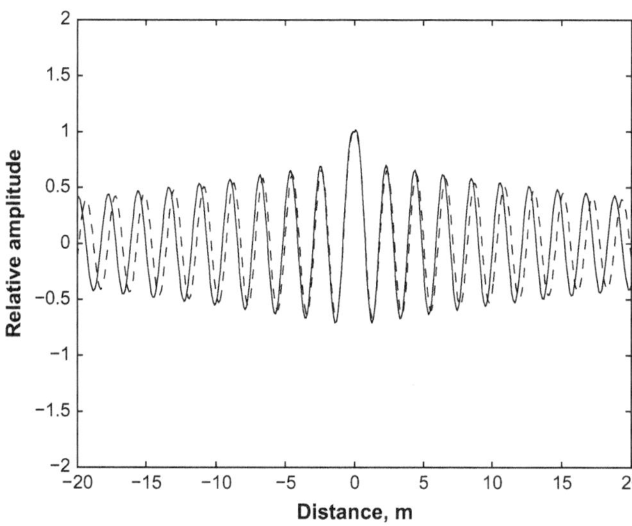

FIGURE 3-17 Instantaneous vibration amplitude versus distance from the load at 500 Hz, showing the part in phase with the excitation point and normalized to the displacement at the excitation point. —, load moving left to right at 50 m/s; – – –, non-moving load

$$EI\frac{\partial^4 u}{\partial x^4} + su + C\frac{\partial u}{\partial t} + m_r'\frac{\partial^2 u}{\partial t^2} = 0 \tag{3.37}$$

Considering harmonic motion at frequency ω, free wave solutions have wavenumber k which satisfies

$$EI\,k^4 + s + \mathrm{i}\omega C - m_r'\,\omega^2 = 0 \tag{3.38}$$

The complex stiffness $s(1 + \mathrm{i}\eta)$ in equation (3.4) has now been replaced by the frequency-dependent term $(s + \mathrm{i}\omega C)$. These two complex stiffnesses are compared in Figure 3.18 for a loss factor of 0.5. The frequency axis is normalized by the frequency at which the two models give identical results ($\omega = s\eta/C$). For the viscously damped system the phase angle of the complex stiffness is 0 at low frequencies and $\pi/2$ at high frequencies, whereas with hysteretic damping it is constant at $\tan^{-1}\eta$. Moreover, for viscous damping the magnitude of the complex stiffness increases due to the dominance of the damping term at high frequency. The two models are equal at only one frequency.

For resonant systems (such as the wheel), where damping only has a significant effect close to resonances, both damping models will give similar results provided that the parameters are chosen to be equivalent at the resonance frequency. However, for a system such as the track, where damping has an effect at all frequencies, it is important to replicate the frequency dependence of the actual system.

Compared with the properties of elastomeric materials, such as used in rail pads, the constant loss factor model is much more realistic (see Section 3.8), although it should strictly be associated with a weak dependence of magnitude on frequency. The properties of ballast, however, are strongly frequency dependent and may in fact be better approximated by a viscous damping model (see Section 3.7).

The mobility of the track is only influenced by the damping around the wave cut-on frequency, ω_0, so a suitable choice of C is to equate it to the required hysteretic damping value at this frequency, i.e. $C = s\eta/\omega_0$. The effect on the mobility of including the viscous damping term in the track model is found to be negligible provided that the damping coefficient C is chosen in this way.

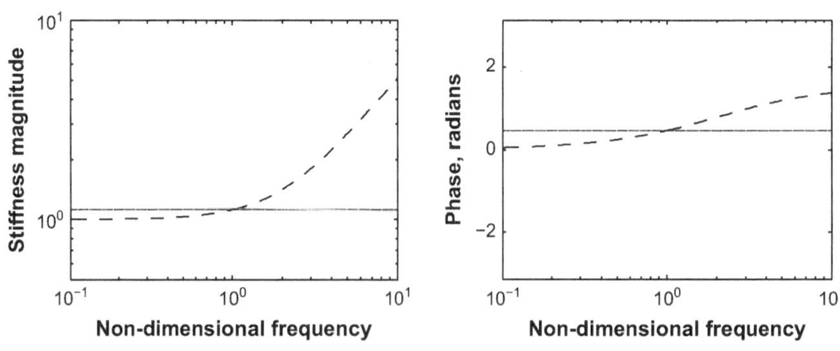

FIGURE 3-18 Magnitude and phase of complex stiffness for different damping models (normalized to frequency at which they are equal). —, hysteretic damping ($\eta = 0.5$); – – –, viscous damping

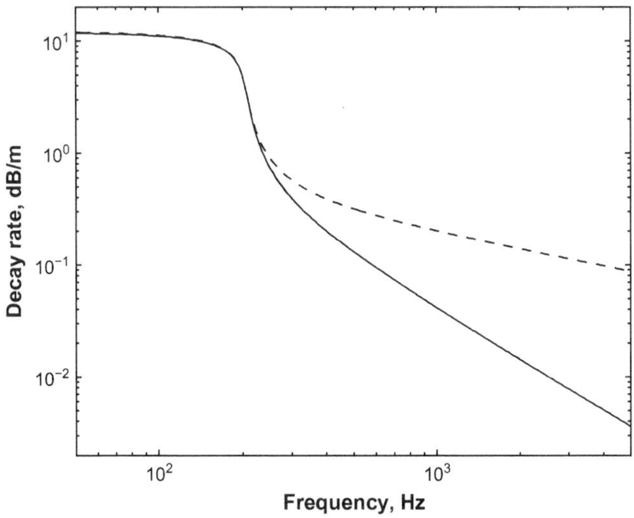

FIGURE 3-19 Decay rates for different damping models. —, hysteretic damping; – – –, viscous damping

The effect on the decay rates is more significant, as shown in Figure 3.19. Below the cut-on frequency of 200 Hz the damping has negligible effect and no significant change occurs. However, above the cut-on frequency the decay rates have a different frequency dependence of $\omega^{-1/2}$ in place of $\omega^{-3/2}$. This can be seen by replacing η by $\omega C/s$ in equation (3.11) giving

$$\mathrm{Im}(k) \approx -k_B\left(\frac{\omega C}{4s(\omega/\omega_0)^2}\right) \qquad \text{for} \quad \omega \gg \omega_0 \tag{3.39}$$

where $k_B \propto \omega^{1/2}$. However, this effect is still less than the influence of the rail damping at high frequencies (compare with Figure 3.13).

3.3 BEAM ON TWO-LAYER SUPPORT

The rails are usually supported on sleepers, with resilient rail pads between the rail and sleeper, see Figure 3.2. The ballast beneath the sleeper provides a further layer of resilience, see Figure 3.1. This system can be modelled as a two-layer support, as shown in Figure 3.20. The support is assumed to be continuous at this stage, that is the pads, sleepers and ballast are 'smeared out' over the length of the rail. Here s_p represents the pad stiffness per unit length, m'_s the sleeper mass per unit length (for one rail) and s_b the ballast stiffness per unit length. If the 'sleepers' are given bending stiffness in the longitudinal direction the same model could be used to represent a slab track, but this is not considered here. (A model of two coupled beams is introduced in Chapter 11 to represent a rail on a bridge.) The introduction of the second layer of stiffness and mass has some interesting effects. This system will

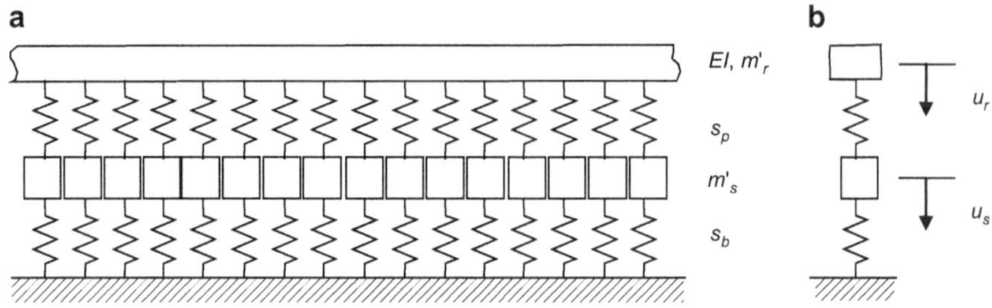

FIGURE 3-20 (a) Beam on two-layer foundation, (b) equivalent mass–spring system

be considered first with no damping as the effects are clearer; damping will then be included in the same way as before using a constant loss factor model.

3.3.1 Undamped system

The analysis of Section 3.2 can be repeated but replacing s by a frequency-dependent support stiffness, $s(\omega)$. From the equations of motion of the simplified system shown in Figure 3.20(b), this is given by

$$s(\omega) = \frac{s_p(s_b - \omega^2 m'_s)}{(s_p + s_b - \omega^2 m'_s)} \tag{3.40}$$

while the ratio of the sleeper displacement, u_s, to that of the rail, u_r, is

$$\tau = \frac{u_s}{u_r} = \frac{s_p}{(s_p + s_b - \omega^2 m'_s)} \tag{3.41}$$

At low frequencies, $\omega \to 0$, these expressions reduce to

$$s(\omega) \to s_0 = \frac{s_p s_b}{(s_p + s_b)} \text{ and } \tau \to \tau_0 = \frac{s_p}{(s_p + s_b)} = \frac{s_0}{s_b} \tag{3.42}$$

where s_0 is the combined stiffness of the two springs in series. If, as is often the case in practice, $s_b \ll s_p$, then $s_0 \approx s_b$ and $\tau \approx 1$. However, for soft pads these approximations are no longer valid and equations (3.42) should be used. At high frequencies, $\omega \to \infty$, the above expressions (3.40) and (3.41) reduce to

$$s(\omega) \to s_p \text{ and } \tau \to \frac{-s_p}{\omega^2 m'_s} \tag{3.43}$$

The frequency-dependent stiffness $s(\omega)$ is shown in Figure 3.21(a) for a typical set of parameters, as listed in Table 3.2, but with no damping at this stage.

Two characteristic frequencies can be identified in Figure 3.21(a), marked by vertical dotted lines:

$$\omega_1 = \sqrt{\frac{s_b}{m'_s}}, \quad \omega_2 = \sqrt{\frac{s_p + s_b}{m'_s}} \tag{3.44}$$

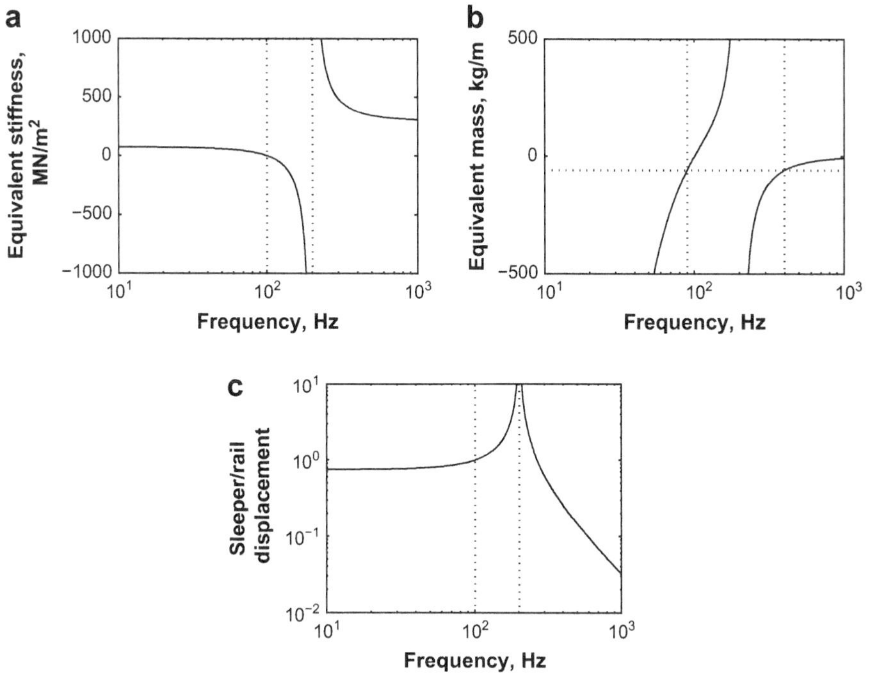

FIGURE 3-21 (a) Frequency-dependent equivalent stiffness, vertical lines: f_1, f_2. (b) Frequency-dependent equivalent mass, vertical lines: f_{c1}, f_{c2}; horizontal line: $-m'_r$. (c) Ratio of sleeper displacement to rail displacement (modulus), vertical lines: f_1, f_2

At ω_1 it can be seen from equation (3.40) that the stiffness $s(\omega)$ becomes zero whereas at ω_2 it is infinite. The frequency ω_1 corresponds to the resonance of the sleeper mass m'_s on the stiffness of the ballast, s_b. The frequency ω_2 corresponds to the resonance of the mass m'_s on the combined stiffness of the ballast and pad with the rail constrained. As will be seen, this latter frequency corresponds to an *anti-*

TABLE 3-2 PARAMETERS USED FOR EXAMPLE CALCULATIONS IN SECTION 3.3

Rail bending stiffness	EI	6.42 MN/m^2
Rail mass per unit length	m_r'	60 kg/m
Damping loss factor of rail	η_r	0.02
Pad stiffness per unit length	s_p	300 MN/m^2
Pad damping loss factor	η_p	0.2
Sleeper mass per unit length of rail (one rail)	m_s'	250 kg/m
Ballast stiffness per unit length (one rail)	s_b	100 MN/m^2
Ballast damping loss factor	η_b	1.0

resonance of the track when excited at the rail. For the present numerical parameter values, these frequencies correspond to $f_1 = 100$ Hz and $f_2 = 200$ Hz.

For frequencies $\omega < \omega_1$ or $\omega > \omega_2$, $s(\omega)$ is positive (stiffness-like) whereas for $\omega_1 < \omega < \omega_2$, $s(\omega)$ is negative (mass-like). Instead of a frequency-dependent stiffness, the support could also be represented as an equivalent (frequency-dependent) mass $m_{eq} = -s(\omega)/\omega^2$. It is shown in this form in Figure 3.21(b).

The ratio τ is shown in Figure 3.21(c). This rises from the value $\tau_0 = 0.75$ (for the present parameters) at low frequencies, passing through $\tau = 1$ at ω_1, to a peak at ω_2, limited in reality by damping, before falling in proportion to ω^{-2} at high frequencies. Thus the sleeper vibrates with a similar amplitude to the rail at low frequencies but is well isolated from the rail at high frequencies. From equations (3.42) and (3.43) it can be seen that it is in phase with the rail at low frequencies but out of phase with it above ω_2.

The equivalent mass of the support m_{eq} is compared in Figure 3.21(b) with $-m'_r$. These two curves can be seen to cross at two frequencies denoted $\omega_{c1} < \omega_1$ and $\omega_{c2} > \omega_2$. These occur approximately at $f_{c1} = 90$ and $f_{c2} = 400$ Hz for the present parameters. At these two frequencies $m'_r + m_{eq} = 0$, or $m'_r\omega^2 - s(\omega) = 0$ and the rail mass is exactly cancelled by the foundation. This corresponds to the condition of cut-on seen for a single layer foundation in Section 3.2, see equation (3.5). These cut-on frequencies are the natural frequencies of the corresponding two-degree-of-freedom system shown in Figure 3.20(b). From the equations of motion of the two-degree-of-freedom system, the natural frequencies satisfy

$$\left|\begin{bmatrix} s_p & -s_p \\ -s_p & s_p + s_b \end{bmatrix} - \omega^2 \begin{bmatrix} m'_r & 0 \\ 0 & m'_s \end{bmatrix}\right| = 0 \tag{3.45}$$

which has solutions

$$\omega_c^2 = \frac{(\omega_0^2 + \omega_2^2)}{2} \pm \sqrt{\frac{(\omega_0^2 + \omega_2^2)^2}{4} - \omega_0^2\omega_1^2} \tag{3.46}$$

where ω_1 and ω_2 are given by equation (3.44) and ω_0 is now given by

$$\omega_0 = \sqrt{\frac{s_p}{m'_r}} \tag{3.47}$$

For the case where $s_b \ll s_p$, $\omega_1 \ll \omega_2$ and the second cut-on frequency is given approximately by

$$\omega_{c2} \approx \sqrt{(\omega_0^2 + \omega_2^2)} = \sqrt{\frac{s_p}{m'_r} + \frac{s_p + s_b}{m'_s}} \approx \sqrt{\frac{s_p}{m'_r}} \tag{3.48}$$

where the final approximation applies if $m'_s \gg m'_r$. This corresponds to the resonance of the rail mass on the pad stiffness. The first cut-on frequency is given by

$$\omega_{c1} \approx \frac{\omega_0\omega_1}{\sqrt{(\omega_0^2 + \omega_2^2)}} = \left(\frac{m'_s}{s_b} + m'_r\left(\frac{1}{s_p} + \frac{1}{s_b}\right)\right)^{-1/2} \approx \sqrt{\frac{s_b}{m'_s + m'_r}} \tag{3.49}$$

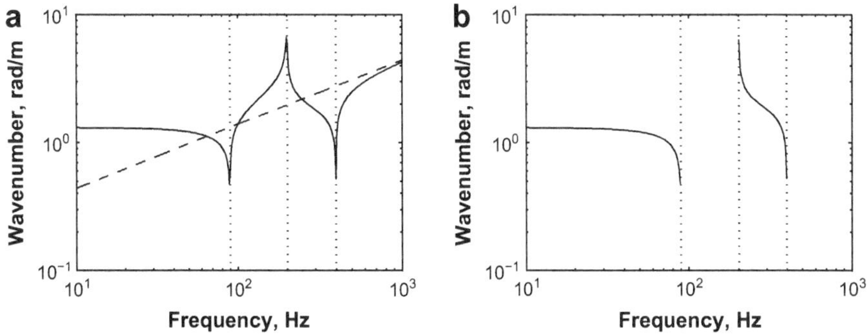

FIGURE 3-22 Wavenumbers for rail on two-layer foundation for example parameters. (a) Real part, (b) imaginary part (negative). Vertical lines: f_{c1}, f_2, f_{c2}. - - -, wavenumber of unsupported rail

where the final approximation applies if $s_b \ll s_p$. This corresponds to the resonance of the combined mass of the rail and sleeper on the ballast stiffness.

The wavenumbers obtained by substituting $s(\omega)$ from equation (3.40) into equation (3.4) are shown in Figure 3.22 for the example parameters of Table 3.2. The frequencies $f_{c1} = \omega_{c1}/2\pi$, $f_2 = \omega_2/2\pi$ and $f_{c2} = \omega_{c2}/2\pi$ are identified by dotted lines. Other wavenumber solutions are given by $i^n k$.

As for the case of a single stiffness support (Figure 3.12), in the first stiffness-like region there is a low frequency 'blocked' region where no waves propagate, the wavenumbers having equal real and imaginary parts. Above the first cut-on frequency, ω_{c1}, free wave propagation occurs with no decay (zero imaginary part). Free wave propagation occurs, moreover, in the whole of the region, $\omega_{c1} < \omega < \omega_2$. Here the mass per unit length of the equivalent beam becomes $m_r' + m_{eq}$ so that above ω_1 there is an increase in the wavenumber compared with that of the free rail (see equation (3.4)), as can be seen in Figure 3.22(a). In this frequency region the sleeper vibration is greater than that of the rail, see Figure 3.21(c). At ω_2 free wave propagation ceases as the support mass (or stiffness) becomes infinite and changes sign.

There follows a second 'blocked' region of complex wavenumbers between ω_2 and ω_{c2}, above which free wave propagation again cuts on. In the region above ω_{c2}, the wavenumber tends to that of the unsupported rail. The sleeper vibration rapidly becomes small compared with that of the rail as frequency increases above ω_2. At $\omega = \omega_{c2}$, for example, from equations (3.40) and (3.48):

$$\tau \approx \frac{-m_r'}{m_s'} \tag{3.50}$$

which for the present parameters is 0.36 at 400 Hz.

3.3.2 Effect of damping

Next, the effect of adding damping to the two layers of the support and to the rail is considered. The damping loss factors used are indicated in Table 3.2.

Figure 3.23 shows the ratio of sleeper vibration to rail vibration, τ. Damping has a strong effect around the peak at ω_2, reducing the sleeper vibration in this region, as

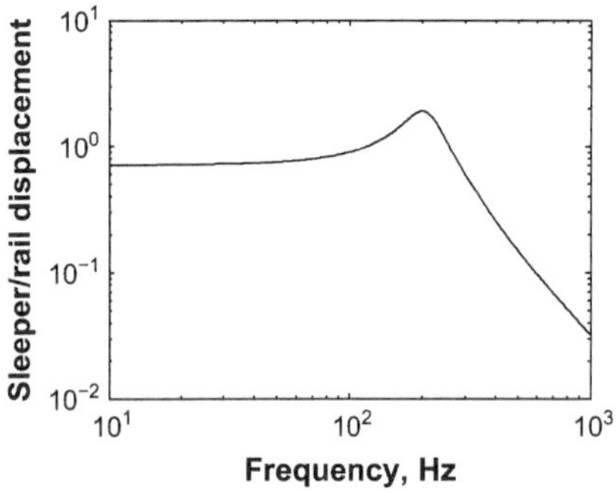

FIGURE 3-23 Ratio of sleeper displacement to rail displacement (magnitude) in the presence of damping

expected. Generally similar behaviour can be seen in the experimental results in Figure 3.9.

Figure 3.24 shows the real and imaginary parts of the wavenumbers in the presence of damping. Similar features can be seen to those in the undamped case, Figure 3.22, but the damping has 'rounded the corners' of the plots. In particular, the 'blocked' regions can still be seen as regions of high imaginary part below 90 Hz and between 200 and 400 Hz. At low frequencies the real and imaginary parts tend to have similar values, although they are no longer equal due to the high damping of the ballast. The imaginary part of the wavenumber falls significantly between 100 and 200 Hz and above 400 Hz, although it is no longer zero.

The decay rates predicted using the two-layer support model for the above parameters are shown in Figure 3.25. At low frequencies the decay rate is around 10 dB/m, as found for the single layer support. After a slight dip between 100 and 200 Hz, corresponding to the first cut-on region, a pronounced peak occurs in the region 200 to 400 Hz, the second 'blocked' region. The effect of adding damping to the rail is only significant at high frequencies, as found for the single layer support.

For comparison, the equivalent results for a single layer support of stiffness equal to that of the rail pad, s_p, are also shown. Above the second cut-on frequency these results tend rapidly to those of the two-layer support, in the region where $s(\omega) \rightarrow s_p$, see Figure 3.21(a). Above 600 Hz for the present parameters (equivalent to $1.5\omega_{c2}$)

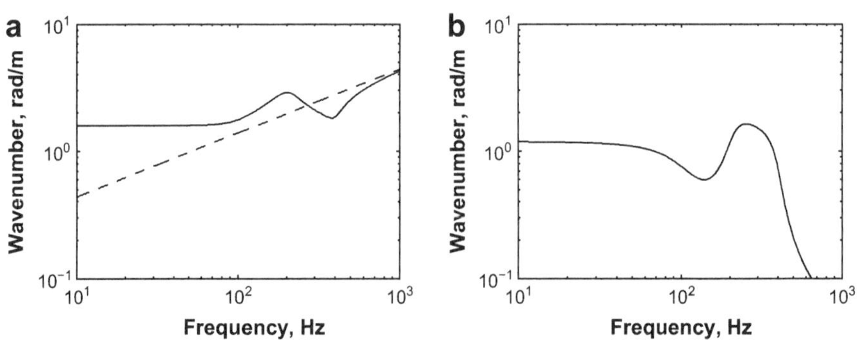

FIGURE 3-24 Wavenumbers for rail on two-layer foundation for example parameters in the presence of damping. (a) Real part, (b) imaginary part (negative). - - -, wavenumber of unsupported rail

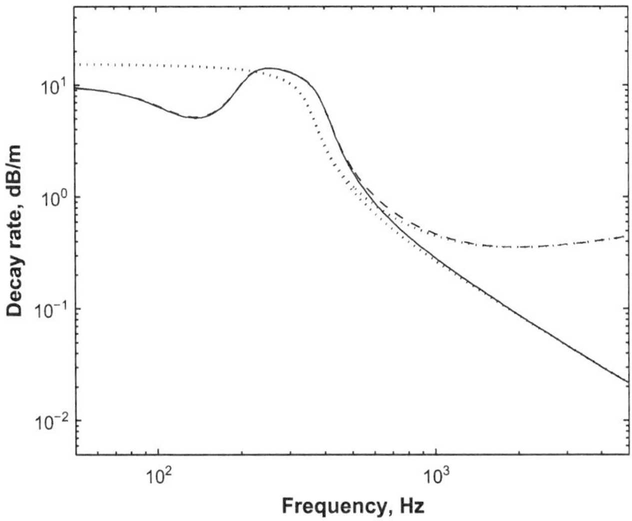

FIGURE 3-25 Decay rates for track with two-layer support. —, $\eta_r = 0.0$; – – –, $\eta_r = 0.02$; ····, corresponding results for single layer support with $s = 300$ MN/m^2

the difference is less than 25% (1 dB in terms of noise) and the details of the support structure below the pad are of only minor significance.

To calculate the track mobility the same procedure can be followed as in Section 3.2. Thus equation (3.18) can be used, but now with the wavenumber appropriate for the two-layer support. Results for the above parameters are shown in Figure 3.26. The single resonance peak seen previously is now replaced by two peaks. These can be interpreted in terms of the resonances of the equivalent two-degree-of-freedom system in Figure 3.20(b). The first peak at around 100 Hz actually corresponds to the first cut-on ω_{c1} (90 Hz in the absence of damping). It is a highly damped peak due to the high damping loss factor of the ballast (here taken as 0.5 but in reality often higher than this; see Figures 3.3 and 3.6). The dip at around 200 Hz corresponds to the anti-resonance, ω_2, at which the sleeper vibrates with a large amplitude between the stiffnesses of the pad and ballast. The second peak at 400 Hz is at the second cut-on frequency, ω_{c2}, corresponding to the second resonance of the two-degree-of-freedom system in Figure 3.20(b), at which the rail and sleeper vibrate out of phase on the stiffness of the pad.

3.3.3 Effect of changing parameters

The effect of varying the pad stiffness is shown in Figure 3.27, which gives the decay rates and the magnitude of the point mobility. The frequencies bounding the blocked region, ω_2 and ω_{c2}, are marked by vertical lines in each case. They can be clearly seen to correspond to the dip and peak in the mobility as well as bounding the main peak in the decay rate. As the pad stiffness increases, this region of high decay rate becomes broader and the decay rate within it increases. Similar effects can be seen in the measured results of Figure 3.10.

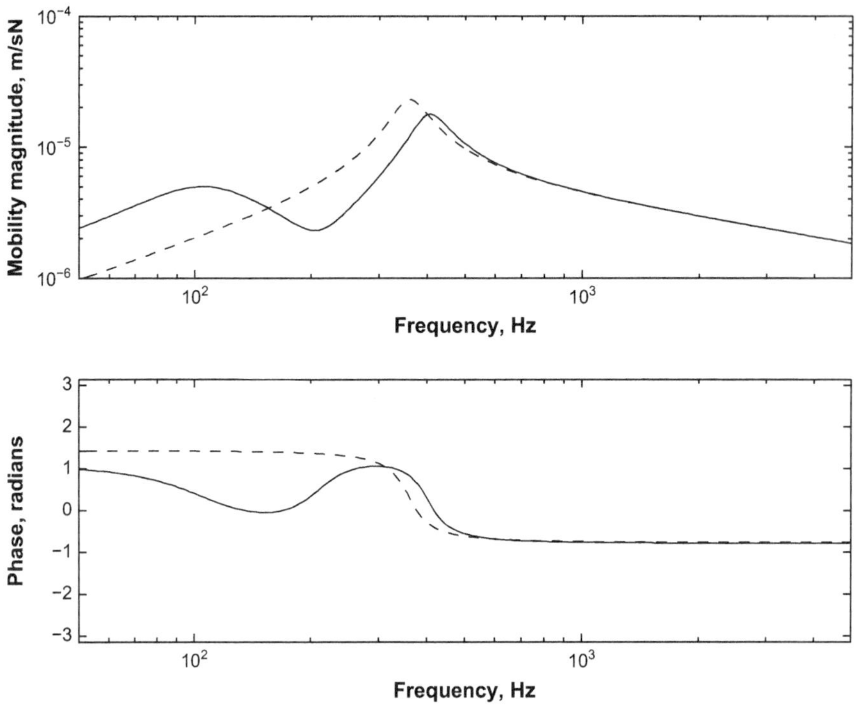

FIGURE 3-26 Point mobility for track. —, with two-layer support; - - -, corresponding result for single layer support with $s = 300\ \text{MN/m}^2$

The effect of varying the sleeper mass is shown in Figure 3.28. Reducing the sleeper mass, for example changing from concrete to wooden sleepers, leads to an increase in the bounding frequencies ω_2 and ω_{c2} but a reduction in the relative bandwidth of the blocked region. This occurs as ω_0 remains constant while ω_2 increases, see equation (3.48).

The effect of varying the ballast stiffness is shown in Figure 3.29. This affects the mobility and decay rate particularly below ω_2. The decay rate at 100 Hz varies between 3 and 10 dB/m for ballast stiffnesses in the range 50 to 200 MN/m^2, although the low frequency asymptote varies much less than this. As discussed in Section 3.5.2 below, the ballast stiffness can vary considerably at a given site, and is also modified by maintenance of the track, e.g. using a track tamping machine.

3.4 TIMOSHENKO BEAM MODEL

3.4.1 Free wave solutions

Euler–Bernoulli beam theory, used in the model up to this point, is based on the assumption that plane sections of the beam remain plane and perpendicular to the neutral axis. When the wavelength of a beam is shorter than about six times its

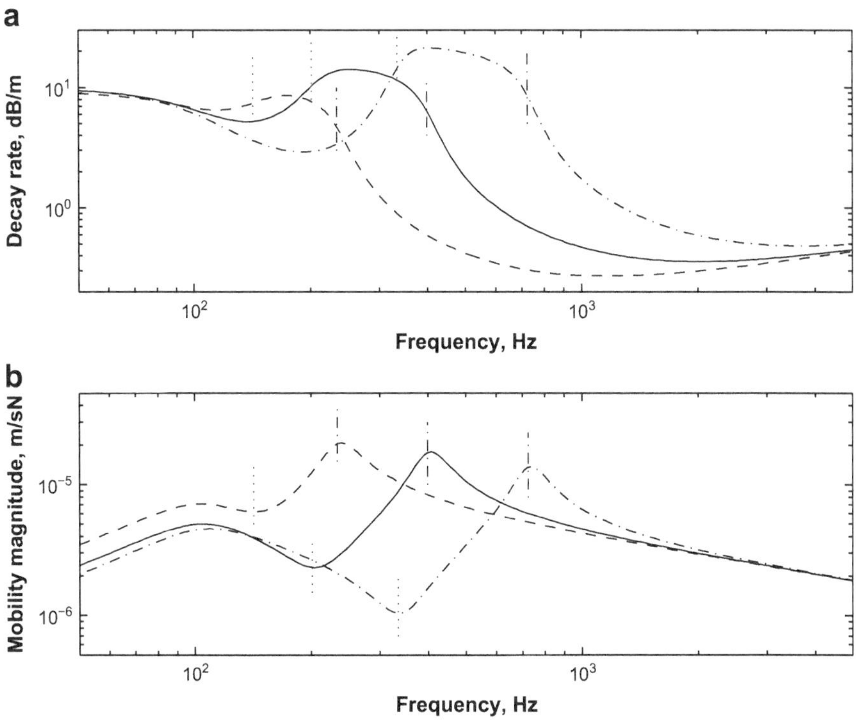

FIGURE 3-27 (a) Decay rates and (b) point mobility for track with two-layer support with various pad stiffnesses. —, $s_p = 300$ MN/m^2; – – –, $s_p = 100$ MN/m^2; – · – ·, $s_p = 1000$ MN/m^2; vertical lines: · · · ·, ω_2; – · – ·, ω_{c2}

height, shear deformation and rotational inertia play a role and should be included in the description of the beam [3.8]. As will be seen, for vertical vibration of a rail, the effects become significant for frequencies above about 500 Hz.

The motion of a Timoshenko beam is described by its deflection u and the rotation of the cross-section relative to the undeformed axis, ϕ (for an Euler–Bernoulli beam where there is no shear deformation it is assumed that $\phi = \partial u/\partial x$). The equations of motion for a Timoshenko beam on an elastic foundation of stiffness s per unit length can be written in the form

$$GA\kappa\frac{\partial}{\partial x}\left(\phi - \frac{\partial u}{\partial x}\right) + su + m'_r\frac{\partial^2 u}{\partial t^2} = F\delta(x)e^{i\omega t} \tag{3.51}$$

$$GA\kappa\left(\phi - \frac{\partial u}{\partial x}\right) - EI\frac{\partial^2 \phi}{\partial x^2} + \rho I\frac{\partial^2 \phi}{\partial t^2} = 0 \tag{3.52}$$

where ρ is the density, G is the shear modulus, A is the cross-sectional area and $\kappa < 1$ is the shear coefficient. Other parameters are defined in Section 3.2. The sleeper mass

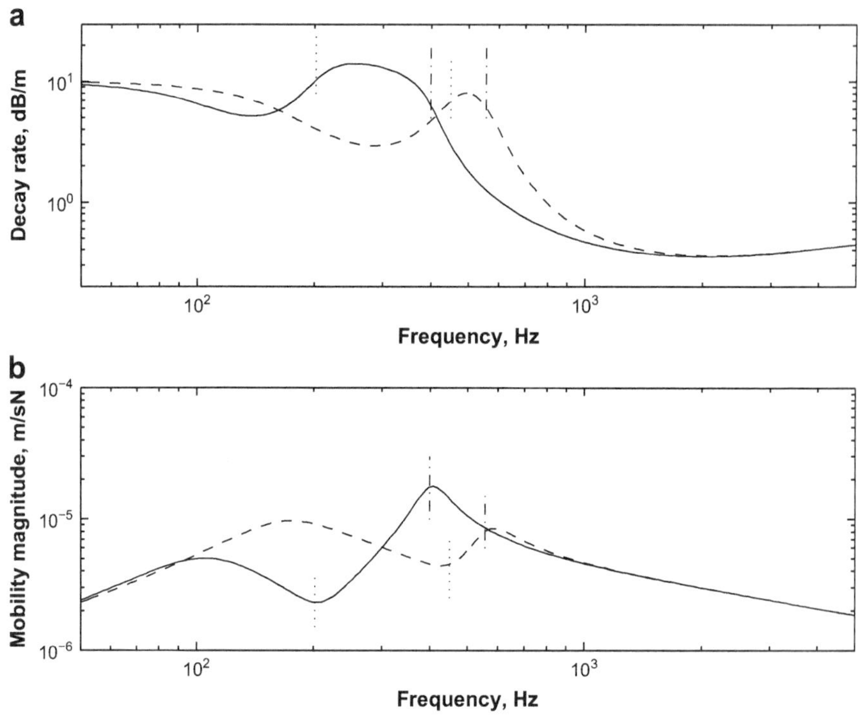

FIGURE 3-28 (a) Decay rates and (b) point mobility for track with two-layer support with various sleeper masses. —, $m'_s = 250$ kg/m; – – –, $m'_s = 50$ kg/m; vertical lines: ····, ω_2; – · – ·, ω_{c2}

etc. can be included in the support stiffness by making $s(\omega)$ frequency dependent according to equation (3.40). Considering harmonic motion at frequency ω, free wave solutions can exist of the form

$$u(x,t) = U\,e^{i\omega t}\,e^{-ikx} \tag{3.53}$$

$$\phi(x,t) = U\,\Psi\,e^{i\omega t}\,e^{-ikx} \tag{3.54}$$

where U is the complex amplitude and $U\Psi$ is the corresponding amplitude of ϕ. Substituting these into equation (3.52) yields

$$\Psi = \frac{ikGA\kappa}{\rho I\omega^2 - GA\kappa - EIk^2} \tag{3.55}$$

which differs for the two waves (propagating and evanescent). Substituting this into equation (3.51) gives a quadratic for k^2:

$$k^4 + C_2(\omega)k^2 + C_3(\omega) = 0 \tag{3.56}$$

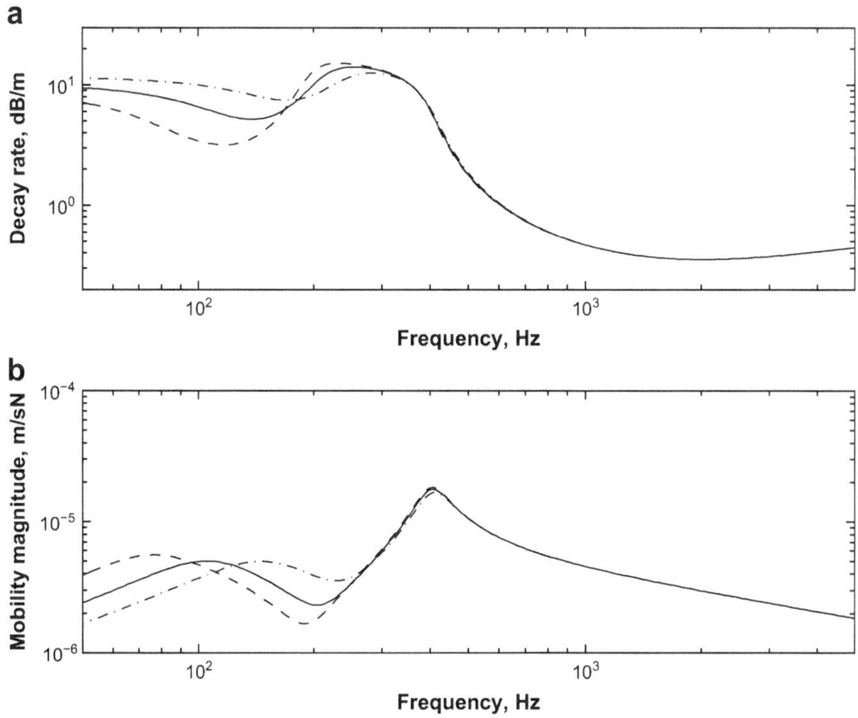

FIGURE 3-29 (a) Decay rates and (b) point mobility for track with two-layer support with various ballast stiffnesses. —, $s_b = 100$ MN/m^2; – – –, $s_b = 50$ MN/m^2; – · – ·, $s_b = 200$ MN/m^2

where

$$C_2(\omega) = \left(\frac{s - m'_r\omega^2}{GA\kappa}\right) - \left(\frac{\rho I\omega^2}{EI}\right) \tag{3.57}$$

and

$$C_3(\omega) = \left(\frac{s - m'_r\omega^2}{EI}\right)\left(1 - \frac{\rho I\omega^2}{GA\kappa}\right) \tag{3.58}$$

Once free wave motion has cut on, equation (3.56) has one positive and one negative solution for k^2 provided that $C_3 > 0$. These correspond to the familiar propagating and evanescent waves, although their wavenumbers differ. However, at

$$\omega = \omega_T = \sqrt{\frac{GA\kappa}{\rho I}} \tag{3.59}$$

the term C_3 becomes equal to zero, which yields $k = 0$. For higher frequencies, $\omega > \omega_T$, both solutions for k^2 are positive, indicating that two propagating waves occur.

3.4.2 Mobility

The solution for the point mobility can be found by solving the boundary conditions at $x = 0$, as before. However, a more efficient method, similar to that used by Grassie [3.12], is to write the displacement in terms of a Fourier transform pair, as given in equations (3.22) and (3.23) in Section 3.2.5. Taking Fourier transforms of equation (3.51) gives

$$(-GA\kappa\, ik(\Psi + ik) + s - m_r'\omega^2)U(k) = F \tag{3.60}$$

which can be rearranged to give

$$\frac{U}{F} = \frac{1}{GA\kappa}\left(\frac{k^2 + C_1(\omega)}{k^4 + k^2C_2(\omega) + C_3(\omega)}\right) \tag{3.61}$$

where C_2 and C_3 are given by equations (3.57) and (3.58) and

$$C_1(\omega) = \frac{GA\kappa}{EI} - \frac{\rho I\omega^2}{EI} \tag{3.62}$$

Finally, taking the inverse Fourier transform, the transfer mobility for a response position x is given by

$$Y(x) = \frac{i\omega u(x)}{F} = \frac{i\omega}{2\pi}\int_{-\infty}^{\infty} \frac{1}{GA\kappa}\left(\frac{k^2 + C_1(\omega)}{k^4 + k^2C_2(\omega) + C_3(\omega)}\right)e^{-ikx}\,dk \tag{3.63}$$

This can be determined using an appropriate contour integration, as for the Euler–Bernoulli beam in Section 3.2.5. For $x \geq 0$, the contour should be closed at infinity in the lower half plane. The poles of the integrand are the free wavenumbers, which are the solutions to equation (3.56). The integral is equal to $-2\pi i$ times the sum of the residues of the poles enclosed by the contour

$$Y(x) = \frac{-i\omega}{2\pi}2\pi i \sum_{\substack{n \text{ with} \\ \mathrm{Im}(k_n)<0}} \mathrm{Res}(k_n) \qquad \text{for} \quad x \geq 0 \tag{3.64}$$

the residues being given by

$$\mathrm{Res}(k_n) = \frac{e^{-ik_nx}}{GA\kappa}\left(\frac{k_n^2 + C_1(\omega)}{4k_n^3 + 2k_nC_2(\omega)}\right) \tag{3.65}$$

As in the previous section, the stiffness s can be replaced by a frequency-dependent stiffness $s(\omega)$. This affects the results in the low frequency region in the same manner as for the Euler–Bernoulli beam.

3.4.3 Results

Example results are given for the two-layer support with the same parameters as above. These and additional parameters are listed in Table 3.3.

TABLE 3-3 PARAMETERS FOR BASELINE TRACK IN SECTION 3.4

		Vertical
Rail bending stiffness	EI	6.42 MN/m^2
Rail mass per unit length	ρA	60 kg/m
Rail shear stiffness	GA	6.17×10^8 N
Rail shear parameter	κ	0.4
Rail rotational inertia	ρI	0.240
Pad stiffness	s_p	300 MN/m^2
Pad damping loss factor	η_p	0.2
Sleeper mass	m_s'	250 kg/m
Sleeper spacing	d	0.6 m
Ballast stiffness	s_b	100 MN/m^2
Ballast loss factor	η_b	1.0

The real parts of the wavenumbers are shown in Figure 3.30(a). This confirms that the Timoshenko beam has an effect above about 500 Hz for vertical motion of the rail. At sufficiently high frequencies the first wavenumber of a Timoshenko beam is approximately that of a shear beam:

$$k \approx \sqrt{\frac{m_r'\omega^2}{GA\kappa}} = \sqrt{\frac{\rho}{G\kappa}}\,\omega \tag{3.66}$$

which has a constant wavespeed $\sqrt{G\kappa/\rho}$, while the wavespeed of the second wave, which cuts on at ω_T, tends to that of a longitudinal wave at high frequencies, $\sqrt{E/\rho}$ (it involves longitudinal motion of the top and bottom of the beam in opposite directions). The wavenumber in equation (3.66) increases in proportion to ω, rather than $\omega^{1/2}$ as found for the Euler–Bernoulli beam. However, from Figure 3.30(a) it can be seen that the slope of the dispersion curve does not approach such a linear relationship until about 5 kHz. When the wavelength in the rail is equal to twice the sleeper spacing distance, $2d$, standing wave patterns ('pinned–pinned mode') can occur, as discussed in more detail in Section 3.5. This corresponds to a wavenumber $k = \pi/d$, which for $d = 0.6$ m is $k = 5.24$ rad/m. For the unsupported Timoshenko beam this occurs at 1070 Hz whereas for the Euler–Bernoulli beam it is much higher at 1425 Hz. The second wave in the Timoshenko beam appears at around 5 kHz.

The decay rates for the track, shown in Figure 3.30(b), are unaffected by the use of a Timoshenko beam at low frequencies, but the higher wavenumbers found at high frequencies lead to higher decay rates. This can be expected from the factor k_B in equation (3.11). In the absence of rail damping the decay rate is proportional to ω^{-1} rather than $\omega^{-3/2}$, and with rail damping the dependence of $\omega^{1/2}$ changes to being proportional to ω.

The point mobility, shown in Figure 3.31, is similar for the Timoshenko and Euler–Bernoulli beams at low frequencies, with divergence again occurring above about 500 Hz. For the Timoshenko beam, the magnitude tends to a constant

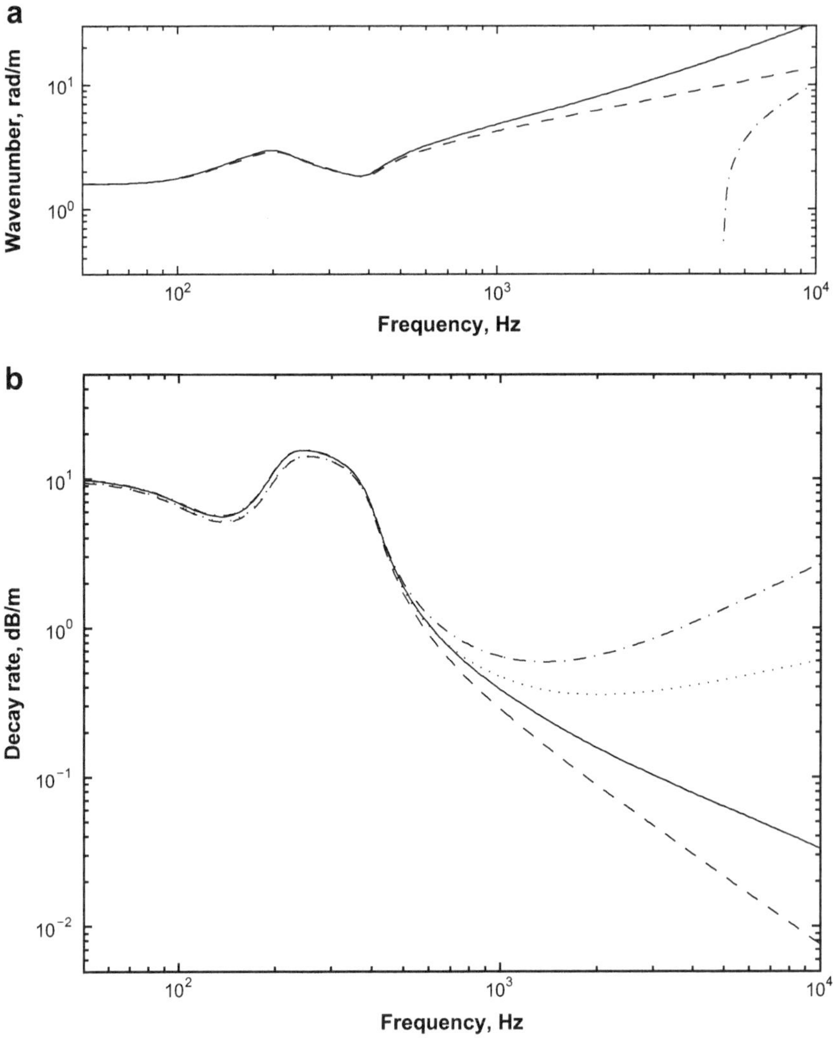

FIGURE 3-30 (a) Wavenumbers (real part) predicted for track. —, using Timoshenko beam; – · – ·, second wave for Timoshenko beam; – – –, using Euler–Bernoulli beam. (b) Decay rates predicted for track. —, using Timoshenko beam with $\eta_r = 0.0$; – · – ·, with $\eta_r = 0.02$; – – –, using Euler–Bernoulli beam with $\eta_r = 0.0$; · · · ·, with $\eta_r = 0.02$

rather than sloping downwards at a rate of $\omega^{-1/2}$, while the phase tends to 0 rather than $-\pi/4$.

Figure 3.32 shows the effect of varying the rail pad stiffness with the rail represented by a Timoshenko beam. Similar results are obtained to those found in Figure 3.27 for the Euler–Bernoulli beam. The low frequency behaviour is mainly determined by the ballast stiffness and so is little affected. The track decay rates are higher than the corresponding results for the Euler–Bernoulli beam for frequencies

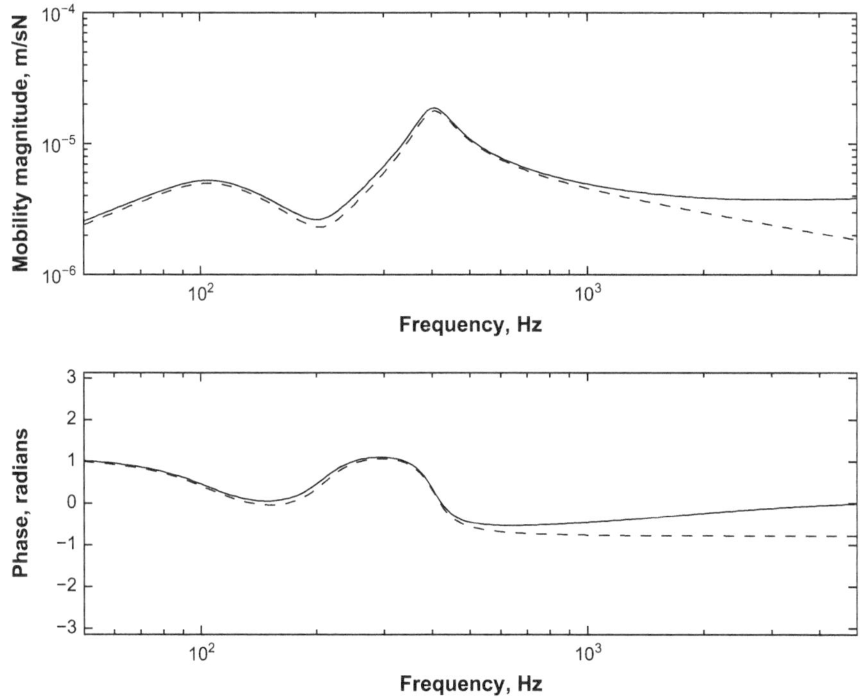

FIGURE 3-31 Point mobility predicted for track. —, using Timoshenko beam; – – –, using Euler–Bernoulli beam

above about 500 Hz. The mobility is also affected at high frequencies but the results in the region where the pad stiffness has most influence are similar to those in Figure 3.27.

The predicted decay rates have been based on a rail loss factor of 0.02. This is chosen to give a good fit to the measurements in Figure 3.10. However, it is known that the rail itself will have a damping loss factor closer to the material damping of steel, which is of the order of 2×10^{-4}. The physical reason for this effect is that the rail foot vibrates with a larger amplitude at high frequencies (see Section 3.6) and this leads to an increase in the damping effect of the pad. Nevertheless, it is convenient in the beam model to assign this extra damping to the rail.

3.5 DISCRETELY SUPPORTED TRACK MODELS

3.5.1 Periodically supported track

The models considered in the previous sections are based on a continuous support below the rail. In practice, the track is not supported continuously but rather on

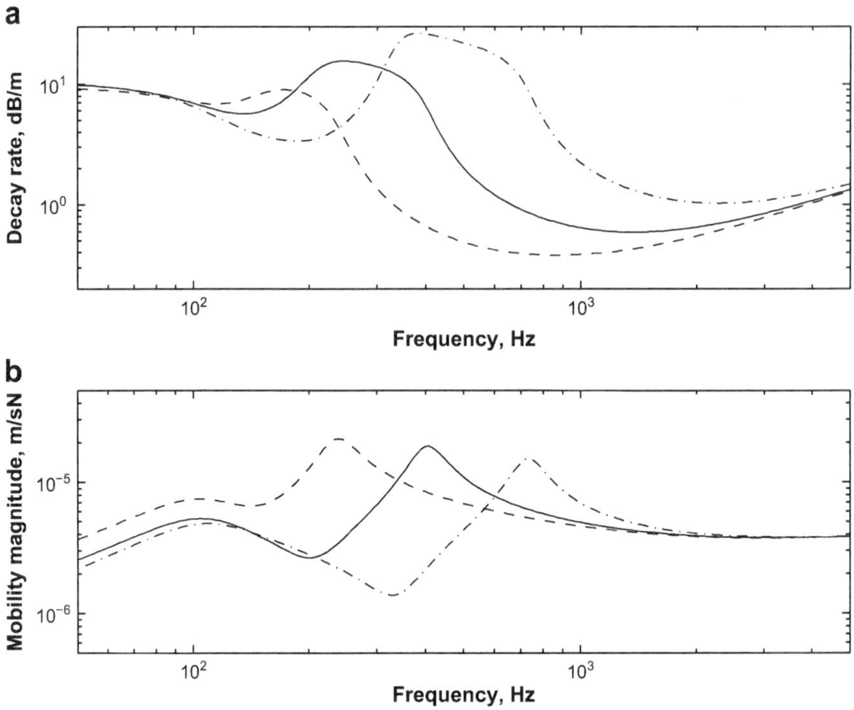

FIGURE 3-32 Results predicted for track with different rail pad stiffnesses using Timoshenko beam with $\eta_r = 0.02$. (a) Decay rates, (b) point mobilities. – – –, $s_p = 100$ MN/m^2; —, $s_p = 300$ MN/m^2; – · – ·, $s_p = 1000$ MN/m^2

discrete sleepers, which are spaced approximately periodically. Models have been developed that allow either for an exactly periodic support or a set of discrete supports that may differ in spacing and stiffness. The following description of a periodically supported rail is based on a model developed by Heckl [3.13, 3.14].

An infinite Timoshenko beam is assumed to be fastened to a finite number of supports, each of dynamic stiffness K and separated by a spacing d. The dynamic stiffness can incorporate the sleeper mass and multiple layers of stiffness in the same way as for the continuous support

$$K(\omega) = \frac{K_p(K_b - \omega^2 m_s)}{(K_p + K_b - \omega^2 m_s)} \tag{3.67}$$

where K_p is the stiffness of the pad, K_b is the stiffness of the ballast and m_s is the mass of the sleeper. Each support is replaced by a reaction force which is proportional to the beam displacement at that point. The response of a *free* Timoshenko beam at position x to a point force of unit amplitude applied at point x' is determined first. This can be found by setting $s = 0$ in the expressions from the previous section (see equations (3.63) and (3.64)). It is convenient to work in terms of receptance rather than mobility:

$$\alpha(x,x') = (u_1 e^{-ik_e|x-x'|} + u_2 e^{-ik_p|x-x'|}) \tag{3.68}$$

where k_p is the wavenumber solution close to the positive real axis and k_e is the solution close to the negative imaginary axis. The terms u_n are given by

$$u_1 = \frac{-i}{GA\kappa}\left(\frac{k_e^2 + C_1}{4k_e^3 + 2k_e C_2}\right), \; u_2 = \frac{-i}{GA\kappa}\left(\frac{k_p^2 + C_1}{4k_p^3 + 2k_p C_2}\right) \tag{3.69}$$

and the wavenumbers are the solutions of

$$k^2 = -\frac{1}{2}C_2 \pm \frac{1}{2}\sqrt{C_2^2 - 4C_3} \tag{3.70}$$

In the absence of damping one of these, k_p, is real and positive; the other, k_e, is negative imaginary. The terms C_i are given by simplifications of those given earlier:

$$C_1 = \frac{GA\kappa}{EI} - \frac{\rho I\omega^2}{EI} \tag{3.71}$$

$$C_2 = -\left(\frac{m_r'\omega^2}{GA\kappa}\right) - \left(\frac{\rho I\omega^2}{EI}\right) \tag{3.72}$$

and

$$C_3 = \left(\frac{m_r'\omega^2}{EI}\right)\left(\frac{\rho I\omega^2}{GA\kappa} - 1\right) \tag{3.73}$$

If the beam is now attached to supports at $x_n = nd$ for integer values of n, a force is applied at each of these points, which is equal to $-Ku(x_n)$, where $u(x)$ is the displacement at position x. Thus the total response to a harmonic point force $Fe^{i\omega t}$ applied at x_c (assumed to be in the range $0 \leq x_c \leq d$) is given by

$$u(x) = F\alpha(x, x_c) - K\sum_{n=-\infty}^{\infty} \alpha(x, x_n)u(x_n) \tag{3.74}$$

There are several ways in which this may be solved. First, it is possible to truncate the sum at a finite number of terms, $n = -N$ to N. Then evaluating $u(x)$ at the support points x_n

$$u(x_m) = F\alpha(x_m, x_c) - K\sum_{n=-N}^{N} \alpha(x_m, x_n)u(x_n) \tag{3.75}$$

allows a matrix equation to be written:

$$([I] + K[\alpha(x_m, x_n)])\{u(x_n)\} = F\{\alpha(x_m, x_c)\} \tag{3.76}$$

where $[\alpha]$ is a square matrix of dimension $2N+1$ consisting of point and transfer receptances of the free beam, $[I]$ is a unit matrix, $\{u\}$ is a vector of the $2N+1$

displacements at the sleeper positions x_n and $\{\alpha\}$ is a vector of $2N+1$ transfer receptances of the free beam from the force point x_c to the positions x_n. Equation (3.76) can be inverted to solve for the displacements $u(x_n)$. These can then be used in equation (3.74) to give the displacement at a general point x. In the above, it is not necessary for K to be identical at each support, or for the points x_n to be equally spaced. The effect of such variations is discussed in Section 3.5.2.

Heckl [3.14] developed a more efficient method for a purely periodically supported track. In a periodic structure Bloch's theorem states that, for free wave propagation, the response in each bay is related to that in neighbouring bays by a constant decay and phase change:

$$u(x + d) = u(x)e^{-\gamma d} \tag{3.77}$$

This property can be used to determine the propagation coefficients γ which describe the phase change and attenuation between adjacent sleeper bays but not, in general, that between any two points along the rail. There are two such propagation coefficients for each direction. The corresponding deflection shapes within a sleeper span depend on the location of the excitation point. More details are given in [3.14–3.16].

Figure 3.33 shows the point mobility of a periodically supported track with the same parameters as in the previous section (and a sleeper spacing, d, of 0.6 m).

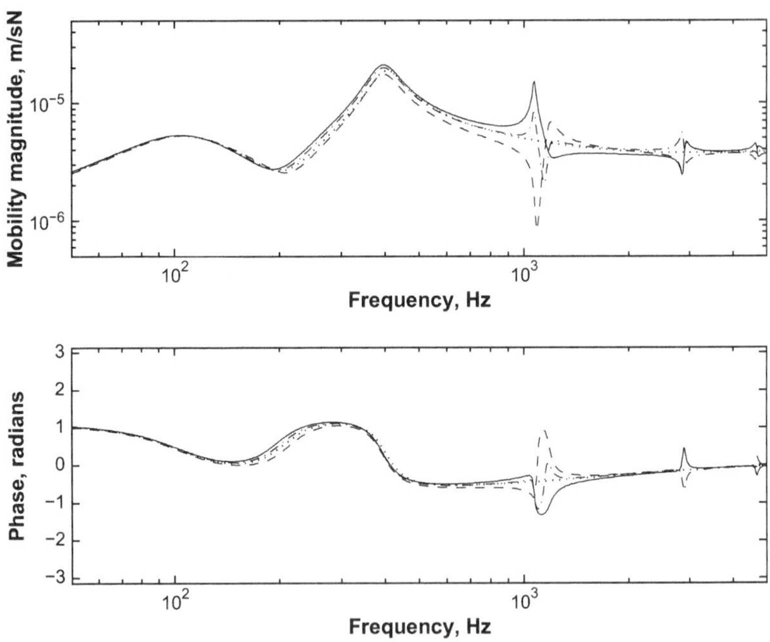

FIGURE 3-33 Vertical point mobility of track with discrete supports, pad stiffness 180 MN/m. —, excitation at mid-span; – – –, excitation above a sleeper; – · – ·, excitation at quarter span; · · · ·, continuous support

Comparing the results with those of the continuously supported model, also shown, it is clear that the response at low frequencies is unaffected but that around 1 kHz there are significant differences. The 'pinned–pinned mode' occurs when a half wavelength corresponds to the support spacing. For an unsupported UIC 60 rail with support distance 0.6 m this occurs at 1070 Hz. At this frequency the mobility at mid-span between two sleepers has a maximum whereas above a sleeper a minimum occurs. In fact, the minimum occurs at a slightly higher frequency than the maximum, the difference in frequency increasing as the pad stiffness increases. The continuously supported model gives a prediction that lies between the results from the discretely supported model.

The effect of periodicity is greater with a stiffer pad. Figure 3.34 illustrates this by showing results for a pad stiffness of 600 MN/m per pad (equivalent to 1000 MN/m^2). The peak in the point mobility at mid-span at 1070 Hz has a larger amplitude than in Figure 3.33 and the dip in the mobility above the sleeper is also deeper. The variations of the mobility within a sleeper span are also significant over a wider range of frequencies.

The corresponding propagation coefficients, γ, of the wave with the lowest decay rate are shown in Figures 3.35 and 3.36. Compared with the wavenumbers of the continuously supported rail, similar behaviour can be seen, apart from the distinct peaks in the real part of γ (the part associated with attenuation). At the pinned–pinned frequency the decay rate is low; in the absence of rail damping it would have

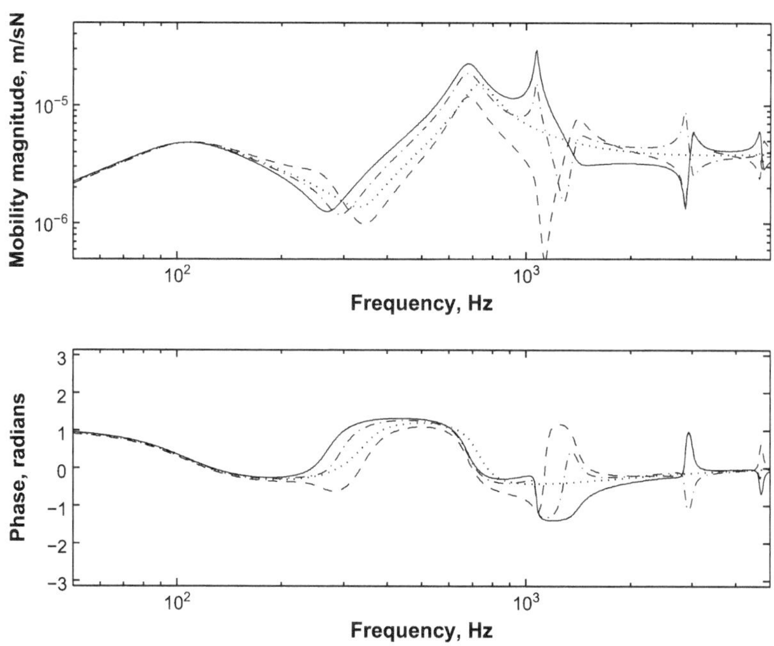

FIGURE 3-34 Vertical point mobility of track with discrete supports, pad stiffness 600 MN/m. —, excitation at mid-span; – – –, excitation above a sleeper; – · – ·, excitation at quarter span; ····, continuous support

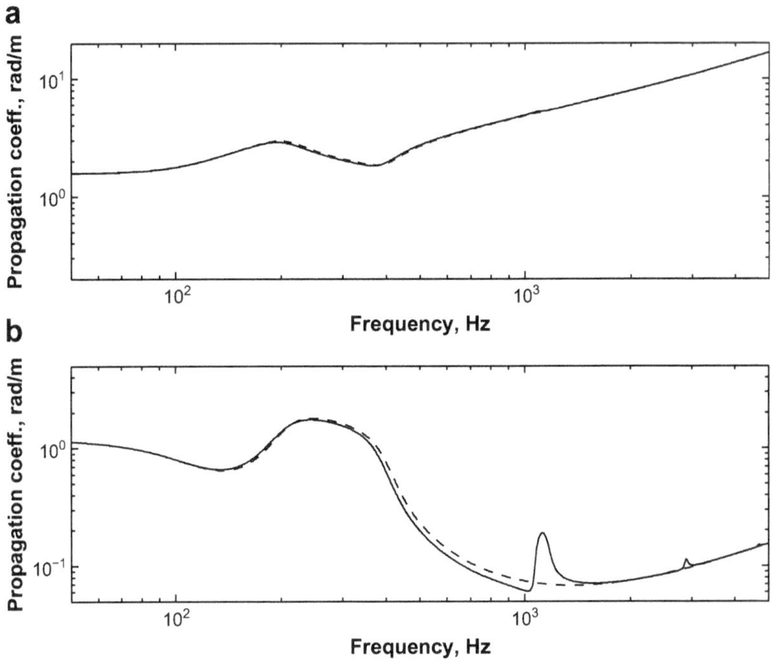

FIGURE 3-35 Propagation coefficients for vertical vibration of track with discrete supports, pad stiffness 180 MN/m. —, discrete supports; – – –, continuous supports. (a) Imaginary (propagating) part, (b) real (decaying) part

a more distinct minimum. However, just above this frequency it rises to a peak which is not found for the continuously supported track. Throughout this peak the phase change is virtually constant at π per sleeper bay (5.24 rad/m). This band is a typical 'blocked band' found in periodic structures [3.17], in which attenuation occurs even in the absence of damping. The width of the blocked zone depends on the stiffness of the support; for soft supports its effect is much less noticeable, whereas with the higher pad stiffness in Figure 3.36 it becomes more significant. The effects of periodicity are much less significant for the lateral direction, for which the support stiffness is much smaller than for the vertical direction. This can be seen in the measured results in Figure 3.4.

The bounding frequencies for the blocked zone of a periodic structure can be calculated as the natural frequencies of a single span of the periodic structure with either symmetric or anti-symmetric boundary conditions at the ends [3.17]. In the present case of a beam, this is either the mode of a simply supported beam, or the mode of a beam with sliding ends attached to half a sleeper at each end. These are shown schematically in Figure 3.37. Equivalently, the latter is the same as the mode of a simply supported beam with the sleeper attached at its mid-point.

In the example shown here these frequencies are 1070 and 1175 Hz for the softer pad, while the latter frequency increases to 1390 Hz for the stiffer pad. The lower of these frequencies, mode (a), is independent of the support stiffness (unless there is some rotational stiffness not considered here), whereas the higher mode, (b) or (c),

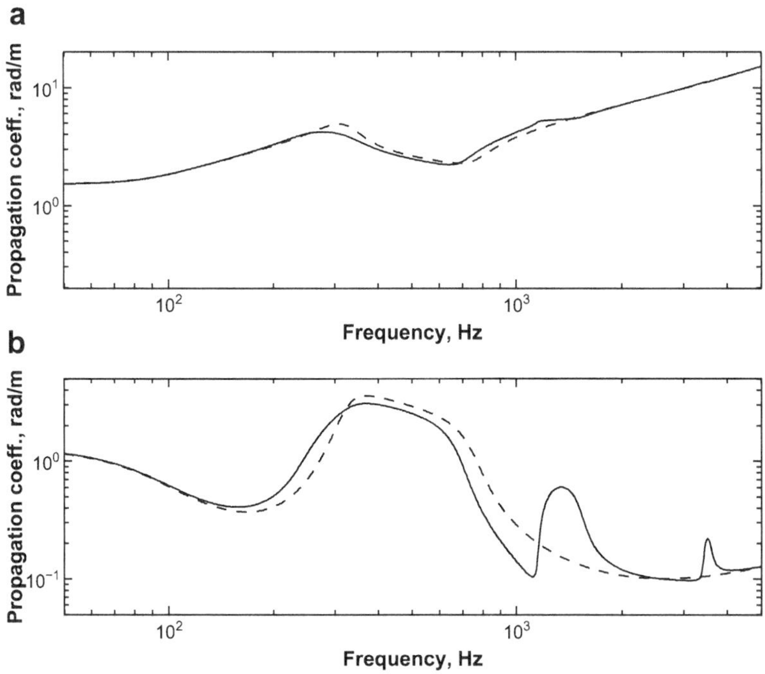

FIGURE 3-36 Propagation coefficients for vertical vibration of track with discrete supports, pad stiffness 600 MN/m. —, discrete supports; – – –, continuous supports. (a) Imaginary (propagating) part, (b) real (decaying) part

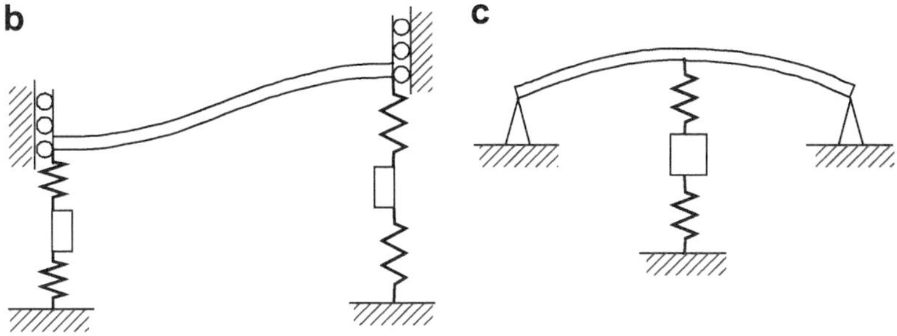

FIGURE 3-37 Modes of single span corresponding to the bounding frequencies of the blocked zone. (a) Anti-symmetric mode, (b) symmetric mode, (c) alternative visualization of mode (b)

increases as the support stiffness increases, in particular that of the pad. At the latter frequency the point mobility above a sleeper has a maximum, as seen in Figures 3.33 and 3.34.

Examples of the spatial distribution of the response are shown in Figures 3.38 to 3.40 for different excitation points. These results are for the softer pad. They show, for different frequencies, the transfer mobility as a function of distance, that is the velocity amplitude at x due to a unit force applied at x_c.

At the first two frequencies shown, behaviour typical of relatively low frequencies is seen. At 100 Hz the decay rate is high, whereas at 500 Hz the decay is much lower. At 100 Hz there is no significant effect of the forcing point location, but at 500 Hz some effect can be seen, as also seen in the point mobility, Figure 3.33.

The other frequencies are chosen to illustrate the pinned–pinned phenomenon. At 1070 Hz a very clear standing wave pattern can be seen with nodes at the sleepers and maxima mid-way between them. At 1170 Hz the maxima are found close to the sleepers. It is interesting to note that, for $x_c = d/4$, the response is strongly asymmetrical in the pinned–pinned region (Figure 3.39(c,d)) with waves propagating in one direction away from the forcing point.

The periodicity effects in a track are clearly significant in causing the point mobility to vary strongly within a sleeper span. However, in terms of noise radiation the key issue is whether the track produces more (or less) noise because of its periodic support than if it were continuously supported.

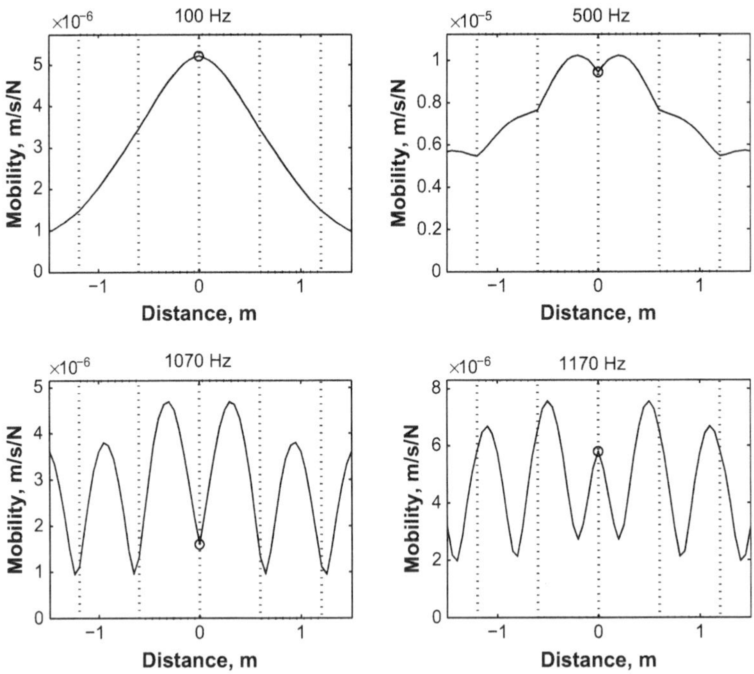

FIGURE 3-38 Rail vertical vibration versus distance predicted using periodic track model, $x_c = 0$ m (above a sleeper). —, transfer mobility; ○, excitation point; ····, sleeper positions

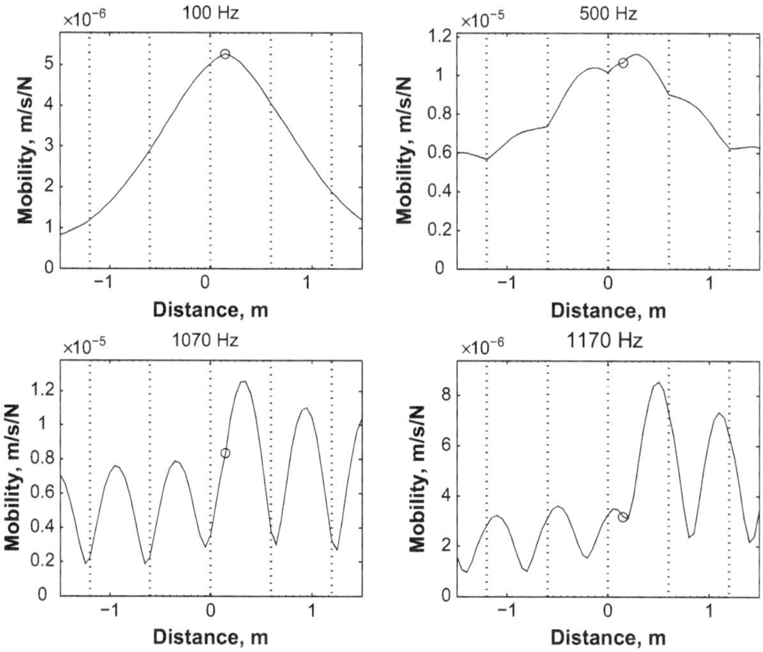

FIGURE 3-39 Rail vertical vibration versus distance predicted using periodic track model, $x_c = 0.15$ m (quarter span). —, transfer mobility; o, excitation point; ····, sleeper positions

To answer this question fully, a wheel/rail interaction model is required. However, a simplified model can be used to give a preliminary answer here. The following steps are performed:

- For a constant amplitude force applied at a number of points within a sleeper span, the average squared velocity is calculated over a sufficient length to capture the decaying waves; in this case ± 20 m is used.
- This is divided by the squared drive point mobility, as the excitation by roughness is approximately a constant displacement, at least for frequencies up to 1 kHz (see Chapter 5).
- This result is then averaged over 12 excitation positions across a sleeper span.

The result is shown in Figure 3.41(a). A refinement to the above is to divide the average response by the square of the sum of the rail, wheel and contact spring mobility ($|Y_r + Y_w + Y_c|^2$ instead of $|Y_r|^2$, see Section 5.2). The result of this is shown in Figure 3.41(b) for a wheel mobility representing a mass of 600 kg and a contact spring of stiffness 1.14×10^9 N/m. The effect of periodicity can be seen to be localized to the region around 1 kHz and for this example is at most 4 dB in the narrow band spectrum. In one-third octave bands the effect is at most 1.4 dB, in the 1 kHz band. Moreover, this is reduced by the effect of random variations in pad stiffness and sleeper spacing, as discussed in Section 3.5.2 below. Therefore for noise radiation the pinned–pinned resonance usually has only a small effect and satisfactory predictions are achieved using a continuously supported track model.

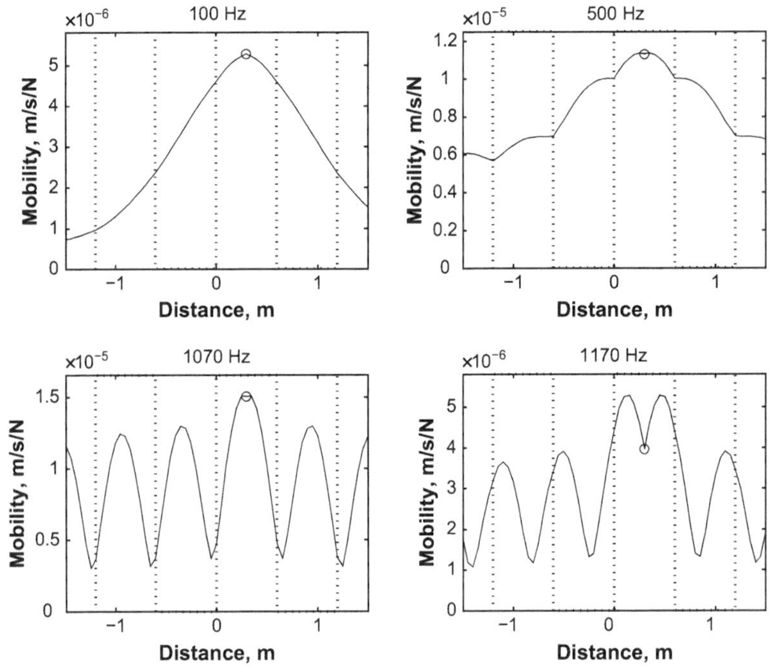

FIGURE 3-40 Rail vertical vibration versus distance predicted using periodic track model, $x_c = 0.3$ m (mid-span). —, transfer mobility, ○, excitation point, ····, sleeper positions

3.5.2 Random sleeper spacing and support stiffness

In reality, neither the sleeper spacing nor the ballast stiffness are constant along the track. Although data are limited, de France [3.18] found that the sleeper spacing on a newly laid section of track had a standard deviation of 39 mm compared with a mean of 628 mm. The ballast stiffness had a much larger relative variation. This was measured dynamically, for frequencies in the range 50–100 Hz with the individual sleepers uncoupled from the rail, and found to have a standard deviation of 25% of the mean value. Other results are given by Oscarsson [3.19] from two sites in Sweden which show a similar trend, although the standard deviations were smaller. These results are summarized in Table 3.4.

The rail pad stiffness may also vary from one pad to another, but no data are available and the variations are likely to be much smaller than those due to the ballast.

In order to quantify the effects of random variations in these parameters, a discretely supported Timoshenko beam with a finite number of supports (set to 80) was used [3.20], as described in Section 3.5.1. Values of sleeper spacing and ballast stiffness for each of these support points were selected according to a probability distribution. Monte Carlo simulations were then performed for a large number of tracks with parameters taken from these distributions.

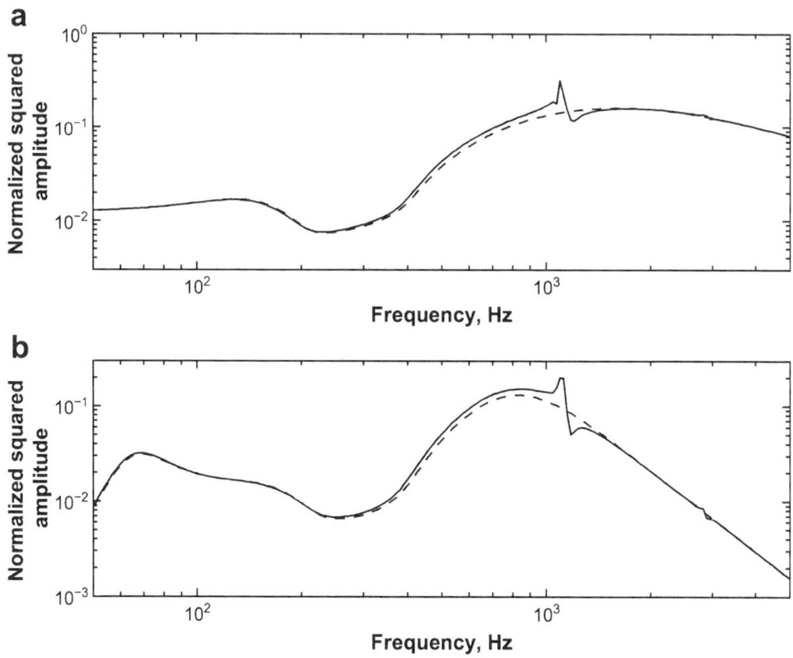

FIGURE 3-41 Squared amplitude of vertical rail vibration averaged over ±20 m and averaged over 12 force positions within a sleeper span. (a) Normalized by rail mobility, (b) normalized by sum of rail, wheel and contact spring mobilities. —, periodic support; – – –, continuous support

Example results from [3.20] are reproduced in Figures 3.42 and 3.43. These are for relatively large variations in both ballast stiffness and sleeper spacing in order to make the effects visible. The standard deviations are 40% and 20% of the mean, respectively, whereas the values in Table 3.4 are considerably smaller than this. The random changes in ballast stiffness only have a significant effect on the track response at low frequencies. The point receptance varies only below 300 Hz for these track parameters. The effect on the track decay rate, which is based on an average over many sleepers, is negligible.

By contrast, the effect of variations in sleeper spacing can be seen in the receptance over the whole frequency range. Nevertheless the decay rate is mainly affected at

TABLE 3-4 **MEAN AND STANDARD DEVIATION OF SLEEPER SPACING AND BALLAST STIFFNESS (FOR WHOLE SLEEPER)**

	Sleeper spacing, mm		Ballast stiffness, MN/m	
	Mean	**Standard deviation**	**Mean**	**Standard deviation**
Southampton [3.18]	628	39 (6%)	300	75 (25%)
Gåsakulla [3.19]	652	17 (3%)	255	16 (6%)
Grundbro [3.19]	650	20 (3%)	186	22 (12%)

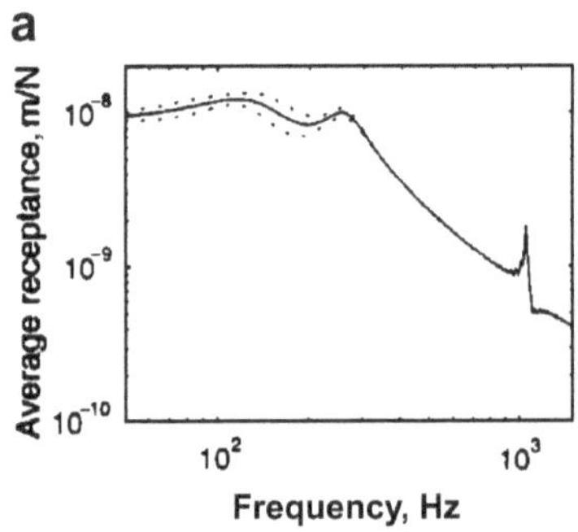

FIGURE 3-42 Results for random ballast stiffness K_b distributed between 0 and 302 MN/m (standard deviation 60 MN/m), $\eta_b = 0.6$, $K_p = 68.8$ MN/m, $\eta_p = 0.25$, $d = 0.6$ m. (a) Average receptance, (b) average decay rate [3.20]

higher frequencies, particularly around the pinned–pinned frequency. No significant peak or dip is found here in either the decay rates or receptance. Consequently, it is shown in [3.20] that, when the sleeper spacing has randomly distributed values, the spatially averaged vibration resembles the result from the continuously supported track model and the peak seen in Figure 3.41 no longer appears.

Figure 3.44 shows results of random sleeper spacing and ballast stiffness for a stiffer pad (five times stiffer). The effect on the track receptance is greater than for the soft pad especially around the pinned–pinned frequency region, but the effect on the average decay rate remains small.

3.6 RAIL CROSS-SECTION DEFORMATION

3.6.1 Finite element model

At high frequencies the rail no longer behaves as a simple beam in which the cross-section remains plane. Instead cross-sectional deformation occurs, particularly

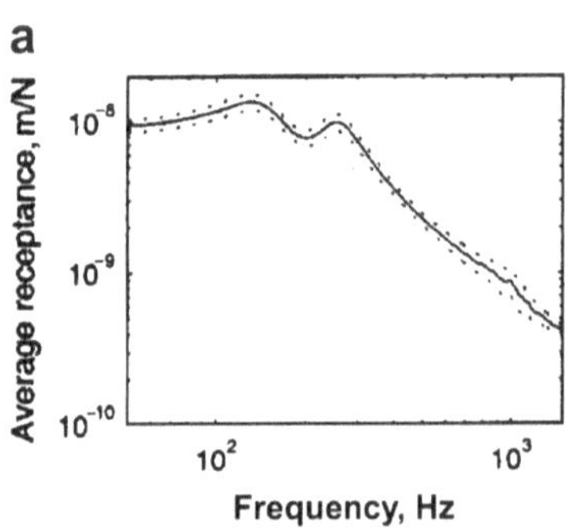

FIGURE 3-43 Results for random sleeper spacing d distributed between 0.3 and 0.9 m (standard deviation 0.11 m), $K_b = 151$ MN/m, $\eta_b = 0.6$, $K_p = 68.8$ MN/m, $\eta_p = 0.25$. (a) Average receptance, (b) average decay rate [3.20]

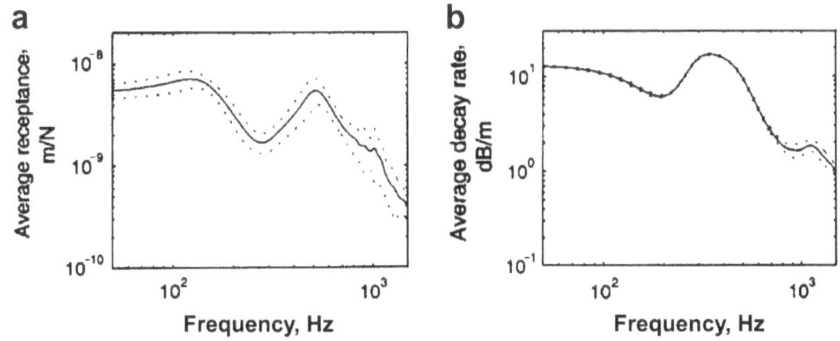

FIGURE 3-44 Results for random sleeper spacing d distributed between 0.3 and 0.9 m, random ballast stiffness K_b distributed between 0 and 302 MN/m, $\eta_b = 0.6$, $K_p = 344$ MN/m, $\eta_p = 0.25$. (a) Average receptance, (b) average decay rate [3.20]

for vibration in the lateral direction where significant web bending occurs above about 1.5 kHz. Also for the vertical direction foot flapping motion occurs progressively above about 2 kHz. Figure 3.45 shows typical finite element results for a finite length of UIC54 rail which illustrate these effects [3.21]. By selecting a finite length of rail and applying boundary conditions at the ends that are either symmetric or anti-symmetric, the modes of the finite rail have an exactly sinusoidal modeshape in the longitudinal direction. They therefore correspond to free waves in an infinite rail. In each case shown the wavelength is the same (the length chosen represents 3/4 of the wavelength for these modes).

Figure 3.46 shows the deformation of the cross-section in the vertical and longitudinal waves for a wider range of frequencies. This shows how the first vertical

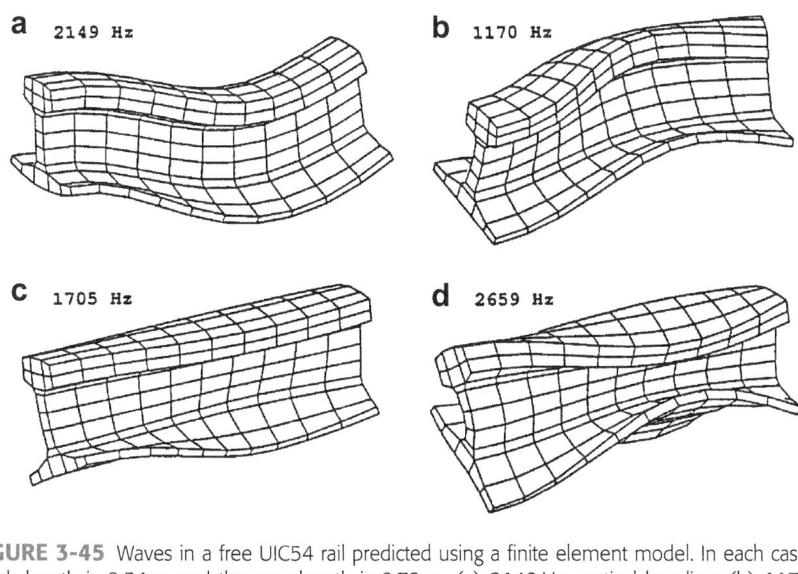

FIGURE 3-45 Waves in a free UIC54 rail predicted using a finite element model. In each case the mesh length is 0.54 m and the wavelength is 0.72 m. (a) 2149 Hz, vertical bending; (b) 1170 Hz, lateral bending; (c) 1705 Hz, torsion; (d) 2659 Hz, web bending [3.21]

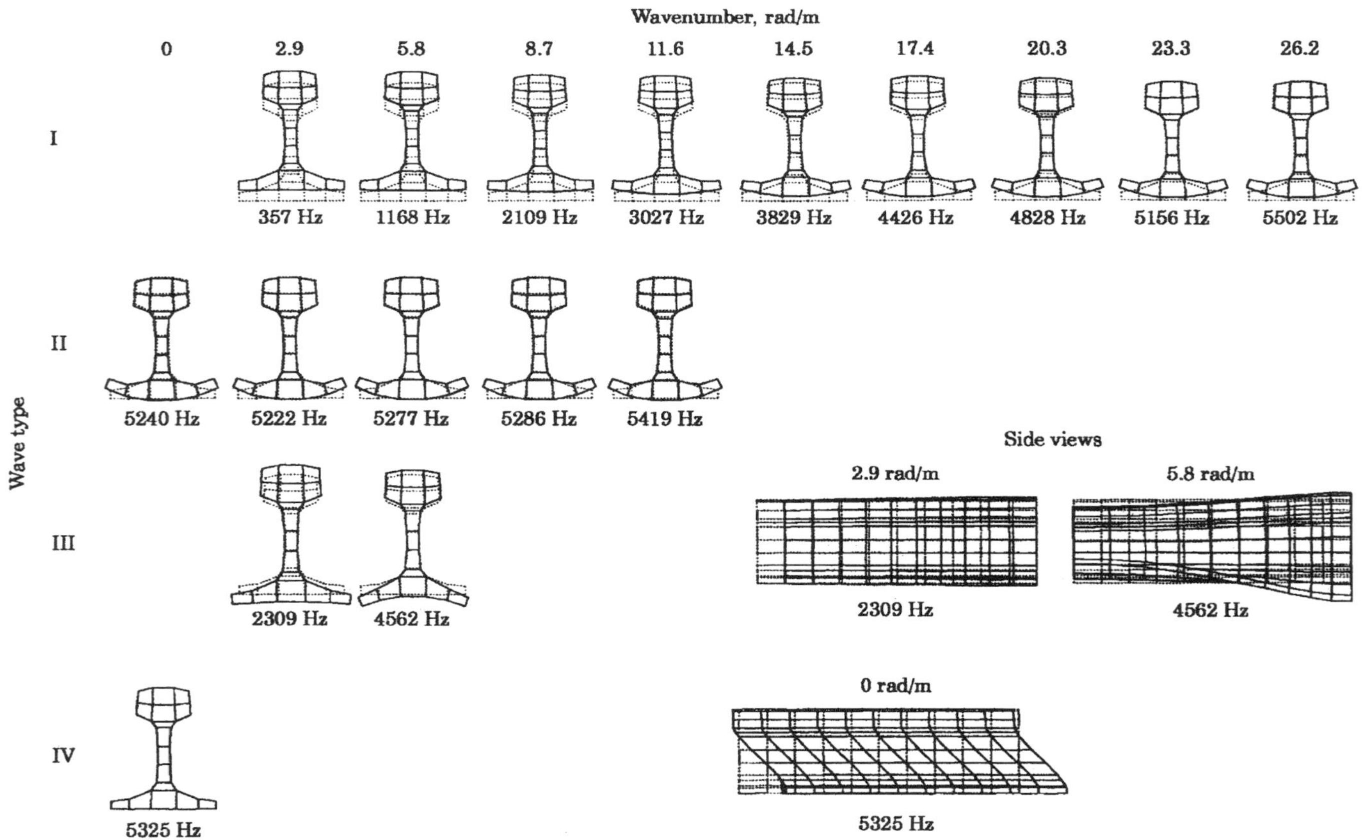

FIGURE 3-46 Normalized cross-section deformation associated with each free vertical and longitudinal wave in a UIC54 rail calculated using finite elements, from [3.21]

wave in the rail involves progressively more foot flapping motion as frequency increases. The higher order foot flapping wave, wave II, cuts on (propagating wave motion begins) at 5.2 kHz. Similarly, a higher order longitudinal wave, wave IV, in which the head and foot move in opposite directions, cuts on at 5.3 kHz.

Corresponding results for the lateral direction are shown in Figure 3.47. Again, four wave types are found in the frequency region considered, below 6 kHz. Two, the lateral bending and torsional waves, commence at 0 Hz (for an unsupported rail). Web bending (wave III) commences at 1.4 kHz and a double web bending (wave IV) cuts on at 4.4 kHz.

3.6.2 Periodic structure – finite element model

The difficulties of using a conventional finite element model are (i) wavenumbers can only be predicted at particular frequencies determined by the length of the model considered, and (ii) the forced vibration of an infinite rail cannot be predicted.

To overcome these difficulties, special spectral finite elements can be used in which the cross-section is modelled with two-dimensional elements that take account of the wavenumber in the third dimension. Gavric [3.22] predicted free waves in a rail using such an approach. The full forced vibration solution can be found from a Fourier transform over the wavenumber, as shown, for example, by Nilsson [3.23] and Sheng *et al.* [3.24].

An alternative finite element-based approach was developed in [3.25]. This used a finite element model of a short slice of rail (an arbitrary length of 10 mm was used) which was then formed into an infinite structure through the use of periodic structure theory (a continuous structure is periodic with arbitrary period). Note, however, that the support structure is considered to be continuous in this model. Such an approach allows a commercial finite element package to be used to generate the matrices of the short slice. The solution method allows the complex wavenumbers of all waves to be found at any arbitrary frequency. There are as many waves as there are degrees of freedom in one cross-section, although many of them are rapidly decaying. Associated with each wavenumber is a wave eigenvector (equivalent to a modeshape). The forced response of the rail can then be found from a sum over the waves.

Example results are shown in Figures 3.48 and 3.49, adapted from [3.16]. These show the vertical and lateral mobilities of the track predicted using this model, corresponding to the measurements in Figures 3.3 and 3.4. For this track, the rails were UIC54 and the pad stiffness was taken as 1300 MN/m vertically and 100 MN/m laterally with a loss factor of 0.25 in each case. The sleeper mass was 122 kg, the sleeper spacing being 0.6 m. The ballast stiffness was 67 MN/m vertically and 34 MN/m laterally with a loss factor of 2.0.

Peaks are found at the cut-on frequencies of the higher order waves, for example at 5 kHz for the vertical direction. The vertical mobility is found to be similar to that from a Timoshenko beam model up to about 2 kHz, but as foot flapping motion becomes more important the response increases. The response of the end of the foot is considerably greater than that of the head above about 2 kHz.

Although the lateral mobility could be represented by a simple beam model, as described in the previous sections, it would be considerably underpredicted by such a model, as shown in Figure 3.49. This is caused at low frequencies by the omission of torsion, and at higher frequencies by the additional flexibility caused by the web

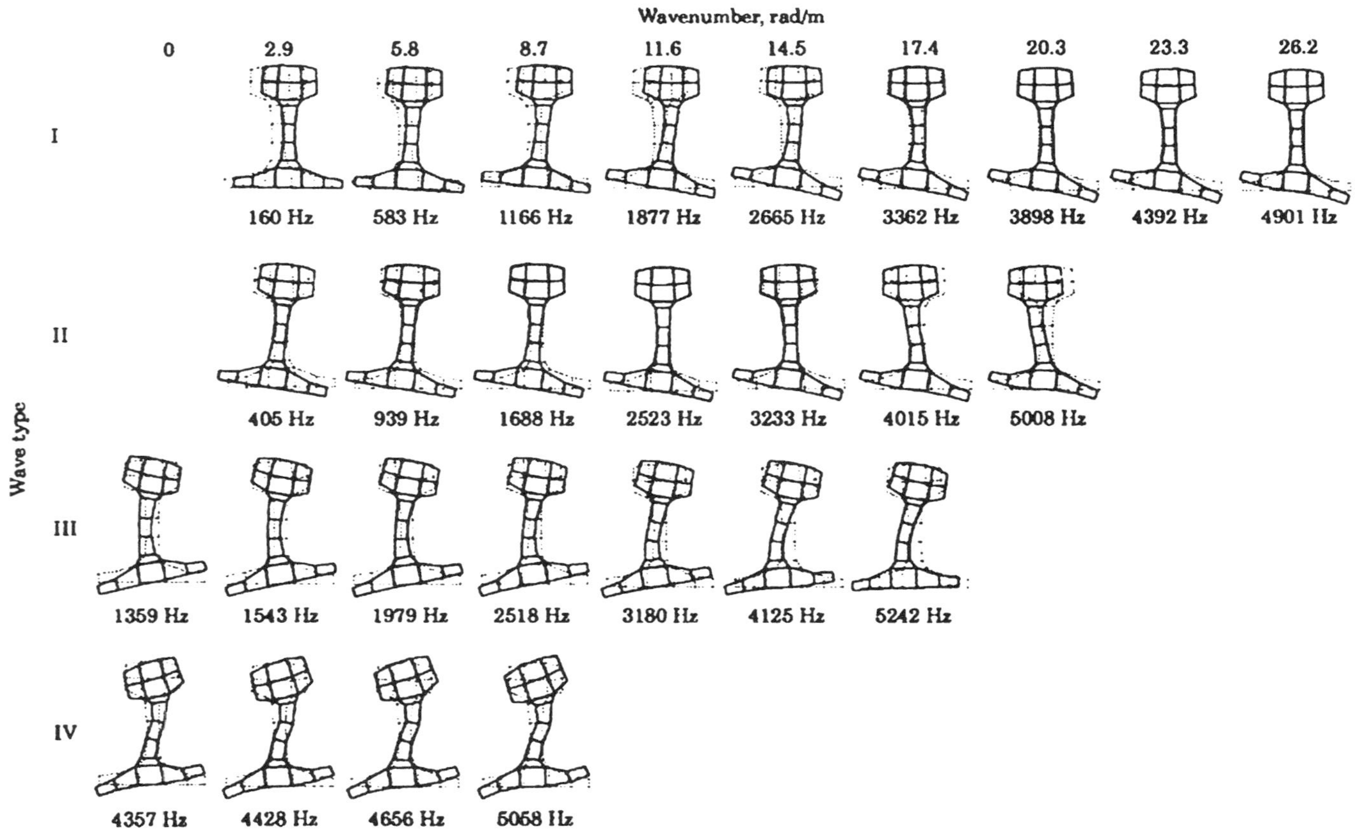

FIGURE 3-47 Normalized cross-section deformation associated with each free lateral wave in a UIC54 rail calculated using finite elements, from [3.21]

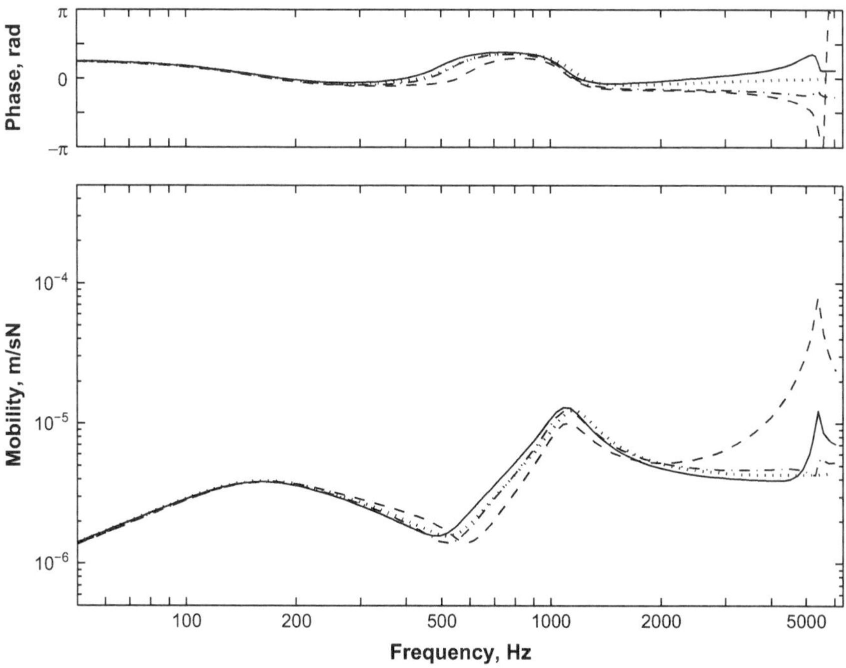

FIGURE 3-48 Point vertical mobility predicted by periodic structure theory for force at the rail head. —, response at rail head; – – –, response at foot end; – · – ·, response at foot centre; ···· Timoshenko beam model

bending motion. The lateral mobility is actually similar at high frequencies to that of a beam representing the head alone. The peak at around 2 kHz corresponds to the cut-on of wave III (web bending) and the peak at 5 kHz to the cut-on of wave IV (double web bending). These were not seen in the measurements in Figure 3.4 as the force and response points were at the centre of the head, rather than at the running surface as here.

3.6.3 Multiple beam models

A simpler approach to modelling the cross-sectional deformation was taken in [3.26] where the vertical motion was represented as a composite beam formed of two Timoshenko beams coupled by a continuous layer of springs, as shown in Figure 3.50(a). The resonance frequency of this system was tuned to the foot flapping frequency (5 kHz) identified from the finite element model, while the mass of the second beam was chosen to represent the mass at the end of the rail foot vibrating as a cantilever.

For the lateral direction it is important to include bending and torsion from low frequencies, web bending (from about 1.5 kHz) and second-order web bending (from about 4 kHz). A model to account for this can be formed of two beams including torsion that are linked by an array of vertical beams representing the web [3.27], see Figure 3.50(b). This omits the twisting stiffness of the

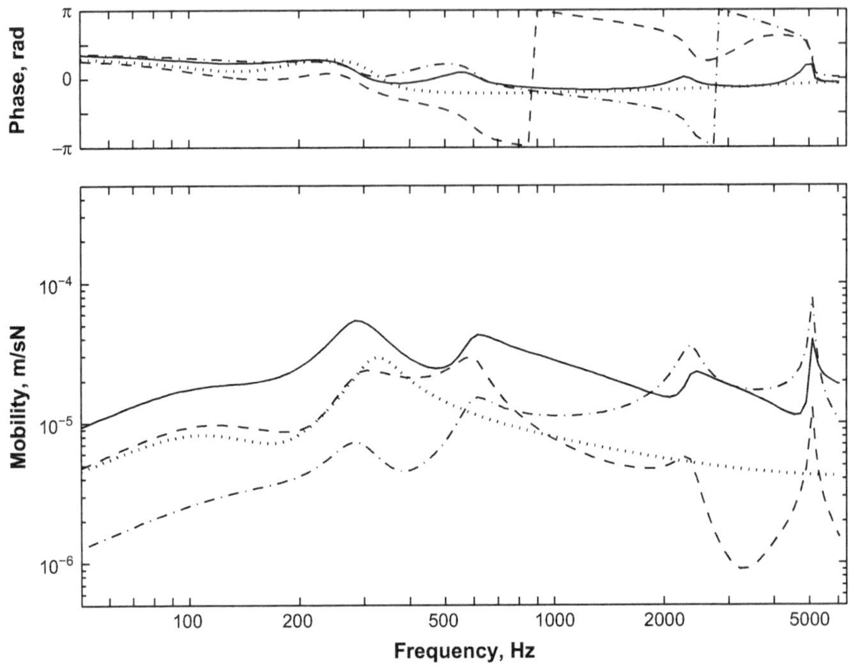

FIGURE 3-49 Point lateral mobility predicted by periodic structure theory for force at the top of the rail head. —, response at rail head; – – –, lateral response at foot; – · – ·, vertical response of foot end; · · · ·, Timoshenko beam model

web but this has been shown to be negligible compared with the other stiffness in the rail.

These models have also been used with periodic supports using a technique similar to that described in Section 3.5.1, see [3.26, 3.28].

3.7 SLEEPER VIBRATION

In all the models discussed throughout this chapter so far, the sleeper has been represented as a mass. This simplification ignores the bending modes that monobloc sleepers have in the frequency range of interest. Grassie [3.29] described a simple Timoshenko beam model for a freely suspended sleeper in which the variation of profile along the length of the sleeper was ignored. Quite reasonable agreement was found with measurements in terms of natural frequencies, apart from the first mode, for which the effect of the non-uniform profile is greatest. However, as will be seen, once the sleeper is supported in ballast, even these differences are of little consequence.

Inclusion of the ballast below the sleeper leads to a slightly more complex model, which is described here. Such a model was initially developed in [3.15, 3.30] in terms of its modes. The model described here is based instead on the forced response. This approach is more straightforward, especially if frequency-dependent ballast

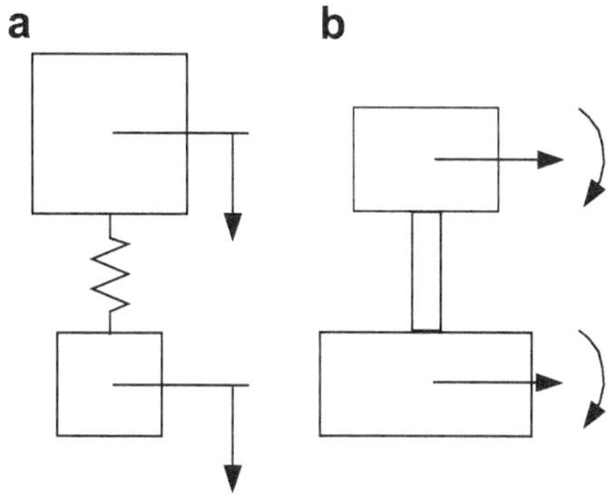

FIGURE 3-50 (a) Double Timoshenko beam model for vertical motion of rail. (b) Multiple beam model for lateral motion of rail

properties are to be included. The ballast stiffness is strongly frequency dependent, as will be discussed in Section 3.7.2.

3.7.1 Model for sleeper vibration

Figure 3.51 shows the sleeper, represented as a finite beam on a continuous elastic foundation. The point force acts at $y = 0$ and various waves are generated. The length is L and the distance from the left-hand end to the force point is y_0.

The equations of motion for a Timoshenko beam on an elastic foundation were given in Section 3.4. The beam now represents the sleeper and the elastic layer the ballast. From this, two wavenumbers, k_p and k_e, are obtained, the roots of equation (3.56), along with their negative counterparts. Above the cut-on frequency k_p is close to real with a positive real part, while k_e is close to imaginary with a negative imaginary part. The beam response amplitude, u, at frequency ω can be written as

$$u(y) = A_1 e^{ik_e y} + A_2 e^{ik_p y} + A_3 e^{-ik_e y} + A_4 e^{-ik_p y} \tag{3.78}$$

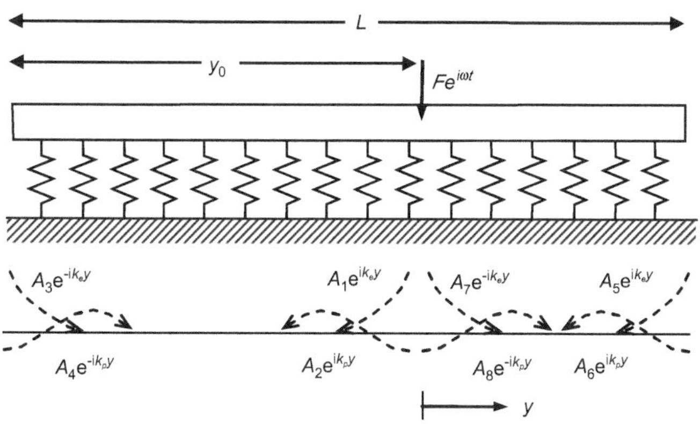

FIGURE 3-51 Sleeper represented as a finite Timoshenko beam on an elastic foundation, excited by a point force at $y = 0$

where y is the coordinate across the track and A_n are amplitudes of the various waves. The natural frequencies of the sleeper can be found as in [3.29] by applying the boundary conditions for free ends: the shear force and bending moment must both be zero at the ends.

Alternatively, the forced response can be determined directly. The beam can be divided into two parts with

$$u_-(y) = (A_1 e^{ik_e y} + A_2 e^{ik_p y} + A_3 e^{-ik_e y} + A_4 e^{-ik_p y}) \qquad \text{for} \quad -y_0 < y < 0 \tag{3.79}$$

$$u_+(y) = (A_5 e^{ik_e y} + A_6 e^{ik_p y} + A_7 e^{-ik_e y} + A_8 e^{-ik_p y}) \qquad \text{for} \quad 0 < y < L - y_0 \tag{3.80}$$

Similarly, the rotation of the cross-section, ϕ, is given by

$$\phi_-(y) = (A_1 \Psi_e e^{ik_e y} + A_2 \Psi_p e^{ik_p y} + A_3 \Psi_e e^{-ik_e y} + A_4 \Psi_p e^{-ik_p y}) \quad \text{for} \quad -y_0 < y < 0 \tag{3.81}$$

$$\phi_+(y) = (A_5 \Psi_e e^{ik_e y} + A_6 \Psi_p e^{ik_p y} + A_7 \Psi_e e^{-ik_e y} + A_8 \Psi_p e^{-ik_p y}) \quad \text{for} \quad 0 < y < L - y_0 \tag{3.82}$$

where Ψ_i are the ratios ϕ/u in each wave according to equation (3.55).

Two boundary conditions apply at each of the beam ends. The shear force is zero at each end:

$$GA\kappa\left(\frac{\partial u}{\partial y} - \phi\right)\bigg|_{y=-y_0} = 0 \quad \text{and} \quad GA\kappa\left(\frac{\partial u}{\partial y} - \phi\right)\bigg|_{y=L-y_0} = 0 \tag{3.83}$$

Similarly the bending moment is zero at each end:

$$EI\frac{\partial \phi}{\partial y}\bigg|_{y=-y_0} = 0 \quad \text{and} \quad EI\frac{\partial \phi}{\partial y}\bigg|_{y=L-y_0} = 0 \tag{3.84}$$

Four more conditions apply at $y = 0$: continuity of displacement, rotation, bending moment and equilibrium of forces:

$$u_-(0) = u_+(0) \tag{3.85}$$

$$\phi_-(0) = \phi_+(0) \tag{3.86}$$

$$EI\frac{\partial \phi_-}{\partial y}\bigg|_{y=0} = EI\frac{\partial \phi_+}{\partial y}\bigg|_{y=0} \tag{3.87}$$

$$GA\kappa\left(\frac{\partial u_-}{\partial y} - \phi_-\right)\bigg|_{y=0} - GA\kappa\left(\frac{\partial u_+}{\partial y} - \phi_+\right)\bigg|_{y=0} = F \tag{3.88}$$

Equations (3.83)–(3.88) can be written as an 8 × 8 matrix equation and solved to find the wave amplitudes A_i. The response at a general position y is given by equation (3.79) or (3.80).

3.7.2 Results

This model is applied to a typical concrete sleeper, the response of which has been measured, [3.31]. The geometry of the sleeper is shown in Figure 3.52. The parameters used for this sleeper are listed in Table 3.5. The cross-sectional area is chosen as the average of the values at the rail seat and at the centre of the sleeper. The density is then chosen to give the correct total mass (300 kg). The second moment of area is the geometrical average of the values at the railseat and the centre of the sleeper, as recommended by Grassie [3.29]. Finally, the equivalent Young's modulus is obtained by fitting predicted natural frequencies to measurements.

In [3.29], measured natural frequencies on 12 different concrete sleepers were used to derive Young's modulus and density values as 5.24×10^{10} N/m^2 and 2570 kg/m^3. These differ slightly from the values used here.

Initially, results are shown for a constant ballast stiffness of 48 MN/m^2, equivalent to 60 MN/m per sleeper end or a continuous stiffness along the direction of the rail of 100 MN/m^2 per rail (as used in the track models in earlier sections). Figure 3.53 shows the predicted point mobility of the sleeper excited at the railseat ($y_0 = 2.0$ m, or 0.5 m from the right-hand end), along with the result from a simple mass model based on half the mass of the sleeper, i.e. 150 kg, and a stiffness of 60 MN/m.

The fundamental resonance of the beam on the foundation can be seen at about 120 Hz in both curves. Above this frequency the beam has a number of bending resonances, not seen for the mass–spring model. The displaced shape at some of these frequencies is shown in Figure 3.54. This shows the real part of the response normalized to that at the excitation point. Note that this is an operating deflection

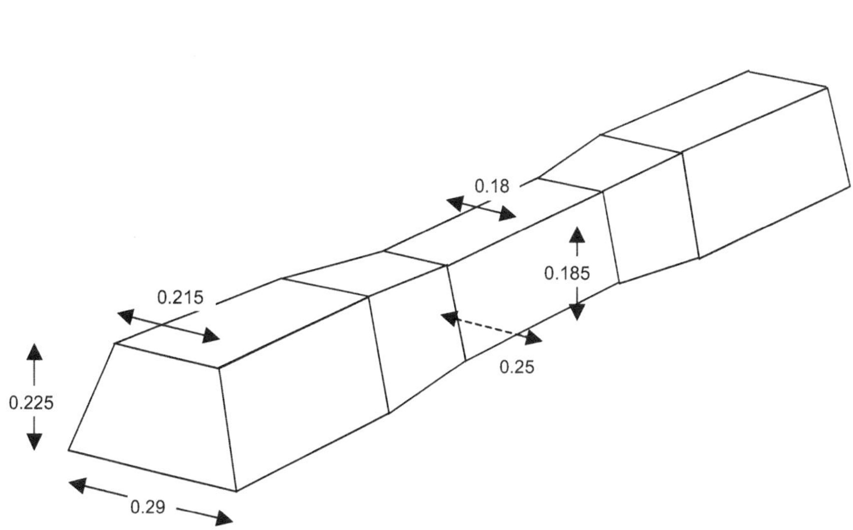

FIGURE 3-52 Concrete sleeper showing dimensions (not to scale)

TABLE 3-5 **PARAMETERS USED TO DESCRIBE A CONCRETE SLEEPER**

Young's modulus (N/m^2)	E	4.3×10^{10}
Shear modulus (N/m^2)	G	1.87×10^{10}
Poisson's ratio	ν	0.15
Sleeper damping loss factor	η_s	0.01
Density (kg/m^3)	ρ	2500
Cross-sectional area at railseat (m^2)	A	0.0568
Cross-sectional area at sleeper centre (m^2)	A	0.0398
Average cross-sectional area (m^2)	A	0.0436
Second moment of area at railseat (m^4)	I_s	2.40×10^{-4}
Second moment of area at sleeper centre (m^4)	I_c	1.13×10^{-4}
Average second moment of area (m^4)	I	1.65×10^{-4}
Shear coefficient	κ	0.83
Length (m)	L	2.5
Excitation point (m)	y_0	2.0
Ballast stiffness (N/m^2)	s	4.8×10^7
Ballast damping loss factor	η	1.0

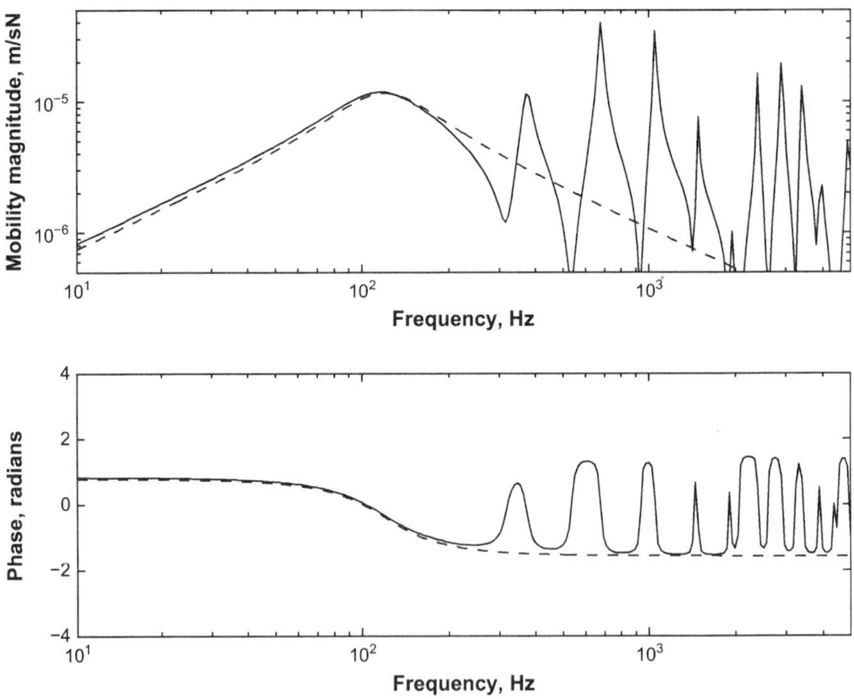

FIGURE 3-53 Point mobility of concrete sleeper: —, predicted using Timoshenko beam on elastic foundation; – – –, equivalent mass–spring system

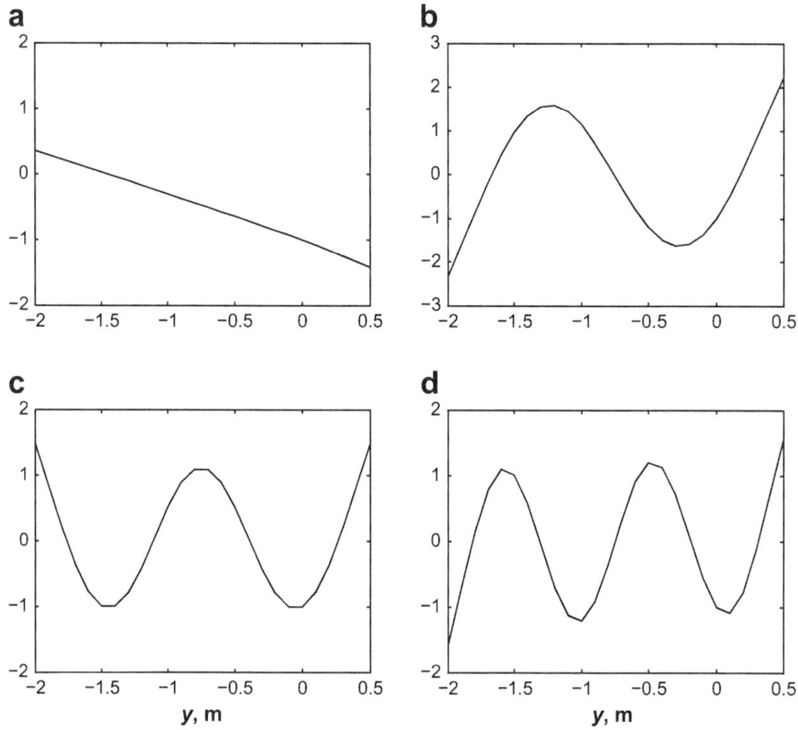

FIGURE 3-54 Displaced shape versus distance y along sleeper at frequencies corresponding to peaks of the mobility, normalized to response at excitation position, $y = 0$. (a) 120 Hz, (b) 370 Hz, (c) 680 Hz, (d) 1050 Hz

shape, which is not the same as a modeshape, since the response will contain contributions from several modes due to the high damping.

Table 3.6 lists the natural frequencies obtained using this sleeper model, both with and without the support stiffness. These have been determined by predicting the frequency response using a very low value of damping and finding the frequencies corresponding to the peaks, to the nearest 1 Hz. It can be seen that the support stiffness has most effect on the natural frequencies of the low frequency modes, and has negligible effect above 1 kHz. The peak at 120 Hz seen in Figure 3.53 actually comprises two modes, a bending resonance at 180 Hz and a rigid body resonance at 100 Hz. They can only be identified separately by reducing the damping. Table 3.6 also shows the effect of increasing the ballast stiffness by a factor of 10. This leads to an increase in all the natural frequencies of the sleeper, with the largest changes occurring for the low frequency modes. At high frequencies the bending stiffness of the sleeper dominates and the support stiffness has negligible effect.

Experimental natural frequencies determined on a concrete sleeper are listed in Table 3.7 [3.31]. These were measured on a freely suspended concrete sleeper (with geometry as in Figure 3.52), and on the same sleeper when installed in the track. These results show similar trends to the predictions, although the exact frequencies differ due to the simplified geometry used in the model. The predicted results for the

TABLE 3-6 PREDICTED NATURAL FREQUENCIES (IN HZ) OF MONOBLOC CONCRETE SLEEPER

	Unsupported	Constant ballast stiffness	Stiffness increased by factor 10	Variable ballast stiffness	Variable stiffness (damped)
Rigid mode	0	100	316	100	184
2 nodes	135	167	341	193	
3 nodes	359	372	473	410	421
4 nodes	669	676	736	718	733
5 nodes	1045	1050	1088	1092	1109
6 nodes	1469	1472	1499	1514	1532
7 nodes	1925	1928	1948	1970	1989

free–free sleeper in Table 3.6 are within 2% of the measured results for all modes except the first. This first bending mode is most affected by the non-uniformity of the actual sleeper geometry which is not included in the model. For higher order modes, the relative increase in natural frequency in Table 3.7 when the sleeper is located in the ballast is greater than that seen in the predictions of Table 3.6. This suggests that the ballast stiffness at higher frequencies is greater than the value used in these predictions.

The model above was based on a constant value for the ballast stiffness. Measurements of ballast stiffness indicate that it is strongly frequency dependent above about 100 Hz. This also manifests itself in a greater effect on the higher order sleeper modes, as noted above. Ballast stiffness results from [3.32] are shown in Figure 3.55 along with an idealized frequency-dependent stiffness, see also Section 11.3.4. (Note that a similar amplitude dependence could be obtained by using a viscous damping model, see Section 3.2.7, although the loss factor would tend to infinity at high frequencies.) Using the idealized stiffness from Figure 3.55 at each frequency in place of the constant value produces the sleeper mobility shown in Figure 3.56. This can be seen to have reduced the height of the peaks and troughs compared with the result in

TABLE 3-7 NATURAL FREQUENCIES AND MODAL LOSS FACTORS FOR THE SAME CONCRETE SLEEPER FREELY SUSPENDED AND IN SITU [3.31]

Number of nodes	Free sleeper		Sleeper in ballast	
	Frequency (Hz)	Loss factor	Frequency (Hz)	Loss factor
2	120	0.0079	186	0.815
3	352	0.0100	428	0.350
4	682	0.0125	745	0.143
5	1036	0.0148	1136	0.094
6	1483	0.0109	1616	0.068
7	1943	0.0200	2077	0.047

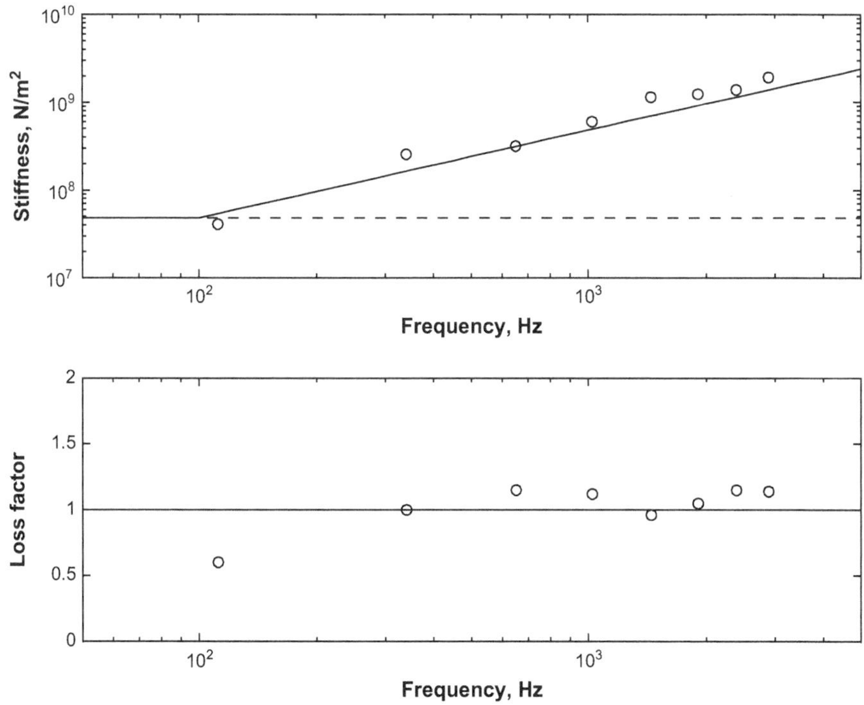

FIGURE 3-55 Ballast stiffness and loss factor. —, idealized frequency-dependent stiffness; – – –, constant stiffness; ○, measured [3.32]

Figure 3.53, particularly for frequencies below about 2 kHz. Again, the result of using a simple mass–spring system representing half the sleeper is also shown.

The natural frequencies obtained using the frequency-dependent ballast stiffness are listed in Table 3.6. Also shown in Table 3.6 are the frequencies corresponding to the peaks in the mobility when damping is included. These can be seen to follow a similar trend to the measured values in Table 3.7, at least in terms of the increase in frequency relative to the free–free case.

Figure 3.57 shows the spatially averaged mobility for each model – the square root of the mean square mobility. This is of direct relevance to the radiated noise (see Chapter 6) as sound power is proportional to the spatially averaged squared velocity. Apart from low frequencies, using the mass–spring model will tend to underestimate the noise from a monobloc sleeper due to the neglect of the bending resonances. On the other hand, using the beam model with a constant ballast stiffness will lead to a considerable overestimate at the beam resonances.

Figure 3.58(a) shows the point mobility at the rail seat and the transfer mobility to the rail seat at the far side of the sleeper. At low frequencies the far side has a much lower amplitude. This can be understood from the displaced shape in Figure 3.54(a), which indicates that the sleeper is rotating about the position of the far rail seat. At higher frequencies, however, the displacements at the two sides of the sleeper are similar.

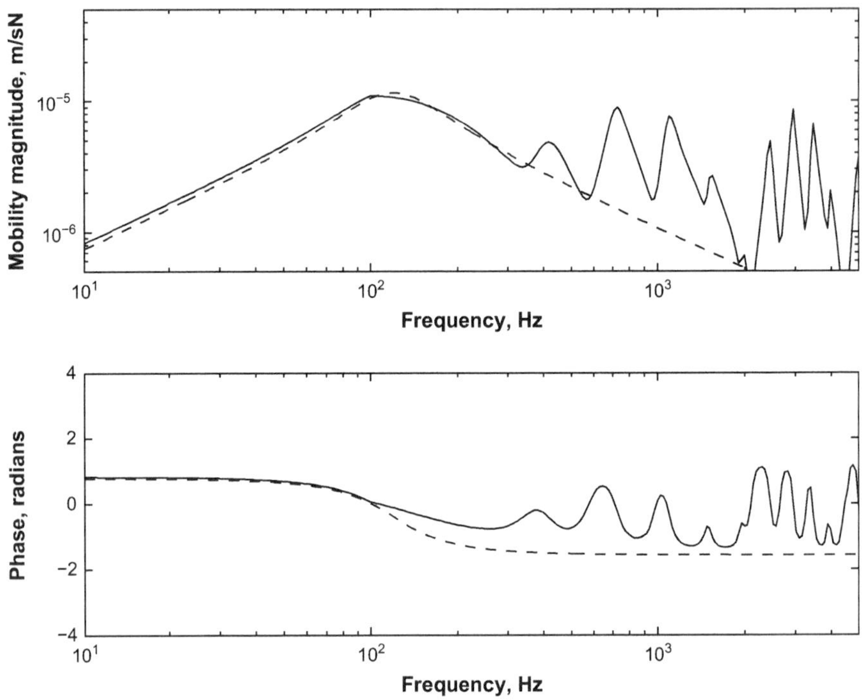

FIGURE 3-56 Point mobility of concrete sleeper predicted using Timoshenko beam on elastic foundation. —, frequency-dependent ballast stiffness; – – –, equivalent mass–spring system

When the ballast is preloaded by the presence of the train its stiffness will increase. Results in [3.32] indicate that the ballast stiffness increases typically by a factor of 2.5 for a preload of 20 kN per sleeper end. Figure 3.58(b) shows corresponding results for this higher value of stiffness.

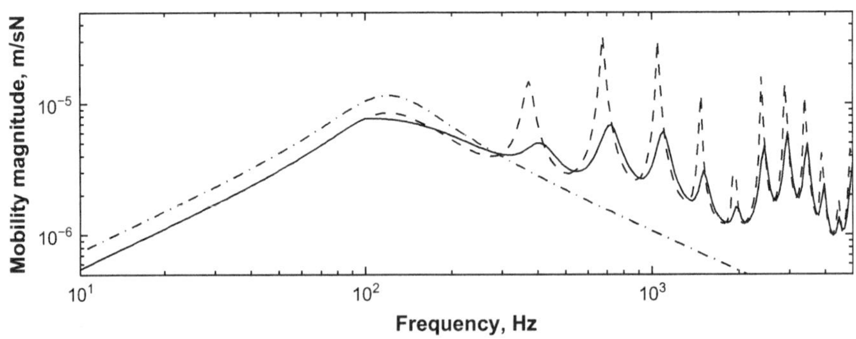

FIGURE 3-57 Spatially averaged response of sleeper. —, frequency-dependent ballast stiffness; – – –, constant ballast stiffness; – · – ·, equivalent mass–spring system

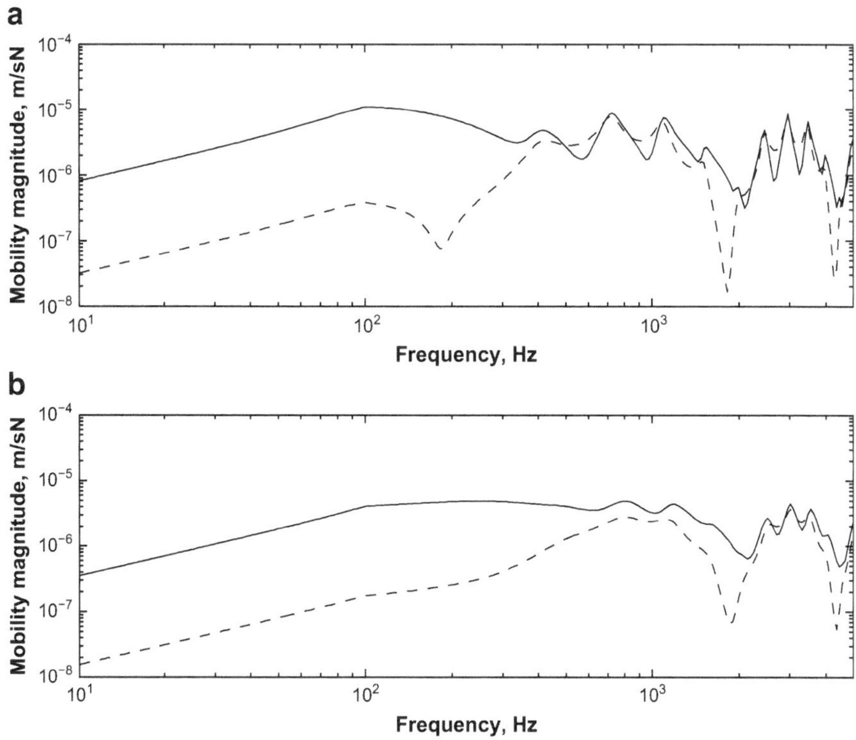

FIGURE 3-58 Response of sleeper for frequency-dependent ballast stiffness. —, at drive point (rail seat); – – –, at far rail seat. (a) Unloaded; (b) ballast stiffness increased by factor of 2.5 to represent preload effect

3.8 RAIL PAD STIFFNESS

The stiffness of a rail pad is usually defined for track engineering purposes in terms of its static or low frequency dynamic behaviour over large strains. However, for predicting noise generation, it is the stiffness at high frequencies and small strains that is required. Elastomeric materials can have a much higher stiffness under such conditions than the conventional static or even low frequency dynamic stiffness. In [3.33] differences of up to a factor of 10 were found.

The rail is held against the sleeper by a spring clip (see Figure 3.2) which is intended to prevent rail roll-over due to large lateral forces at the wheel/rail interface. Typically these clips have a large static deflection under the nominal preload – for example, a Pandrol clip typically has a deflection of 10 mm for a static preload of 10 kN (on each side of the rail). The corresponding stiffness is therefore of the order of 2 MN/m for the two clips combined, which is much lower than the stiffness of even a soft rail pad. The main effect of the clips is to apply a preload and this should be taken into account when measuring the dynamic stiffness. Nevertheless, the lateral stiffness of the clip may be significant. For example, in [3.33] a DE spring clip was found to correspond to a lateral stiffness of 60 MN/m in parallel with the pad.

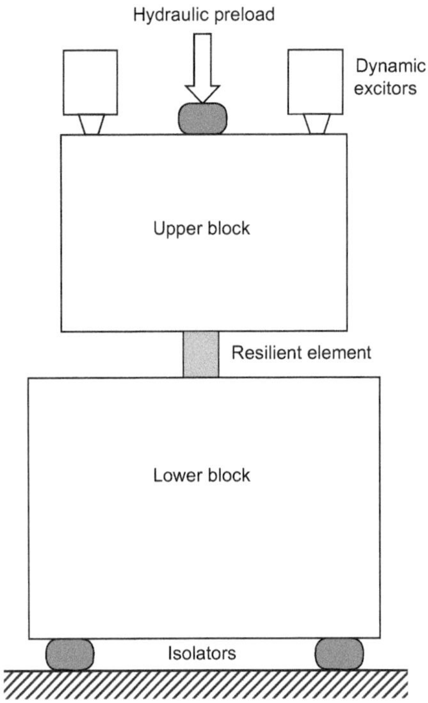

FIGURE 3-59 Schematic diagram of test apparatus used for indirect measurements of dynamic stiffness

Measurements of the stiffness of a number of rail fastener systems are described in [3.33, 3.34]. These measurements used an indirect method in which the resilient element is placed between two large masses which are isolated from the rest of the test rig (see Figure 3.59). A static preload can also be applied. The upper mass is excited dynamically and the response of both masses is measured using accelerometers [3.34]. Valid measurements can be obtained over about a decade of frequency, here 100–1000 Hz.

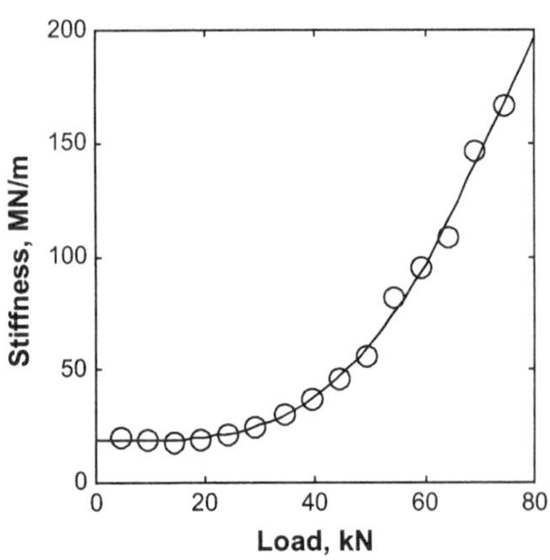

FIGURE 3-60 Incremental static stiffness of 10 mm studded rubber rail pad, from [3.34]

Figure 3.60 shows the incremental static stiffness of a 10 mm studded rubber pad [3.34], i.e. the gradient of the load–deflection curve. At low values of preload the stiffness is quite low and fairly constant but above the 20 kN preload applied by the clip the stiffness of the pad increases considerably with the preload. Corresponding dynamic stiffness results are shown in Figure 3.61 [3.34]. These results are mildly frequency dependent but the dependence on preload is similar to that of the static stiffness. For this pad the dynamic stiffness is about 3.5 times the static incremental stiffness.

The phase angle shown in Figure 3.61 can be seen to be approximately constant with frequency, justifying the use of the hysteretic damping model for the rail pad. The corresponding loss factor, equal to the tangent of the phase angle, is around 0.14.

Further measured stiffnesses and damping loss factors are given in [3.33] for a number of rail fastener systems. A range of values derived from fitting track models to measured track mobilities is given in Table 3.8. These are intended to be taken as generic values. The corresponding damping loss factors are usually found to be in the range 0.2 to 0.25 [3.2], whereas measurements on individual pads yield values in the range 0.1 to 0.15 [3.33, 3.34]. As can be seen, the stiffness of rail pads can vary over a wide range, from around 60 MN/m, found on systems intended for use on slab tracks, up to 1300 MN/m noted in Figure 3.3, and even greater. In the presence of the preload of the train the values given here will increase by a further factor of 1.5 to 2 [3.33]. Much softer systems are also used for isolation from ground-borne noise, as will be discussed in Chapter 13.

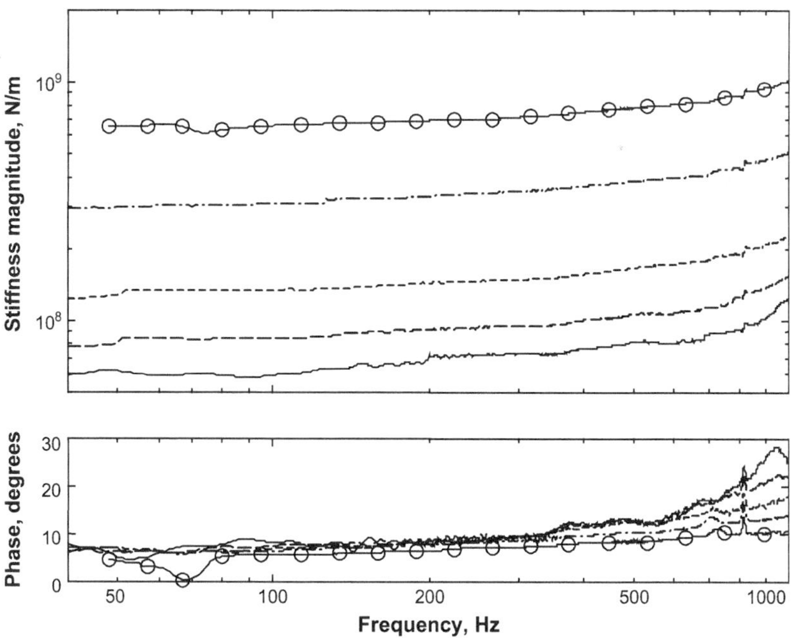

FIGURE 3-61 Magnitude and phase of complex stiffness of 10 mm studded rubber rail pad for various preloads (—, 20 kN; – – –, 30 kN; - - - -, 40 kN; – · – ·, 60 kN; o—o, 80 kN), from [3.32]

TABLE 3-8 TYPICAL VALUES OF RAIL PAD STIFFNESS DERIVED FROM FITTING TRACK MODELS TO MEASURED TRACK MOBILITIES (UNLOADED) [3.2, 3.31]

	Vertical stiffness	Lateral stiffness
System for slab track installation (Germany)	60 MN/m	50 MN/m
Embedded rail (per m of rail, Netherlands)	75 MN/m^2	25 MN/m^2
Baseplate for use on bridges (Netherlands)	85 MN/m	15 MN/m
10 mm studded pad (UK)	120 MN/m	40 MN/m
9 mm ribbed pad (France)	200 MN/m	50 MN/m
Medium stiffness pad (Germany)	350 MN/m	50 MN/m
4.5 mm cork/rubber pad (Netherlands)	1000 MN/m	20 MN/m
4.5 mm ribbed pad (France)	1300 MN/m	100 MN/m

REFERENCES

3.1 C. Esveld. *Modern Railway Track*, 2nd edition. MRT Productions, Zaltbommel, 2001.

3.2 N. Vincent and D.J. Thompson. Track dynamic behaviour at high frequencies. Part 2: experimental results and comparisons with theory. *Vehicle System Dynamics Supplement*, **24**, 100–114, 1995.

3.3 C.J.C. Jones, D.J. Thompson, and R.J. Diehl. The use of decay rates to analyse the performance of railway track in rolling noise generation. *Journal of Sound and Vibration*, **293**, 485–495, 2006.

3.4 M.H.A. Janssens, M.G. Dittrich, F.G. de Beer, and C.J.C. Jones. Railway noise measurement method for pass-by noise, total effective roughness, transfer functions and track spatial decay rates. *Journal of Sound and Vibration*, **293**, 1007–1028, 2006.

3.5 M.J. Brennan and N.S. Ferguson. Vibration control, Chapter 12. In: F. Fahy and J. Walker (eds.), *Advanced Applications of Acoustics, Noise and Vibration*. E&FN Spon, London, 2004.

3.6 D.J. Thompson, C.J.C. Jones, T.X. Wu and G. de France. The influence of the non-linear stiffness behaviour of railpads on the track component of rolling noise. *Proceedings of the Institution of Mechanical Engineers, Part F (Journal of Rail and Rapid Transit)*, **213**, 233–241, 1999.

3.7 K. Knothe and S.L. Grassie. Modelling of railway track and vehicle/track interaction at high frequencies. *Vehicle System Dynamics*, **22**, 209–262, 1993.

3.8 K.F. Graff. *Wave Motion in Elastic Solids*. Dover Publications, New York, 1991.

3.9 L. Cremer, M. Heckl, and E.E. Ungar. *Structure-borne Sound*, 2nd edition. Springer, 1988.

3.10 H. Priestley. Introduction to Complex Analysis. Oxford University Press, 2nd edition, 2003.

3.11 L. Fryba. Vibration of Solids and Structures Under Moving Loads. Thomas Telford, 3rd edition, 1999.

3.12 S.L. Grassie, R.W. Gregory, D. Harrison, and K.L. Johnson. The dynamic response of railway track to high frequency vertical excitation. *Journal of Mechanical Engineering Science*, **24**, 77–90, 1982.

3.13 M.A. Heckl. Railway noise – can random sleeper spacings help? *Acustica*, **81**, 559–564, 1995.

3.14 M.A. Heckl. Coupled waves on a periodically supported Timoshenko beam. *Journal of Sound and Vibration*, **252**, 849–882, 2002.

3.15 D.J. Thompson, M.H.A. Janssens and F.G. de Beer. TWINS: Track-Wheel Interaction Noise Software, theoretical manual (version 3.0), TNO report HAG-RPT-990211, 1999.

3.16 D.J. Thompson and N. Vincent. Track dynamic behaviour at high frequencies. Part 1: theoretical models and laboratory measurements. *Vehicle System Dynamics Supplement*, **24**, 86–99, 1995.

3.17 D.J. Mead. Wave propagation and natural modes in periodic systems: I. Mono-coupled systems. *Journal of Sound and Vibration*, **40**, 1–18, 1975.

3.18 G. de France. Railway track: effect of rail support stiffness on vibration and noise. MSc Dissertation, ISVR. University of Southampton, 1998.

3.19 J. Oscarsson. Dynamic train-track interaction: variability attributable to scatter in the track properties. *Vehicle System Dynamics*, **37**, 59–79, 2002.

3.20 T.X. Wu and D.J. Thompson. The influence of random sleeper spacing and ballast stiffness on the vibration behaviour of railway track. *Acustica* united with *Acta Acustica*, **86**, 313–321, 2000.
3.21 D.J. Thompson. Experimental analysis of wave propagation in railway tracks. *Journal of Sound and Vibration*, **203**, 867–888, 1997.
3.22 L. Gavric. Computation of propagative waves in free rail using a finite element technique. *Journal of Sound and Vibration*, **185**, 531–543, 1995.
3.23 C-M. Nilsson. Waveguide finite elements applied on a car tyre. PhD thesis. KTH, Stockholm, 2004.
3.24 X. Sheng, C.J.C. Jones, and D.J. Thompson. Responses of infinite periodic structures to moving or stationary harmonic loads. *Journal of Sound and Vibration*, **282**, 125–149, 2005.
3.25 D.J. Thompson. Wheel-rail noise generation, Part III: rail vibration. *Journal of Sound and Vibration*, **161**, 421–446, 1993.
3.26 T.X. Wu and D.J. Thompson. A double Timoshenko beam model for vertical vibration analysis of railway track at high frequencies. *Journal of Sound and Vibration*, **224**, 329–348, 1999.
3.27 T.X. Wu and D.J. Thompson. Analysis of lateral vibration behavior of railway track at high frequencies using a continuously supported multiple beam model. *Journal of the Acoustical Society of America*, **106**, 1369–1376, 1999.
3.28 T.X. Wu and D.J. Thompson. Application of a multiple-beam model for lateral vibration analysis of a discretely supported rail at high frequencies. *Journal of the Acoustical Society of America*, **108**, 1341–1344, 2000.
3.29 S.L. Grassie. Dynamic modelling of concrete railway sleepers. *Journal of Sound and Vibration*, **187**, 799–813, 1995.
3.30 M.H.A. Janssens and D.J. Thompson. Improvements of ballast and sleeper description in TWINS. Step 2: development and implementation of theoretical models. TNO report TPD-HAG-RPT-960108, 1996.
3.31 A.M. David and D.J. Thompson. Measurements of the vibration and sound radiation of a concrete railway sleeper. ISVR Contract Report 98/22, September 1998.
3.32 N. Frémion, J.P. Goudard and N. Vincent. Improvements of ballast and sleeper description in TWINS. Step 1: experimental characterization of ballast properties. Vibratec report 072.028a, July 1996.
3.33 D.J. Thompson and J.W. Verheiij. The dynamic behaviour of rail fasteners at high frequencies. *Applied Acoustics*, **52**, 1–17, 1997.
3.34 D.J. Thompson, W.J. van Vliet, and J.W. Verheiij. Developments of the indirect method for measuring the high frequency dynamic stiffness of resilient elements. *Journal of Sound and Vibration*, **213**, 169–188, 1998.

CHAPTER 4

Wheel Vibration

4.1 INTRODUCTION

As has been seen in Chapter 2, the vibration of the wheels is an important source of rolling noise, along with the vibration of the track. In this chapter the dynamic behaviour of a railway wheel is introduced. As well as rolling noise, it is of importance for curve squeal (Chapter 9).

Most railway wheels are axi-symmetric, having a constant cross-section around the azimuthal direction. As an example of a typical wheel structure, the cross-section of a standard UIC freight wheel with a diameter of 0.92 m is shown in Figure 4.1. Many wheels are now constructed as a single piece structure rather than having a separate tyre, but the term 'tyre' will be retained here to refer to the region that runs on the rail. For mainline railways this usually has an overall width of 135 mm including the flange, whereas the thinner web region, connecting the tyre to the hub, is typically only 20–25 mm wide at its narrowest point. At the hub, the wheel is connected to the axle, usually by a press fit. The running surface is machined to a well-defined transverse profile required for stable running. The flange is provided for protection against derailment. The curved shape of the web shown in Figure 4.1 is typical of many tread-braked wheels and is used to allow thermal expansion during braking of the wheel tread.

4.2 WHEEL MODES OF VIBRATION

4.2.1 Characterization of modes

All finite structures have a series of resonances and associated natural frequencies. The extent to which these resonances characterize the response depends on the damping of the structure. Damping comes from a number of sources. Material damping of metal structures is generally very low and in built-up structures is usually of less importance than damping from joints, boundaries, etc.

While many engineering structures are heavily damped and only respond with a dull thud when tapped, a railway wheel has very light damping so that its vibration is particularly strongly characterized by its resonances. This is due in large part to its axi-symmetric nature, supported at its centre, which allows modes of vibration to occur with negligible motion at the point of connection to the axle. Similar structures are a wine glass or a church bell.

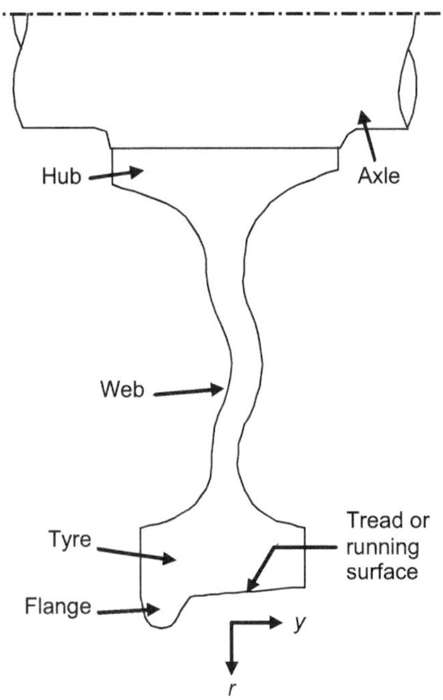

FIGURE 4-1 Cross-section of UIC 920 mm freight wheel

To characterize the modes of vibration of structures, it is usual to indicate the number of node lines (stationary points) in different directions. For a wheel, similarly to a flat disc, out-of-plane (axial) modes correspond to various numbers of nodal diameters, n, possibly in combination with a number of nodal circles, m. Some of these mode shapes are illustrated in Figure 4.2. The mode with n nodal diameters and m nodal circles will be denoted mode (n,m). If there are n nodal diameters, the motion can be described essentially in the form $A \cos n\theta$ or $A \sin n\theta$ where θ is the circumferential angular coordinate and A is a function of the radial and axial coordinates, (r,y). For $n = 0$, as $\sin n\theta = 0$ and $\cos n\theta = 1$, the motion is axi-symmetric, i.e. independent of θ. For $n > 0$ there are actually two independent modes of vibration at each resonance frequency, corresponding to the sine and cosine mode shapes. However, for an axi-symmetric structure the angular position of the node lines is actually arbitrary, and will be determined by the location of the excitation. If the wheel is not perfectly symmetric there are preferred locations for these nodal diameters and the pair of modes occur at slightly different frequencies (see box on page 100).

In addition, a flat disc can sustain in-plane motion in the form of either radial modes or circumferential modes, also with n nodal diameters and m nodal circles. For a wheel, the in-plane modes with one or more nodal circles ($m \geq 1$) do not occur in the frequency range of interest, so there is only a single set of radial modes to be considered, shown in Figure 4.2. Similarly, the circumferential modes (not shown in Figure 4.2) can be categorized by their n nodal diameters. These sets of modes will therefore be denoted (n,R) and (n,C).

Clearly, as seen from Figure 4.1, a railway wheel is not a flat disc – the cross-section is non-uniform and is not symmetric. Even a wheel with a straight web is

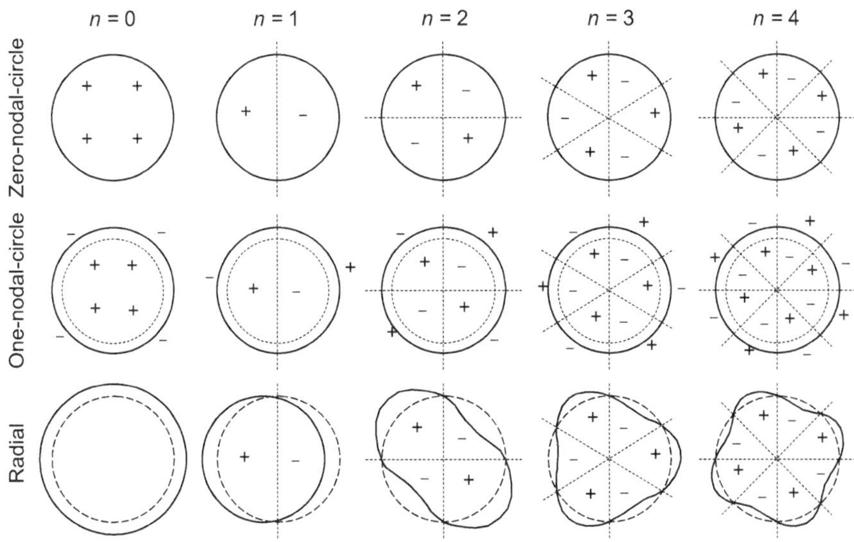

FIGURE 4-2 Examples of modeshapes of a wheel, +/– indicate the relative phase of the motion in each area. —, deformed shape; – – –, undeformed shape; · · · ·, node lines

asymmetric due to the presence of the flange. Due to the lack of symmetry, in-plane and out-of-plane motions become coupled. Where such motions are coupled this means that pairs of modes occur, each containing both motions but with one of them dominant in each. Of the axial modes, the one-nodal-circle modes are closest in frequency to the radial modes and so experience the greatest degree of coupling. Thus the one-nodal-circle axial modes contain some radial motion and the radial modes contain some axial motion. This can make it difficult to identify the modes.

It should also be noted that wheel modes with $n = 1$ (1 nodal diameter) are coupled to bending motion of the axle, while $n = 0$ axial modes are coupled to extension of the axle and $n = 0$ circumferential modes are coupled to torsion of the axle.

4.2.2 Numerical models

Although various analytical models for wheel vibration have been produced, such as a ring or a disc [4.1, 4.2], the complex geometry means that more reliable results are obtained by using the finite element method. Various types of finite element mesh can be used. The most computationally efficient FE method is to use axi-symmetric (sometimes called axi-harmonic) elements in which only the cross-section is modelled using two-dimensional elements and a separate calculation is performed for each 'harmonic' (number of nodal diameters) required [4.3].

Solid (three-dimensional) elements can also be used to model the wheel [4.4]. This type of model is particularly useful for wheels that are not actually axi-symmetric, for example with a doubly curved web. With this type of model it is only

Illustration of effect of axisymmetry

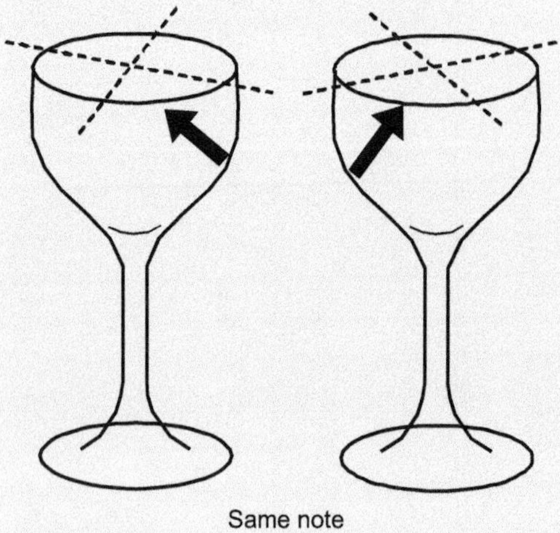

Take a wine glass and place it on a table. Tap it gently at the rim, for example with a pencil, and it will produce a note. Tap it at any other position around the rim and it should produce the same note. The forcing point determines the position of the node lines.

Now take a china cup or mug or a beer glass with a handle and place it on the table.

Tap it radially (towards the centre) on the handle and it will produce a note. Tap it radially at a position that is 90° or 180° from the handle and it will produce the same note because the principal mode has two nodal diameters. The position of the node lines is fixed by the mass of the handle.
However, tap it at a position 45° from the handle and the note will be higher. In this case the handle is at the node line and its mass does not contribute to the mode. (Tap the handle laterally and it will again produce the higher note.)

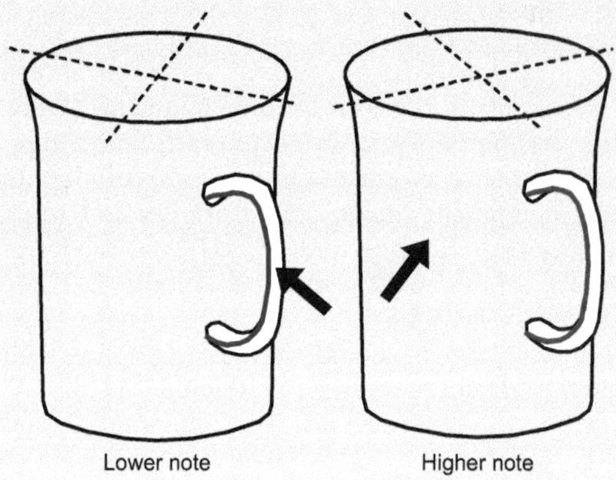

necessary to model a quarter wheel by introducing two perpendicular planes of symmetry, with either symmetry or anti-symmetry boundary conditions on these two planes. Modes with even values of n are obtained from the combination of symmetry/symmetry boundary conditions while those with odd values of n are obtained from symmetry/anti-symmetry boundary conditions. To obtain the $n=0$ circumferential modes the anti-symmetry/anti-symmetry condition is required and if the wheel is not axi-symmetric the fourth combination is also required.

Earlier FE models used a combination of plate and beam elements in order to reduce computational effort [4.3], but this is no longer relevant as computing power has increased.

To obtain modes that include motion of the axle, a complete model of the wheelset is required (symmetry at the centre of the axle between the two wheels can be invoked). However, this model is only required for modes with $n = 0$ and 1 as, for higher values of n, the axle motion is negligible. Nevertheless, some care is required to merge modes from different models.

The usual precautions in using finite element methods are required, for example the need to use sufficient elements, usually at least six per structural wavelength taking into account the frequency range of interest, for which reference is made to standard texts, e.g. [4.5].

Figure 4.3 shows examples of mode shapes obtained for a UIC 920 mm standard freight wheel [4.4]. These have been calculated without the axle present, but instead a rigid constraint was applied at the inner edge of the hub. This gives a very good approximation to the modes of a wheel with $n \geq 2$, which, as will be seen, are the most important modes for rolling noise generation. The first row of modes all look very similar in cross-section, essentially resembling a cantilever with maximum deflection at the tread. However, it should be remembered that these each have a different value of n, so in the circumferential direction their mode shapes differ, as indicated in Figure 4.2. These are clearly axial modes with no nodal circle. They occur strongly in curve squeal (see Chapter 9) but hardly appear at all in rolling noise.

The second and third sets look similar to each other due to the coupling between radial and one-nodal-circle axial motion. The second row has been labelled 'radial' and the third 'one-nodal-circle' as these appear to be the dominant motions: look particularly at the running surface which is rotating more in the modes of the third row. However, it is not always easy to distinguish these modes. Indeed, in some situations the second set may be predominantly radial at low values of n and predominantly axial at high values and vice versa for the third set.

The coupled radial and one-nodal-circle axial modes are the most important for rolling noise due to their large radial component at the tread, which is well excited by vertical forces, and their large out-of-plane motion of the web which radiates sound well. For a straight-webbed wheel the coupling between radial and axial motion is much weaker, as shown in the example in Figure 4.4. Such a wheel is therefore often quieter. Straight webs can only be used where there are no tread brakes, as the curved web is intended to allow thermal expansion of the wheels under braking.

The two-nodal-circle axial modes and the circumferential modes occur at higher frequencies and are of little importance in rolling noise as they have smaller radial components at the tread.

The natural frequencies calculated by the finite element model for each of these wheels are plotted in Figure 4.5. Agreement with experiment is typically within around 3–4%, apart from modes with $n = 1$ where the omission of the axle in the FE model leads to large differences. However, it should be noted that the design tolerance on a railway wheel is often several millimetres in the width of the web, so that nominally identical wheels may differ considerably in natural frequency. Moreover,

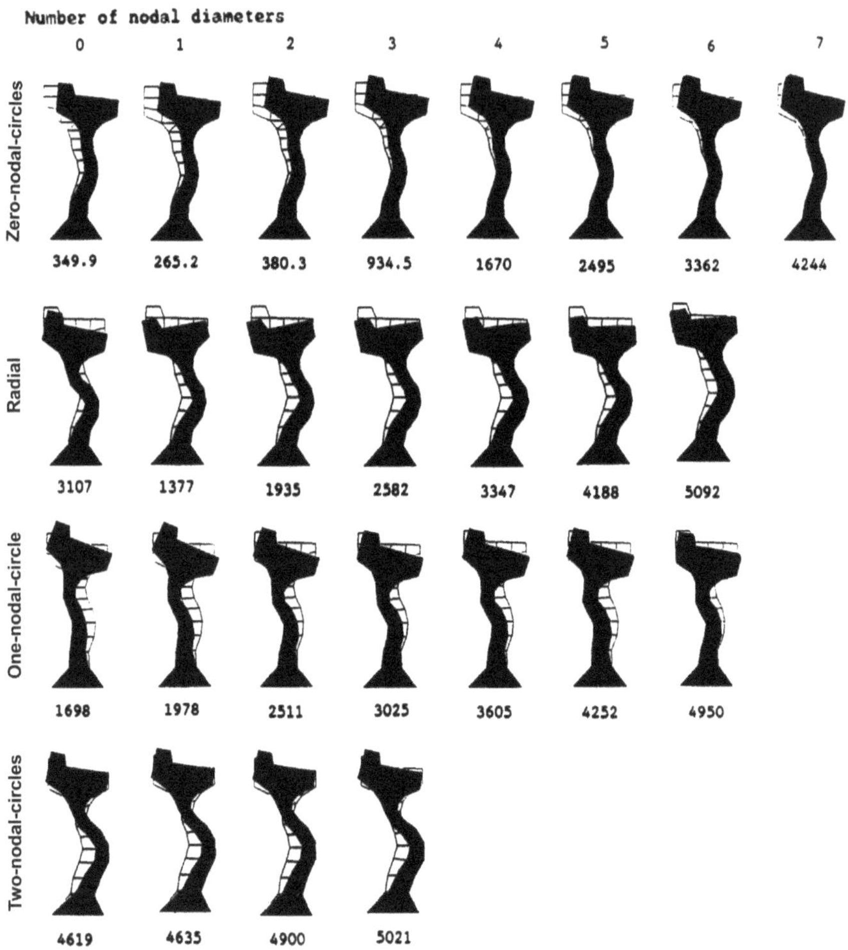

FIGURE 4-3 Modes of vibration of a UIC 920 mm standard freight wheel shown in cross-section and natural frequencies in Hz [4.4]

as the wheel is used, the tread is gradually worn away, further material being removed by regular turning (reprofiling) so that worn wheels may be typically 50 mm smaller in diameter than new ones before they are replaced. This also has quite a large effect on the natural frequencies. In one example the natural frequencies of the zero-nodal-circle modes were reduced by up to 10% while those of the one-nodal-circle modes were increased by up to 13%. The radial modes are affected less with reductions of up to 5%.

4.2.3 Modal damping

Unlike natural frequencies and mode shapes, which can be predicted using numerical methods, damping factors must usually be measured. The modal damping of

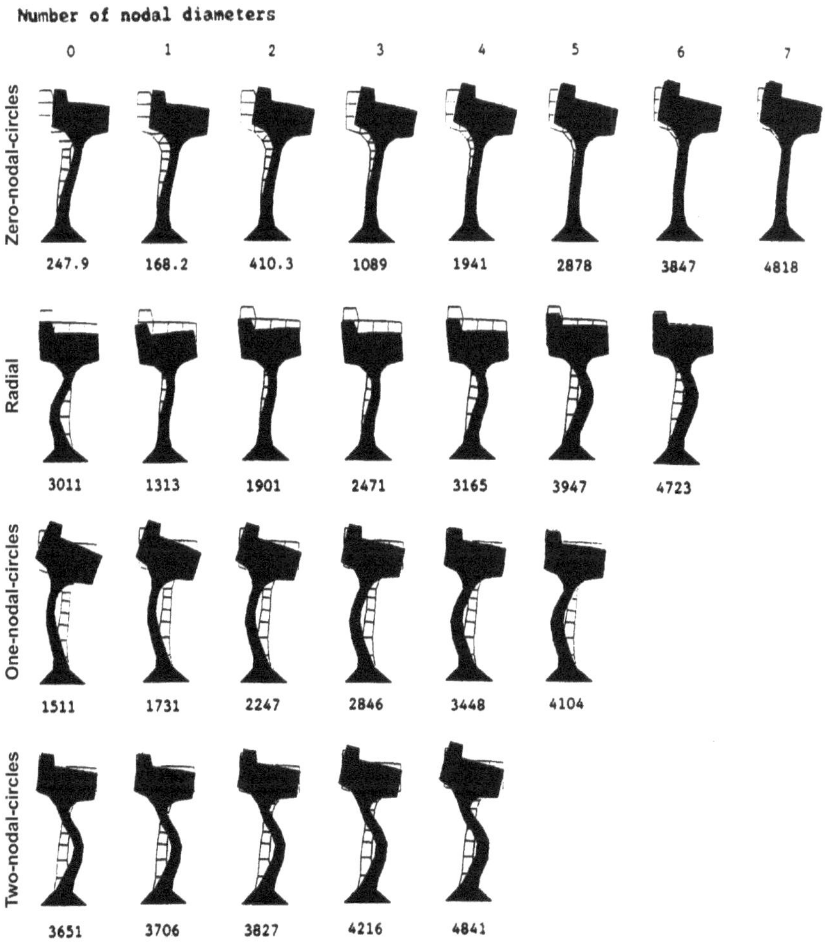

FIGURE 4-4 Modes of vibration of NS Intercity wheel shown in cross-section and natural frequencies in Hz [4.4]

wheels is generally very small, so some care is required in determining it (see Section 4.3.3 below).

Figure 4.6 shows some damping ratios measured on five different monobloc wheels (from [4.4]). The damping ratio is the ratio of the modal damping to critical damping (see also Section 4.3.1 below). No systematic differences were found between the five types of wheel. The modes with $n \geq 2$, shown in the first four groups, have average values of 0.8×10^{-4} for modes (n,0) and (n,R) and 1.3×10^{-4} for modes (n,1) and (n,2). Results for modes (0,0) and (1,0) are less consistent but show considerably higher damping ratios, of the order of 10^{-3}. However, the modes with $n = 1$ could often not be found at all, suggesting that higher damping values apply in other cases. The radial mode with $n = 0$ occurring at about 3 kHz has an intermediate value of damping between 10^{-3} and 10^{-4}. For two of the wheels

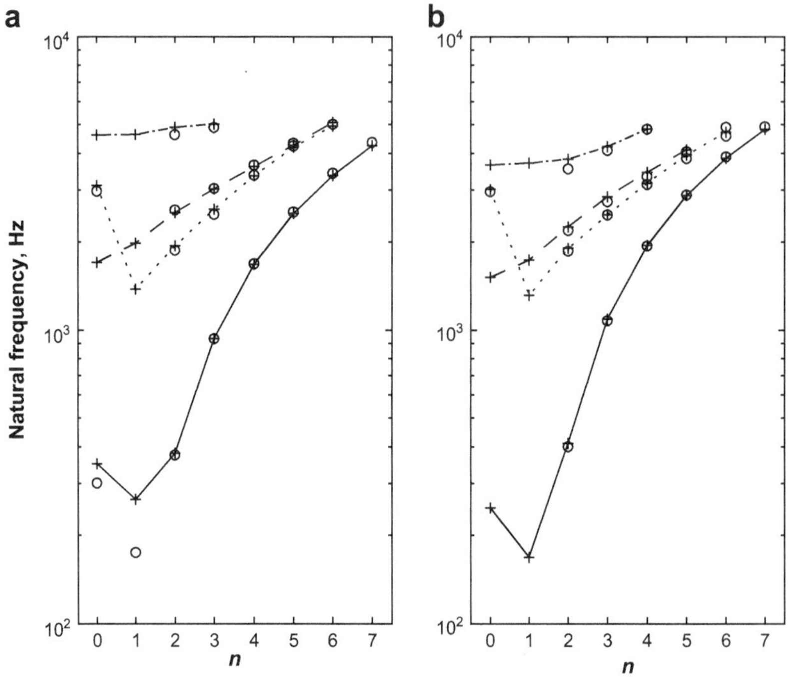

FIGURE 4-5 Natural frequencies of railway wheels predicted using finite elements (+) and measured (○). (a) UIC 920 mm freight wheel and (b) NS Intercity wheel. —, zero-nodal-circle; · · · ·, radial; – – –, one-nodal-circle; – · – ·, two-nodal-circle

a circumferential mode at about 4.5 kHz was also identified and found to have a very low damping ratio.

For rolling noise calculations the exact value of modal damping is not critical. Reasonable predictions can therefore be made using nominal damping ratios of

10^{-3} for modes with $n = 0$,
10^{-2} for modes with $n = 1$,
10^{-4} for modes with $n \geq 2$.

4.3 FREQUENCY RESPONSE

4.3.1 Theory

The frequency response of a wheel can be found from its modes of vibration. To introduce this, consider first a damped single degree-of-freedom system, a mass M on a spring K and damper C, as shown in Figure 4.7. As found in many vibration text books (e.g. [4.6]), the complex displacement amplitude u of the

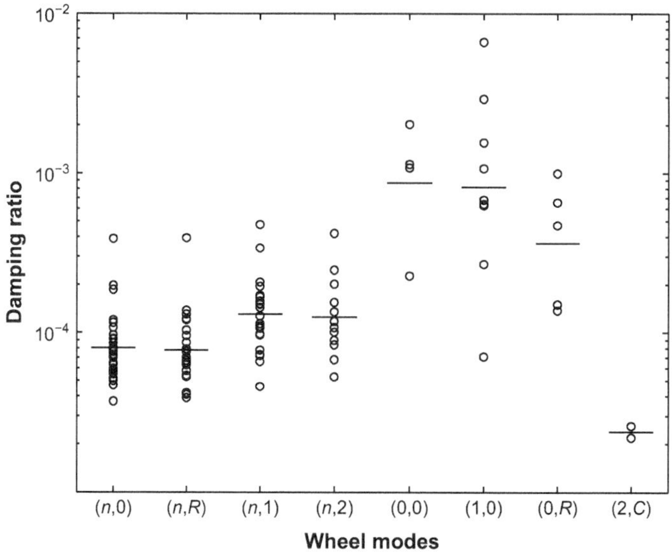

FIGURE 4-6 Measured damping ratios for five types of railway wheel. ○, individual modes; —, logarithmic average. The first four sets are for $n \geq 2$

mass can be found from the equation of motion at circular frequency ω, which can be written as

$$-\omega^2 Mu + i\omega Cu + Ku = F \tag{4.1}$$

where a time dependence of $e^{i\omega t}$ is implicit. Writing $\omega_n = (K/M)^{1/2}$ for the natural frequency, and $\zeta_n = C/2(KM)^{1/2}$ for the damping ratio, this can be expressed as

$$M(\omega_n^2 - \omega^2 + 2i\zeta_n \omega \omega_n)u = F \tag{4.2}$$

Thus the mobility is given by

$$Y = \frac{i\omega u}{F} = \frac{i\omega}{M(\omega_n^2 - \omega^2 + 2i\zeta_n \omega \omega_n)} \tag{4.3}$$

FIGURE 4-7 Damped single degree-of-freedom system

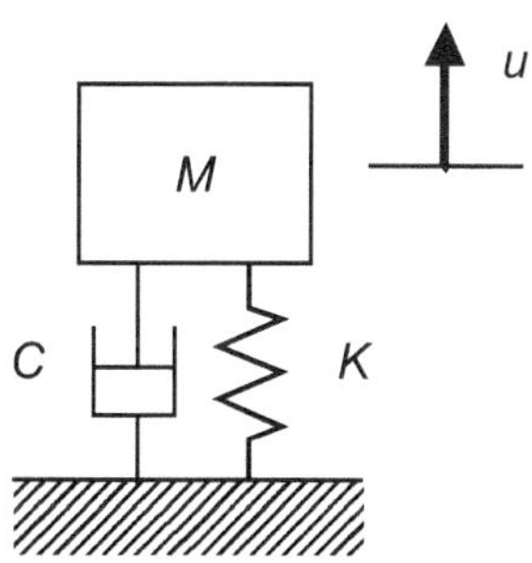

For a more general structure, if the normal modes are known, the frequency response function between a force at a location k and the velocity response at a location j can be found from a sum of the response in each mode

$$Y_{jk} = \sum_n \frac{i\omega\psi_{jn}\psi_{kn}}{m_n(\omega_n^2 - \omega^2 + 2i\zeta_n\omega\omega_n)} \tag{4.4}$$

where ψ_{jn} is the modeshape amplitude of mode n at location j and m_n is the modal mass, which is a normalization factor for the nth modeshape [4.7]. ζ_n is the modal damping ratio. For a continuous system there is an infinite number of modes, but the series can be truncated by ignoring modes with natural frequency greater than some suitable limit, which should usually be greater than the maximum frequency of interest.

4.3.2 Frequency response of wheelset

The general expression in equation (4.4), the modal summation, can be used to determine the frequency response of a wheel. The modeshape amplitudes and corresponding modal masses can be obtained from a finite element calculation along with the natural frequencies. The only parameter that causes some difficulty is the damping, which must usually be obtained from measurements, or using estimates based on experience of measurements on other wheels, such as listed in Section 4.2.3 above.

For an unconstrained flexible body the rigid body modes must also be included in the modal summation, equation (4.4). Even though the flexible wheel modes may be determined reasonably well by ignoring the axle and constraining the wheel at the inner edge of the hub, the rigid body modes of the whole wheelset *must* still be included in order to determine the frequency responses. There are six rigid modes of a free structure but for lateral and vertical motion it is sufficient to include three of these: translation in the vertical and lateral directions and rotation in the vertical/lateral plane.

Figures 4.8 and 4.9 show the mobility of the UIC 920 mm standard freight wheel in the radial and axial directions at the wheel tread. Curves are shown for models based on the modes of the wheel alone (constrained at the hub) and for the full wheelset.

The radial mobility is essentially mass-like at low frequencies. This corresponds to the unsprung mass of the wheelset (modified by the fact that excitation is on one wheel so that rotation of the wheelset also occurs). At around 500 Hz an anti-resonance occurs (i.e. a minimum in the frequency response) and above this frequency the mobility is stiffness controlled, rising to a first peak at between 1.5 and 2 kHz. At high frequencies a series of strong resonance peaks occur, which are the radial and one-nodal-circle axial modes as discussed in Section 4.2. Inclusion of the axle in the model leads to the introduction of a number of more highly damped modes which have little direct effect on the level of the mobility but which cause the anti-resonance to shift.

The mode at around 1150 Hz with $n = 1$, which appears as a distinct peak in the wheel-only model, is affected in practice by coupling with the axle and therefore appears too strongly in the wheel-only model. This can lead to incorrect noise predictions. A way around this with a wheel-only model is to assign this mode a very

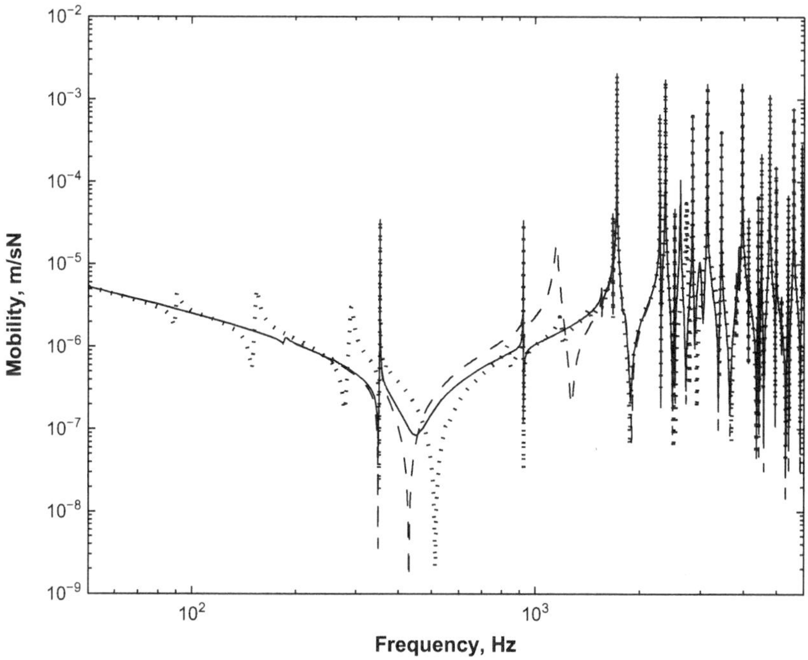

FIGURE 4-8 Radial mobility of UIC 920 mm standard freight wheelset. – – –, wheel-only model; · · · ·, full wheelset model; —, wheel-only model with higher damping in mode at 1150 Hz

high value of damping, say 1.0 [4.8]. As shown in the figure this gives reasonable agreement with the full wheelset model.

The axial mobility, Figure 4.9, is dominated by peaks associated with the zero-nodal-circle axial modes (see Figure 4.3). The highest of these are for $n \geq 2$ which have lower damping than for $n < 2$. Here, the inclusion of the axle has little effect on the mobility except at low frequency, below about 300 Hz, where a series of axle bending modes are present.

Figure 4.10 shows the effect of truncation of the modal summation for the case of the axial mobility. Results for the radial mobility show similar trends. Curves are shown in which the series is truncated at 5, 7 and 10 kHz as well as the largest set of modes available, which extends up to a maximum resonance frequency of 13 kHz. Differences are minimal below 3 kHz. Above this frequency, the main differences are in the location of the anti-resonances, so that even truncating the series at 6 kHz is satisfactory for frequencies up to about 5 kHz. Especially around the resonances it is found that the effect of truncation is negligible. This is a consequence of the light damping of the wheel.

4.3.3 Frequency resolution

The damping of wheel modes is very light so that the half power bandwidth is small, typically only a few Hz. The half power bandwidth is the frequency bandwidth at which the amplitude is 3 dB, or a factor of $1/\sqrt{2}$, below that at the peak. It is given by

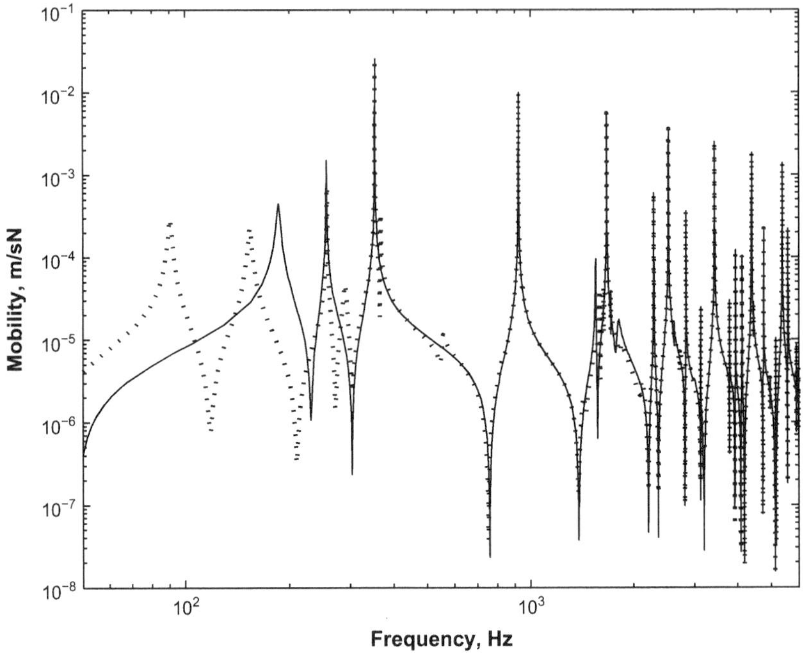

FIGURE 4-9 Axial mobility of UIC 920 mm standard freight wheelset. —, wheel-only model; · · · ·, full wheelset model

$$\Delta f = 2\zeta_n f_n \tag{4.5}$$

where $f_n = \omega_n/2\pi$. For $\zeta_n = 10^{-4}$ this gives a bandwidth of 0.1 Hz at 500 Hz and 1 Hz at 5000 Hz.

To ensure that the resonances are correctly represented in the mobility, at least two calculation points are required within this bandwidth. This suggests that the frequency resolution should be of the order of 0.1 Hz. Conversely, in regions well away from resonances the mobility changes only slowly with frequency and such a resolution would require far more calculation points than is necessary. Therefore, it is most efficient in calculations to use a variable frequency resolution which is finest around the resonances (and anti-resonances) but is much coarser well away from them. This can be achieved by varying the spacing between successive points interactively such that

- the natural frequencies of all resonances are included,
- the amplitude does not change more than a specified amount (e.g. 5 dB) between successive points,
- the phase does not change by more than a specified amount between successive points (e.g. 45°) to ensure that anti-resonances are not overlooked,
- the frequency spacing between adjacent points is not greater than a specified amount (e.g. 50 Hz).

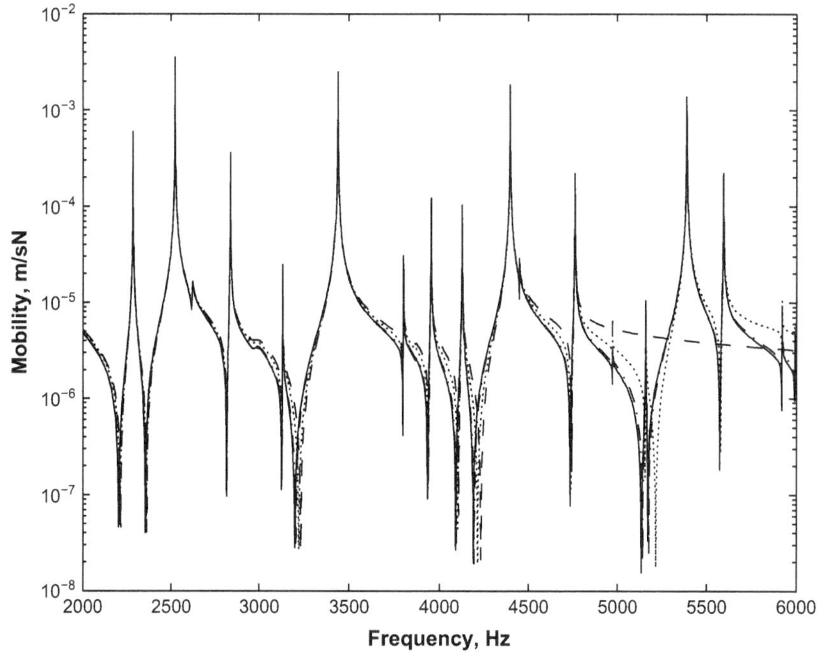

FIGURE 4-10 Axial mobility of UIC 920 mm standard freight wheelset. – – –, all modes up to 5 kHz; · · · ·, all modes up to 6 kHz; – · – ·, all modes up to 10 kHz; —, all modes up to 13 kHz

In the examples shown in Figures 4.8 and 4.9 (in which mobilities in all six degrees of freedom were generated simultaneously) almost 1800 frequency points were required to cover the range 50 to 5000 Hz, which is only 4% of those that would be needed for 0.1 Hz spacing throughout.

Similar considerations apply to the measurement of wheel mobility. If the bandwidth is insufficient, the resonance peaks will be incorrectly measured. With a random excitation, the coherence will then drop at the resonances. Equivalently, for an impact excitation, the impulse response will last considerably longer than the analysis window. For example, for a 1 Hz bandwidth the analysis time is 1 s. The corresponding measurement time T is related to the frequency resolution Δf by $T = 1/\Delta f$ [4.9]. A measure of the length of the impulse response is the reverberation time (time taken for the vibration to decay by 60 dB), given by

$$T_{60} = \frac{1.1}{\zeta_n f_n} \tag{4.6}$$

This gives about 20 s for $\zeta_n = 10^{-4}$ at 500 Hz. Indeed reverberation times as large as 90 s have been measured [4.4]. A very fine zoom analysis or a Fourier transform with a very large number of points is therefore required to measure the peaks correctly.

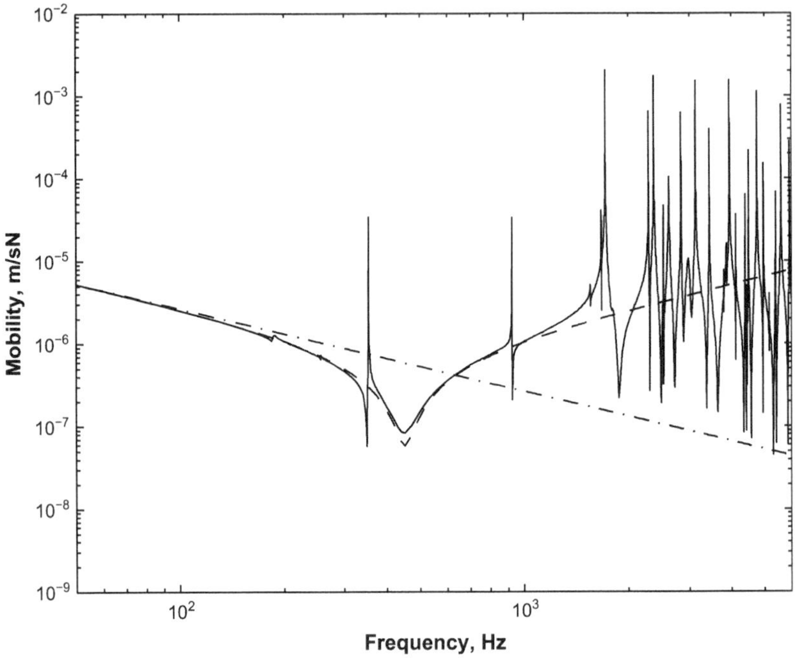

FIGURE 4-11 Radial mobility of UIC 920 mm standard freight wheelset. —, modal summation (from Figure 4.8); – · – ·, mass of 600 kg; – – –, mass and modal stiffness

4.4 SIMPLE MODELS FOR WHEEL MOBILITY

4.4.1 Mass–spring models

In some situations, where mainly the track response is required, the radial wheel mobility can be represented by a simpler model. The low frequency behaviour has been seen to be mass controlled. Figure 4.11 shows the modal summation result from Figure 4.8 along with the equivalent mass line. This corresponds to 600 kg, which is half the wheelset mass.

This is a satisfactory approximation up to about 200 Hz. At higher frequencies the anti-resonance at around 500 Hz must also be modelled. This can be represented by a spring in series with the mass,[1] as shown in Figure 4.12:

$$Y = \frac{i\omega}{K} - \frac{i}{M\omega} \tag{4.7}$$

This result is also shown in Figure 4.11, where the value of K is chosen to ensure that the anti-resonance frequency occurs at 450 Hz, as in the figure, giving $K = 4.8 \times$

[1] Note that this differs from the system of equations (4.1–4.3) which has the mass and spring in parallel. The latter is the more usual model, in which the mass is excited directly with the bottom of the spring grounded. The system in Figure 4.12 has an anti-resonance where that in Figure 4.7 has a resonance.

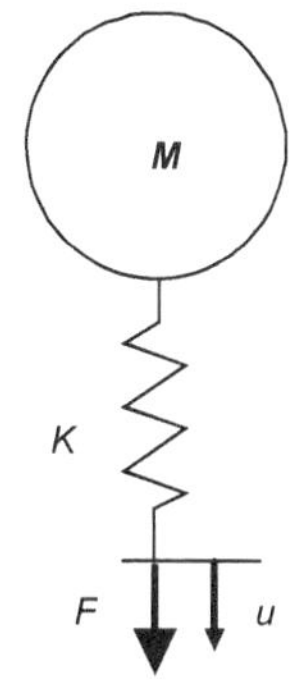

FIGURE 4-12 Simple model of wheelset consisting of its mass M and stiffness K

10^9 N/m. K is additionally made complex $\tilde{K} = K(1 + i\eta)$, with a loss factor $\eta = 0.1$. Apart from two sharp resonance peaks (the (2,0) and (3,0) modes at 380 Hz and 930 Hz, respectively) this gives a good approximation for frequencies up to about 1.5 kHz. At higher frequency it is clear that this model does not represent the modal behaviour of the wheel, but it nevertheless provides an adequate representation of the frequency-average behaviour, which can be used in modelling the track response. It will be noted in Chapter 5 that the contact spring also isolates the wheel from the track at higher frequency, so that detail of the wheel mobility is less important in determining the response of the track at high frequency, although it still has a small influence.

4.4.2 Beam model

Various alternative models have been considered to represent the dynamic behaviour of the wheel. In [4.10], Remington modelled the radial vibration of the wheel as a mass at low frequencies, finding that an appropriate mass was that of the wheel plus 1/3 of the axle. Above about 1 kHz, where many modes occur, an infinite beam was used, with the same cross-section as the tyre.

Remington later developed a model for the radial wheel response (expressed as impedance) using Love's equations for the in-plane vibration of a ring, modified to include the compressional stiffness of the web and the rigid body mass of the wheel [4.1]. For the axial vibration Love's equations for the out-of-plane vibration of a ring were used including torsion as well as bending.

In order to gain more insight, a simple model for the vibration of a wheel can be developed in terms of a finite beam on an elastic foundation as shown in Figure 4.13 (see also Chapter 3). This is developed from an idea by Boennen used to represent a brake disc [4.11]. It is similar to Remington's ring model, but ignores the effects of curvature for simplicity.

The beam represents the wheel tyre and the elastic foundation represents the web. By appropriate choice of parameters either radial or axial motion may be represented. Beam curvature is ignored but the beam is assumed to be closed at its ends $x = \pm L/2$. Thus the waves A_3 and A_4 travel to the right and when they reach the right-hand end $x = L/2$ they continue at the left-hand end $x = -L/2$. Similarly, waves A_1 and A_2 are transmitted from the left-hand end to the right-hand end. Thus the response to a harmonic force at $x = 0$ can be written in terms of these propagating and near-field waves:

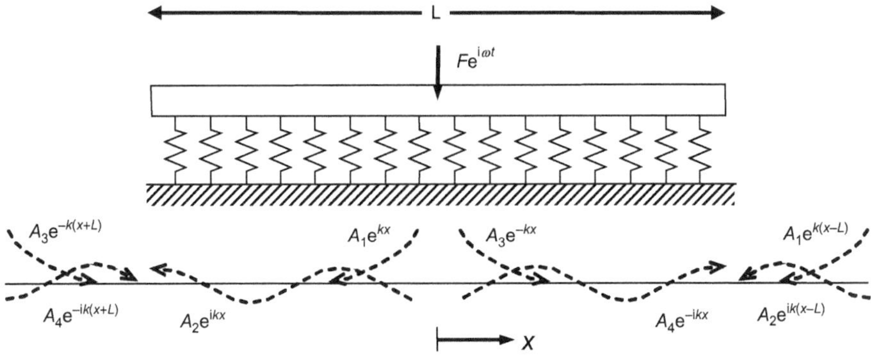

FIGURE 4-13 Waves generated in a closed finite beam by a point force at $x = 0$

$$\begin{aligned} u_-(x) &= A_1 e^{kx} + A_2 e^{ikx} + A_3 e^{-k(x+L)} + A_4 e^{-ik(x+L)} \quad &\text{for} \quad x \leq 0 \\ u_+(x) &= A_1 e^{k(x-L)} + A_2 e^{ik(x-L)} + A_3 e^{-kx} + A_4 e^{-ikx} \quad &\text{for} \quad x \geq 0 \end{aligned} \tag{4.8}$$

where the wavenumber k depends on frequency and is given in Section 3.2. The boundary conditions at the ends of the beam are automatically satisfied by this form of solution. Four boundary conditions are required at $x = 0$: continuity of displacement, rotation and bending moment and force balance:

$$(A_1 - A_3)(1 - e^{-kL}) + (A_2 - A_4)(1 - e^{-ikL}) = 0 \tag{4.9}$$

$$(A_1 + A_3)(1 - e^{-kL}) + i(A_2 + A_4)(1 - e^{-ikL}) = 0 \tag{4.10}$$

$$(A_1 - A_3)(1 - e^{-kL}) - (A_2 - A_4)(1 - e^{-ikL}) = 0 \tag{4.11}$$

$$\frac{F}{EIk^3} = (A_1 + A_3)(1 - e^{-kL}) - i(A_2 + A_4)(1 - e^{-ikL}) \tag{4.12}$$

Hence the point mobility is found as

$$Y(\omega) = \frac{\omega}{4EIk^3}\left(\frac{-i(1 + e^{-kL})}{(1 - e^{-kL})} + \frac{(1 + e^{-ikL})}{(1 - e^{-ikL})}\right) \tag{4.13}$$

Modes occur when $kL = 2n\pi$ for integer values of n and the second term in the mobility tends to infinity (in practice limited by damping). At these natural frequencies, the response is dominated by the A_2 and A_4 waves and has the form $\cos 2n\pi x/L$, a standing wave formed by interference between the two propagating waves.

Damping can be added by making E complex with the form $E(1 + i\eta)$ where η is the damping loss factor; k consequently also becomes complex. A rigid body mode with mass M can also be added to the above beam mobility to give the correct low frequency behaviour of a free wheelset.

TABLE 4.1 PARAMETERS USED TO REPRESENT THE WHEEL USING CLOSED BEAM MODEL

	Axial	Radial
I (m^4)	1.03×10^{-5}	3.80×10^{-5}
A (m^2)	0.0169	0.0203
E (N/m^2)	2.1×10^{11}	2.1×10^{11}
ρ (kg/m^3)	7850	7850
s (N/m^2)	1.0×10^{8}	2.0×10^{10}
M (kg)	600	600
L (m)	2.76	2.76
η (beam)	2×10^{-4}	2×10^{-4}
η (support)	2×10^{-3}	2×10^{-3}

Choosing appropriate parameters to represent the wheel, listed in Table 4.1, the resulting natural frequencies are plotted in Figure 4.14. Not surprisingly, the modes with $n = 0$ and 1 are not well predicted using this model. In the FE results (and the measurements, see Figure 4.5) the $n = 0$ mode is actually higher in frequency than the $n = 1$ mode in both cases, whereas this cannot be obtained from the beam model. The difference is especially great for the radial motion, where the $n = 0$ mode occurs

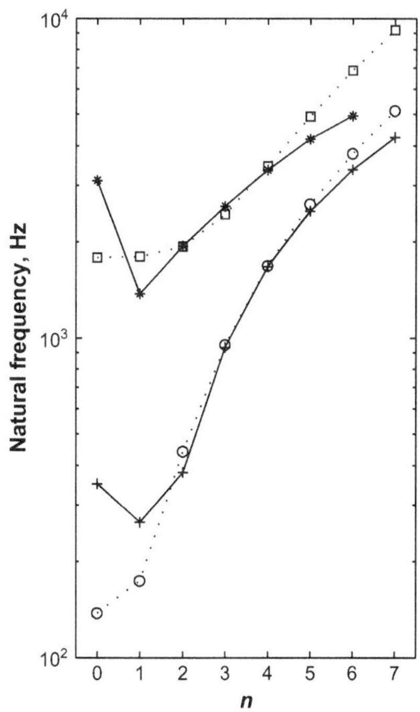

FIGURE 4-14 Natural frequencies of UIC 920 mm freight wheel predicted using finite elements (—) and using closed beam model (····). +, ○, axial modes $m = 0$; *, □, radial modes

at about 3 kHz. The $n = 0$ radial mode of a ring should occur at the 'ring frequency', given by

$$f_r = \frac{c_L}{L} = \frac{1}{L}\sqrt{\frac{E}{\rho}} \tag{4.14}$$

where c_L is the longitudinal wave speed in the tyre and L is the circumference length. However, even this equation only gives a frequency of 1870 Hz; this effect should be added to that of the foundation stiffness.

Despite these differences for $n = 0$ and 1, by a suitable choice of parameters the modes with $n \geq 2$ and above can all be arranged to be within 15% of the finite element results, which is good for such a simple model. Modes with $n \geq 6$ are affected by shear deformation and would require a Timoshenko beam formulation for increased accuracy.

Figure 4.15 shows the mobility for the radial and lateral directions predicted using the beam model. Especially for the radial direction, the effect of adding a rigid body mode to represent the motion of the entire wheel at low frequency can be clearly seen by the mass-controlled region below 500 Hz. The foundation stiffness is also much more important for the radial direction since the web will be much stiffer radially than axially. Despite the simplicity of this model, reasonable agreement can be seen

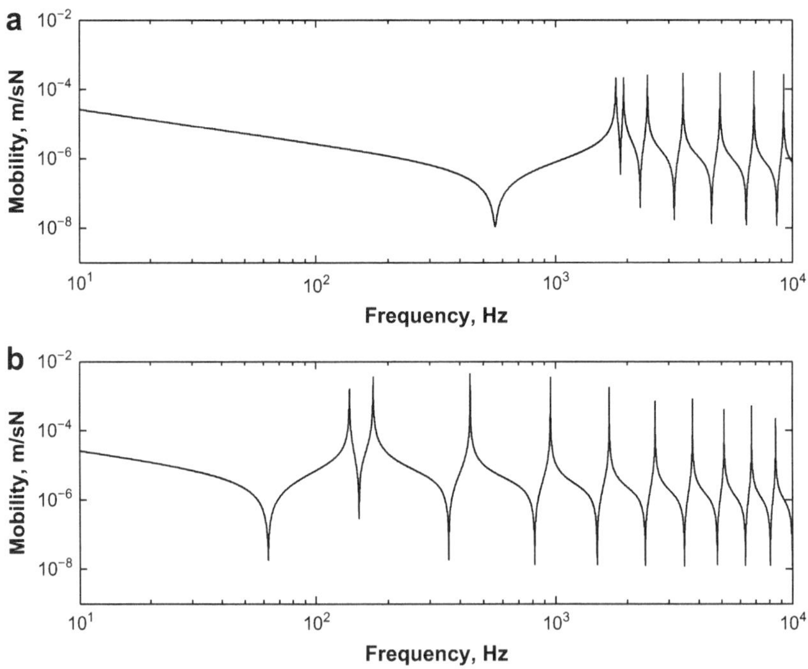

FIGURE 4-15 Mobilities of UIC 920 mm freight wheel predicted using closed beam model. (a) Radial, (b) axial

in comparison with the results of the FE model in Figures 4.8 and 4.9. The main differences are caused by the $n = 0$ and $n = 1$ modes which have incorrect frequencies and should be given higher damping. The damping has been adjusted by giving the support layer a higher loss factor than the beam, but this also tends to increase the modal damping of the $n = 2$ modes.

4.5 EFFECTS OF WHEEL ROTATION

When the wheel rolls along the track at speed V, as shown in Figure 4.16(a), the forcing point effectively moves around its circumference at angular speed $\Omega = V/r_0$ as shown in Figure 4.16(b). This causes the natural frequencies to split into pairs when viewed from the forcing point [4.12]. Before considering the full model based on FE, the above beam model can be used to illustrate the phenomenon more simply.

4.5.1 Beam model

If the force in Figure 4.13 moves to the right at speed V, this can also be modelled by keeping the force fixed and moving the beam to the left. As a result of this motion, the wavespeeds in the positive and negative directions change (see Section 3.2.6). The four wavenumber solutions $k_n = \mathrm{i}^n k_p$ (with k_p the propagating wavenumber in the static case) are modified to become $k_n \to \mathrm{i}^n k_p(1 + \varepsilon)$, with

$$\varepsilon = i^n \frac{V}{c_g} \tag{4.15}$$

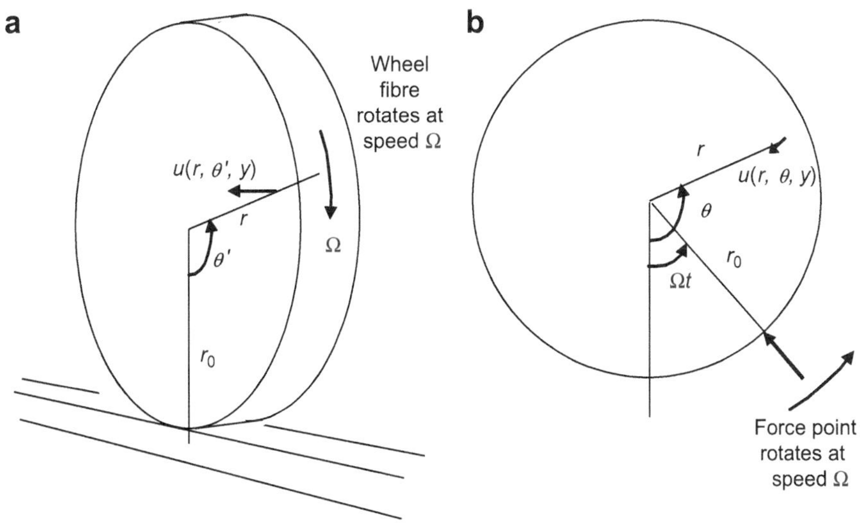

FIGURE 4-16 (a) Coordinates with reference point fixed at forcing point, (b) coordinates with reference point fixed in wheel where c_g is the group velocity associated with k (i.e. $\partial\omega/\partial k$). (Recall that k_2 and k_4 are close to purely real whereas k_1 and k_3 are close to imaginary.)

This gives a modified set of boundary conditions which can be written as a 4 × 4 matrix which

$$\begin{bmatrix} -(1-e^{ik_1L}) & -(1-e^{ik_2L}) & (1-e^{-ik_3L}) & (1-e^{-ik_4L}) \\ ik_1(1-e^{ik_1L}) & ik_2(1-e^{ik_2L}) & -ik_3(1-e^{-ik_3L}) & -ik_4(1-e^{-ik_4L}) \\ k_1^2(1-e^{ik_1L}) & k_3^2(1-e^{ik_2L}) & -k_3^2(1-e^{-ik_3L}) & -k_4^2(1-e^{-ik_4L}) \\ -ik_1^3(1-e^{ik_1L}) & -ik_2^3(1-e^{ik_2L}) & ik_3^3(1-e^{-ik_3L}) & ik_4^3(1-e^{-ik_4L}) \end{bmatrix} \times \begin{Bmatrix} A_1 \\ A_2 \\ A_3 \\ A_4 \end{Bmatrix} = \begin{Bmatrix} 0 \\ 0 \\ 0 \\ F/EI \end{Bmatrix} \tag{4.16}$$

This can be solved to find A_i and hence the mobility as before. The result is shown in Figure 4.17 for radial and axial directions for a speed of 160 km/h (44 m/s). Compared with Figure 4.15, each resonance peak for $n \geq 1$ is split into two peaks. This can also be seen from equation (4.18), as the matrix is singular when the exponentials containing k_2 or k_4 are equal to unity

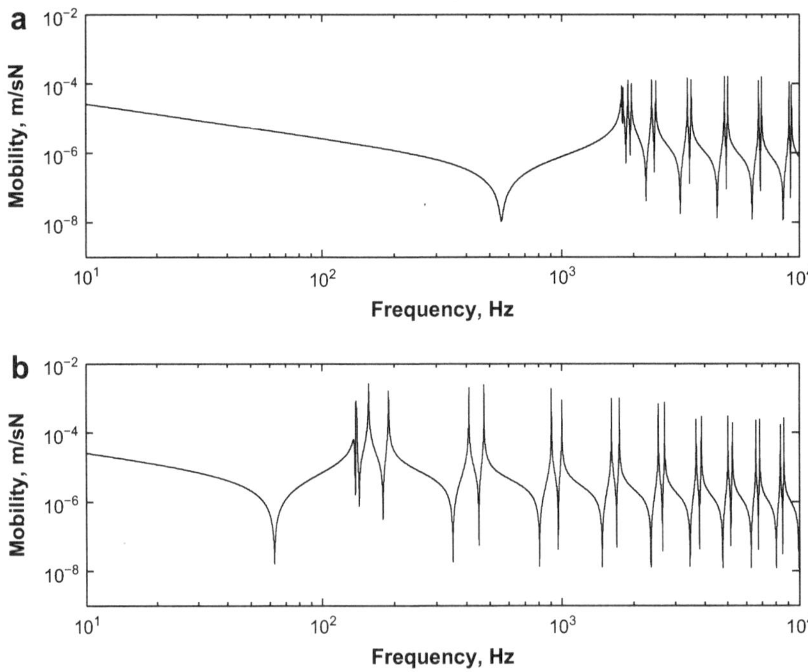

FIGURE 4-17 Mobilities of UIC 920 mm freight wheel predicted using closed beam model for a speed of 160 km/h. (a) Radial, (b) axial

$$e^{ik_2L} = 1, \text{giving } k_2 = -k\left(1 - \frac{V}{c_g}\right) = \frac{2\pi n}{L} \tag{4.17}$$

and

$$e^{-ik_4L} = 1, \text{giving } k_4 = k\left(1 + \frac{V}{c_g}\right) = \frac{2\pi n}{L} \tag{4.18}$$

The first of these conditions is met at a frequency slightly higher than the original resonance and the second at a slightly lower frequency. In each case only a single propagating wave is excited into resonance, and no standing wave is formed. At the lower peak the forward travelling wave is excited; at the higher peak the backward travelling wave is excited.

4.5.2 Theory

Returning now to the more general case of the modal summation described in Section 4.3, the effects of wheel rotation are developed in [4.12]. The main results are given here. Both gyroscopic effects and centrifugal stiffening are neglected.

For a harmonic force at angular frequency ω, the displacement can be written as a sum of terms over all modes as in Section 4.3 above [4.12]. For each mode, however, this response consists of two waves which are rotating in opposite directions. When observed in the frame of reference that is rotating with the wheel, these occur *at different frequencies*, $\omega \pm n\Omega$, where n is the number of nodal diameters in the mode and $\Omega = V/r_0$ is the rotational velocity, where r_0 is the wheel radius. The response vector $\underline{u}$ is given by (the real part of):

$$\underline{u}(r,\theta,y,t) = \sum_{m,n}\left\{ \underline{\Psi}_{mn}(r,y)T_{mn}\frac{e^{i((\omega+n\Omega)t-n\theta)}}{2m_{mn}d_+(\omega)} + \underline{\Psi}^*_{mn}(r,y)T^*_{mn}\frac{e^{i((\omega-n\Omega)t+n\theta)}}{2m_{mn}d_-(\omega)}\right\} \tag{4.19}$$

where (r, θ, y) are the radial, circumferential and axial coordinates of the response point, see Figure 4.16, m_{mn} is the modal mass, * denotes complex conjugate and $d_\pm$ are given by

$$d_\pm(\omega) = \omega_{mn}^2 - (\omega \pm n\Omega)^2 + 2i\zeta_{mn}(\omega \pm n\Omega)\omega_{mn} \tag{4.20}$$

where ω_{mn} is the natural frequency and ζ_{mn} is the modal damping ratio. The denominators of equation (4.19) approach zero when $d_\pm(\omega) \to 0$, i.e. when

$$\omega = \omega_{mn} \mp n\Omega. \tag{4.21}$$

These are the natural frequencies of the rotating wheel as seen from the excitation point. They occur below and above the corresponding natural frequency of the stationary wheel, in the same way as found for the beam in Section 4.5.1.

The terms $\underline{\Psi}_{mn}$ and T_{mn} are defined as follows. The mode shape can be divided into two sets of coordinates: type *I*, the axial and radial displacements and the rotation about the circumferential direction, and type *II*, the circumferential displacement and the other two rotations. Then for 'even' modes the full mode shapes can be written as

$$\underline{\varphi}^{e}_{mn}(r,\theta,y) = \underline{\psi}^{I}_{mn}(r,y)\cos n\theta + \underline{\psi}^{II}_{mn}(r,y)\sin n\theta \tag{4.22}$$

and the odd modes as

$$\underline{\varphi}^{o}_{mn}(r,\theta,y) = \underline{\psi}^{I}_{mn}(r,y)\sin n\theta - \underline{\psi}^{II}_{mn}(r,y)\cos n\theta \tag{4.23}$$

The corresponding amplitudes are combined by writing

$$\Psi_{mn} = \psi^{I}_{mn} + i\psi^{II}_{mn} \tag{4.24}$$

The forcing terms have been divided into type *I* and *II* coordinates and written as

$$T_{mn} = \underline{F}^{I}\cdot\underline{\psi}^{I}_{mn}(r_0,y_0) - i\,\underline{F}^{II}\cdot\underline{\psi}^{II}_{mn}(r_0,y_0) \tag{4.25}$$

where r_0 and y_0 are the coordinates of the forcing point.

Transforming the response into the frame of reference in which the point of application of the force is stationary, the coordinates are (r, θ', y) with $\theta' = \theta - \Omega t$ and the response is given by

$$\underline{u}(r,\theta',y,t) = e^{i\omega t}\sum_{m,n}\left\{\underline{\Psi}_{mn}(r,y)T_{mn}\frac{e^{-in\theta'}}{2m_{mn}d_{+}(\omega)} + \underline{\Psi}^{*}_{mn}(r,y)T^{*}_{mn}\frac{e^{in\theta'}}{2m_{mn}d_{-}(\omega)}\right\} \tag{4.26}$$

from which it can be seen that the two waves now exist at the same frequency (the excitation frequency). However, as ω is varied, they are still 'resonant', that is their response is maximum, at different frequencies, as given by equation (4.21). Equation (4.26) allows the mobility to be determined for the rotating wheel, as seen at the contact patch for degrees of freedom j and k:

$$Y_{jk}(\omega) = i\omega\sum_{m,n}\left\{\frac{\psi_{mnj}(r,y)\psi_{mnk}(r_0,y_0)}{2m_{mn}}\left(\frac{\varepsilon_{jk}e^{-in\theta'}}{d_{+}(\omega)} + \frac{\varepsilon^{*}_{jk}e^{in\theta'}}{d_{-}(\omega)}\right)\right\} \tag{4.27}$$

where the force is at $\theta' = 0$ and the response at θ', and

$$\varepsilon_{jk} = \begin{cases} 1 & \text{if } j \text{ and } k \text{ are both of type } I \text{ or both of type } II \\ -i & \text{if } j \text{ is of type } I \text{ and } k \text{ is of type } II \\ i & \text{if } j \text{ is of type } II \text{ and } k \text{ is of type } I \end{cases} \tag{4.28}$$

Figure 4.18 shows the radial mobility of the UIC 920 mm standard freight wheelset for a train speed of 160 km/h (44 m/s). Comparison with Figure 4.8 shows that modes with $n > 0$ are separated into dual peaks, according to equation (4.21). Similarly, Figure 4.19 shows the result for axial vibration. Again each resonance with

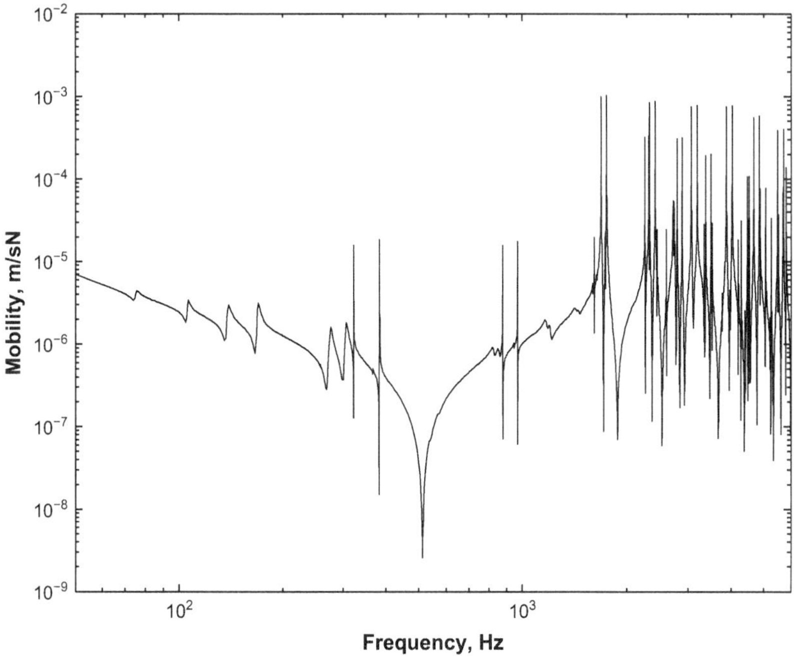

FIGURE 4-18 Radial mobility of UIC 920 mm standard freight wheelset for a train speed of 160 km/h as seen at the point of contact with the rail

$n > 0$ is separated into two peaks. For this speed, the rotational frequency is given by $\Omega = 44/0.46 = 96$ rad/s (equivalent to 15.4 Hz). Thus, for example, the $n = 2$ mode at 353 Hz is divided into two peaks at 322 and 384 Hz.

If, instead of a harmonic force, the excitation is a broad-band random process, such as that due to roughness (see Chapter 5), the velocity response at position j is given in terms of its spectral density, which is given in the frame of reference rotating with the wheel by

$$S_{v_j}(\omega) = \omega^2 \sum_n \left\{ \left| \sum_k H_k(\omega - n\Omega) \sum_m \frac{\psi_{mnj}(r,y)\psi_{mnk}(r_0,y_0)\varepsilon_{jk}}{2m_{mn}(\omega_{mn}^2 - \omega^2 + 2i\zeta_{mn}\omega\omega_{mn})} \right|^2 \times S_r(\omega - n\Omega) \right\} + \omega^2 \sum_n \left\{ \left| \sum_k H_k(\omega + n\Omega) \times \sum_m \frac{\psi_{mnj}(r,y)\psi_{mnk}(r_0,y_0)\varepsilon_{jk}^*}{2m_{mn}(\omega_{mn}^2 - \omega^2 + 2i\zeta_{mn}\omega\omega_{mn})} \right|^2 S_r(\omega + n\Omega) \right\} \tag{4.29}$$

where $S_r(\omega)$ is the spectral density of the roughness and H_k is a transfer function from roughness to the contact force in the direction k obtained using the method described in Chapter 5. It can be seen that the response at frequency ω is caused by

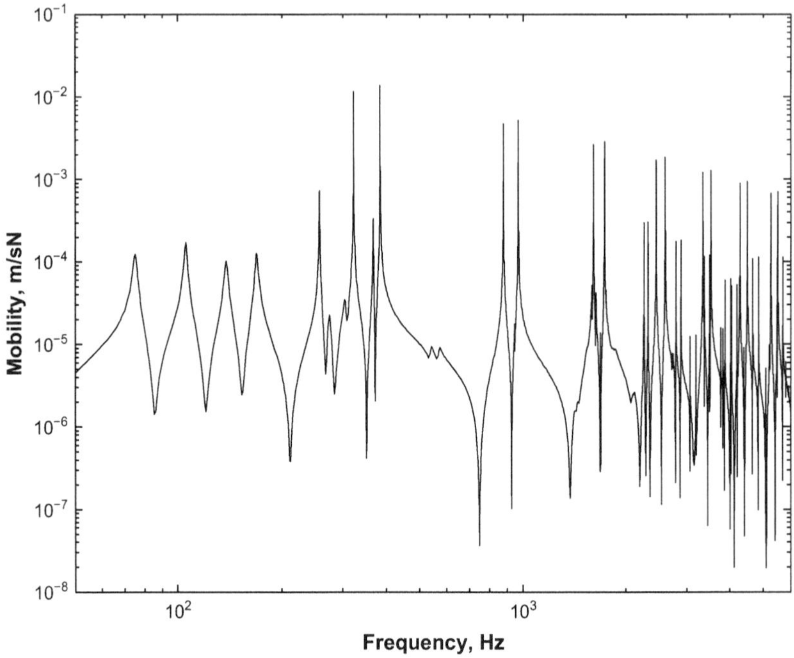

FIGURE 4-19 Axial mobility of UIC 920 mm standard freight wheelset for a train speed of 160 km/h as seen at the point of contact with the rail

roughness excitation at the frequencies $\omega \pm n\Omega$, depending on the modal order n as well as the direction of travel of the waves.

4.5.3 Form of the response

It should be remembered that for each natural frequency ω_{mn} with $n > 0$ there are actually two independent modes. While they are normally represented in terms of cosine and sine components (see equations (4.22, 4.23)), an equally valid representation is in terms of two contra-rotating waves, as in the beam model in Section 4.5.1. The two forms are mathematically equivalent as, for any angle ϕ

$$Ae^{i\phi} + Be^{-i\phi} = (A + B)\cos\phi + i(A - B)\sin\phi \tag{4.30}$$

Normally, resonance of a structure is considered to occur when two waves travelling in opposite directions interfere to form a standing wave pattern, also known as the phase-closure principle. However, for a rotating wheel the two waves travelling in opposite directions are excited by statistically independent forces (roughness components at different frequencies) and thus resonance (maximum response) can occur in each of them separately. Here the phase-closure principle should be applied to a wave rotating in a single direction without any need to consider reflections or interference.

Since the force components at different excitation frequencies are not correlated, the response of the two contra-rotating waves is consequently uncorrelated, even though both waves have their maximum response at a response frequency of approximately ω_{mn}, as seen in the denominators in equation (4.29). Therefore the response of a wheel to random forcing cannot contain a fixed pattern of nodal diameters but is a random combination of two independent waves. The time-averaged response at any circumferential position is the same as at any other, so that equation (4.29) does not depend on θ or θ'. This conclusion only holds for random excitation; for sinusoidal excitation as in curve squeal, it might be expected that only one of the two waves is excited but again there should be no node lines.

4.6 EXPERIMENTAL RESULTS

4.6.1 Response during rolling

Figure 4.20 shows measured results of the axial vibration of the wheel web and tyre for four speeds [4.13] and the overall axial acceleration levels at various positions on the wheel are compared [4.14]. The wheel here is a 1.06 m diameter 'Commonwealth' wheel, the same wheel as used in the analysis presented in Figures 2.11 and 2.12 in Chapter 2.

It can be seen that the vibration levels on the web are considerably higher than on the tyre. The reason for this is as follows. The modes of vibration seen on the tyre are the zero-nodal-circle ($m = 0$) axial modes ($n = 2, 3, 4, \ldots$) as seen in the lateral mobility. However, those seen most clearly in the web response are the one-nodal-circle ($m = 1$) axial modes and the radial modes. These are the modes seen in the radial mobility (e.g. Figure 4.8). They have higher modal responses than the zero-nodal-circle modes. It is their high amplitudes which lead to the high wheel radiation identified in Figure 2.12 for the 1.25 kHz band and above. As will be seen in Chapter 5, the major excitation occurs via the radial direction at the contact point (due to the roughness). This means that the one-nodal-circle and radial modes are excited because they occur strongly in the radial mobility. The web is then also brought into vibration, and depending on the mode shapes, can have a higher amplitude than at the contact patch, thus acting as a sort of mechanical amplifier. Its large area radiates sound effectively.

4.6.2 'Rolling damping'

A feature of the above measured data is that the peaks associated with each wheel resonance are much broader than seen in the mobilities. This implies a much higher 'damping' is present than for a free wheel. As will be seen in Chapter 5, this effect occurs because of the coupling with the rail. It means that in order to reduce the wheel vibration by damping treatments, the added damping must be sufficiently high to overcome this 'rolling damping' as well as the material damping of the free wheel. It is also the reason that it is not critical exactly what values of modal damping ζ_n are used in the calculation of rolling noise.

Table 4.2, from [4.15], lists modal damping ratios extracted from measurements on a running wheel. These tests were carried out on wheels on which a sinusoidal roughness profile had been machined. By slowly increasing the running speed the

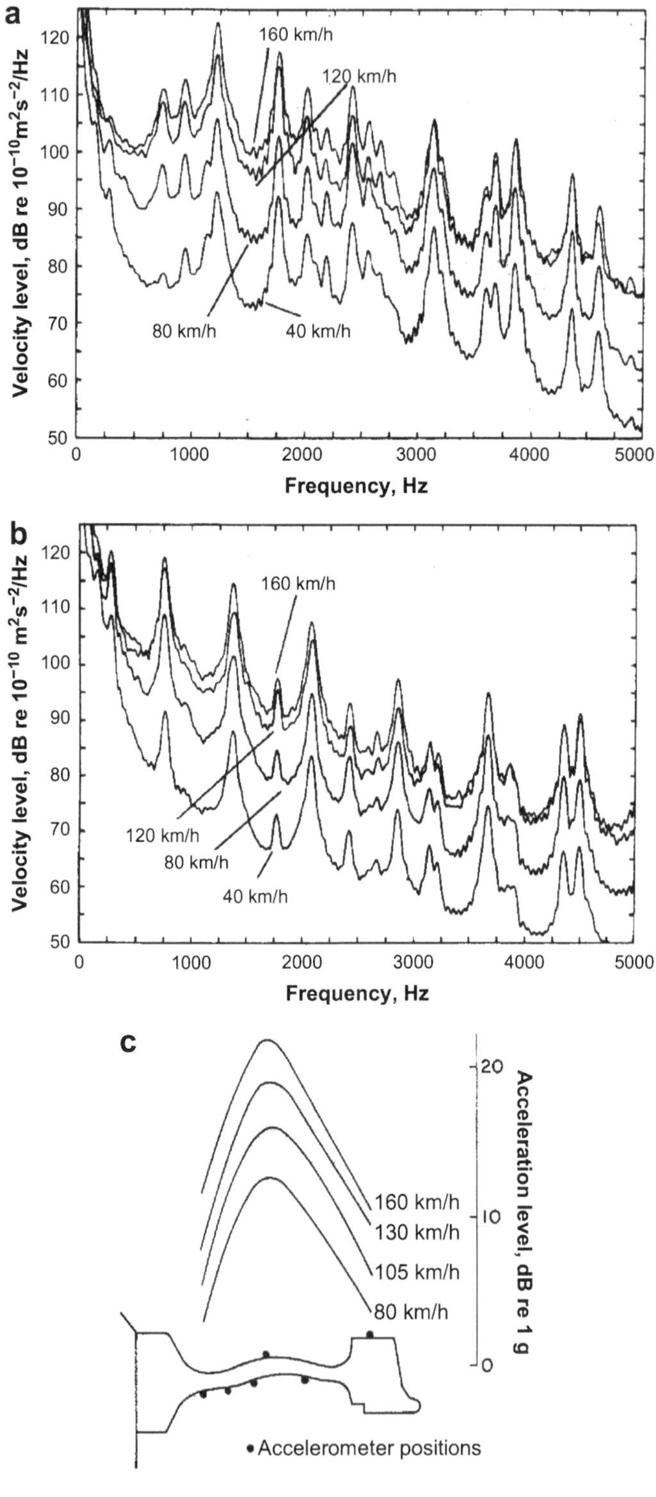

FIGURE 4-20 Measured response of a BR 'Commonwealth' 1.06 m diameter wheel during rolling, (a) axial on tyre, (b) axial on web, (c) amplitude distribution [4.13, 4.14]

TABLE 4.2 **DAMPING OF WHEEL MODES DURING ROLLING [4.15]**

n	Mode	Natural frequency, Hz	Measured damping ratio
2	radial	1722	0.009
2	one-nodal-circle	2387	0.004
3	radial	2426	0.003
0	radial	2761	0.003
4	one-nodal-circle	3707	0.0013

frequency of excitation was varied. The equivalent modal damping was estimated from the half power bandwidth of the peaks. The damping ratios found, between 0.0013 and 0.009, are much greater than the damping of the free wheel which is around 10^{-4} (see Figure 4.6). The implications for damping treatments will be considered further in Chapter 7.

4.7 NOISE FROM BOGIE AND VEHICLE SUPERSTRUCTURE

4.7.1 Introduction

As well as noise from the wheel vibration it is possible that the bogie and vehicle superstructure also vibrate and radiate noise. However, this has been neglected throughout this book. This section provides a brief justification for this.

Within the Silent Freight project (described in Chapter 7), extensive investigations were carried out into the contribution of the bogie and the vehicle superstructure to the radiated noise from a freight wagon. An SNCF 'tombereau' wagon was studied, because it was perceived as a noisy wagon type. It was fitted with standard Y25 bogies.

Several methods of investigation were used in parallel, including theoretical modelling [4.16] and experimental analysis [4.17]. The latter used a combination of field experiments at 100 km/h and static measurements of vibro-acoustic transfer functions to derive the various contributions to the noise during a train pass-by. Using an indirect technique [4.18] estimates were derived of the forces acting on the bogie above each wheel bearing and on the vehicle body at the bogie pivot points.

4.7.2 Bogie

A finite element model was constructed of the bogie frame [4.16] and the responses to vertical and lateral forces at the top of the primary suspension were determined. The element density was sufficient to yield valid results up to 500 Hz.

Figure 4.21 compares measured and predicted point accelerances. They are shown in one-third octave bands for ease of comparison. The finite element model gives generally good predictions in the frequency range for which it is valid. In the frequency range 100–1000 Hz, the response to a lateral force can be seen to be around 20 dB (a factor of 10) higher than that due to the vertical force.

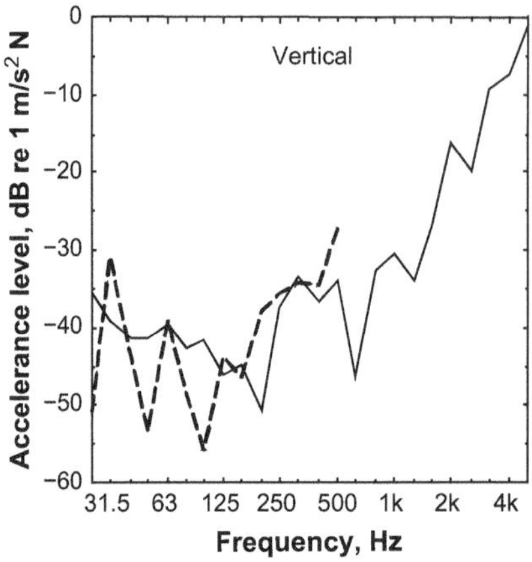

FIGURE 4-21 Measured and predicted input accelerances of Y25 bogie frame for vertical and lateral forces

The radiated sound power level for a unit force was estimated by calculating the normal vibration velocity at many points on the bogie frame and using the technique described in Chapter 6.

Combining this with estimates of the forces acting at the top of the suspension, obtained from the field measurements [4.17], the contribution of the bogie to the radiated noise could be determined. The contribution from the lateral force was found to be greater than that due to the vertical force, even though the vertical force spectrum is greater. Figure 4.22 compares the various spectra in the form of sound pressure at 7.5 m from the track. The two estimates of the noise from the bogie are consistent and show that the bogie contributes 20 dB less noise energy than the total

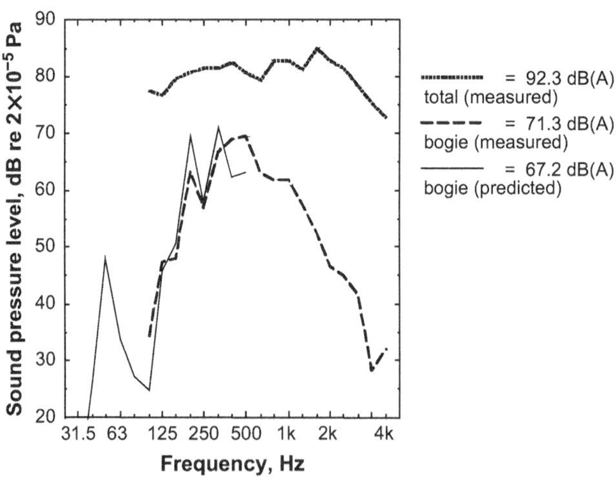

FIGURE 4-22 Contributions to the pass-by noise from the bogie at 7.5 m estimated from measured transfer functions and predicted from finite element model

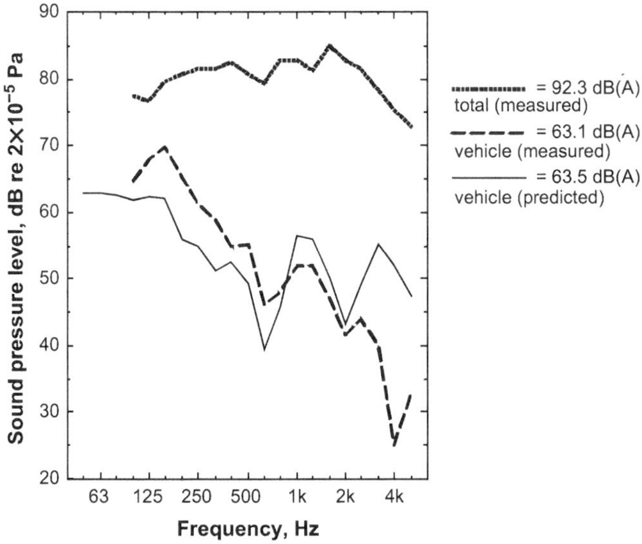

FIGURE 4-23 Contributions to the pass-by noise from the vehicle body estimated from measured transfer functions and predicted from SEA model

(i.e. about 1%), although between 200 and 500 Hz, the contribution is within about 10 dB of the total noise.

4.7.3 Vehicle body

A similar experimental technique was used to identify the forces acting on the vehicle body at the bogie pivot point [4.17]. The vehicle body is too large a structure to represent using finite elements over the frequency range of interest, so for this a statistical energy analysis approach was used. Figure 4.23 shows the predicted sound pressure level at 7.5 m from the track due to the vehicle body, again compared with the total measured noise and the estimate based on measured transfer functions. The vehicle body was found to contribute about 30 dB less than the total (0.1% of the sound energy), while at low frequencies, the contribution is within about 10–15 dB of the total noise.

4.7.4 Discussion

It is clear that the contribution of vibrations of the bogie frame and vehicle body to the wayside noise was negligible for the case studied. This was confirmed by additional measurements carried out on two other wagon types, a bogie tank wagon and a container flat wagon, both with the same bogie type. It is notable that measurement techniques such as microphone arrays do not have sufficient amplitude resolution to measure such low amplitude sources accurately in the presence of louder sources (the wheels and rails).

The conclusion of the extensive programme of investigation was therefore that the dominant source is the wheel/rail region. However, this should be qualified by noting that bogie and suspension designs differ widely. Where no primary suspension is present, the bogie frame can be expected to contribute more significantly to the radiated noise than the results shown here.

REFERENCES

4.1 P.J. Remington. Wheel/rail rolling noise, I: Theoretical analysis. *Journal of the Acoustical Society of America*, **81**, 1805–1823, 1987.

4.2 H. Irretier. The natural and forced vibrations of a wheel disc. *Journal of Sound and Vibration*, **87**, 161–177, 1983.

4.3 D.J. Thompson. Wheel-rail noise generation, Part II: Wheel vibration. *Journal of Sound and Vibration*, **161**, 401–419, 1993.

4.4 D.J. Thompson and M.G. Dittrich. Wheel response and radiation – laboratory measurements of five types of wheel and comparisons with theory. ORE Technical Document DT248 (C163), Utrecht, June 1991.

4.5 M. Petyt. *Introduction to Finite Element Vibration Analysis*. Cambridge University Press, Cambridge, 1990.

4.6 S.S. Rao. *Mechanical Vibrations*, 4th edition. Prentice Hall, Upper Saddle River, NJ, 2003.

4.7 D.J. Ewins. *Modal Testing: Theory and Practice*. Research Studies Press, Letchworth, 1984.

4.8 C.J.C. Jones and D.J. Thompson. Extended validation of a theoretical model for railway rolling noise using novel wheel and track designs. *Journal of Sound and Vibration*, **267**, 509–522, 2003.

4.9 D.E. Newland. *An Introduction to Random Vibrations and Spectral Analysis*. Longman, London, 1975.

4.10 P.J. Remington. Wheel/rail noise – Part 1: Characterization of the wheel/rail dynamic system. *Journal of Sound and Vibration*, **46**, 359–379, 1976.

4.11 D. Boennen and S.J. Walsh. Investigation of analytical beam and annular plate models for automotive disc brake vibration. IMechE Braking Conference, Leeds, 2006.

4.12 D.J. Thompson. Wheel-rail noise generation, Part V: Inclusion of wheel rotation. *Journal of Sound and Vibration*, **161**, 467–482, 1993.

4.13 D.J. Thompson. Theoretical modelling of wheel-rail noise generation. *Proceedings of the Institution of Mechanical Engineers*, **205**, Part F *(Journal of Rail and Rapid Transit)*, 137–149, 1991.

4.14 B. Hemsworth. Vibration of a rolling wheel – preliminary results. *Journal of Sound and Vibration*, **87**, 189–194, 1983.

4.15 D.J. Thompson, N. Vincent, and P.E. Gautier. Validation of a model for railway rolling noise using field measurements with sinusoidally profiled wheels. *Journal of Sound and Vibration*, **223**, 587–609, 1999.

4.16 N.S. Ferguson. Modelling the vibrational characteristics and radiated sound power for a Y25 type bogie and wagon. *Journal of Sound and Vibration*, **231**, 791–803, 2000.

4.17 F. de Beer and J.W. Verheij. Experimental determination of pass-by noise contributions from the bogies and superstructure of a freight wagon. *Journal of Sound and Vibration*, **231**, 639–652, 2000.

4.18 M.H.A. Janssens, J.W. Verheij, and D.J. Thompson. The use of an equivalent forces method for quantifying structural sound transmission in ships. *Journal of Sound and Vibration*, **226**, 305–328, 1999.

CHAPTER 5

Wheel/Rail Interaction and Excitation by Roughness

5.1 INTRODUCTION

As already discussed in Chapter 2, the most important form of noise generated at the wheel/rail interface is rolling noise caused by the roughness of the wheel and rail running surfaces. As given in equation (2.1), when the train runs at speed V, over undulations of the surfaces with wavelength λ, these produce vibrations, and hence noise, at a frequency f given by

$$f = \frac{V}{\lambda} \tag{5.1}$$

Typical wavelengths of roughness relevant to rolling noise are between about 5 and 500 mm. The roughness at such wavelengths has amplitudes in the range from tens of microns at long wavelengths to less than a micron at short wavelengths. Thus the amplitudes are of the order of 10^{-4} times the wavelength. They are not visible on what appears to be a smooth surface. It may seem surprising that such small amplitudes can lead to high noise levels, but it should be remembered that the displacement amplitudes of a typical acoustic field are similarly small at audio frequencies.

Longer wavelength roughness excites lower frequencies and is therefore relevant to ground vibration (see Chapters 12 and 13) and ride comfort. Track roughness with wavelengths greater than around 1 m tends to be caused by undulations in the track bed or in the straightness of the rails. Such a wavelength range is measured routinely in loaded conditions by track recording coaches and used to determine the need for maintenance such as tamping of the ballast [5.1]. On the wheel, the primary out-of-roundness occurs at a wavelength equal to the wheel circumference (typically about 3 m), which is caused by wheel eccentricity on its bearings. This may have a peak-to-trough amplitude of 1 mm or more. The first few harmonics of this wavelength are also important and caused by wheel shape defects [5.2].

At the wavelengths of relevance to rolling noise, roughness may be caused by manufacturing variability in the rail section or variable wear of the wheel and rail surfaces. Severe quasi-periodic roughness, known as corrugation, may also develop in some situations, see Figure 2.5. This may have peak-to-trough amplitudes of typically 50 μm at a wavelength of 50 mm; at longer wavelengths the amplitudes can be greater than this.

Shorter wavelength roughness, often termed micro-roughness, with wavelengths less than 1 mm, is not directly relevant to noise generation, but is important for adhesion (traction and braking) and electrical conductivity and should therefore not be too low.

In this chapter the basic model of the excitation of the wheel/rail system by roughness is introduced. The wheel/rail interaction model is described in Section 5.2. As well as models for the wheel and rail mobilities, as developed in the previous chapters, this requires models for the mobilities of the contact zone itself which are introduced in Section 5.3. The effect of the contact patch between wheel and rail in filtering the roughness is described in Section 5.4 followed by a discussion of the measurement and analysis of surface roughness in Sections 5.5 and 5.6. Other mechanisms of excitation are summarized in Section 5.7.

5.2 WHEEL/RAIL INTERACTION MODEL

5.2.1 Vertical excitation model

The wheel/rail system, shown in Figure 5.1, can be represented by two dynamic systems connected at a point and excited by a relative displacement between them. A third system, the contact spring, is connected in parallel with the others. In this model the motion of the wheel along the rail can be ignored and replaced by a 'moving excitation' in which the roughness 'strip' is pulled through the gap between wheel and rail. Initially, only vertical vibration will be considered.

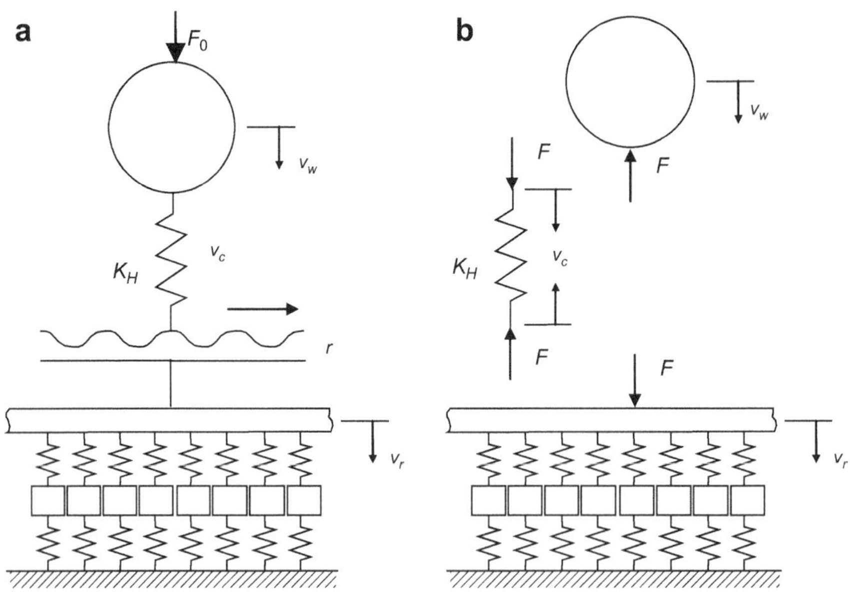

FIGURE 5-1 Schematic diagram of the wheel/rail system. (a) Excitation by roughness r. (b) Dynamic forces acting in vertical direction

If the rail is excited by a vertical harmonic force $Fe^{i\omega t}$ of circular frequency ω and complex amplitude F, its velocity amplitude v_r is given by

$$v_r = Y_r F \tag{5.2}$$

where Y_r is the vertical rail mobility, as discussed in Chapter 3. Quantities are defined as positive downwards. Similarly, the same force acts *upwards* on the wheel so that the (downwards) wheel velocity v_w is given by

$$v_w = -Y_w F \tag{5.3}$$

where Y_w is the wheel mobility. The contact spring (discussed in more detail in Section 5.3.1) has a mobility Y_c so that if v_c is the *relative* velocity across the contact spring (positive for compression of the spring):

$$v_c = Y_c F = \frac{i\omega F}{K_H} \tag{5.4}$$

where K_H is the linearized (Hertzian) contact stiffness. The contact spring can be located on the wheel side of the roughness or the rail side or divided between the two; the result is the same. Introducing a roughness of amplitude r at circular frequency ω (positive for an asperity) to ensure that contact is maintained, the various velocities are related by

$$v_r = i\omega r + v_w - v_c \tag{5.5}$$

where $i\omega r$ is the roughness *velocity* amplitude. The frequency $\omega = 2\pi f$ is determined from equation (5.1). Combining equations (5.2)–(5.5) gives the force amplitude:

$$F = \frac{i\omega r}{Y_r + Y_w + Y_c} \tag{5.6}$$

from which the velocity amplitudes can be obtained as

$$v_r = \frac{i\omega r Y_r}{Y_r + Y_w + Y_c} \tag{5.7}$$

$$v_w = \frac{-i\omega r Y_w}{Y_r + Y_w + Y_c} \tag{5.8}$$

Note that the force is a derived quantity, dependent on the combined dynamic properties of the wheel, track and contact spring. The implications of this will be explored further in Section 5.2.5.

In general, the responses of wheel and rail are coupled and depend on all three mobilities in the denominator of equations (5.7) and (5.8). However, where one of these mobilities has a larger magnitude than the others, this system will have a response that is similar to the roughness. In simple terms, a roughness asperity can push the wheel upwards, the rail downwards or compress the contact spring, depending on the relative size of the mobilities – see Figure 5.2. The three mobilities in the above equations were compared in Figure 2.16 for a typical wheel and track

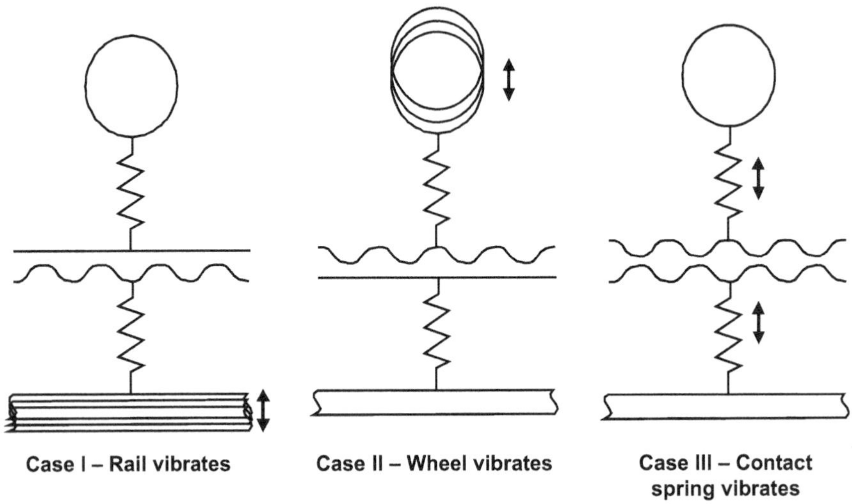

FIGURE 5-2 Schematic indication that roughness may excite the rail, the contact spring or the wheel depending on their frequency response functions

(the UIC 920 mm freight wheel of Figure 4.3 and the track represented by the parameters of Table 3.3).

The interaction force, from equation (5.6), is shown in Figure 5.3, for the mobilities of Figure 2.16. This is shown normalized by the roughness amplitude, r. This is equal to the inverse of the sum of the mobilities (divided by $i\omega$). From this it

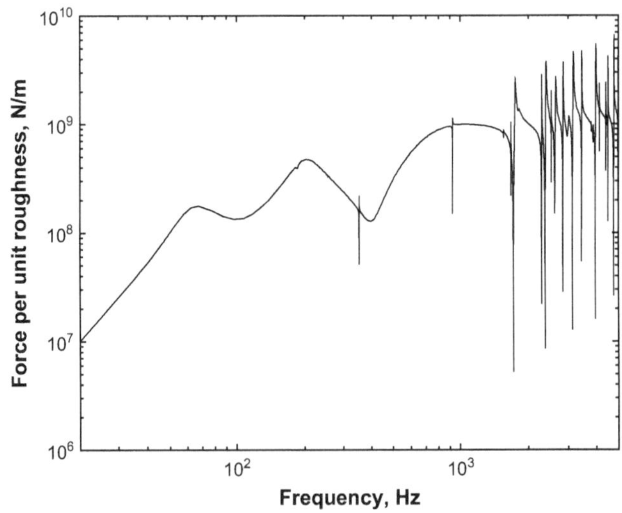

FIGURE 5-3 Contact force for unit roughness based on vertical interaction only and mobilities of Figure 2.16

can be seen that the force has a sharp minimum at each of the high frequency wheel resonances (radial modes) due to the dominance of the wheel mobility in the denominator of equation (5.6) at these frequencies. The broad peak in the force at around 200 Hz corresponds to an anti-resonance in the track mobility. Other features will be explained below.

The corresponding responses, v_r and v_w, estimated using equations (5.7) and (5.8), are plotted in Figure 5.4, each normalized by $i\omega r$. At low frequencies the wheel moves with an amplitude approximately equal to the roughness, $|v_w/i\omega r| \approx 1$, whereas between about 70 and 1000 Hz the rail amplitude is similar to the roughness, $|v_r/i\omega r| \approx 1$. In order to explain the features of these graphs, in the following section various frequency regions are considered in which one of the mobilities is greater than the others, allowing approximations can be made.

5.2.2 Approximations to the vertical excitation model

In Section 4.4 it was shown that the wheel can be represented by its unsprung mass, M_w, for frequencies up to around 200 Hz, giving

$$Y_W = \frac{-i}{\omega M_W} \tag{5.9}$$

For the track, the Timoshenko beam model given in Section 3.4 has been used here. At low frequencies its mobility is stiffness controlled. Extending equation (3.20) to include damping, it can be written as a complex stiffness $\tilde{K}_T = K_T(1 + i\eta_T)$ where η_T is the damping loss factor. This can be given approximately by $\eta_T = 3\eta/4$, where η is the loss factor of the support stiffness. For a two-layer support system η is dominated by the loss factor of the ballast.

At low frequencies, the wheel mobility is large and mass controlled, equation (5.9), whereas both the rail and the contact spring are stiffness controlled with a mobility proportional to $i\omega$. Consequently, below about 70 Hz, the wheel has the largest of the three mobilities and equations (5.7) and (5.8) can be approximated to

$$v_r \approx i\omega r \frac{Y_r}{Y_W} \approx -i\omega r \frac{\omega^2 M_W}{\tilde{K}_T} \tag{5.10}$$

$$v_W \approx -i\omega r \tag{5.11}$$

Thus the wheel moves with the roughness amplitude (velocity $i\omega r$), whereas the track motion is much smaller (as can be seen in Figure 5.4). The track motion is approximately in phase with the wheel motion, that is as the wheel runs down into a roughness trough the track deflects downwards beneath it; as the wheel reaches a roughness peak the track rises beneath it. This is a consequence of the fact that Y_W and Y_r have almost opposite phases (see Figure 2.16); if both were stiffness controlled, for example, then they would vibrate out of phase with each other, being pushed apart by a roughness asperity.

There is a frequency between about 50 and 100 Hz at which the mobilities of the wheel mass and the combined stiffness of the track and contact spring are equal and opposite:

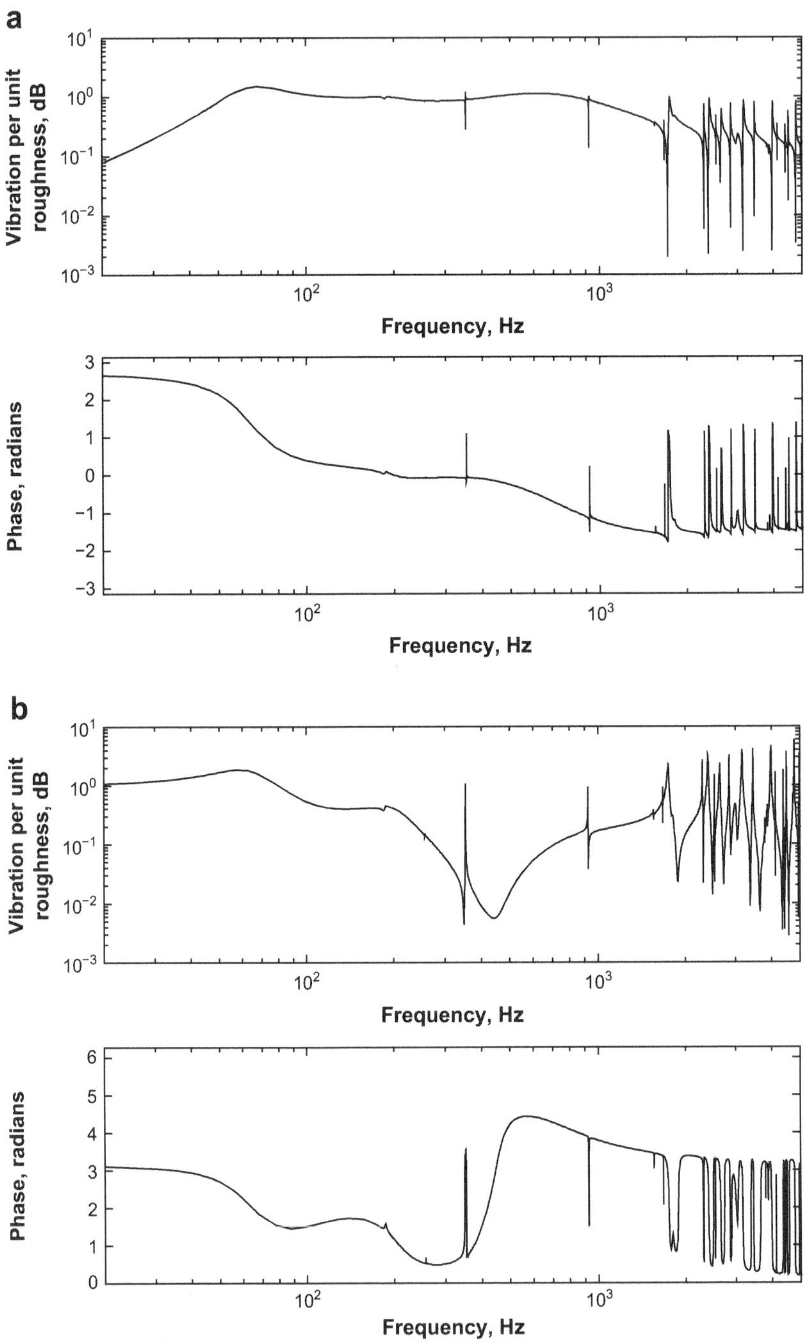

FIGURE 5-4 Vertical vibration at the wheel/rail contact for unit roughness based on vertical interaction only and mobilities of Figure 2.16. The phase is shown relative to the roughness. (a) Rail, (b) wheel

$$\mathrm{Im}\left(Y_w + Y_r + Y_c\right) = 0, \tag{5.12}$$

giving

$$\omega = \sqrt{\frac{K_T + K_H}{M_w}} \tag{5.13}$$

This is sometimes referred to as a resonance of the coupled wheel/track system at which the interaction force, and hence the system response, has a maximum. This can be seen as a peak in the contact force at 70 Hz in Figure 5.3. Note that the contact stiffness is usually small compared with the track stiffness at these frequencies, typical values for K_H being between about 1.1×10^9 and 1.4×10^9 N/m while $K_T = 1.4 \times 10^8$ N/m (for $s = 10^8$ N/m^2). Since the phase of the track mobility does not correspond purely to a stiffness, in particular due to the influence of ballast damping which can be quite high, this resonance is usually quite well damped. At this frequency, the denominator in equations (5.7) and (5.8) is given by

$$Y_w + Y_r + Y_c \approx Y_w + Y_r = \frac{-i\omega}{K_T} + \frac{i\omega}{K_T\left(1 + i\eta_T\right)} = \frac{\omega\eta_T}{K_T\left(1 + i\eta_T\right)} \tag{5.14}$$

The vibration amplitudes are therefore

$$v_r \approx \mathrm{i}\omega r \frac{K_T\left(1 + i\eta_T\right)}{\omega\eta_T} \frac{i\omega}{K_T\left(1 + i\eta_T\right)} = \frac{-\omega r}{\eta_T}, \tag{5.15}$$

$$v_w \approx -\mathrm{i}\omega r \frac{K_T\left(1 + i\eta_T\right)}{\omega\eta_T}\left(\frac{-i\omega}{K_T}\right) = -\frac{\omega r}{\eta_T}\left(1 + i\eta_T\right) \tag{5.16}$$

If the damping is not too large, these are approximately equal in magnitude, as found in Figure 5.4. They are also in phase with each other (90° out of phase with the roughness velocity, $\mathrm{i}\omega r$). If the damping η_T is lower, both amplitudes will be considerably greater than the input roughness. This can occur for slab track systems or directly fastened track on bridges or in tunnels. However, for ballasted track, which has a higher damping loss factor, the amplification is relatively small.

Above this resonance frequency, the track mobility becomes the largest of the three and equations (5.7) and (5.8) can be approximated to

$$v_r \approx \mathrm{i}\omega r \tag{5.17}$$

$$v_w \approx -\mathrm{i}\omega r \frac{Y_w}{Y_r} \approx \mathrm{i}\omega r \frac{K_T}{\omega^2 M_w} \tag{5.18}$$

Thus the rail now moves with the roughness amplitude, the result in Figure 5.4(a) being close to unity, whereas the wheel motion is much smaller due to its high inertia.

Above about 500 Hz the wheel mobility becomes stiffness controlled while at higher frequencies, it is dominated by its radial modes, as described in Sections 4.2 and 4.3. Moreover, at higher frequencies, the track mobility is no longer stiffness controlled; when waves propagate freely in the rail its mobility has a phase corresponding to a combination of mass and damping (see Chapter 3). Nevertheless, it remains higher in magnitude than either the wheel or the contact stiffness mobilities until about 1 kHz. Thus, the rail has the highest mobility between about 70 and 1000 Hz, and its response is approximately equal to the roughness in this region, as seen in Figure 5.4(a).

It is important to note that modifying the track mobility in this frequency region, where it is greater than that of the wheel, has no effect on the magnitude of the track response due to roughness. However, it does affect the contact force, equation (5.6), and thereby the response of the wheel. Similarly, increasing the unsprung mass of the wheel would increase the track vibration below the coupled wheel/track resonance frequency but would not affect the wheel response there. Above this resonance frequency it would reduce the wheel response but would not affect the rail response.

Above 1 kHz the contact stiffness has the highest mobility of the three in the denominator of equations (5.6)–(5.8), at least away from wheel resonances. Then, the track vibration is given by

$$v_r = \mathrm{i}\omega r \frac{Y_r}{Y_c} \tag{5.19}$$

Since the track mobility is approximately independent of frequency (for a Timoshenko beam, see Chapter 3) whereas $|Y_c| \propto \omega$ (see equation (5.4)) it is found that $|v_r/\mathrm{i}\omega r|$ rolls off at a rate of approximately $1/\omega$ at high frequency, as seen in Figure 5.4(a). The wheel vibration is only significant near wheel resonances, where

$$v_w \approx -\mathrm{i}\omega r \tag{5.20}$$

Peaks with $|v_w/\mathrm{i}\omega r| > 1$ occur just above each wheel natural frequency. This is considered in more detail in Section 5.2.4. The rail vibration also shows significant features near the wheel natural frequencies due to the dips and peaks seen in the contact force.

5.2.3 Multi-degree-of-freedom excitation model

In practice, although the roughness acts in the vertical direction, the wheel and rail are not only coupled in the vertical direction but also in other directions. Up to six coordinates (three displacements and three rotations) can be included. Apart from the vertical direction, the lateral direction is the most important to include. The coordinate system was introduced in Chapter 3 (see box on page 39) with the axis system x–y–z (x along the track, y in the lateral direction and z vertically downwards). A local axis system 1–2–3 is introduced here, with 3 normal to the contact, 2 in the transverse direction and 1 along the track. The wheel and rail mobilities require rotating from the x–y–z coordinate system into the 1–2–3 coordinate system. The notation used here is consistent with that generally used in studies of creepage (e.g. [5.3]). However, it differs from that used in the TWINS model for rolling noise [5.4], where x is vertical and z longitudinal.

The model for coupling in the vertical direction given by equations (5.2)–(5.5) can be extended to cover multi-degree-of-freedom coupling. Equation (5.2) can be written in matrix form:

$$\{v_r\} = [Y_r]\{F\} \tag{5.21}$$

where $[Y_r]$ is a symmetric matrix of rail mobilities at frequency ω, $\{v_r\}$ is a vector of up to six velocities and $\{F\}$ a vector of up to six forces. Similarly, the same forces act on the wheel to give

$$\{v_w\} = -[Y_w]\{F\} \tag{5.22}$$

where $[Y_w]$ is the wheel mobility matrix. The contact zone can also be represented by a mobility matrix, $[Y_c]$, as will be discussed further in Section 5.3:

$$\{v_c\} = [Y_c]\{F\} \tag{5.23}$$

Roughness only acts in the normal (3) direction, so the continuity of velocity takes the form

$$\{v_r\} = \begin{Bmatrix} 0 \\ \mathrm{i}\omega r \\ 0 \end{Bmatrix} + \{v_w\} - \{v_c\} \tag{5.24}$$

where all but the 3 element of the 'roughness' vector is zero. Combining equations (5.21)–(5.24) gives the force vector

$$\{F\} = [Y_r + Y_w + Y_c]^{-1} \begin{Bmatrix} 0 \\ \mathrm{i}\omega r \\ 0 \end{Bmatrix} \tag{5.25}$$

from which the velocity amplitudes can be obtained:

$$\{v_r\} = [Y_r][Y_r + Y_w + Y_c]^{-1} \begin{Bmatrix} 0 \\ \mathrm{i}\omega r \\ 0 \end{Bmatrix} \tag{5.26}$$

$$\{v_w\} = -[Y_w][Y_r + Y_w + Y_c]^{-1} \begin{Bmatrix} 0 \\ \mathrm{i}\omega r \\ 0 \end{Bmatrix} \tag{5.27}$$

These are the multi-dimensional versions of equations (5.6)–(5.8). Example results are shown in Figure 5.5 for coupling in vertical and lateral directions. Comparing these with the results in Figure 5.4, it can be seen that the vertical response is not greatly affected by the addition of coupling in other directions. However, it is found that the lateral response is considerably modified by the addition of coupling in the lateral direction. In Figure 5.5 the lateral responses of the wheel and rail are seen to be similar up to around 1 kHz due to the coupling through the creep force.

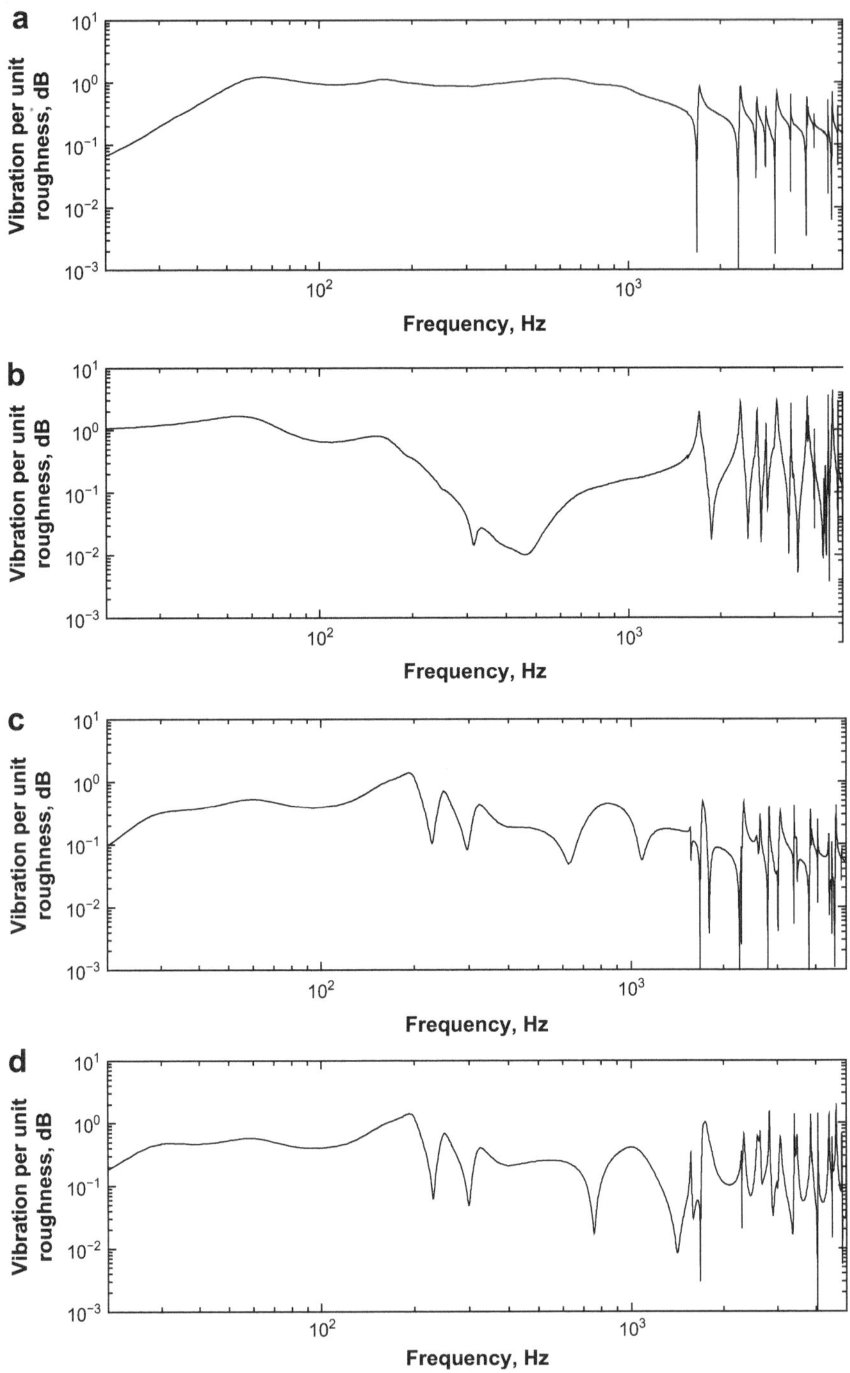

FIGURE 5-5 Vertical vibration at the wheel/rail contact for unit roughness based on interaction in vertical and lateral directions. (a) Rail vertical, (b) wheel vertical, (c) rail lateral, (d) wheel lateral

5.2.4 Damping of a rolling wheel

As seen in Section 4.6.2, the wheel experiences a much higher effective damping when rolling on the track than when it is suspended freely. To explain this phenomenon, it is sufficient to consider interaction in the vertical direction only. Although the rotation of the wheel has an effect in terms of splitting the wheel resonances (see Section 4.5), for simplicity a non-rotating wheel is considered in the present analysis.

In the vicinity of a wheel resonance, the wheel mobility is dominated by this single mode and can be approximated from equation (4.4) as

$$Y_w \approx \frac{i\omega\psi_n^2}{m_n(\omega_n^2 - \omega^2 + 2i\zeta_n\omega\omega_n)} \tag{5.28}$$

where ψ_n is the vertical modeshape amplitude of mode n at the contact point, m_n is the modal mass and ζ_n is the modal damping ratio. Close to the wheel natural frequency, $\omega \approx \omega_n$, this can be approximated further as

$$Y_w \approx \frac{D_n}{2(\zeta_n + i\varepsilon)} \tag{5.29}$$

where $\omega = \omega_n(1 + \varepsilon)$ for some small ε and

$$D_n = \frac{\psi_n^2}{m_n\omega_n} \tag{5.30}$$

At the resonance itself, $\varepsilon = 0$ and $Y_w = D_n/2\zeta_n$. In the high frequency region where wheel radial resonances occur, the track mobility based on a Timoshenko beam has a phase close to 0 and a magnitude approximately independent of frequency (see Section 3.4). This can be expressed in terms of an equivalent damper:

$$Y_r \approx \frac{1}{C_{eq}} \tag{5.31}$$

The vertical contact spring has a mobility given by equation (5.4). Substituting these into equation (5.8), the wheel response for a unit roughness is given by

$$\frac{v_w}{i\omega r} = \frac{-D_n}{D_n + 2(\zeta_n + i\varepsilon)\left(\dfrac{1}{C_{eq}} + \dfrac{i\omega_n}{K_H}\right)} \tag{5.32}$$

At the wheel resonance frequency, $\varepsilon = 0$, this ratio is approximately equal to unity. Expanding the denominator

$$D_n + 2(\zeta_n + i\varepsilon)\left(\frac{1}{C_{eq}} + \frac{i\omega_n}{K_H}\right) = D_n - \frac{2\omega_n\varepsilon}{K_H} + \frac{2i\omega_n\zeta_n}{K_H} + \frac{2i\varepsilon}{C_{eq}} + \frac{2\zeta_n}{C_{eq}} \tag{5.33}$$

it can be seen that for $\varepsilon < 0$ the real part of the denominator is greater than D_n, so that the ratio in equation (5.32) has a magnitude less than 1. At $\varepsilon = (D_n K_H)/(2\omega_n)$, which corresponds to a frequency just above the wheel resonance, the real part of the denominator is approximately zero. This is a resonance of the *coupled system* at which the wheel response has a maximum. The height of this maximum is given by

$$\left.\frac{v_W}{\mathrm{i}\omega r}\right|_{\max} = \frac{iD_n}{2\left(\dfrac{\varepsilon}{C_{eq}} + \dfrac{\omega_n \zeta_n}{K_H}\right)} \tag{5.34}$$

In the absence of the damper C_{eq} this would be determined by the wheel damping and would have a half-power bandwidth of $\Delta\omega = 2\zeta_n\omega_n$, identical to the bandwidth of the resonance in the wheel mobility. However, due to the presence of C_{eq}, and provided that ε is not too small, the peak is given by

$$\left.\frac{v_W}{\mathrm{i}\omega r}\right|_{\max} \approx \frac{iD_n C_{eq}}{2\varepsilon} \tag{5.35}$$

and the half-power bandwidth can be found as

$$\Delta\omega = \frac{D_n K_H^2}{C_{eq}\omega_n} \tag{5.36}$$

This can be expressed as an equivalent damping ratio of the wheel in contact with the rail:

$$\zeta_{\text{roll}} = \frac{D_n K_H^2}{2C_{eq}\omega_n^2} = \frac{\psi_n^2 K_H^2}{2m_n C_{eq}\omega_n^3} \tag{5.37}$$

Some examples are shown in Figure 5.6. These are calculated for the following parameters which are typical of the wheel/rail contact: $K_H = 10^9$ N/m, $C_{eq} = 3 \times 10^5$ Ns/m, $\psi_n^2/m_n = 1/200$, $\zeta_n = 10^{-4}$ and various notional natural frequencies. The corresponding mobilities are shown in Figure 5.7. In each case it can be seen that the normalized wheel response is unity at its natural frequency,

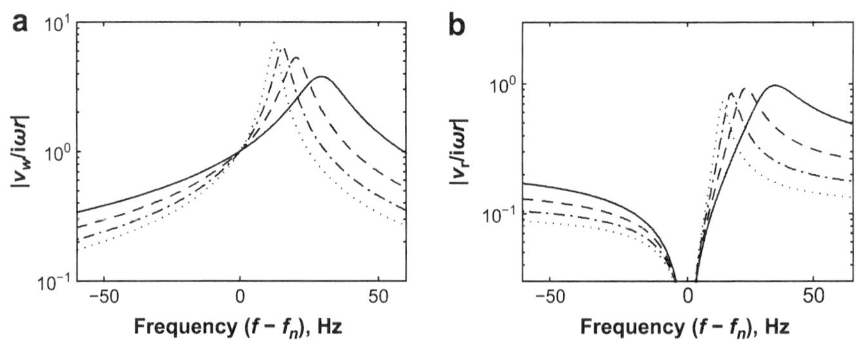

FIGURE 5-6 Wheel and rail vibration normalized to roughness for various nominal wheel natural frequencies. (a) Wheel, (b) rail. —, 2 kHz; – – –, 3 kHz; – · – ·, 4 kHz; · · · ·, 5 kHz

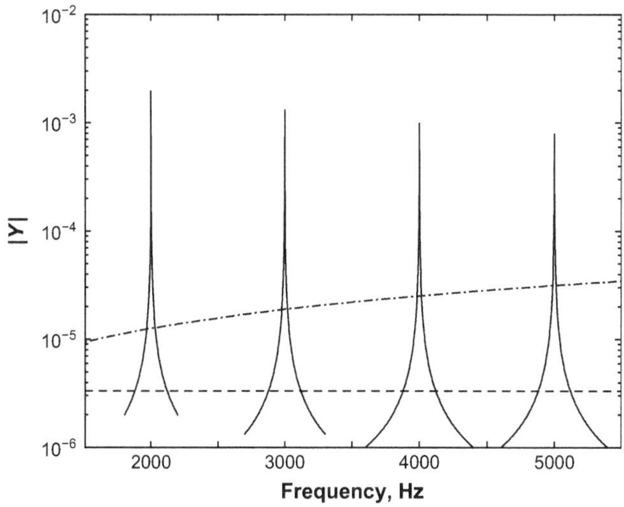

FIGURE 5-7 Example vertical mobilities of wheel, rail and contact spring used to illustrate rolling damping. —, wheel; – – –, rail; – · – ·, contact spring

where the rail response has a minimum. The peak in the wheel response occurs at a frequency greater than the natural frequency and this peak has a bandwidth much greater than the corresponding peak in the mobility. This indicates that the coupled system has an equivalent damping which is much greater than the wheel damping. The rail response also has a peak at the same frequency as the wheel, although its maximum response is 10–20 dB smaller.

It is clear from the above analysis that the additional damping effect experienced by the wheel is caused by the coupling with the rail, the mobility of which is damping controlled. However, it is interesting to note that increasing the equivalent damper describing the track, C_{eq}, would lead to a *reduction* in ζ_{roll} in equation (5.37). This is due to the fact that the damper is in series with the contact spring. When the value of C_{eq} is too large the damper becomes locked and the spring takes all the deflection.

5.2.5 Note on the interaction force

As discussed in Section 5.2.1, the wheel/rail interaction force is determined by the roughness spectrum divided by the sum of the wheel, rail and contact spring mobilities. Consequently, the interaction force spectrum in Figure 5.3 contains narrow band features at high frequencies due to the lightly damped wheel resonances. The force reaches a minimum at the wheel natural frequencies to ensure that $|v_w/i\omega r| = 1$. The force spectrum then has peaks at frequencies above the wheel natural frequencies corresponding to the resonances of the coupled system.

As a result of these features, it is important to note that an interaction force spectrum calculated for one wheel cannot be used to determine the response of another. Unless the dips in the force spectrum align with the peaks in the wheel mobility, unrealistic lightly damped resonance peaks will occur in the predicted wheel response. Moreover, even if the correct force spectrum is used, similar

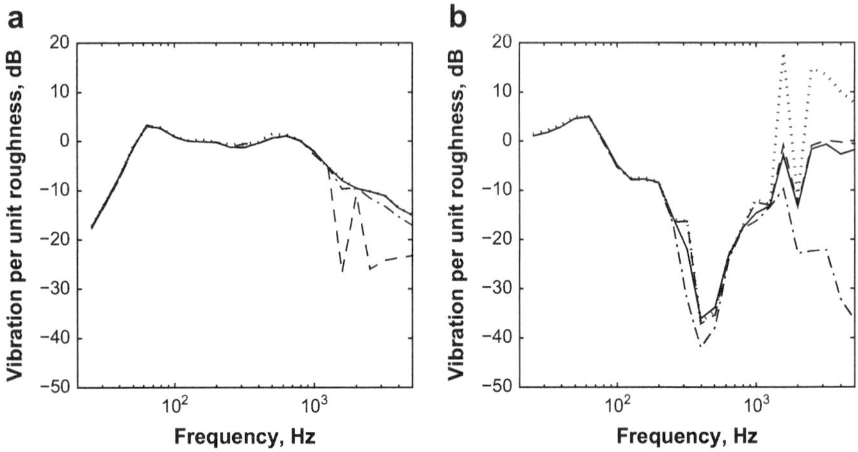

FIGURE 5-8 Predictions of (a) rail vibration and (b) wheel vibration, normalized to roughness input, from full narrow-band model (—) and from one-third octave model with different wheel mobilities. – – –, from average mobility; – · – ·, from average impedance; · · · ·, using interaction force converted into one-third octave bands

problems will occur unless the force spectrum is determined with a very fine frequency resolution around these resonances.

To illustrate this, some results are presented in Figure 5.8 in one-third octave bands. As a reference, the solid line gives the rail and wheel vibration calculated in narrow frequency steps, from Figure 5.4, and subsequently converted to one-third octave band spectra. Three other results are given which have been calculated directly using one-third octave band data. None of them agree with the reference curve for both wheel and rail.

The dotted line in Figure 5.8 has been produced by converting the interaction force from Figure 5.3 into one-third octave bands. Multiplying this force by the wheel or rail mobilities gives an estimate of the response. Clearly the rail response is estimated very well using this approach since its mobility changes only slowly with frequency. However, the wheel response is overestimated by up to 20 dB in the region where wheel resonances occur.

The dashed line is obtained using one-third octave band wheel and rail mobilities directly in equations (5.7) and (5.8). The wheel response is estimated quite closely by this approach, although there are differences of around 2 dB at high frequencies. However, the rail response is underestimated in the high frequency region by 10–20 dB.

Due to the large dynamic range of the wheel mobility, it is difficult to choose a suitable value that represents the mobility within a one-third octave band. While an average of the mobility modulus (as used in producing the dashed line) tends to emphasize the peaks of the mobility, an average of the impedance would tend to emphasize the peaks of impedance, i.e. the dips of the mobility. Using such a one-third octave band mobility, the results shown by the chain line in Figure 5.8 were obtained. This gives a reasonably good estimate of the rail response, underestimated by about 2 dB at high frequencies, but the wheel response is much too low.

Similar problems, although less severe, would occur at lower frequencies if, for example, the force calculated for one track was used to calculate the response of a track with a different support stiffness. In such a case the force peaks corresponding to the coupled wheel/track resonance and the track anti-resonance would no longer occur at the correct frequencies, leading to errors in the response calculation.

In conclusion, therefore, it is essential to consider the interaction force as an 'internal variable' in the calculation, and not to use the force calculated in one situation to determine the response in another.

5.3 CONTACT ZONE MOBILITIES

5.3.1 Hertz contact spring

The contact stiffness at the wheel/rail contact is caused by local elastic deformation which creates a contact area, the size of which increases as the normal load increases. Consequently the load–deflection relation is non-linear.

Hertz first defined the stiffness between contacting bodies described by their radii of curvature at the contact, the simplest case being the contact between a sphere and a plane [5.5]. If the wheel has radii of curvature R_w in the rolling direction and R_{wt} in the transverse direction (positive for convex), as shown in Figure 5.9, and the rail has transverse radius R_{rt} and radius R_r in the rolling direction (usually infinite) the contact will exist over an ellipse. This has semi-axes a in the rolling direction and b in the transverse direction given by

$$a = \sigma_1 \left(\frac{3F_0 R_0}{2E'} \right)^{1/3} \tag{5.38}$$

$$b = \sigma_2 \left(\frac{3F_0 R_0}{2E'} \right)^{1/3} \tag{5.39}$$

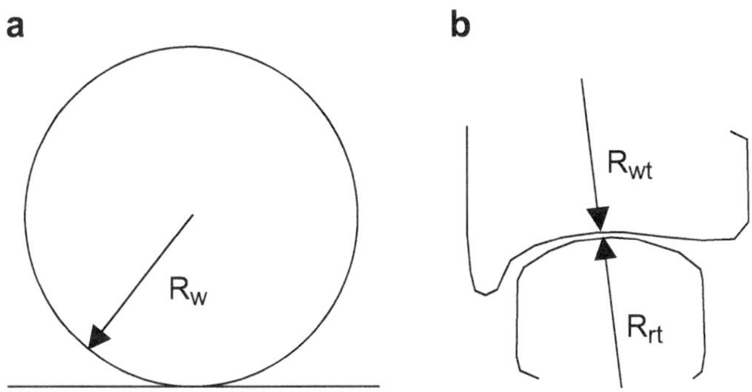

FIGURE 5-9 Wheel and rail radii of curvature at the contact, R_w, R_{rt} and R_{rt}. R_r, the rail radius corresponding to R_w, is infinite. (a) Side view, (b) cross-sectional view. R_{wt} is negative for a concave surface

where F_0 is the normal load, E' is the plane strain elastic modulus, $E' = E/(1-\nu^2)$, with both bodies assumed to have the same material properties (E is Young's modulus and ν is Poisson's ratio) and R_0 is an effective radius of curvature of the surfaces in contact given by

$$\frac{1}{R_0} = \frac{1}{2}\left(\frac{1}{R_w} + \frac{1}{R_{wt}} + \frac{1}{R_r} + \frac{1}{R_{rt}}\right) \tag{5.40}$$

The formulation used here is based on that given by, e.g., Timoshenko and Goodier [5.6] and Harris [5.7] although with altered notation. (Note that Johnson [5.5] uses a slightly different formulation where a composite elastic modulus E^* is defined which, for identical materials, is $E'/2$ and an equivalent radius R_e is introduced.) The parameters σ_i are given by [5.6, 5.7]

$$\sigma_1 = \left(\frac{2g^2\mathbf{E}(e)}{\pi}\right)^{1/3} \tag{5.41}$$

$$\sigma_2 = \left(\frac{2\mathbf{E}(e)}{\pi g}\right)^{1/3} \tag{5.42}$$

where $g = a/b$, $e = (1 - 1/g^2)$ and $\mathbf{E}$ is the complete elliptic integral of the second kind. Values of σ_i are listed in Table 5.1 against θ, which is defined by

$$\cos\theta = -\frac{R_0}{2}\left(\frac{1}{R_w} - \frac{1}{R_{wt}} + \frac{1}{R_r} - \frac{1}{R_{rt}}\right) \tag{5.43}$$

so that $\theta = 90°$ corresponds to a circular contact. This is related to g according to

$$\cos\theta = \frac{(g^2+1)\mathbf{E}(e) - 2\mathbf{K}(e)}{(g^2-1)\mathbf{E}(e)} \tag{5.44}$$

where $\mathbf{K}$ is the complete elliptic integral of the first kind. The approach of the two bodies, u_0, due to the load F_0 is similarly given by

$$u_0 = \frac{\xi}{2R_0}\left(\frac{3F_0R_0}{2E'}\right)^{2/3} \tag{5.45}$$

where ξ is given by

$$\frac{\xi}{2} = \frac{2\mathbf{K}(e)}{\pi}\left(\frac{\pi}{2g^2\mathbf{E}(e)}\right)^{1/3} \tag{5.46}$$

which is also listed in Table 5.1.

Although the relation between the approach distance and the load in equation (5.45) is non-linear, this expression can be linearized for small displacement amplitudes about a nominal approach distance u_0 or vertical load F_0. Thus the incremental contact stiffness is given by

TABLE 5.1 CONTACT PARAMETERS

g = a/b	cos θ	σ_1	σ_2	$\xi/2$
1.0	0	1	1	1
1.259	0.1711	1.1257	0.8942	0.9934
1.585	0.3329	1.2754	0.8047	0.9741
2.00	0.4781	1.4536	0.7285	0.9436
2.51	0.6022	1.6652	0.6629	0.9036
3.16	0.7036	1.9160	0.6059	0.8566
3.98	0.7836	2.2121	0.5557	0.8048
5.01	0.8446	2.5609	0.5110	0.7503
6.31	0.8900	2.9708	0.4708	0.6949
7.94	0.9231	3.4514	0.4345	0.6398
10.0	0.9467	4.0141	0.4014	0.5861
12.59	0.9634	4.6721	0.3711	0.5346
15.85	0.9750	5.4410	0.3433	0.4859
20.0	0.9831	6.3387	0.3177	0.4401
25.1	0.9886	7.3864	0.2941	0.3974
31.6	0.9923	8.6088	0.2722	0.3580
39.8	0.9949	10.0346	0.2521	0.3217
50.1	0.9966	11.6976	0.2334	0.2885
63.1	0.9977	13.6370	0.2161	0.2582
79.4	0.9985	15.8984	0.2001	0.2307
100.0	0.9990	18.5353	0.1854	0.2058

$$\frac{1}{K_H} = \frac{du_0}{dF_0} = \frac{2}{3F_0^{1/3}}\frac{\xi}{2R_0}\left(\frac{3F_0R_0}{2E'}\right)^{2/3} = \frac{\xi}{2}\left(\frac{2}{3E'^2F_0R_0}\right)^{1/3} \tag{5.47}$$

The effect of the non-linearity of the contact spring is discussed in Chapter 10 which deals with impact noise.

The contact between two static bodies has a compliance in the transverse direction, which Mindlin [5.8] showed to be approximately the same as for the vertical case. The transverse contact stiffness is thus given by

$$\frac{1}{K_L} = \frac{\chi}{K_H} = \chi\frac{\xi}{2}\left(\frac{2}{3E'^2F_0R_0}\right)^{1/3} \tag{5.48}$$

where χ varies between about 1 and 1.4 depending on the aspect ratio of the contact patch, $g = a/b$. This is given approximately by

$$\chi \approx 1 + \frac{\nu}{1-\nu}\left(\frac{1}{4} + \frac{1}{\pi}\tan^{-1}g\right)^{1/3} \tag{5.49}$$

where ν is Poisson's ratio. It should be noted that K_L differs for the longitudinal and lateral directions as the appropriate aspect ratio is g or $1/g$, respectively.

5.3.2 Rolling contact and creepage

During rolling a small relative motion occurs between the two rolling bodies. The relative velocity normalized by the rolling velocity is known as the creepage. This gives rise to a reaction force known as the creep force. Kalker [5.3] developed the theory of rolling contact in detail. Although the lateral creepage is of more relevance for rolling noise, it is more instructive to begin with the longitudinal creepage.

During rolling, the wheel does not move at exactly the velocity of the train. For example, if traction forces are applied, the wheel tends to rotate slightly faster than it would in an undriven state. A particle on the wheel surface therefore passes through the contact zone at a higher average speed than a corresponding particle on the rail. If v_{w1} is the velocity of the wheel at the contact and v_{r1} is that of the rail, the relative slip velocity is given by $v_{c1} = v_{w1} - v_{r1}$. This relative slip velocity divided by the mean rolling velocity V is defined as the longitudinal creepage, $\gamma_1 = v_{c1}/V$. According to the definitions in Figure 5.10, v_{c1} and hence γ_1 are negative when the wheel is under traction (as the wheel slips backwards). Note that the quasi-static values of v_{w1} and v_{r1} are negative. For steady rolling, the longitudinal force F_1 (which is positive forwards on the rail and backwards on the wheel) is

$$F_1 = Gc^2 C_{11} \gamma_1 \tag{5.50}$$

where $G = E/2(1+\nu)$ is the shear modulus, $c^2 = ab$, a and b are given by equations (5.38) and (5.39) and C_{11}, the creep coefficient, is a function of the aspect ratio,

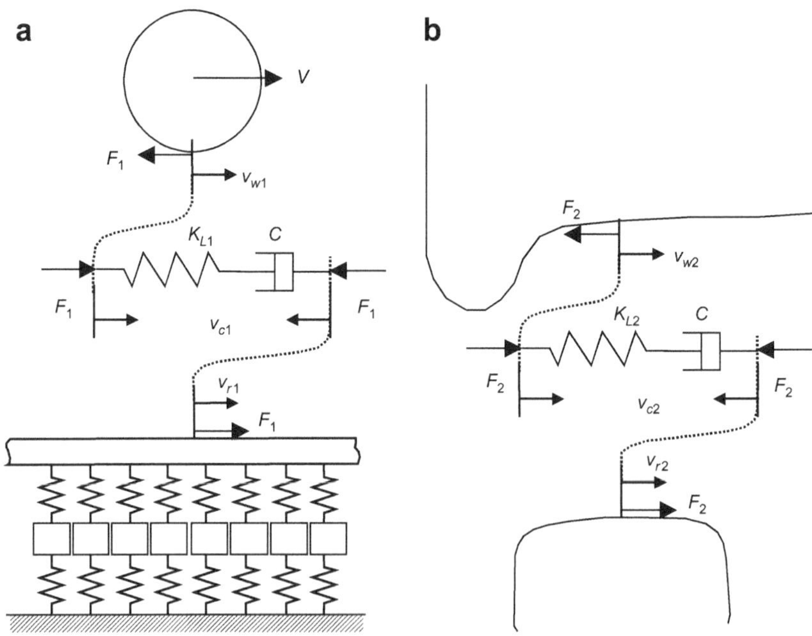

FIGURE 5-10 Sign conventions at wheel/rail interface. (a) Longitudinal direction, (b) lateral direction

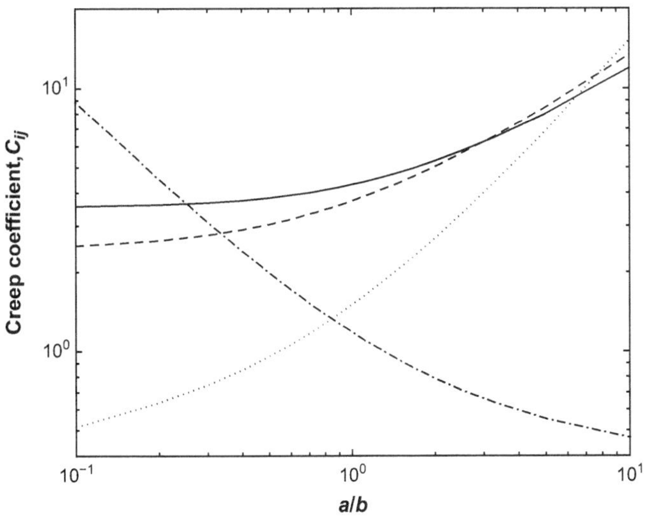

FIGURE 5-11 Creep coefficients. —, C_{11}; – – –, C_{22}; · · · ·, C_{23}; – · – ·, C_{33}

g, and the Poisson's ratio, ν. These values are tabulated by Kalker [5.3] and are plotted in Figure 5.11 for a Poisson's ratio of 0.3.

The lateral creep force is similarly related to the lateral creepage, γ_2, but with an additional term due to the spin creepage ω_3 (relative rotational velocity about the normal to the contact plane, divided by the mean rolling velocity):

$$F_2 = Gc^2 C_{22}\gamma_2 + Gc^3 C_{23}\omega_3 \tag{5.51}$$

where $\gamma_2 = v_{c2}/V$ and $\omega_3 = v_{c6}/V$. A spin moment F_6 may also be generated:

$$F_6 = -Gc^3 C_{23}\gamma_2 + Gc^4 C_{33}\omega_3 \tag{5.52}$$

The coefficients C_{pq} are again tabulated by Kalker [5.3] and are plotted in Figure 5.11 for a Poisson's ratio of 0.3.

For a harmonic lateral force of amplitude F_2 and relative velocity of amplitude v_{c2} (ignoring spin), the above can be expressed as a mobility coupling the wheel and rail:

$$Y_{c22} = \frac{v_{c2}}{F_2} = \frac{V}{Gc^2 C_{22}} \tag{5.53}$$

This mobility is real and independent of frequency so that it can be represented by a damper coupling the wheel and rail. The damping coefficient is simply $1/Y_{c22}$. Similar mobilities can also be derived for other coordinate directions [5.9]. Where a non-zero steady creepage is present this will introduce non-zero off-diagonal terms into the contact mobility matrix [5.9]

5.3.3 Saturation of creepage

While the creepage describes the overall relative motion of the wheel relative to the rail, in the contact zone itself there are simultaneously regions of adhesion, where there is no slip, and regions where there is micro-slip. This is shown schematically in Figure 5.12(a). On entering the contact zone from the front (the right), two particles on the wheel and rail lock together and move towards the rear. The gross relative motion of the two bodies causes elastic deformation to build up in the wheel and rail until the transverse stresses reach the local friction limit, μp, where p is the normal pressure, defined by Hertz as being elliptical, and μ is the local friction coefficient, as shown in Figure 5.12(b). Once this limit has been exceeded, micro-slip can occur between the two surfaces. Note that the creepage represents the overall relative motion between the two bodies, whereas the micro-slip is local to points within the contact zone.

The extent of the adhesion zone reduces as the creepage increases. In the extreme case where the slip zone covers the whole contact area, the transverse force is limited to μF_0. This is known as saturation. At small values of creepage, the creep force can

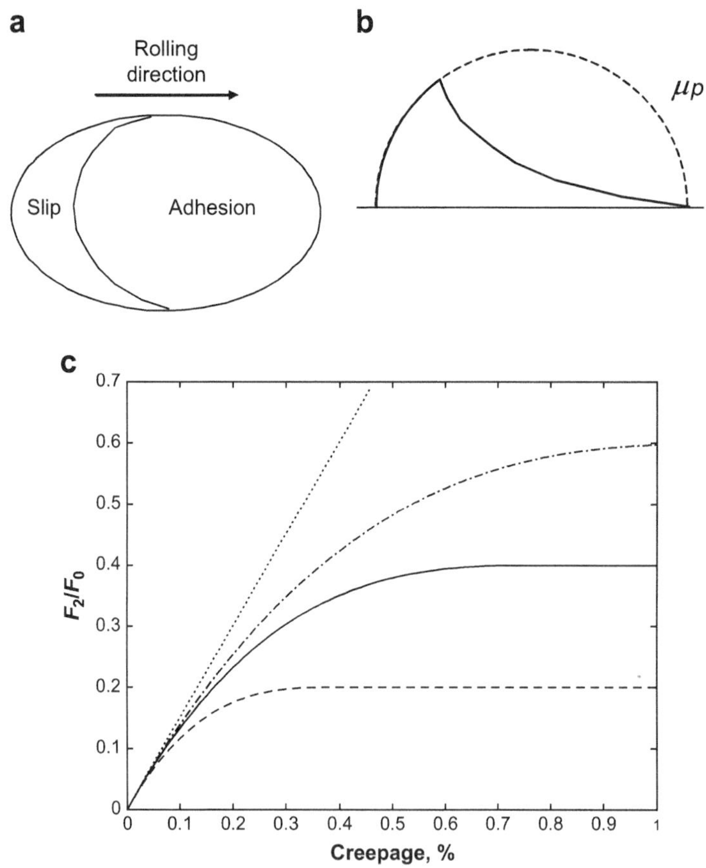

FIGURE 5-12 (a) Adhesion and slip regions in the contact zone. (b) Corresponding transverse stresses. (c) Typical creep force–creepage relationships calculated using equation (5.54) with $a = b = 0.005$ m, $C_{22} = 3.73$, $G = 81$ GN/m^2, $F_0 = 50$ kN. – – –, $\mu = 0.2$; —, $\mu = 0.4$; – · – ·, $\mu = 0.6$; · · · ·, linear theory

be described by Kalker's linear theory given in the previous section. At larger values of creepage a non-linear theory is required. Kalker [5.3] gives an exact theory which requires numerical solution, as well as an approximate solution known as FASTSIM.

A simple approximate model for the unsaturated creepage was given by Vermeulen and Johnson [5.10]:

$$F_i = \begin{cases} \mu F_0 \left(\Gamma_i - \frac{1}{3}\Gamma_i^2 + \frac{1}{27}\Gamma_i^3 \right) & \text{for} \quad \Gamma_i < 3 \\ \mu F_0 & \text{for} \quad \Gamma_i > 3 \end{cases} \tag{5.54}$$

where F_i is the longitudinal or lateral creep force and Γ_i is a *normalized creep* given by

$$\Gamma_i = \frac{G C_{ii} a b}{\mu F_0} \gamma_i \qquad \text{for} \quad i = 1, 2 \tag{5.55}$$

in which a and b are the semi-axis lengths of the contact patch, G is the shear modulus of the wheel and rail material (assumed identical) and C_{ii} is the creep coefficient described above. This model, while not exact, gives a good approximation to the behaviour in the region where linear creep theory no longer applies.

Some typical creep force/creepage curves are shown in Figure 5.12(c). These are calculated for lateral creep using equation (5.54) for a circular contact patch, $a = b = 5$ mm, and a normal load of 50 kN. Linear theory, equation (5.51), gives a result which is independent of μ at small creepage, but as the creepage increases the lateral force tends to μF_0.

Shen, Hedrick and Elkins [5.11] extended this approach to the case of combined longitudinal and lateral forces. The forces found from linear theory, F_{1l} and F_{2l}, are combined to give a resultant, $F_l = \sqrt{F_{1l}^2 + F_{2l}^2}$. Then a reduction factor

$$\delta = 1 - \frac{1}{3}\left(\frac{F_l}{\mu F_0}\right) + \frac{1}{27}\left(\frac{F_l}{\mu F_0}\right)^2 \quad \text{for} \quad F_l \leq 3\,\mu F_0 \tag{5.56}$$

is determined and is applied to give reduced forces $F_1 = \delta \cdot F_{1l}$ and $F_2 = \delta \cdot F_{2l}$. This approach allows the inclusion of spin through equation (5.51) but is only valid for small values of spin since the effect of the spin moment on the saturation is ignored.

At higher values of creepage the creep force may reduce in amplitude, due to the dependence of the friction coefficient on sliding velocity (it is often considered that 'dynamic' friction is lower than 'static' friction). It is this falling amplitude at high creepage that is believed to be the main reason for the unstable dynamic behaviour leading to squeal noise, see Chapter 9.

5.3.4 High frequency creepage relations

At high frequencies the above expressions for creepage and contact stiffness are no longer valid.

The vertical contact stiffness is virtually unaffected by the rolling process as any inertial effects will be small. Nevertheless, Sheng *et al.* [5.12] found that, by modelling two half-spaces in contact for a harmonic load, the contact spring should be replaced by a complex stiffness. This has a small frequency-dependent loss factor given approximately by $\eta_H \approx 3.3 \times 10^{-6} f$ where f is the frequency. This accounts for the radiation of power into the half-spaces.

For the lateral contact mobility some means must be found to combine the contact stiffness, which applies at high frequency, with the creep forces, which apply at low frequency. Knothe and Gross-Thebing [5.13] developed frequency-dependent creep coefficients C^*_{pq} by studying non-steady rolling contact. In fact, for C_{11} and C_{22}, their results can be reduced to a form that is equivalent to adding the transverse contact stiffness in series with a damper that represents the creep force term [5.9]

$$Y_{c22} = \frac{v_{c2}}{F_2} = \frac{V}{Gc^2 C^*_{22}} = \frac{V}{Gc^2 C_{22}} + \frac{i\omega}{K_{L2}} \tag{5.57}$$

where K_{L2} is the transverse stiffness applying in the 2-direction (see equation (5.48)). Hence, C^*_{22} can be expressed as

$$\frac{1}{C^*_{22}} = \frac{1}{C_{22}} + \frac{i\omega Gc^2}{VK_{L2}} = \frac{1}{C_{22}} + \frac{i2\pi Gc^2}{\lambda K_{L2}} \tag{5.58}$$

where λ is the wavelength (V/f). Clearly, at low frequency (or long wavelength) this is dominated by C_{11} whereas at high frequency the second term (the transverse contact stiffness) becomes important. A similar expression can be found for C^*_{11} (K_{L2} is replaced by K_{L1}). For C^*_{23} and C^*_{33}, however, the expressions are more complex [5.9, 5.13].

5.4 CONTACT FILTER EFFECT

In the interaction model described above, the surface roughness is assumed to excite the wheel/rail system at the contact point. It is clear that this roughness may be present on the wheel or the rail with the same effect. In each case it introduces a relative displacement between the wheel and rail. Where significant roughness is present on both surfaces, it can safely be assumed that these are incoherent: the wheel roughness is essentially periodic with the wheel circumference, whereas the rail roughness is traversed by many wheels of different circumference and is essentially random in nature. To determine the combined response due to wheel and rail roughness, therefore, their spectra can be added in terms of mean-square values.

As seen above, the contact zone between wheel and rail exists over an area. According to Hertz theory [5.5] this is an ellipse with semi-axes a and b, although in practice the geometry can be more complex. Typical dimensions ($2a$, $2b$) are 10–15 mm. Roughness with wavelengths that are short compared with this length in the rolling direction, $2a$, tends to be attenuated in its excitation of the wheel/rail system. This effect is known as the 'contact filter'. Moreover, the contact exists over a finite width and where the roughness profile differs across the width of the contact, its effect will tend to be averaged out.

Remington [5.14] developed an analytical model for the contact filter effect. For a circular contact patch of radius a, the filter transfer function is given by

$$|H(k)|^2 = \frac{4}{\alpha} \frac{1}{(ka)^2} \int_0^{\tan^{-1}\alpha} J_1^2(ka \sec\psi) d\psi \tag{5.59}$$

where $k = 2\pi/\lambda$ is the roughness wavenumber in the longitudinal direction, λ being the wavelength, and α is a parameter that describes the degree of correlation between roughness across the width of the contact zone at a given wavenumber. Large values of α imply poor correlation. It is unclear what value of α should be used, so a range of values was considered in [5.14].

The above function can also be approximated by a simple filter function:

$$|H(k)|^2 = \left(1 + \frac{\pi}{4}(ka)^3\right)^{-1} \quad (5.60)$$

These are compared in Figure 5.13. The filter from equation (5.59) rolls off at around $ka = 1$ and has a minimum at around $ka = \pi$ (i.e. $\lambda = 2a$) and integer multiples of this, irrespective of the value of α. Cancellation can be expected at this value of ka as the wavelength corresponds to the contact patch length. As α is increased the filter effect becomes greater although the dips become less pronounced. The simple filter function of equation (5.60) provides a reasonable fit to the curve for $\alpha = 2$ for values of ka up to about 2 and gives approximately the same slope at higher values of ka, but it does not contain the dips associated with the contact patch length.

A numerical method was developed in [5.15] in which the contact zone between a rigid wheel and rail is approximated by a series of 'distributed point-reacting springs' (DPRS). The stiffness behaviour of these springs must be chosen to be non-linear, with the force proportional to the square root of the local deflection, in order to give the correct overall relationship between force and deflection. Moreover, the radii of curvature of the wheel and rail have to be modified in order to ensure the correct contact area is maintained.

The original DPRS model in [5.15] produced a 'blocked force' induced by the roughness passed between the rigid wheel and rail. To avoid excessive forces or loss of

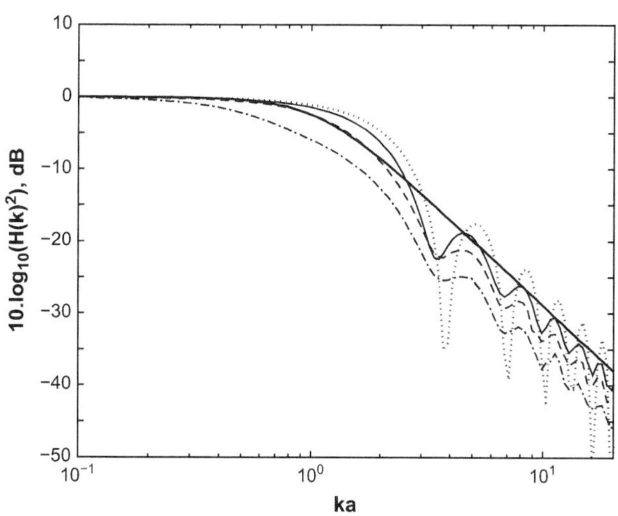

FIGURE 5-13 Contact filter effect from analytical models. —, simplified filter, equation (5.60); from equation (5.59): ····, $\alpha = 0.1$; —, $\alpha = 1$; – – –, $\alpha = 2$; – · – ·, $\alpha = 5$

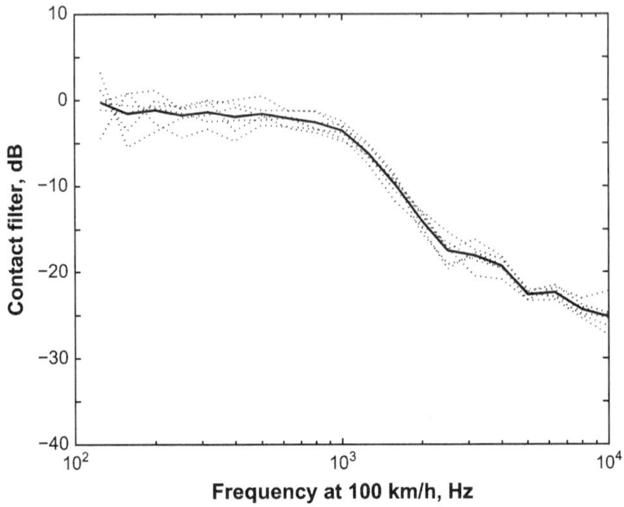

FIGURE 5-14 Contact filter from DPRS model for 460 mm wheel radius, 50 kN load. ····, results from roughness of six wheels; —, average

contact due to large amplitudes at long wavelengths, a high pass filter was built into the model by representing the wheel as a mass supported on a damper. This model was extended in [5.16] to give an equivalent roughness. In this form the results were shown to be relatively independent of the choice of parameters, even below the cut-off frequency of the high pass filter.

The filter effect from this model is presented in Figure 5.14. These results were determined using six sets of particularly detailed wheel roughness measurements, from [5.17]. These were obtained on 25 parallel lines at 2 mm intervals across the running surface, and sampled at 0.5 mm intervals around the perimeter. The average of the roughness spectra from the four or five measurement lines passing through the contact zone have been subtracted from the results of the DPRS model to determine the filter effect. This is then averaged over the six sets of data. It is plotted against frequency for a particular train speed (100 km/h). Up to about 800 Hz (35 mm wavelength) the filter effect is negligible (average values of −1 dB are not significantly different from 0, given the small number of measurements and the variation between them). At higher frequencies, the contact filter reduces the effective roughness due to averaging over the length of the contact patch.

The filter effect depends on the wheel radius and the normal load used in the calculation. Figure 5.15(a) presents results from the DPRS model in one-third octave bands assuming different wheel radii. Again these are obtained from the average result from the six sets of wheel roughness. In each case the rail transverse radius has been assumed to be 0.3 m and the wheel assumed to have a flat (coned) profile. Changing the wheel radius alters the length of the contact patch and hence the wavelength at which the filter rolls off.

The smaller wheel radii have a shorter contact patch ($a = 3.6$ mm for 180 mm radius instead of 5.7 mm for 460 mm radius), by a factor of 1.6. It would therefore be expected that the contact filter would take effect at frequencies about two one-third

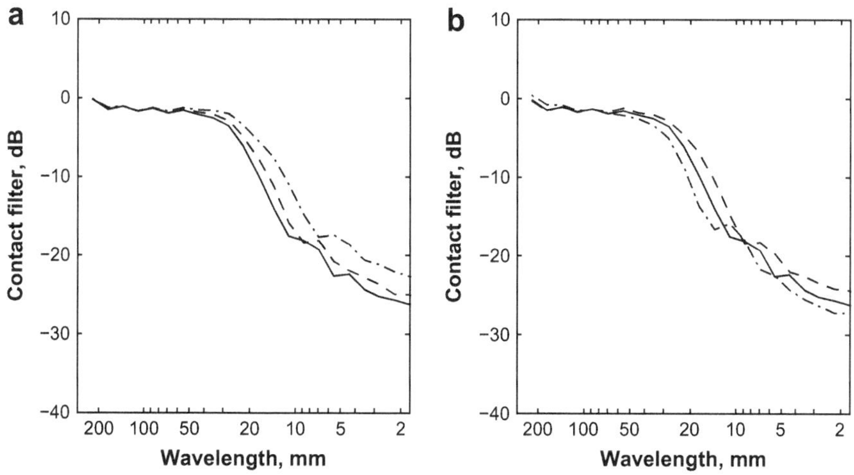

FIGURE 5-15 Contact filter effect from DPRS model. (a) Due to various wheel radii, 50 kN load: —, 460 mm radius wheel ($a = 5.7$ mm); – – –, 327 mm radius wheel ($a = 4.8$ mm); ····, 180 mm radius wheel ($a = 3.6$ mm). (b) Due to various wheel loads, 460 mm radius: —, 50 kN ($a = 5.7$ mm); – – –, 25 kN ($a = 4.5$ mm); –·–·, 100 kN ($a = 7.2$ mm) [5.17]

octaves higher for the smaller wheel. This can be seen to be approximately the case. Dips are found in each case at wavelengths of around $2a$ and at a.

The effect of the wheel load is illustrated in Figure 5.15(b) for a constant radius of 460 mm. Again, the contact patch length determines the wavelength at which the filter rolls off (see box below).

Figure 5.16 shows these results plotted against the non-dimensional parameter ka, where a is the semi-axis length in the rolling direction. The various calculated results

Effect of normal load

If the wheel load increases, does the noise increase? Intuition suggests it will. However, rolling noise actually reduces as the wheel load increases.

If the normal load, F_0, applied to a wheel is increased, there are two effects:

- the contact stiffness is increased (equation (5.47)),
- the contact patch length $2a$ is increased (equation (5.38)) and hence the contact filter rolls off at a lower frequency.

The first effect tends to increase the noise slightly whereas the second effect reduces it. The overall effect is a reduction in noise. This has been quantified as a reduction of approximately 2 dB in A-weighted noise level when the wheel load is increased from 23.5 to 64 kN [5.41]. Similar results have also been found experimentally.

Other effects, however, will increase as the wheel load increases. These include the effects of the moment excitation (Section 5.7.1), moving load (Section 5.7.4) and parametric excitation (Section 5.7.5) which all increase in proportion to the wheel load. In addition, the noise due to impacts will increase considerably, see Chapter 10.

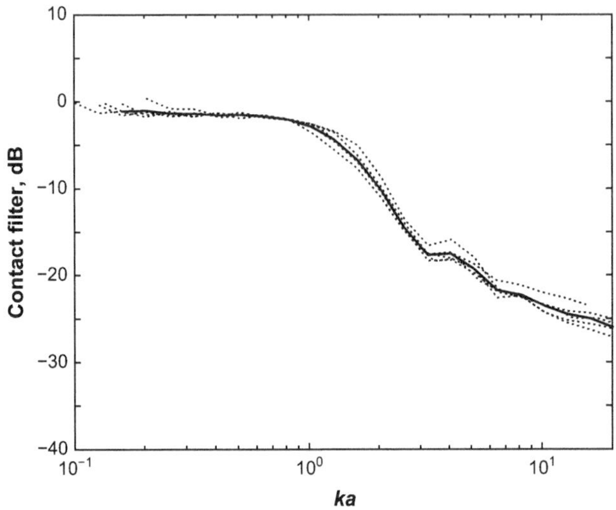

FIGURE 5-16 Contact filter from DPRS model for various wheel radii and normal loads each averaged over six wheels. ····, results from Figure 5.15; —, average

collapse to a single curve. Their average is shown by the thick line. These numerically determined contact filter values are listed in Table 5.2. In practice, rather than use the DPRS model, which requires detailed roughness measurements, these values of contact filter can be used directly to modify measured roughness spectra obtained from a single measurement line.

These average results from the DPRS model are compared with results from the analytical filter of equation (5.59) and the simplified model of equation (5.60) in Figure 5.17. In each case they are plotted against *ka*. The analytical model gives similar results to the numerical DPRS method when a value of $\alpha = 2$ is used but the roll-off is lower than that of the analytical filter above about $ka = 3.2$ (corresponding to 2.5 kHz for the parameters used in Figure 5.14). This is believed to be due to the fact that the DPRS model takes account of the variation of the normal load across the contact zone. Similarly, the simple filter of equation (5.60) can be used up to about $ka = 6.5$ (5 kHz for the parameters used in Figure 5.14) although there are differences of up to 5 dB from the DPRS result at around $ka = 3$. The results from equation (5.60) are included in Table 5.2 at low values of *ka* where the DPRS results are believed to be less reliable; it is recommended to use these values in place of the DPRS results.

In [5.18] a two-dimensional version of the DPRS model was developed, referred to as a mattress model. This allows the contact filter to be implemented for a single line of roughness measurement. It can also be implemented directly in a time-stepping model. However, the high frequency roll-off is slightly less than that predicted using the three-dimensional model.

5.5 MEASUREMENT OF ROUGHNESS

The wavelength range of roughness required in order to predict noise can be established from the relation $f = V/\lambda$, equation (5.1). The frequency range of

TABLE 5.2 **CONTACT FILTER EFFECT DERIVED FROM DPRS MODEL**

Frequency (Hz) (for $a = 5.69$ mm, $V = 100$ km/h)	λ, mm (for $a = 5.69$ mm)	λ/a	ka	Filter effect, dB (values in brackets from equation (5.60))
125	220	38.8	0.162	−1.1 (0)
160	175	30.8	0.204	−1.0 (0)
200	140	24.5	0.257	−1.4 (0)
250	110	19.4	0.323	−1.4 (−0.1)
315	88	15.4	0.407	−1.6 (−0.2)
400	70	12.3	0.512	−1.5 (−0.4)
500	55	9.74	0.645	−1.8 (−0.8)
630	44	7.74	0.812	−2.1 (−1.4)
800	35	6.15	1.02	−2.9 (−2.5)
1000	28	4.89	1.29	−4.4
1250	22	3.88	1.62	−6.7
1600	17.5	3.08	2.04	−10.2
2000	14	2.45	2.57	−14.6
2500	11	1.94	3.23	−17.7
3150	8.8	1.54	4.07	−17.6
4000	7.0	1.23	5.12	−19.3
5000	5.5	0.974	6.45	−21.8
6300	4.4	0.774	8.12	−22.3
8000	3.5	0.615	10.2	−23.5
10 000	2.8	0.488	12.9	−24.5
12 500	2.2	0.388	16.2	−24.5
16 000	1.8	0.308	20.4	−26.1

relevance in rolling noise depends on the train speed but is usually considered to extend from around 100 Hz to 5000 Hz. At higher frequencies the contact filter ensures that the noise contribution is reduced. At low frequencies aerodynamic noise is expected to be dominant. Moreover, low frequencies are significantly attenuated by the A-weighting filter and are thus usually given less importance.

Table 5.3 summarizes the wavelengths or relevance for different train speeds. This implies that, to cover the range 100 to 5000 Hz for this range of train speeds, wavelengths between 2 and 900 mm are required. However, in practice the lower frequencies are of less relevance, particularly for the higher train speeds, while a noise spectrum at 40 km/h has less content above 2 kHz. Therefore a realistic range is around 5 to 500 mm. In practice the measurement range is often limited further by the measuring instrument.

There are two main types of measuring instrument used for rail roughness. The first type comprises a beam or straight-edge, mounted above the rail, examples of which are shown in Figure 5.18. A displacement transducer is moved along the beam and measures the height of the rail surface relative to the beam. Corrections can be made for undulations in the beam straightness by measuring on a calibrated flat surface [5.19]. The main limitation of such equipment is its length. Several systems

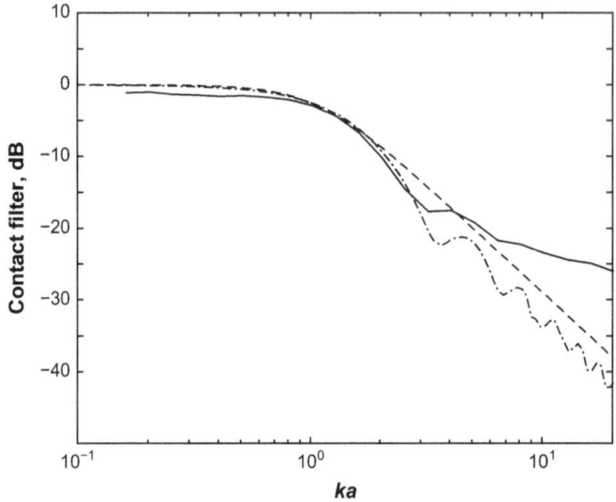

FIGURE 5-17 Contact filter effect: —, from DPRS model averaged over various wheel radii and loads; – – –, simplified filter, equation (5.60); – · – ·, from analytical filter, equation (5.59) for $\alpha = 2$

have been produced with a length of 1.2 m, and this is a minimum length in the proposed international standards. Greater lengths are impractical. Unless multiple measurements are concatenated, which is a difficult and rather unreliable process, spectral analysis is limited to a base wavelength of 1.2 m. In order to obtain one-third octave band resolution, the maximum wavelength band is usually considered to be 100 mm (see Section 5.6.1). While this is satisfactory for train speeds of 40 km/h, at 300 km/h the minimum frequency covered would be 800 Hz (see Table 5.3).

The alternative type of equipment consists of a portable trolley that runs along the railhead, e.g. [5.20], examples of which are shown in Figure 5.19. By using an accelerometer to measure the profile, an inertial reference is used and the beam can be dispensed with. This allows long sections of rail to be measured relatively rapidly. Distance along the rail is measured using a tachometer pulse and sampling can be synchronized with this. A steady speed is required to allow conversion of acceleration to displacement, but this is achievable using a hand-operated trolley which

TABLE 5.3 EXAMPLES OF ROUGHNESS WAVELENGTHS (IN MM) FOR VARIOUS FREQUENCIES AND TRAIN SPEEDS

	50 Hz	100 Hz	250 Hz	500 Hz	1000 Hz	2500 Hz	5000 Hz
40 km/h	230	110	45	23	11	4.5	2.3
80 km/h	450	230	90	45	22	9.0	4.5
100 km/h	570	290	110	57	29	11	5.7
160 km/h	900	450	180	90	45	18	8.9
320 km/h	1800	900	360	180	90	36	18

FIGURE 5-18 Examples of rail roughness measurement equipment based on a 1.2 m beam. (a) RM1200E developed by Muller-BBM; (b) ØDS system

FIGURE 5-19 Examples of rail roughness measurement equipment based on a trolley. (a) Corrugation Analysis Trolley (CAT); (b) DeltaRail trolley

eliminates vibration from the motor present in earlier versions and the need to carry large batteries.

Wheel roughness can be measured more simply using a displacement transducer such as a linear variable differential transformer (LVDT) mounted on a bracket attached to the rail. The main requirement is to lift the wheel clear of the rail to allow it to turn freely. The wheel is turned either using an electric motor or by hand, in which case a tachometer pulse is required for distance synchronization. Powered wheels can also be measured in principle but are more difficult to turn.

Each surface should ideally be measured on a number of parallel lines to cover the whole width of the likely contact patch. In practice, for monitoring purposes often only a single line in the centre of the visible running band is measured, or possibly three lines spaced 5 or 10 mm apart. Separate equipment exists to measure wheel and rail transverse profiles and should ideally be used to help determine the likely running line for particular wheel and rail combinations.

In 1986 ORE C163 conducted a comparison of seven rail roughness measurement systems [5.21] known unofficially as the rail roughness world championships. Each was used to measure six rail sections on the Derby to Burton-on-Trent railway line in the UK that varied from very smooth to badly corrugated. None of these measurement systems is currently used in the same form so the detailed results are of limited value. Nevertheless, it is worth noting that all systems identified corrugation amplitudes and wavelengths reliably but at low levels of roughness there was considerable spread in the data. In the absence of an absolute datum it is difficult to use such comparisons to draw firm conclusions. However, noise measurements were also carried out on (smooth) disc-braked wheels which revealed a range of over 20 dB and were consistent with at least some of the roughness systems. The noise spectra were shown in Figure 2.7 along with results from one of the roughness systems, a trolley-based system developed by Cambridge University for British Rail which is a forerunner of the two systems shown in Figure 5.19.

More recently, a comparison was made of a number of systems as part of a 'road test' of a new standard for the measurement of rail roughness [5.22]. The main purpose of this exercise was to compare how different measurement teams interpreted the draft standard in a practical situation. It also allowed a check to be made that the standard leads to a consistent estimate of roughness spectrum when used by different measurers with different instruments. It was not concerned with testing instruments or measurement technology as such.

Comparisons were carried out at two test sites. Seven measurement instruments of five designs were included. These consisted of both LVDT-based 1.2 m straight edge instruments and accelerometer-based 'trolleys'. When used to measure the same prescribed line on the rail head the agreement between all instruments within each one-third octave band was about ±2 dB across the whole spectrum from 250 to 3.15 mm wavelength bands.

In order to test the noise of a vehicle in a meaningful way, some control is required over the rail roughness (and the track structure) so that ideally the noise measurements are independent of the test site. Various proposals have been developed, including ISO3095:2005 [5.23] and the EU Technical Specifications for Interoperability (TSI), [5.24, 5.25]. Each of these contains a requirement for the rail roughness to be measured at the test site and for it to be below a certain limiting spectrum. It should be emphasized that these limits only apply to such vehicle 'type' tests. There is no requirement for a railway infrastructure owner to maintain their

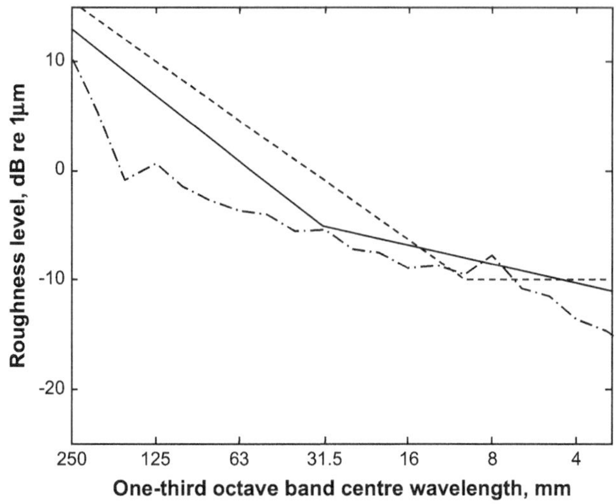

FIGURE 5-20 Limits for rail roughness applying during vehicle type testing. — in TSI [5.24, 5.25]; - - -, in ISO 3095:2005 [5.23]; – · – ·, measured 'average European' rail roughness spectrum from [5.26]

track to any particular roughness and in that sense these are not limits. Nevertheless they have come to be seen as a measure of good practice.

The various limit curves are presented in Figure 5.20. They have each been developed in order to minimize the effect of rail roughness upon a noise measurement, while representing an achievable spectrum. Details of compliance criteria are complex; the reader is referred to the appropriate procedures [5.23–5.25]. In addition, Figure 5.20 shows a typical rail roughness spectrum identified in [5.26] as corresponding to the 'average' European roughness.

Various methods are used to monitor roughness indirectly, using, for example, axlebox acceleration [5.27] or noise under a measurement coach [5.28]. In addition, an indirect method has been developed to determine the combined (wheel plus rail) filtered roughness spectrum from rail vibration measurements [5.29].

5.6 PROCESSING OF ROUGHNESS DATA

5.6.1 Frequency analysis

Roughness data are most usefully presented in terms of spectra. Some examples of one-third octave band spectra have already been shown. Due to the random nature of the roughness signal the use of one-third octave band analysis is preferred over narrow-band analysis due to its smoothing effect. Another advantage is that the one-third octave band levels remain unaltered when the train speed is changed (only the frequency axis is shifted), whereas the magnitude of a power spectral density (in squared amplitude per Hz) is affected by changes in train speed.

One-third octave spectra can be estimated from narrow-band spectra by summing the energy of the narrow-band spectral lines contained within the one-third octave band. Normally, as a 'rule of thumb', at least three such lines should be contained within the band. The statistical reliability of a spectral estimate can be determined in terms of the product, BT, of the frequency bandwidth, B, and the analysis time, T [5.30]. For an analysis length of L, or time $T = L/V$, the bandwidth of the narrow-band spectrum is $B = 1/T$ giving a BT product of 1. Thus, normally, narrow-band spectra are determined using an average over a number of samples. For a one-third octave band containing N narrow-band lines, the bandwidth is approximately N/T giving a BT product of N.

For a one-third octave band with centre frequency f, the bandwidth is $0.23f$. To ensure three narrow-band spectral lines within the band, $3/T \leq 0.23f$ or $f \geq 13/T$. For an analysis length L of 1.2 m this is (almost) satisfied at a wavelength of 100 mm (1/12 of the analysis length). If this is applied as a limit to the wavelength it restricts the frequency range covered by a rail roughness measurement using a 1.2 m long device, especially for high train speeds, see Table 5.3. Nevertheless, it has been noted that omission of these longer wavelength bands has a relatively small effect on the overall A-weighted sound level [5.31].

The BT product can be increased by averaging over a number of measurements at a test site. This may allow the restriction of three lines in a one-third octave band to be relaxed, enabling the frequency range to be extended downwards somewhat.

The choice of centre wavelengths for one-third octave band analysis is dictated by the intended use of the roughness data. One-third octave roughness data is normally used to produce corresponding noise predictions and should therefore ideally be provided at wavelengths corresponding to the standard frequencies. This means that the wavelengths required depend on the train speed. Unfortunately, as found in Table 5.3, for many common train speeds the wavelengths obtained are not the familiar one-third octave band values but actually correspond to values virtually mid-way between standard values. In the author's opinion, therefore, these values would be preferable. However, proposals for standardization of roughness analysis (e.g. [5.23]) have tended to concentrate on the use of 'standard' values (10, 12.5, 16, 20, etc.) as wavelengths in mm (or cm) which means that the resulting spectra are of direct use only at the less usual speeds of 55, 70, 88, 110, 140 km/h, etc. For intermediate speeds the one-third octave band spectra have to be interpolated.

In processing roughness data, account must be taken of several features that are peculiar to roughness signals. First, discrete features known as 'pits and spikes' may be present in the data. These can have a large influence on the spectra but would in most cases have minimal influence on the wheel/rail system; they must therefore be removed. Second, the shape of the spectrum, with much greater amplitudes associated with low frequencies than those at high frequencies, introduces particular problems. These two aspects are considered in the next subsections.

5.6.2 Pits and spikes

To see the effect of pits in the rail surface consider the signal shown in Figure 5.21(a). This comprises a broad-band random roughness component and a large discrete feature. The frequency spectrum of such a signal can be decomposed by superposition into the spectra of the two components, as shown in Figure 5.21(b). The roughness has a falling spectrum with increasing frequency whereas the discrete feature, similar to an impulse, has a flat spectrum up to a frequency at which the width of the feature is roughly equal to one wavelength. Thus a small hole (a 'pit') in

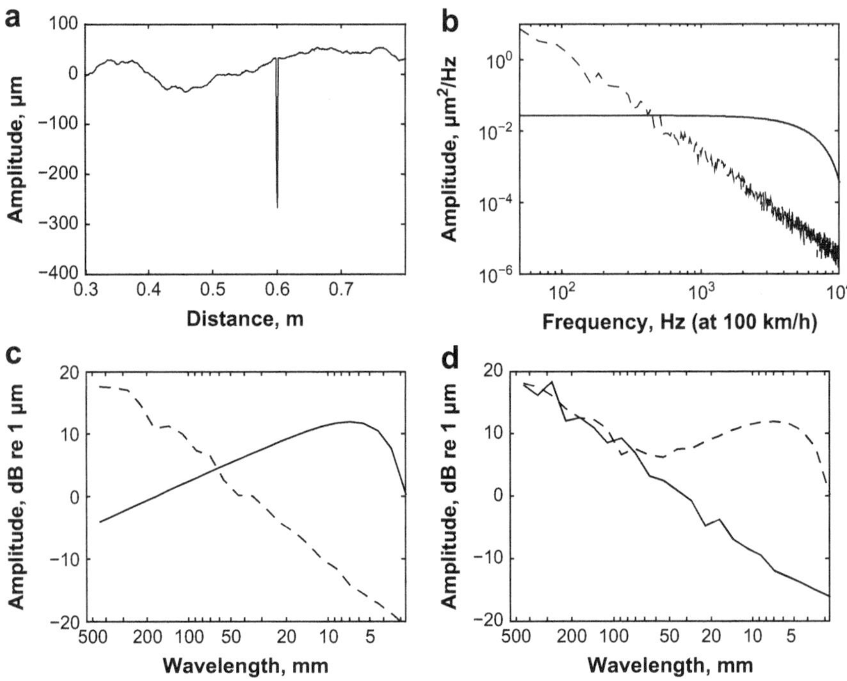

FIGURE 5-21 The effects of a 'pit' on the rail roughness spectrum. (a) Spatial domain, (b) narrow-band spectra, (c) one-third octave band spectra, – – –, random roughness; —, pit. (d) Effect of using 'curvature analysis' on the rail roughness containing a pit: – – –, spectrum of raw data; —, spectrum after applying curvature analysis

the rail surface with a width of 3 mm, as shown, here will have a flat spectrum for wavelengths down to about 3 mm (frequencies up to 10 kHz at 100 km/h). The amplitude of this spectral component depends on the depth of the hole, and therefore also on the radius of curvature of the sensor used to measure it. The one-third octave band spectra in Figure 5.21(c) show similar trends but their slopes are modified by the fact that the bandwidth of a one-third octave band increases in proportion to the centre frequency. For convenience these are plotted against wavelength.

When a wheel runs along this rail, its finite curvature means that it cannot reach the bottom of this discrete feature; it will 'jump' from one side to the other, possibly dropping slightly if the hole is wide enough. The wheel will therefore not 'see' most of this part of the roughness. Such pits are quite common on rail and wheel surfaces (see e.g. Figure 2.4) and they should be excluded from the data if realistic predictions of noise are to be obtained from this roughness measurement. It is not possible to exclude them using a frequency domain filter – instead a technique in the time (or spatial) domain is required.

If the feature is a bump on the rail surface (a 'spike'), the wheel will 'see' it and have to rise over it. Nevertheless, such spikes are often present in roughness data because of dirt on the running surface or electrical noise and should therefore be removed for a correct roughness signal. Care is required in interpreting such features.

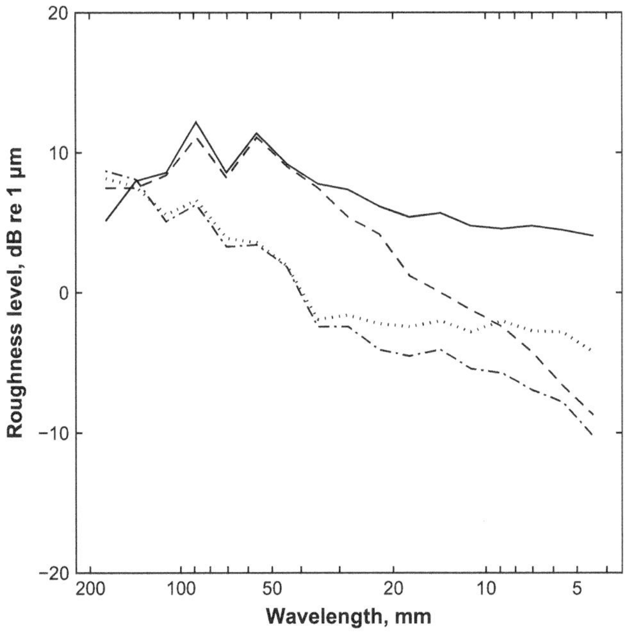

FIGURE 5-22 Effect of removing pits on wheel roughness spectra of two wheels. Wheel with cast-iron block brakes before removal of pits (—) and after (– – –); unbraked wheel before removal of pits (····) and after (– · – ·) from [5.32, 5.33]

The need for such an analysis was first identified in [5.32]. A number of schemes have been devised for dealing with 'pits and spikes'. In [5.32] a simple 'curvature analysis' was performed in which the roughness was resampled using a fictitious curved probe with the same radius as the wheel. This is lowered onto the profile at each longitudinal position until it first touches. The height by which it is lowered is the corresponding result for this location, even if the point of contact is not directly below the probe centre. Example results are shown in Figure 5.21(d) for the simulated data and for the two measured wheel roughness samples of Figure 2.4 in Figure 5.22 (from [5.33]).

The DPRS model described in Section 5.4 also deals directly with pits and spikes although it requires data from multiple parallel measurements across the running surface and so is not often applicable.

5.6.3 Long wavelengths and trend removal

It has been seen that typical roughness spectra contain large amplitudes at long wavelengths and small amplitudes at short wavelengths. Since a practical measurement always involves finite length samples, the contribution of wavelengths longer than the sample length must be eliminated. These components have two effects on results: (i) long wavelengths tend to spill over into shorter wavelength components due to truncation of the sample length, and (ii) linear trends lead to a discontinuity between the end and the start of the sample. Both features are also present in regular time series data, but are much more severe in roughness data due to the shape of the spectrum.

Consider as a first example a wavelength of twice the length of the sample:

$$r(x) = R_0 \sin\left(\frac{\pi x}{L}\right) \tag{5.61}$$

where L is the sample length and $0 \leq x \leq L$. Fourier analysis decomposes this into multiples of the sample length giving

$$r(x) = A_0 + \sum_{n=1}^{\infty} A_n \cos\left(\frac{2n\pi x}{L}\right) + \sum_{n=1}^{\infty} B_n \sin\left(\frac{2n\pi x}{L}\right) \tag{5.62}$$

where the coefficients are given by

$$A_0 = \frac{1}{L}\int_0^L r(x)dx; \quad A_n = \frac{2}{L}\int_0^L r(x)\cos\left(\frac{2n\pi x}{L}\right)dx;$$
$$B_n = \frac{2}{L}\int_0^L r(x)\sin\left(\frac{2n\pi x}{L}\right)dx \tag{5.63}$$

Evaluating these integrals gives

$$A_0 = \frac{2R_0}{\pi}; \quad A_n = \frac{-4R_0}{(4n^2 - 1)\pi} \quad \text{for } n > 0; \quad B_n = 0 \tag{5.64}$$

Thus a single long wavelength is decomposed into many shorter wavelengths, with amplitudes that decrease roughly in proportion to $1/n^2$. Converted into one-third octave bands this spectrum decreases at −30 dB/decade as shown in Figure 5.23. The frequency scale is chosen to correspond to a train speed of 100 km/h. For a half-sine of amplitude 1 mm over a length 1.2 m, the one-third octave band level at 100 Hz is 22 dB re 1 μm and at 500 Hz is 1 dB re 1 μm. These levels are comparable with actual roughness signals, indicating the importance of avoiding this effect.

Second, if a linear trend is present, since the discrete Fourier transform assumes implicitly that the signal is periodic beyond the sample window, a step exists at the end of the window. Thus if

$$r(x) = R_0\left(\frac{x}{L} - \frac{1}{2}\right) \tag{5.65}$$

for $0 \leq x \leq L$, evaluating the integrals as in equation (5.63), gives

$$A_n = 0; \; B_n = -\frac{R_0}{n\pi} \tag{5.66}$$

The result is also shown in Figure 5.23. From this it can be seen that the frequency dependence is $1/n$, or in one-third octave bands −10 dB/decade. A discontinuity of 1 mm will give a level of 27 dB re 1 μm at 500 Hz, which is larger even than the level due to typical corrugation amplitudes.

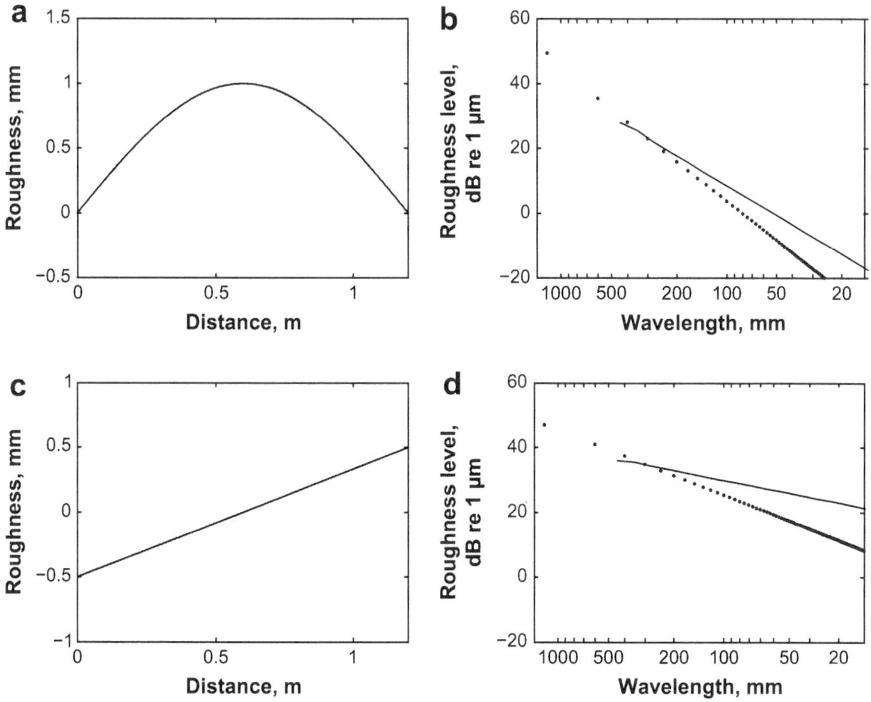

FIGURE 5-23 (a) A half-sine of amplitude 1 mm; (b) corresponding roughness spectrum; (c) a ramp of amplitude 1 mm; (d) corresponding roughness spectrum. •••••, harmonic components (equations (5.64, 5.66)); —, one-third octave band spectrum

Clearly, these effects must be removed from the data. The usual method in signal analysis is to apply a window function to the data [5.30]. Figure 5.24 shows the effects of (i) a Hanning (1 − cosine) window and (ii) a Tukey window consisting of an 80% flat section and a cosine window at the two ends. The latter is much less successful in removing the effect of the trends than the former, but even the Hanning window has negligible effect on the half-sine result at 50 Hz (550 mm). Both windows actually amplify the component at 1.2 m (23 Hz), although this is not shown in the figure. The problem with using a Hanning window is that it gives much more emphasis to the central region of the window and much less to the ends of the sample. This is acceptable if the roughness signal is a stationary random signal or if overlapping samples are measured, but often neither is the case.

Although windowing can be used relatively successfully to reduce these effects, it is often preferable to remove trends directly.

5.6.4 Single number metrics for roughness

There has been some interest in the development of a single number measure for roughness in the same way as the A-weighted level is used for sound. The purpose is to give a single number defining how rough or smooth a particular

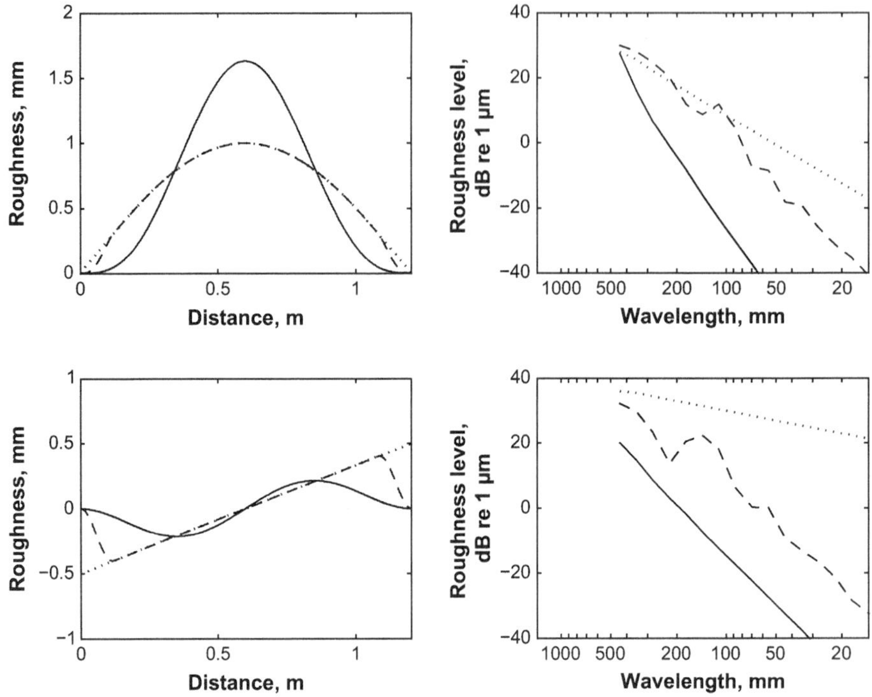

FIGURE 5-24 Effect of windowing, left on roughness amplitude, right on one-third octave band spectrum. Upper: half-sine of amplitude 1 mm, lower: a ramp of amplitude 1 mm. —, Hanning window, – – –, 80% flat plus cosine, · · · ·, no window

track or wheel is. As such, it should provide a weighting of different frequency components that is similar to the weighting of components in the noise spectrum. Due to the typical shape of roughness spectra, with large amplitudes at low frequencies, a simple linear sum or even an A-weighted roughness level, would overemphasize the low frequency components. Instead, the weighting should take a form which is similar to the transfer function between roughness and noise. An example of such a transfer function, which depends on the wheel and track design, is shown in Figure 5.25. This applies to the roughness after application of the contact filter. Note that this transfer function depends on frequency, whereas roughness spectra and the contact filter depend on wavelength. Thus the weighting to be applied to a roughness spectrum at a given wavelength will be different for different train speeds [5.33]. For these reasons single number values for roughness have not been widely adopted.

5.7 OTHER EXCITATION MECHANISMS

Throughout this chapter, the assumption has been made that the wheel/rail system is excited by a relative displacement induced by the surface roughness. While this, indeed, is the predominant excitation mechanism, as demonstrated by the good

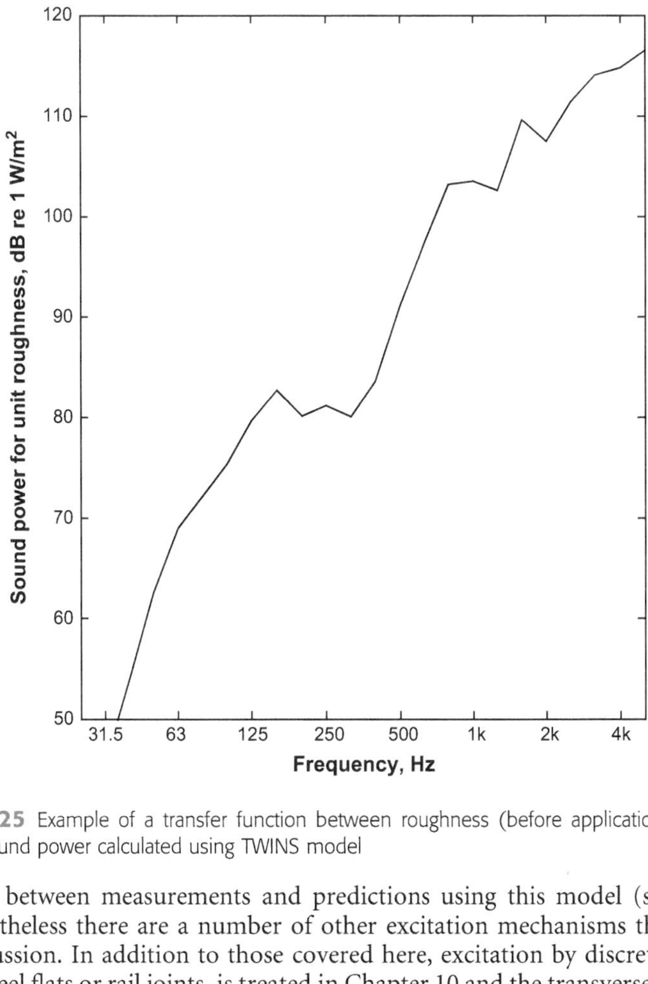

FIGURE 5-25 Example of a transfer function between roughness (before application of contact filter) and sound power calculated using TWINS model

agreement between measurements and predictions using this model (see Section 6.7), nevertheless there are a number of other excitation mechanisms that require some discussion. In addition to those covered here, excitation by discrete features, such as wheel flats or rail joints, is treated in Chapter 10 and the transverse excitation causing curve squeal is discussed in Chapter 9.

5.7.1 Moment excitation

The lateral position of the nominal contact point between the wheel and rail depends on the lateral position of the wheelset on the track and the transverse profiles of wheel and rail. Examples of wheel and rail profiles are shown in Figure 5.26. The lines joining the two profiles indicate the nominal point of contact on the two bodies for various lateral wheelset positions. The nominal contact position can be seen to move quite large distances across the running surfaces for small changes in wheelset position, particularly for conforming profiles (i.e. where $R_{rt} \approx -R_{wt}$). This occurs in particular at two lateral positions of the wheelset. These are, in fact, the stable positions, at which the wheelset is most likely to run [5.17].

As the wheel runs along the rail, the transverse profiles can fluctuate due to the presence of roughness. Thus it is possible, for a constant wheelset position, for the

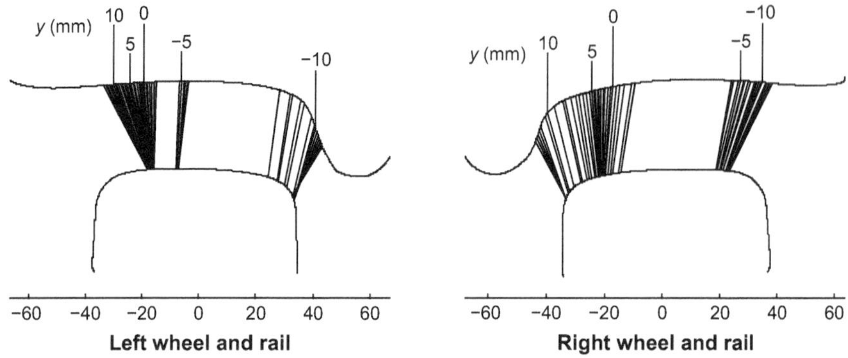

FIGURE 5-26 Transverse profile of a worn wheel profile (sinter block tread braked) in combination with a new UIC54 rail profile showing wheel/rail contact positions as a function of lateral wheelset position [5.16]

contact position to fluctuate laterally across the railhead and wheel tread. This forms a moment excitation of the system that can be an additional source of excitation of the wheel/rail system [5.34]. If the centre of the contact patch moves laterally a distance y_c relative to the *nominal* contact position, a moment $M = y_c F_0$ is induced about the rolling direction with respect to the original position, where F_0 is the nominal normal load. If the distance y_c varies, the corresponding variation in this moment excites the wheel/rail system. Its magnitude is greatest for large static wheel loads, F_0.

Example results are given in Figure 5.27(a), from [5.16]. These were obtained using the DPRS roughness processing model (see Section 5.4). The moment spectrum is presented for various fixed values of wheel transverse radius and a rail radius of 300 mm. As the profiles become more conforming the moment excitation increases considerably. In contrast, the corresponding (vertical) roughness excitation is only slightly changed [5.16].

Using these results along with a model for excitation of the wheel/rail system due to a moment [5.9, 5.34], the sound radiation can be estimated. Figure 5.27(b) compares this with the noise due to roughness excitation. The moment excitation is not significant for a coned wheel profile. However, the conforming profile increases the noise due to moment excitation by 16 dB so that it is only 7 dB(A) less than that due to the conventional roughness excitation. These results are all for a wheel load of 50 kN. As the wheel load is increased, the moment excitation increases, particularly for frequencies below about 1 kHz [5.16].

Results for other sets of wheel roughness data resulted in moment excitation noise levels that vary between 2 and 15 dB(A) below that of the roughness excitation [5.16]. It can therefore be concluded that moment excitation can be a significant additional mechanism for conforming profiles and in extreme cases may be comparable to conventional roughness excitation.

5.7.2 Multiple wheels on a rail

The models considered in this chapter have been based on the interaction between a single wheel and the track. It is assumed that the response of the rail to multiple wheels can be determined using the superposition principle. However,

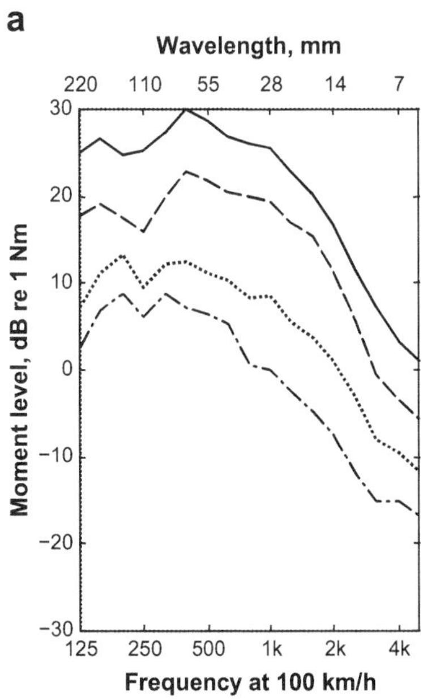

FIGURE 5-27 (a) Equivalent moment level from measured roughness and different transverse radii. ——, $R_{wt} = -330$ mm; – – –, $R_{wt} = -400$ mm; ····, coned wheel; – · – ·, $R_{wt} = +200$ mm; (b) Sound power level due to roughness and moment excitation, for a cast-iron braked wheel at 100 km/h. – – –, due to roughness, 118.1 dB(A); ····, due to moment for coned wheel, 94.8 dB(A); ——, due to moment for conforming wheel ($R_{wt} = -330$ mm), 111.2 dB(A) [5.16]

when multiple wheels are present on a rail, there are some other effects that have to be considered.

First, it has to be considered whether the multiple wheels exciting the rail can be treated as incoherent sources. In [5.35] it is shown that this is a valid assumption provided that the vibration or noise is analysed in frequency bands with a bandwidth of at least 0.8 V/L, where V is the train speed and L is the minimum distance between wheels. For typical parameters and one-third octave analysis, this is satisfied for frequencies over 20 Hz but for low frequency ground vibration it may be necessary to account for the interaction between the forces at each wheel.

More importantly, wave interference effects occur in the rail at frequencies where the track decay rates are low. This is modelled in [5.35] using the concept of a single 'active' wheel (excited by roughness) and a series of 'passive' wheels (with no roughness present) coupled to a rail. The total vibration can be determined using the principle of superposition by taking each wheel in turn as the active wheel. The passive wheels can be represented by a simple mass–spring model as described in Section 4.4.1, as the high frequency modal behaviour is isolated from the rail by the contact spring.

These wave reflections modify the point mobility of the rail, leading to fluctuations of up to ±50%, particularly for soft rail pads. This can be important for roughness growth models, see, e.g., [5.36]. However, the effect on the average rail vibration and noise, considered in [5.37], is found to be insignificant, with only small differences in one-third octave band levels and differences of less than 1 dB in overall noise levels.

5.7.3 Correlation between the two wheels on an axle or two rails in the track

Throughout the models in Chapters 3 to 5 only a single rail and wheel have been considered. Where the bending of a whole sleeper is included (Section 3.7), it has been assumed that the excitation is on one rail only. The response due to excitation on the two rails can be determined by assuming that the roughness on both sides is uncorrelated. Thus the sound powers, or mean-square responses, due to excitation on the two sides can be added.

There is a lack of data on the correlation between the roughness on two rails of a track or two wheels of a wheelset. However, it appears a reasonable assumption that they can be treated as uncorrelated at short wavelengths. Longer wavelength defects, such as those associated with the spacing between sleepers, rail joints, wheel circumference, etc., are more likely to be correlated. It seems reasonable to assume that wavelengths shorter than about 0.6 m (typical sleeper spacing) can be treated as uncorrelated between the two sides, whereas those longer than this should be treated as correlated inputs. The exceptions to this are discrete features such as wheel flats or rail joints, which are likely to be strongly correlated between the two sides even for quite short wavelengths.

Measurements of track geometry at very long wavelengths exist from track measuring coaches. These measurements are usually separated into vertical alignment and 'cross level', that is the height of one rail relative to the other. These components of excitation are relevant to ground-borne vibration (see Chapter 12) and low frequency vibration of bridges.

5.7.4 Moving load

Each wheel exerts a static load on the track, causing a deflection that is localized around the wheel. For a static load F_0 acting on a beam of bending stiffness EI supported by a stiffness s per unit length, the deflection at a distance x from the load is given by

$$u(x) = \frac{F_0\beta}{2s} e^{-\beta|x|}(\cos\beta|x| + \sin\beta|x|) \tag{5.67}$$

where $\beta = (s/4EI)^{1/4}$. Figure 5.28(a) shows an example of the deflection under a series of wheels for a track with $s = 100$ MN/m^2 and $EI = 6.4$ MN/m^2. The wheels are each assigned a load of 50 kN. The vehicle length is chosen as 20 m, the bogie wheelbase as 1.8 m and the distance between bogie centres as 14.6 m.

As the train moves along the track this vibration signature passes any given point on the track so that the rail at that point vibrates vertically. Figure 5.28(b) shows the vibration spectrum for a train speed of 100 km/h and Figure 5.28(c) shows the result for 300 km/h. The shape of these vibration spectra due to the quasi-static deflection depends on the wheel spacing within the vehicles.

These spectra are compared with a notional roughness spectrum of 23 dB re 1 μm at 1 m wavelength, falling at −20 dB/decade. The corresponding rail

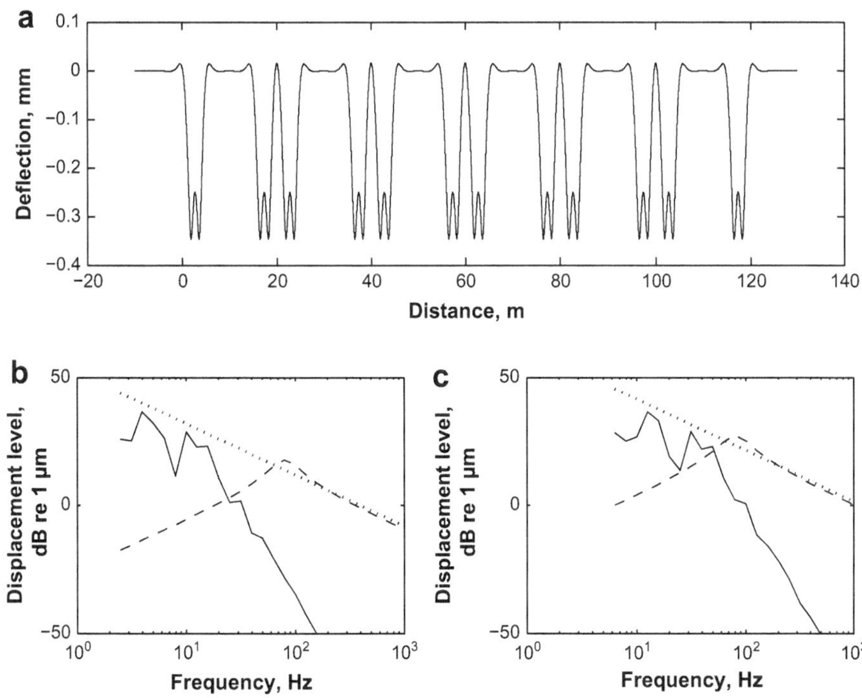

FIGURE 5-28 (a) Static deflection of track under a train of six vehicles with wheel load 50 kN. (b) Spectra of displacement for train speed 100 km/h: —, quasi-static deflection; – – –, rail vibration due to roughness; · · · ·, assumed roughness spectrum. (c) As for (b) at 300 km/h

vibration due to this roughness is also shown. It is similar to the roughness itself above the coupled wheel/track resonance but is much lower than the roughness at low frequencies (see equation (5.10)). These results are determined for a wheel mass of 600 kg.

From these results, it is clear that the quasi-static rail deflection is dominant up to about 25 Hz at 100 km/h, and up to about 50 Hz at 300 km/h. At higher frequencies the vibration due to roughness excitation is considerably larger than that due to the quasi-static deflection.

The quasi-static deflection has no effect on the wheel vibration.

5.7.5 Parametric excitation

Consider a perfectly smooth wheel and rail where the support stiffness provided by the rail, K, varies with distance x. For relatively low frequency variations the wheel can be represented simply by a mass, M_w, and a static load, F_0. Writing the response of the wheel as a quasi-static deflection of the track, u_0, and a dynamic component, u, this satisfies

$$M_w\ddot{u} + K(x)(u_0 + u) = F_0 \tag{5.68}$$

If the stiffness consists of a steady value K_T, and a small sinusoidal variation, $\delta K \sin(2\pi x/\lambda)$, this can be written as

$$M_w \ddot{u} + K_T(u_0 + u) + \delta K u_0 \sin\left(\frac{2\pi x}{\lambda}\right) = F_0 \tag{5.69}$$

where products of small quantities have been neglected. Noting that $K_T u_0 = F_0$:

$$M_w \ddot{u} + K_T u = -\left(\frac{\delta K}{K_T}\right) F_0 \sin\left(\frac{2\pi x}{\lambda}\right) \tag{5.70}$$

The steady state response at frequency ω, corresponding to the wavelength λ, has amplitude

$$u = -\left(\frac{\delta K}{K_T}\right) \frac{F_0}{K_T - M_w \omega^2} \tag{5.71}$$

This differs from the relative displacement excitation due to roughness in that both the wheel and rail move with the same amplitude (apart from the deflection of the contact spring which has been neglected here). However, this response has a coupled resonance at the same frequency as that given in equation (5.13).

A particular source of this 'parametric excitation' occurs at the 'sleeper passing' frequency. Since the rail usually has discrete supports, the track 'stiffness' varies periodically within each sleeper span leading to additional excitation at the sleeper passing frequency. For a sleeper span of 0.6 m this frequency corresponds to 46 Hz at 100 km/h, rising to 138 Hz at 300 km/h.

These variations within a sleeper span can be seen in Figures 3.33 and 3.34 in terms of the mobility. The quasi-static stiffness, which can be found from the low frequency asymptote of these graphs, only varies by about 4% between mid-span and above a sleeper for these parameters, whereas the mobility varies much more at higher frequencies. Figure 5.29 shows the relative variation in mobility over a sleeper span for the track parameters of Figure 3.34 (ballast stiffness 60 MN/m, pad stiffness 600 MN/m). Results are shown at three frequencies, the first being representative of the quasi-static stiffness. The second frequency is chosen as the rail-on-pad resonance of this track, at which the mobility varies by almost a factor of 2. The third is the pinned–pinned frequency at which the mobility varies by a factor of 16.

The simplest way of including the excitation due to the sleeper-passing frequency is to add it to the roughness spectrum at the sleeper-passing wavelength. However, the equivalent roughness is different for the wheel and the rail. Equating equation (5.71) to equations (5.10), (5.16) and (5.18) the equivalent roughness for the rail is

$$r_{eq} = \left(\frac{\delta K}{K_T}\right) \frac{F_0}{M_w \omega^2} \tag{5.72}$$

whereas for the wheel, from equations (5.11), (5.17) and (5.19) the equivalent roughness is

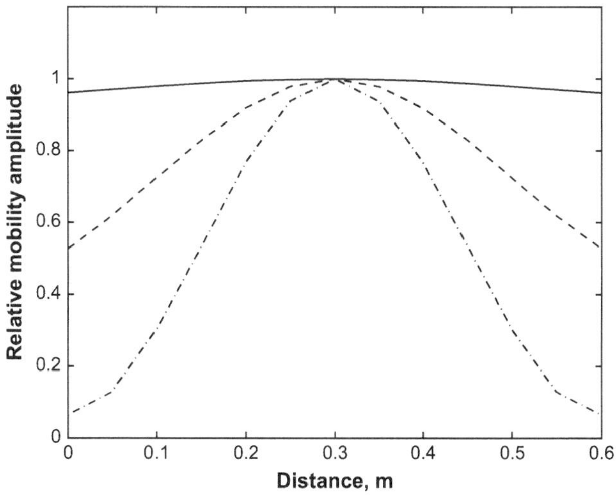

FIGURE 5-29 Variation of point mobility modulus across a sleeper span, normalized to the value at mid-span. —, quasi-static; – – –, 680 Hz; – · – ·, 1070 Hz

$$r_{eq} = -\left(\frac{\delta K}{K_T}\right)\frac{F_0}{K_T} \tag{5.73}$$

For the track parameters considered above, the track stiffness, K_T, is 1.4×10^8 N/m. The relative variation $\delta K/K_T$ within a sleeper span has an amplitude of $\pm 2\%$. For a static wheel load of 50 kN this leads to an equivalent roughness amplitude (for the wheel) of ± 7 μm. The equivalent roughness seen by the track depends on frequency but is similar to this at the coupled wheel/track system resonance frequency. The amplitude of the oscillation increases as the ballast stiffness increases; for example, for a ballast stiffness of 180 MN/m the track mobility varies by $\pm 6\%$ within a sleeper span, equivalent to a displacement of ± 21 μm. It also increases in direct proportion to the wheel load.

The variations in mobility at higher frequencies can lead to other parametric excitation effects. These can either be investigated using a numerical time-domain model or an analytical approach such as that adopted in [5.38–5.40].

5.7.6 Non-linear effects

Finally, it should be remembered that the models in this chapter are linear. As the Hertzian contact spring is actually non-linear, equation (5.45), for large amplitude defects of the wheel or rail the linearized stiffness can no longer be used. Consequently, the frequency-domain approach used here has to be replaced by a time-domain approach. This is discussed in relation to impact noise in Chapter 10.

REFERENCES

5.1 C. Esveld. *Modern Railway Track*, 2nd edition. MRT Publications, Zaltbommel, 2001.

5.2 A. Johansson. Out-of-round railway wheels – assessment of wheel tread irregularities in train traffic. *Journal of Sound and Vibration*, **293**, 795–806, 2006.

5.3 J.J. Kalker. *Three Dimensional Elastic Bodies in Rolling Contact.* Kluwer Academic Publishers, Dordrecht, 1990.

5.4 D.J. Thompson, M.H.A. Janssens, and F.G. de Beer. *TWINS: Track-Wheel Interaction Noise Software, theoretical manual (version 3.0). TNO report HAG-RPT-990211.* TNO Institute of Applied Physics, Delft, 1999.

5.5 K.L. Johnson. *Contact Mechanics.* Cambridge University Press, Cambridge, 1985.

5.6 S.P. Timoshenko and J.N. Goodier. *Theory of Elasticity,* 3rd edition. McGraw-Hill, 1982.

5.7 T.A. Harris. *Roller Bearing Analysis,* 4th edition. John Wiley & Sons, New York, 2001.

5.8 R.D. Mindlin. Compliance of elastic bodies in contact. *Journal of Applied Mechanics,* **71**, 259–268, 1949.

5.9 D.J. Thompson. *Wheel-rail noise: theoretical modelling of the generation of vibrations.* PhD thesis, University of Southampton, 1990.

5.10 P.J. Vermeulen and K.L. Johnson. Contact of nonspherical elastic bodies transmitting tangential forces. *Trans ASME Journal of Applied Mechanics,* 338–340, June 1964.

5.11 Z.Y. Shen, J.K. Hedrick and J.A. Elkins. A comparison of alternative creep–force models for rail vehicle dynamic analysis. *Proceedings of 8th IAVSD Symposium,* Cambridge MA, Swets and Zeitlinger, Lisse, 591–605, 1983.

5.12 X. Sheng, D.J. Thompson, and C.J.C. Jones. *Modelling rail roughness growth on tangent tracks.* ISVR Technical Memorandum No. 929, University of Southampton, March 2004.

5.13 K. Knothe and A. Gross-Thebing. Derivation of frequency dependent creep coefficients based on an elastic half-space model. *Vehicle System Dynamics,* **15**, 133–153, 1986.

5.14 P.J. Remington. Wheel/rail noise, Part IV: Rolling noise. *Journal of Sound and Vibration,* **46**, 419–436, 1975.

5.15 P. Remington and J. Webb. Estimation of wheel/rail interaction forces in the contact area due to roughness. *Journal of Sound and Vibration,* **193**, 83–102, 1996.

5.16 D.J. Thompson. The influence of the contact zone on the excitation of wheel/rail noise. *Journal of Sound and Vibration,* **267**, 523–535, 2003.

5.17 D.J. Thompson and P.J. Remington. The effects of transverse profile on the excitation of wheel/rail noise. *Journal of Sound and Vibration,* **231**, 537–548, 2000.

5.18 R.A.J. Ford and D.J. Thompson. Simplified contact filters in wheel/rail noise prediction. *Journal of Sound and Vibration,* **293**, 807–818, 2006.

5.19 G. Hölzl, M. Redmann, and P. Holm. Entwicklung eines hochempfindlichen Schienenoberflächenmeßgeräts als Beitrag zu weiteren möglichen Lärmminderungsmaßnahmen im Schienenverkehr. *Eisenbahn Technische Rundschau,* **39**, 685–689, 1990.

5.20 S.L. Grassie, M. Saxon, and J.D. Smith. Measurement of longitudinal rail irregularities and criteria for acceptable grinding. *Journal of Sound and Vibration,* **227**, 949–964, 1999.

5.21 *ORE Question C163, Report RP9. Wheel/rail contact noise – an experimental comparison of systems for measuring the rail roughness.* Utrecht, 1988.

5.22 CEN, Railway applications – Noise emission – Road test of draft standard for rail roughness measurement prEN 15610:2006, draft Technical Report, prCEN/TR 1-1:2008:2, 2008.

5.23 International standard. *Railway applications – Acoustics – Measurements of noise emitted by railbound vehicles.* International Standards Organization, ISO 3095, 2005.

5.24 Directive 2001/16/EC of the European Parliament and of the Council of 19 March 2001 on the interoperability of the trans-European conventional rail system. *Official Journal of the European Communities,* 2 April 2001.

5.25 Commission Decision 2002/735/EC concerning the Technical Specification for Interoperability (TSI) relating to the rolling stock subsystem of the trans-European high-speed rail system. *Official Journal of the European Communities* 12.9.2002 L245/402–506, 2002.

5.26 A.E.J. Hardy. Draft proposal for noise measurement standard for ERRI committee C163. Report RR-SPS-97–012 Issue 1, BR Research – published through ERRI, 1997.

5.27 E. Verheijen. A survey on roughness measurements. *Journal of Sound and Vibration,* **293**, 784–794, 2006.

5.28 B. Asmussen, H. Onnich, R. Strube, L.M. Greven, S. Schröder, K. Jäger, and K.G. Degen. Status and perspectives of the 'specially monitored track'. *Journal of Sound and Vibration,* **293**, 1070–1077, 2006.

5.29 M.H.A. Janssens, M.G. Dittrich, F.G. de Beer, and C.J.C. Jones. Railway noise measurement method for pass-by noise, total effective roughness, transfer functions and track spatial decay. *Journal of Sound and Vibration*, **293**, 1007–1028, 2006.
5.30 K. Shin and J.K. Hammond. *Fundamentals of Signal Processing for Sound and Vibration Engineers.* John Wiley & Sons, 2008.
5.31 R.J. Diehl and P. Holm. Roughness measurements – have the necessities changed? *Journal of Sound and Vibration*, **293**, 777–783, 2006.
5.32 . *ORE Question C163, Report RP7. Wheel/rail contact noise – comparative study between theoretical model and measured results.* Utrecht, 1986.
5.33 D.J. Thompson. On the relationship between wheel and rail roughness and rolling noise. *Journal of Sound and Vibration*, **193**, 149–160, 1996.
5.34 D.J. Thompson. Wheel-rail noise generation, Part IV: Contact zone and results. *Journal of Sound and Vibration*, **161**, 447–466, 1993.
5.35 T.X. Wu and D.J. Thompson. Vibration analysis of railway track with multiple wheels on the rail. *Journal of Sound and Vibration*, **239**, 69–97, 2001.
5.36 T.X. Wu and D.J. Thompson. An investigation into rail corrugation due to micro-slip under multiple wheel/rail interactions. *Wear*, **258**, 1115–1125, 2005.
5.37 T.X. Wu and D.J. Thompson. The effects on railway rolling noise of wave reflections in the rail and support stiffening due to the presence of multiple wheels. *Applied Acoustics*, **62**, 1249–1266, 2001.
5.38 A. Nordborg. Wheel/rail noise generation due to nonlinear effects and parametric excitation. *Journal of the Acoustical Society of America*, **111**, 1772–1781, 2002.
5.39 T.X. Wu and D.J. Thompson. On the rolling noise generation due to wheel/track parametric excitation. *Journal of Sound and Vibration*, **293**, 566–574, 2006.
5.40 X. Sheng, C.J.C. Jones, and D.J. Thompson. Responses of infinite periodic structures to moving or stationary harmonic loads. *Journal of Sound and Vibration*, **282**, 125–149, 2005.
5.41 T. Kitagawa and D.J. Thompson. Comparison of wheel/rail noise radiation on Japanese railways using the TWINS model and microphone array measurements. *Journal of Sound and Vibration*, **293**, 496–509, 2006.

CHAPTER 6

Sound Radiation from Wheels and Track

6.1 INTRODUCTION

This chapter discusses the radiation of sound by vibrations of the wheels and track. The final two sections cover the propagation of sound to the wayside and validation tests carried out to verify the overall rolling noise model.

Sound can be generated by various different mechanisms, but these can mostly be grouped into two main categories:

- sound radiation from *structural vibrations* – the vibration of a solid structure causes the air around it to vibrate and hence produces sound, e.g. a loudspeaker or a drum;
- sound produced by *unsteady aerodynamic flow* – thus turbulence and air flow over solid objects also produce sound, e.g. jet noise, turbulent boundary layer noise, exhaust noise and fan noise.

In this chapter, the discussion is limited to the first of these two mechanisms, in particular the sound radiated by wheels and track, the vibrations of which have been described in the previous chapters. The same models can be used for the sound radiation of wheels during curve squeal (Chapter 9) and a similar approach can be used for bridge noise (Chapter 11). Aerodynamic noise is considered separately in Chapter 8.

The sound field produced by a vibrating structure can be expressed in terms of the distribution of *sound pressure*. This depends on the distance and orientation of the receiver relative to the source. It is also useful to consider the total sound produced by a source, which can be described by its *sound power*. This does not depend on a receiver location.

The sound power, W, radiated by a vibrating object in a particular frequency band can be written as [6.1]

$$W = \rho_0 c_0 S \langle \overline{v^2} \rangle \sigma \tag{6.1}$$

where S is the surface area of the vibrating structure and $\langle \overline{v^2} \rangle$ is the squared velocity normal to the surface in the frequency band of interest, which is averaged both over time ($\overline{\ }$) and over the surface area ($\langle\ \rangle$). For sinusoidal motion of complex amplitude v, the mean-square corresponds to $\overline{v^2} = \frac{1}{2}|v^2|$. ρ_0 is the density of air and c_0 is the speed of sound, which take the values 1.2 kg/m^3 and 343 m/s respectively at 20°C. Their product is known as the characteristic specific acoustic impedance of air.

The factor σ in equation (6.1) is known as the *radiation ratio* or *radiation efficiency*. In decibels, $L_\sigma = 10 \log_{10} \sigma$ is known as the *radiation index*. The radiation ratio expresses the ratio of the sound power actually produced to that which would be produced in an idealized case producing plane waves. This reference case can be considered as a piston in a tube; if the piston has the same area S and mean-square velocity amplitude $\overline{v^2}$ as the actual source, it radiates a sound power $\rho_0 c_0 S \overline{v^2}$. Equivalently, the reference situation can also be considered to be an infinite flat surface, all vibrating in phase with mean-square velocity $\overline{v^2}$. This also produces plane waves and a sound power $\rho_0 c_0 \overline{v^2}$ per unit area.

The radiation ratio, σ, is determined by the size and shape of the vibrating structure. At low frequencies, where the wavelength of sound is large compared with the size of the vibrating object, the radiation ratio is generally small. At high frequencies, where the wavelength is much smaller than the object, each part of the surface can radiate sound independently of the rest, as if it were part of an infinite surface, and the radiation ratio tends to unity. The distribution of vibration on the structure, in particular the wavelength of its vibrating modes, can also affect the radiation ratio, leading to acoustic cancellation in situations where the acoustic wavelength is longer than the structural wavelength. However, the radiation ratio is independent of the amplitude of vibration, this being included separately in equation (6.1).

Predictions of the radiation ratio can be obtained for simple cases using analytical methods. In more complex situations numerical methods can be used, such as the boundary element method (BEM). For certain types of source, use can also be made of the Rayleigh integral technique [6.2]. In this, a vibrating surface is visualized as part of an infinite flat surface (an infinite 'baffle') and the sound pressure at any location can be found by integrating the contribution from each part of the surface. The sound power can be found by integrating the squared pressure over a hemisphere in the far field. Such a technique has been used successfully, for example, in determining the radiation ratio of rectangular flat plates, where the effect of acoustic short-circuiting across the plate is important [6.3]. However, it does not take into account any interaction between the sound field generated by the front and rear of the structure, or the effect of complex geometrical shape.

As well as the sound power, the *directivity* can also be important. This describes the proportion of sound radiated in particular directions, which is thus also independent of the distance from the source. The mean-square sound pressure at some distance r from a compact source, emitting power W into free space, can be written as

$$\overline{p^2} = \frac{\rho_0 c_0 W}{4\pi r^2} D(\theta, \phi) \tag{6.2}$$

where $D(\theta,\phi)$ is the *directivity factor* [6.1, 6.4]. This depends on the direction of the receiver from the source, which can be defined by two angles, the elevation, θ, and the azimuth, ϕ. An omnidirectional source has simply $D = 1$. From equation (6.2) the sound pressure due to a point source can be seen to be inversely proportional to the distance r, i.e. the sound pressure level reduces by 6 dB per doubling of distance. This equation effectively separates the dependence on distance from the variation with direction. This separation is valid provided that the receiver is in the far field, i.e. not closer than an acoustic wavelength and further away than the characteristic source dimension. In decibels, $\mathrm{DI} = 10\log_{10} D$ is the *directivity index.*

6.2 SIMPLE MODELS FOR SOUND RADIATION

6.2.1 Point sources

It is useful to consider first the sound radiation from simple sources, before proceeding to study the radiation from more complex structures, such as wheels and rails. One of the simplest acoustic sources consists of a pulsating sphere, which represents a monopole. This has a radiation ratio given by [6.5]

$$\sigma = \frac{(ka)^2}{1+(ka)^2} \tag{6.3}$$

where a is the radius of the sphere and k is the acoustic wavenumber, $k = \omega/c_0$. At low frequencies $\sigma \approx (ka)^2$ which is proportional to f^2, where f is the frequency, i.e. the radiation index increases at a rate of 20 dB/decade of frequency. At high frequencies σ tends to unity. The corresponding directivity factor is simply $D = 1$.

Similarly, the radiation from a circular piston of radius a mounted in an infinite baffle can be found using the Rayleigh integral [6.5]. This gives a radiation ratio of

$$\sigma = 1 - \frac{2J_1(2ka)}{2ka} \tag{6.4}$$

where J_1 is the Bessel function of order 1. For low frequencies this is given approximately by

$$\sigma \approx \frac{(ka)^2}{2} \quad \text{for} \quad ka \ll 1 \tag{6.5}$$

Thus, comparing equations (6.3) and (6.5), a baffled disc of radius a is equivalent to a pulsating sphere of radius $a/\sqrt{2} \approx 0.71a$. The results for the circular piston and the pulsating sphere are compared in Figure 6.1. It may be noted that a piston in a baffle is equivalent to a thin pulsating disc, as a symmetry condition exists in the plane of the baffle (the region behind the baffle has arbitrary sound field). The surface area of the sphere is $S = 4\pi a^2$ whereas the disc has an area of $S = 2\pi a^2$ including both sides. This difference in area accounts for the factor of 2 between equations (6.3) and (6.5).

The directivity factor of a piston mounted in a baffle is [6.5]

$$D = \frac{I(\theta)}{\int_0^{\pi/2} I(\theta)\sin\theta d\theta} \quad \text{with} \quad I(\theta) = \left(\frac{2J_1(ka\sin\theta)}{ka\sin\theta}\right)^2 \tag{6.6}$$

where I is the normalized sound intensity at angle θ. The elevation $\theta = 0$ on the axis of the piston and $\theta = \pi/2$ on the baffle. D is plotted in Figure 6.2 in decibel form for various values of ka. As can be seen, the radiation is omnidirectional at low frequencies but becomes very directional at high frequencies, in contrast to that for a pulsating sphere which remains omnidirectional for all frequencies.

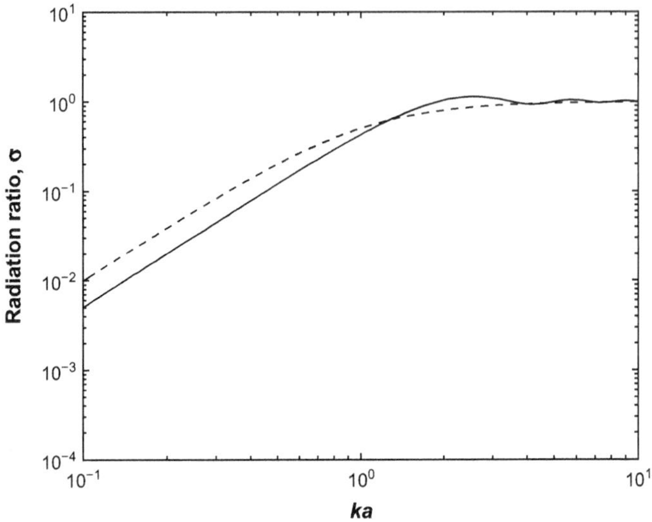

FIGURE 6-1 Radiation ratios for simple monopole sources of radius a. —, disc in infinite baffle; - - -, pulsating sphere

An oscillating sphere has a radiation ratio given by [6.5]

$$\sigma = \frac{(ka)^4}{4 + (ka)^4} \tag{6.7}$$

At low frequencies this is lower than the result for a pulsating sphere, equation (6.3), due to cancellation between the sound radiation from the front and back of the

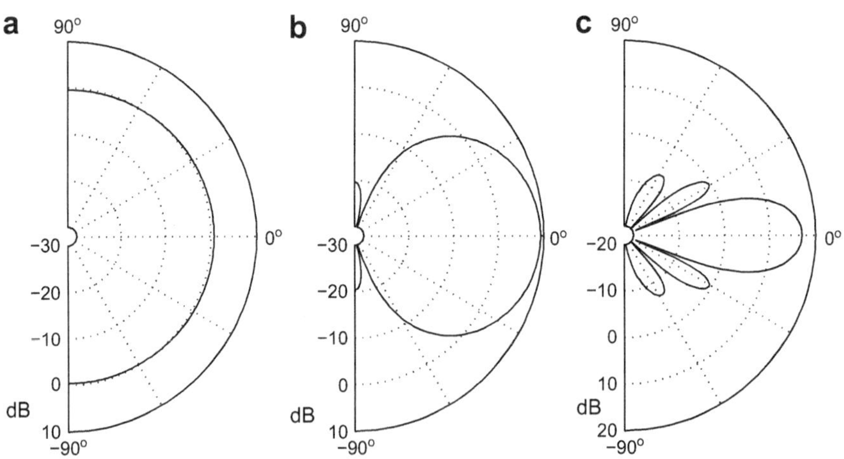

FIGURE 6-2 Directivity index of radiation from a circular piston mounted in an infinite baffle. (a) $ka = 1$, (b) $ka = 4$, (c) $ka = 10$

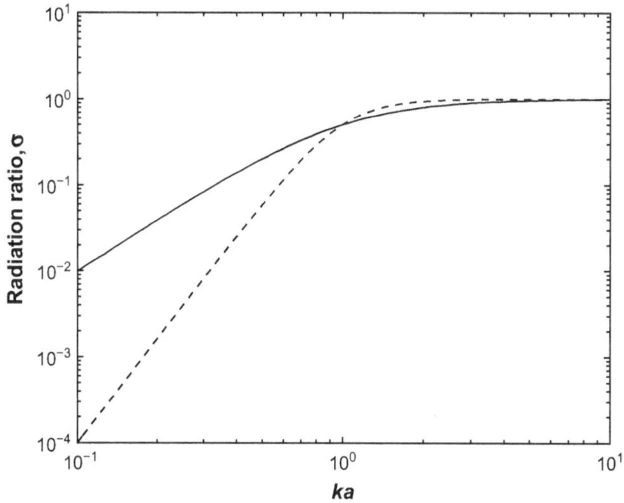

FIGURE 6-3 Radiation ratios for simple sources of radius a. —, pulsating sphere; – – –, oscillating sphere

source. The radiation ratio increases in proportion to f^4, or a rate of 40 dB/decade of frequency. At high frequency it again tends to unity. The results for pulsating and oscillating spheres are compared in Figure 6.3. At high frequencies, above $ka = 1$, both curves tend to unity, whereas at low frequencies their different slopes can be seen. It should be noted that the spatially averaged normal velocity of the oscillating sphere $\langle \overline{v^2} \rangle = \frac{1}{3}\overline{v_0^2}$ where v_0 is the velocity of oscillation. Because of this, other normalizations are sometimes used for σ based on $\overline{v_0^2}$, for which σ does not tend to unity in the high frequency limit.

The corresponding directivity for the oscillating sphere is

$$D(\theta) = 3\cos^2\theta \tag{6.8}$$

which is shown in Figure 6.4. The factor of 3 comes from the fact that the average of the directivity factor over a solid angle of 4π should be unity. The maximum value of $D = 3$ (or DI $= +5$ dB) occurs on the axis in the direction of motion.

The pulsating sphere represents a monopole while the oscillating sphere corresponds to a dipole. A dipole can also be represented by two closely spaced monopoles pulsating 180° out of phase with each other. For two spheres of radius a (with $ka \ll 1$) separated by a distance d, the radiation ratio is

$$\sigma = \left(1 - \frac{\sin kd}{kd}\right)(ka)^2 \approx \frac{(kd)^2}{6}(ka)^2 \qquad (\text{for} \quad kd \ll 1) \tag{6.9}$$

A quadrupole source, which can be represented by two dipoles oscillating in opposite directions to each other, would have a slope proportional to the sixth power of ka at low frequency. Higher order multipoles also exist and have correspondingly steeper slopes at low frequency.

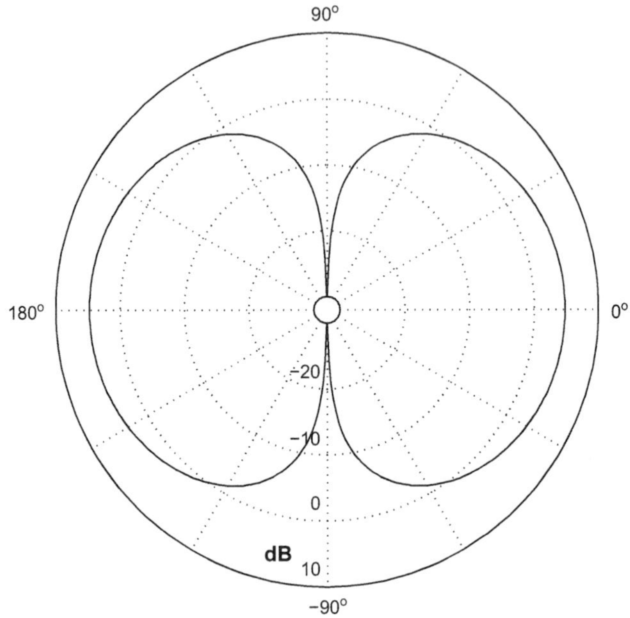

FIGURE 6-4 Directivity index of an oscillating sphere

A thin disc of radius a which is freely suspended and is oscillating also has a dipole-type radiation. At low values of ka it has a radiation ratio equivalent to an oscillating sphere of radius $(4/3\pi)^{1/3}a \approx 0.75a$ [6.5].

6.2.2 Line sources

Sources which are extended in one direction can be represented as line sources. The general solution for a cylindrical source is given in [6.6]. For example, for an infinitely long pulsating cylinder of radius a, this leads to a radiation ratio of

$$\sigma = \frac{2}{\pi ka\left|H_1^{(2)}(ka)\right|^2} \tag{6.10}$$

where H is a Hankel function which reduces to

$$\sigma \approx \frac{1}{2}\pi ka \qquad \text{for} \quad ka \ll 1 \tag{6.11}$$

Similarly, an oscillating cylinder has

$$\sigma = \frac{1}{\pi\left|kaH_0^{(2)}(ka) - H_1^{(2)}(ka)\right|^2} \tag{6.12}$$

which reduces to

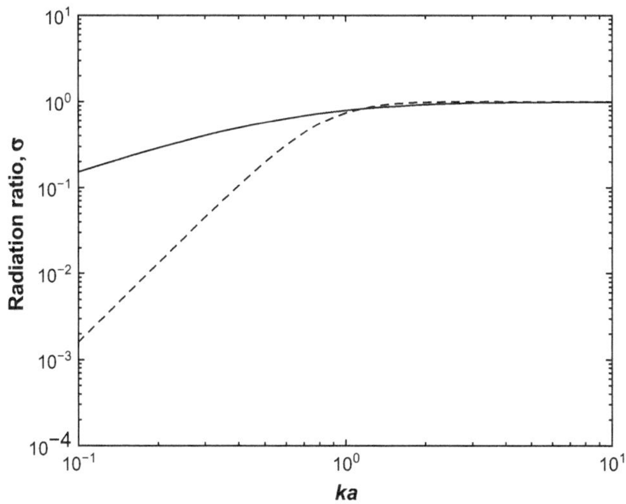

FIGURE 6-5 Radiation ratios for cylindrical sources of radius a. —, pulsating cylinder; – – –, oscillating cylinder

$$\sigma \approx \frac{1}{2}\pi(ka)^3 \qquad \text{for} \quad ka \ll 1 \tag{6.13}$$

The spatially averaged normal velocity of the oscillating cylinder $\langle \overline{v^2} \rangle = \frac{1}{2}\overline{v_0^2}$ where v_0 is the velocity of oscillation. The radiation ratios of a pulsating and an oscillating cylinder are shown in Figure 6.5. It can be noted that the slope of these relations is different from the corresponding point sources, equations (6.3) and (6.7). The line monopole (pulsating cylinder) has a radiation ratio proportional to f at low frequency, whereas a line dipole (oscillating cylinder) has a frequency dependence of f^3. The resulting sound fields are two-dimensional. A general line source has

$$\overline{p^2} = \frac{\rho_0 c_0 W'}{2\pi r} D(\theta) \tag{6.14}$$

where W' is the sound power per unit length emitted by the source. Thus the sound pressure level reduces with 3 dB per doubling of distance. The directivity factor depends only on the angle θ around the cylinder. For a pulsating cylinder $D(\theta) = 1$, whereas for an oscillating cylinder

$$D(\theta) = 2\cos^2\theta \tag{6.15}$$

The factor of 2 ensures that the average of the directivity factor over an angle of 2π is unity.

6.2.3 Bending waves

When the radiating structure vibrates in bending modes with a wavelength that is short compared with the wavelength in air, an acoustic short-circuiting effect occurs.

Since bending waves have a frequency-dependent wavespeed, a frequency exists at which the bending wavelength and acoustic wavelength are equal. This is known as the critical frequency and for a flat plate of thickness h is given by

$$f_c = \frac{c_0^2}{2\pi h}\left(\frac{12\rho(1-\nu^2)}{E}\right)^{1/2} \tag{6.16}$$

where c_0 is the speed of sound in air and the plate has Young's modulus E, density ρ and Poisson's ratio ν. As will be seen, for railway wheels, rails and sleepers the critical frequency is mostly sufficiently low that no acoustic short-circuiting occurs. The radiation from flat plates sustaining bending waves is considered further in Chapter 11.

6.3 WHEEL RADIATION

6.3.1 Radiation ratio

The modes of vibration of a railway wheel have been described in Chapter 4. For a typical wheel radius of 0.46 m, the condition $ka = 1$ corresponds to a frequency of 120 Hz. It can therefore be expected that its radiation ratio will be close to unity for most frequencies of interest. Moreover, as it is relatively thick, the structural wavelengths are not particularly short and acoustic short-circuiting effects only occur at relatively low frequencies.

In [6.7] Remington simply used $\sigma = 1$ for the wheel radiation ratio (it was actually stated in the form $\sigma = 2$, as the surface area S in the equation was taken only as the area of one side of the wheel). Earlier, in [6.8], he also used an expression for the radiation ratio of a rigid unbaffled disc, which gave a lower radiation at low frequency particularly below about 100 Hz.

This approach does not take account of the vibration distribution due to the mode shape. Other early models of the sound radiation from railway wheels [6.9, 6.10] made use of the Rayleigh integral technique [6.2] but this does not give the correct radiation ratio at low frequencies. For example, for a simple $n = 0$ mode (i.e. with 0 nodal diameters) the Rayleigh integral would predict the radiation ratio as that of a monopole, where the true result is that of a dipole, as discussed below.

The boundary element method (BEM) [6.11] can be used in order to take account of complex geometry. This is a numerical technique in which the vibrating surface is represented by a mesh of elements and the sound field is calculated by solving the Helmholtz integral equation in discrete form. Unlike the acoustic finite element method, it readily allows for infinite domains. Fingberg [6.12] used the boundary element method to predict the sound radiation from a wheel in its various normal modes. Making use of the axisymmetry of the wheel, the wheel was represented only by its cross-section. It was shown that the radiation ratio can be significantly smaller at low frequencies than that predicted by the Rayleigh integral method. At high frequencies, where the radiation ratio is close to unity, variations of around 2 dB were found. Experimental validation was performed on a model wheel at scale 1:5.

A similar BEM approach was used in [6.13] but the radiation ratio due to vibration in each normal mode was calculated for a range of frequencies rather than just at the corresponding natural frequency. These results will be summarized here.

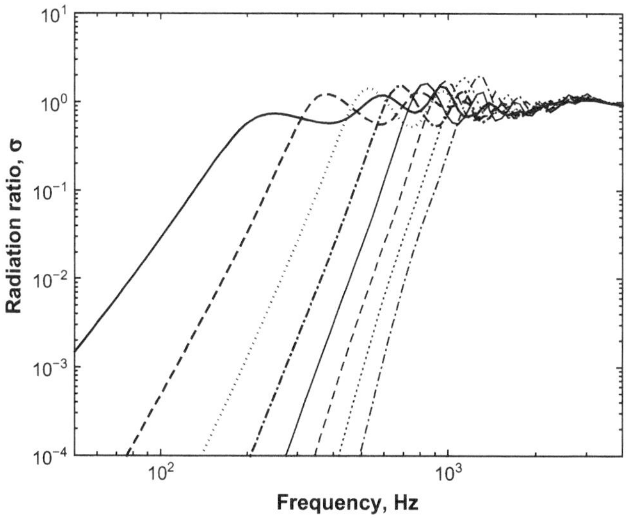

FIGURE 6-6 Radiation ratio of zero-nodal-circle axial modes of 920 mm freight wheel with various numbers of nodal diameters, n. Thick lines: —, $n = 0$; – – –, $n = 1$; ····, $n = 2$; – · –– ·, $n = 3$; thin lines: —, $n = 4$; – – –, $n = 5$; ····, $n = 6$, – · – ·, $n = 7$

The modes of vibration of a wheel were calculated first using the finite element method, see Chapter 4. The radiation ratios obtained for the zero-nodal-circle axial mode shapes with different numbers of nodal diameters, n (see Figure 4.3) are shown in Figure 6.6. In each case, the radiation ratio rises sharply at low frequency, reaching unity between about 250 and 1250 Hz, and then oscillating at higher frequencies.

The result for the $n = 0$ mode has a frequency dependence at low frequency of f^4, reaching unity at about 250 Hz. For other modes of the wheel, the slope of the low frequency part of the curve increases with increasing n. For $n = 1$ it increases at a rate of f^6, for $n = 2$ at f^8, etc. Thus in general it follows $f^{2(n+2)}$.

In order to explain these slopes, Figure 6.7 shows a number of wheel modes schematically. The upper row shows a disc set in an infinite baffle. The $n = 0$ mode is a simple volume source which acts as a monopole; the $n = 1$ mode has regions either side of the node line that are vibrating out of phase with each other, thus forming a dipole at low frequencies, where the wavelength is long compared with the disc dimensions. Similarly, the $n = 2$ mode has two positive and two negative source regions which together form a quadrupole at low frequencies.

In the second row of Figure 6.7 axial modes are shown of a wheel in free field. Here, in addition to the positive and negative source regions seen for the baffled disc, the front and rear of the wheel correspond to source regions of opposite polarity. Thus the $n = 0$ mode now forms a dipole, the $n = 1$ mode a quadrupole, etc. The $n = 0$ mode has similar behaviour to the oscillating sphere, see Figure 6.3, but the result is lower; the correct behaviour can be found by using a reduced radius of 0.3 m in equation (6.7). A similar phenomenon has already been noted for an oscillating disc.

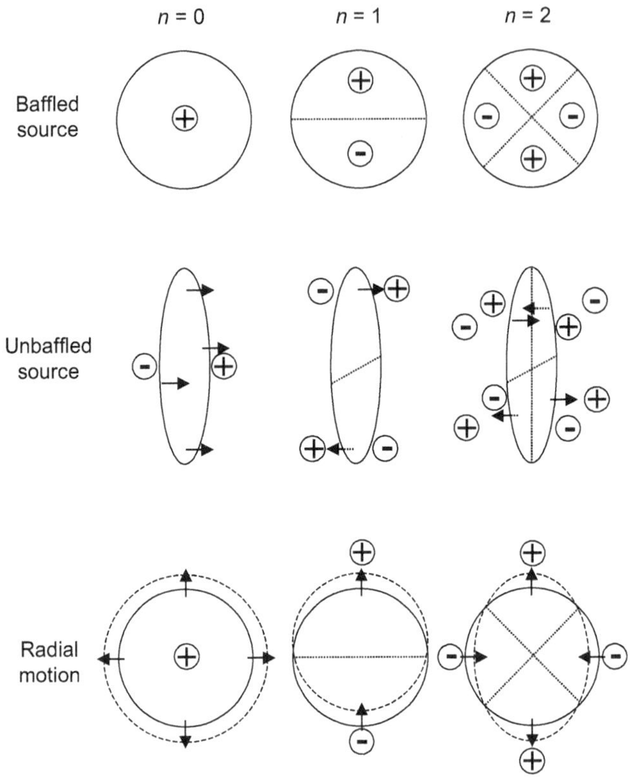

FIGURE 6-7 Indication of the equivalent multipoles for various wheel modes. Top: modes of a disc set in a rigid baffle; middle: modes of an unbaffled disc; bottom: radial modes

The frequency dependence of the radiation ratio for these various modes can thus be explained in terms of the characteristic frequency dependence of simple sources. As the mode order, n, increases, the order of the multipole increases. The frequency at which σ becomes equal to unity also increases as n is increased. This can be related to the size of the component simple sources on the wheel surface, which reduce as the wavelength around the wheel reduces.

It may be noted that the wavelengths in the wheel are such that the critical frequency is very low and no acoustic short-circuiting occurs. The wheel web has a typical thickness of 25 mm and a flat steel plate of this thickness would have a critical frequency of 500 Hz (equation (6.16)). The zero-nodal-circle axial modes have wavelengths that are determined by the wheel tyre which has a thickness of 135 mm and is even stiffer than the web.

Figure 6.8 shows corresponding BEM results calculated for one-nodal-circle axial modes. In these mode shapes the maximum axial motion is in the web region; the tyre has very little axial motion (see Figure 4.3). These results are very similar to those for the zero-nodal-circle axial modes, although for frequencies below about 500 Hz they are approximately a factor of 2 lower. Apart from this factor, the

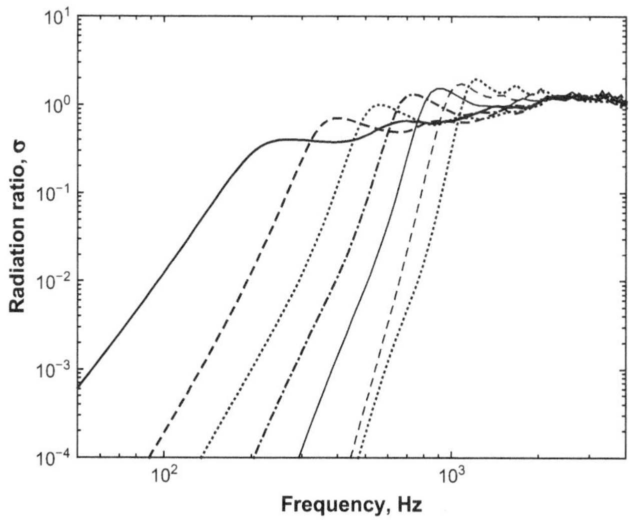

FIGURE 6-8 Radiation ratio of one-nodal-circle axial modes of 920 mm freight wheel with various numbers of nodal diameters, n. Thick lines: —, $n = 0$; – – –, $n = 1$; ····, $n = 2$; – · – ·, $n = 3$; thin lines: —, $n = 4$; – – –, $n = 5$; ····, $n = 6$

number of nodal diameters, n, is more important in determining the radiation ratio than the deformed shape of the cross-section.

Due to the curved shape of the wheel web, the predominantly radial modes of this wheel are found to contain considerable axial motion, similar in form to the one-nodal-circle modes, see Figure 4.3. Their radiation ratios are consequently rather complex. The radiation ratio is therefore presented for a purely radial motion of the tyre, in which the web is assumed not to vibrate [6.13]. This result is shown in Figure 6.9. Here the radiation ratio does not reach unity until close to 1 kHz. At very low frequencies the slope of each curve is less than those in Figure 6.6. For the $n = 0$ motion the low frequency slope is found to correspond to f^2. This motion is a radial in-phase pulsation of the whole tyre, which can be seen to be equivalent to a monopole, see Figure 6.7. For $n = 1$ motion, the tyre oscillates vertically, which thus corresponds to a dipole and the slope of the curve is f^4. For each value of n, the order of the multipole approximation at low frequency is one less than for the corresponding axial motion and the slope is $f^{2(n+1)}$. However, there is also a region where the curves appear to drop below this trend. This can be associated with partial cancellation between the inner and outer faces of the tyre.

Although these results have been given for each mode shape for a wide range of frequencies, allowing the radiation due to forced vibration to be determined from them, it is worth noting that most modes of vibration of the wheel occur in the region where the radiation ratio is close to unity. Figure 6.10 shows the results calculated using BEM for the (measured) natural frequencies of each mode. These are all within 1 dB of unity except for the low order zero-nodal-circle axial modes below 1 kHz. The latter have radiation indices of -1.7, -18.5, -7.8 and -2.6 dB for $n = 0, 1, 2, 3$.

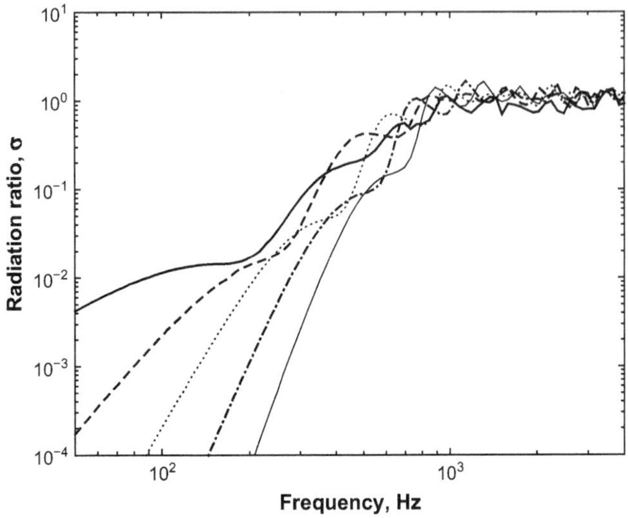

FIGURE 6-9 Radiation ratio for radial motion of the tyre of 920 mm freight wheel with various numbers of nodal diameters, n. Thick lines: —, $n = 0$; – – –, $n = 1$; ····, $n = 2$; – · – ·, $n = 3$; thin line: —, $n = 4$

In [6.13] boundary element results were presented for a series of notional wheels of different diameters and web thicknesses, from which the dependence of the radiation ratio on various parameters could be established. This allowed simple engineering formulae to be developed to describe the radiation ratio of a wheel in

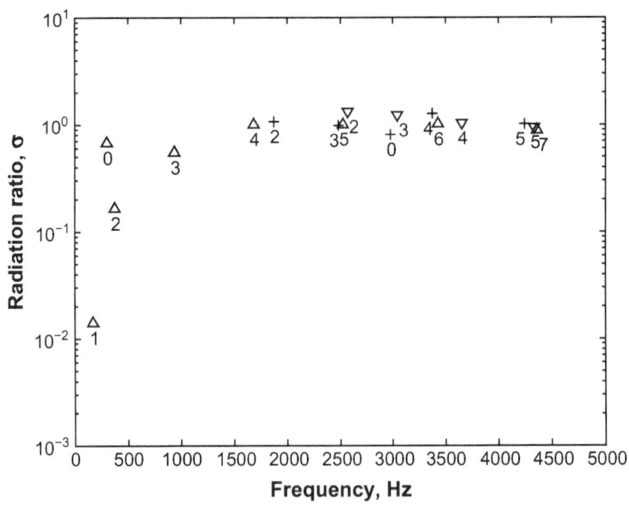

FIGURE 6-10 Radiation ratio for modes of 920 mm freight wheel at resonance frequencies. Numbers indicate n, number of nodal diameters. Δ, zero-nodal-circle axial modes; ∇, one-nodal-circle axial modes; +, radial modes

terms of basic geometrical parameters that can be implemented into a wheel/rail noise prediction model.

6.3.2 Directivity of wheel radiation

The radiation ratio allows the sound power to be predicted from a knowledge of the vibration amplitude. To obtain the sound pressure at a given receiver location, the directivity is also required. In practice, however, the precise details of the directivity are usually unimportant, provided that the average sound level during the passage of a train can be predicted reliably. Similarly, for interior noise, the airborne sound transmission into the vehicle is determined by the overall sound field in a region under the vehicle rather than the sound pressure at any particular location under the floor. In addition, the directivity of the sound radiated from a wheel will be affected by the local environment such as the bogie frame and vehicle body and by turbulence. For these reasons, a detailed knowledge of the directivity is of little practical relevance and it is sufficient to obtain an indication of the approximate nature of the directivity.

In reference [6.14], measurements were presented of the sound field around five different types of railway wheel, excited at their natural frequencies. Microphones were located on a semi-circular frame of radius 1.5 m at intervals of 5°, as shown in Figure 6.11. The wheel was excited radially on the tyre, on the opposite side to the microphones. The floor was covered in absorptive material.

Example results are presented here in the form of the directivity factor for several modes of an NS Intercity 920 mm diameter wheel. This wheel has an almost straight web so that its radial and axial motions are less well coupled than those of wheels such as the standard freight wheel, as shown in Figure 4.4 (compare with Figure 4.3). Figure 6.12 shows results for radial modes and Figure 6.13 shows those for one-nodal-circle axial modes. The sound pressure was generally found to have a minimum on or close to the axis of the wheel for both types of mode. This is due to

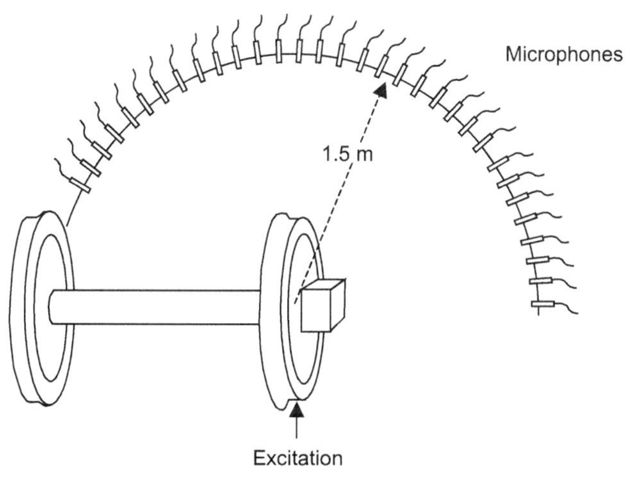

FIGURE 6-11 Location of microphones around a wheel to measure directivity

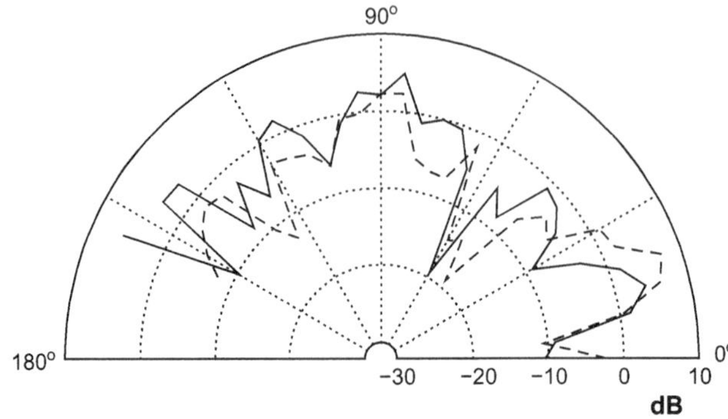

FIGURE 6-12 Directivity of an NS Intercity wheel for radial modes. The wheel axis lies at the angle 0°, angles greater than 90° represent positions behind the wheel. ——, $n = 3$, 2470 Hz; – – –, $n = 5$, 4060 Hz, adapted from [6.14]

cancellation between the contributions from different parts of the mode shape. Conversely, a maximum was obtained on-axis for the $n = 0$ mode [6.14]. However, apart from this, it can be seen that there is no significant directivity, on average, for the radial modes. In contrast, the axial modes have a sound pressure level above the wheel tread (90°) that is about 10 dB lower on average than in the axial direction (0°).

Results for other wheels with a curved web in reference [6.14] were found to resemble the radial mode results of Figure 6.12 for both radial and axial modes, due to the higher coupling of axial and radial motion. It is therefore more appropriate to calculate the radiation separately from radial and axial motions, rather than according to the type of modes, as already noted for the radiation ratio.

For radial motion the sound field can be adequately approximated by an omnidirectional field, whereas for the axial motion a dipole distribution is more appropriate since this has a maximum in the axial direction and a minimum in the plane of

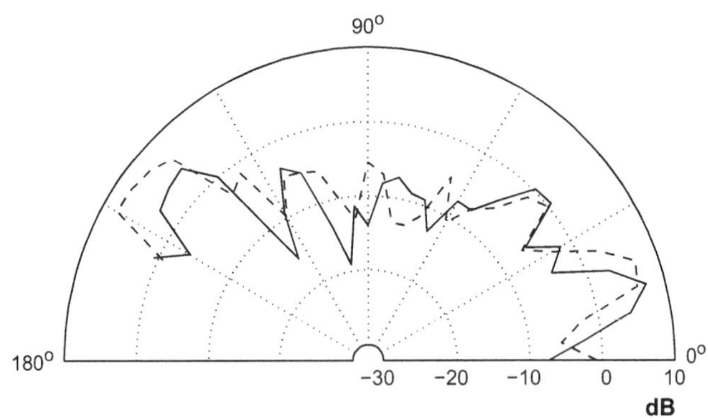

FIGURE 6-13 Directivity of an NS Intercity wheel for one-nodal-circle axial modes. The wheel axis lies at the angle 0°, angles greater than 90° represent positions behind the wheel. ——, $n = 3$, 2740 Hz; – – –, $n = 5$, 3850 Hz, adapted from [6.14]

the wheel. The corresponding directivity is given by equation (6.8) and plotted in Figure 6.4. It is also possible to fit the directivity using a combination of monopole and dipole functions [6.15] but this has only marginal additional benefit if the objective is to predict the average sound pressure level during the passage of a train or vehicle. The prediction of sound pressure levels is discussed further in Section 6.6.

6.4 RAIL RADIATION

6.4.1 Radiation ratio for two-dimensional case

A rail effectively has a two-dimensional geometry. If it could be assumed that the rail vibration is independent of distance along the rail, the sound field would also be two-dimensional and could be calculated efficiently using a two-dimensional boundary element model. The validity of this assumption will be considered in Section 6.4.6. First, the results of using such a two-dimensional model are presented here. The radiation ratio for a UIC60 rail vibrating vertically or horizontally in a free field is shown in Figure 6.14.

At low frequency these results are proportional to f^3, where f is the frequency, corresponding to a line dipole (see equation (6.13)). The size of the equivalent oscillating cylinder can be derived by fitting equation (6.13) to these results, as shown. The results for the lateral motion correspond closely to a cylinder of diameter 172 mm, equal to the rail height. However, for the vertical direction, the radiation is

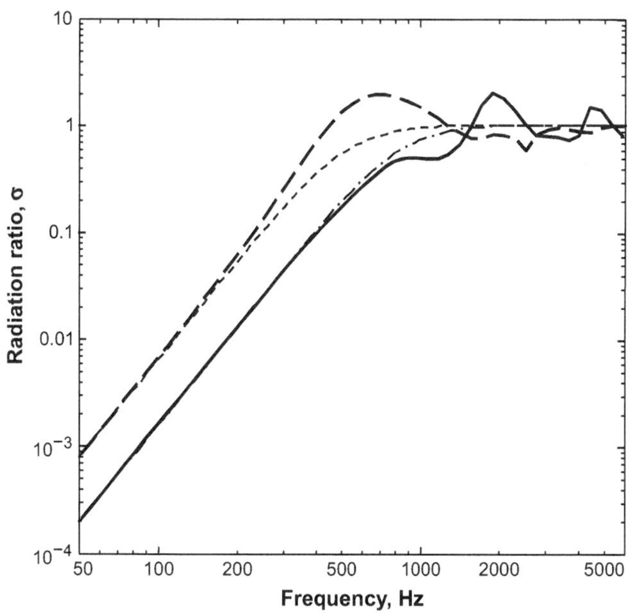

FIGURE 6-14 Radiation ratio of UIC 60 rail section modelled using two-dimensional BEM. —, vertical motion; – – –, horizontal motion. Equivalent cylinders: – · – ·, vertical motion (radius 54 mm); - - -, horizontal motion (radius 86 mm)

considerably less than would be obtained for a cylinder of diameter equal to the rail width (150 mm). This is caused by partial cancellation between the radiation from the upper and lower surfaces of the head and foot. The equivalent cylinder has a radius of only 53 mm.

At around 1 kHz peaks and troughs appear in the radiation ratio curve, which correspond to constructive and destructive interference caused by phase differences in the sound from different parts of the rail section. In this frequency region the wavelength of sound in air is similar to the characteristic dimensions of the rail cross-section. At about 700 Hz a strong peak occurs in the lateral radiation ratio, which does not occur for the equivalent cylinder. Similarly, at 1 kHz a dip occurs for the vertical direction. At this frequency the wavelength in air is approximately equal to twice the distance between the radiating faces of the top of the rail-head and the rail foot. The radiation from these two surfaces in the vertical direction thus interferes destructively at this frequency. At 2 kHz, where the distance equals a whole wavelength, the sound interferes constructively and a peak is observed.

6.4.2 Effect of ground plane

The above results have been obtained for a rail radiating in free field conditions. In practice, the rail is located close to the ground. Depending on the assumptions made, this can have a significant effect on the radiation ratio at low frequencies.

Two extreme cases can be considered, shown in Figure 6.15. First, if the rail is actually mounted on a rigid ground plane, the bottom of the rail foot cannot radiate sound. This is the case on a slab track if the rail is supported on a continuous pad. As the top and bottom areas of the rail section are no longer equal, a net monopole component is present and the radiation ratio is increased at low frequencies, as shown in Figure 6.16. The radiation ratio is proportional to f at low frequencies.

On the other hand, if the rail is located close to the rigid ground but is not in contact with it, the dipole source due to vertical motion will be reflected in the ground plane and together the source and its image will form a quadrupole. The radiation is therefore reduced at low frequency, as shown in Figure 6.16. However, between 500 and 1000 Hz the order of these three results is reversed and a peak appears for the case above a rigid ground, caused by a standing wave occurring under the rail foot.

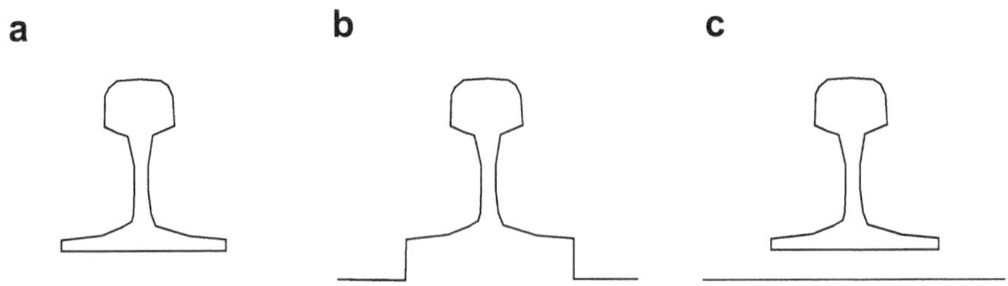

FIGURE 6-15 Two-dimensional boundary element models of a rail. (a) In free space; (b) on rigid ground; and (c) located above a rigid ground

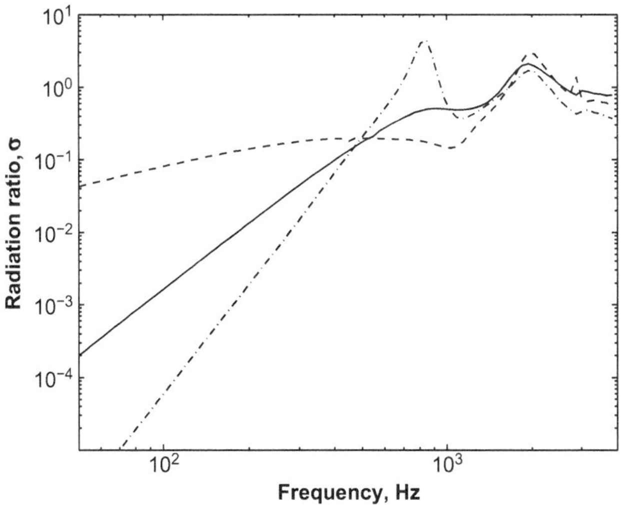

FIGURE 6-16 Sound radiation from vertical vibration of a rail in the presence of a rigid ground calculated using two-dimensional boundary elements. —, in free field; – – –, attached to the surface; – · – ·, just above rigid surface

For a rail in ballasted track, the situation is more complex. Part of the rail will be supported on the rail pads directly above the sleepers while the remainder is located a small distance above the ground. However, this ground is not rigid but rather is partially absorptive. It is therefore not a simple matter to determine the most appropriate model for this situation. Despite these uncertainties, in ballasted track the rail component is masked by the sound radiation from the sleeper for frequencies below about 400 Hz, see Figure 2.14, and it is not critical which model is used for the rail.

6.4.3 Directivity in two dimensions

In [6.16] some measurements were presented of the directivity from the rail in the vertical plane. The rail was excited by an impact hammer vertically or laterally and the sound pressure was measured at a series of angles at 1.5 m from the rail, directly opposite the excitation point. Example results for three one-third octave bands are given in Figure 6.17 for vertical excitation and in Figure 6.18 for lateral excitation. These are presented here in the form of a directivity factor. As with the wheel, the results do not show any strong directivity, although the vertical motion tends to radiate slightly more strongly in the vertical direction and the lateral motion more strongly in the lateral direction.

It should be remembered that the sound radiation from the rail will also be affected by the presence of the vehicle above the track so that these directivities will be considerably modified during a train passage. Therefore a simplified calculation model has been adopted in [6.17] consisting of a line dipole directivity for lateral vibration, see equation (6.15), and omnidirectional sound radiation for the vertical motion.

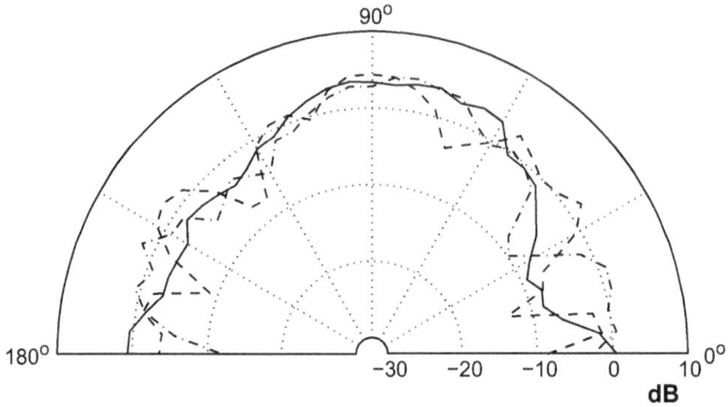

FIGURE 6-17 Directivity measured from rail vibration excited by vertical force. ——, 500 Hz; – – –, 1000 Hz; – · – ·, 2000 Hz, adapted from [6.16]

6.4.4 Directivity in three dimensions

In order to find the directivity of sound radiation in the direction along the track, it is necessary to take account of variations of the rail vibration along its length. As seen in Chapter 3, waves propagate in the rail with a frequency-dependent wavenumber and decay rate. In order to calculate the radiation from such a wave, a model can be used based on a line array of simple acoustic sources to represent the rail [6.18]. This is a valid approach for the radiation from the rail provided that cross-sectional deformation effects can be neglected. The sources must have a separation that is less than a quarter of an acoustic wavelength at the frequency of interest (and less than a quarter of a structural wavelength). The source strength of each source is chosen according to the amplitude and phase of the rail vibration at that location.

The calculations presented here are based on the vertical track vibration modelled using a Timoshenko beam with a pad stiffness of 300 MN/m^2, a sleeper mass of 250 kg/m and other parameters as in Table 3.3. This is excited by a point force at $x = 0$; vibration is transmitted symmetrically to either side. The complex vibration

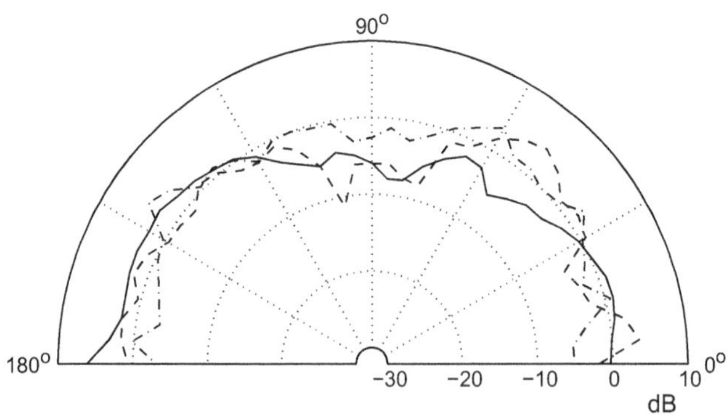

FIGURE 6-18 Directivity measured from rail vibration excited by lateral force. ——, 500 Hz; – – –, 1000 Hz; – · – ·, 2000 Hz, adapted from [6.16]

amplitude is determined at many points along the rail and used to define the source strengths and phases of a set of point sources located along the axis of the rail. The sound field due to these point sources is then calculated in terms of the pressure and particle velocity at a number of receiver positions, from which the intensity can also be determined. The sources are modelled here as dipoles, although similar results are obtained using monopoles.

Figure 6.19 shows the sound pressure level at various distances from the rail plotted against distance along the track for four example frequencies. The magnitude of the rail vibration is also shown on the same graph. The sound pressure and velocity levels have been plotted to arbitrary reference values to allow them to be plotted together. The highest sound pressure level curve (at least at $x = 0$) corresponds to a distance of 0.5 m from the rail, the subsequent ones to 1 m, 2 m and 4 m.

At 125 Hz the track vibration consists only of near-field waves and the decay rate is high, at around 10 dB/m, see also Figure 3.30. The source is therefore localized to a region close to the forcing point and the sound radiation decreases at approximately 6 dB per doubling of distance from the source. At 250 Hz the decay rate

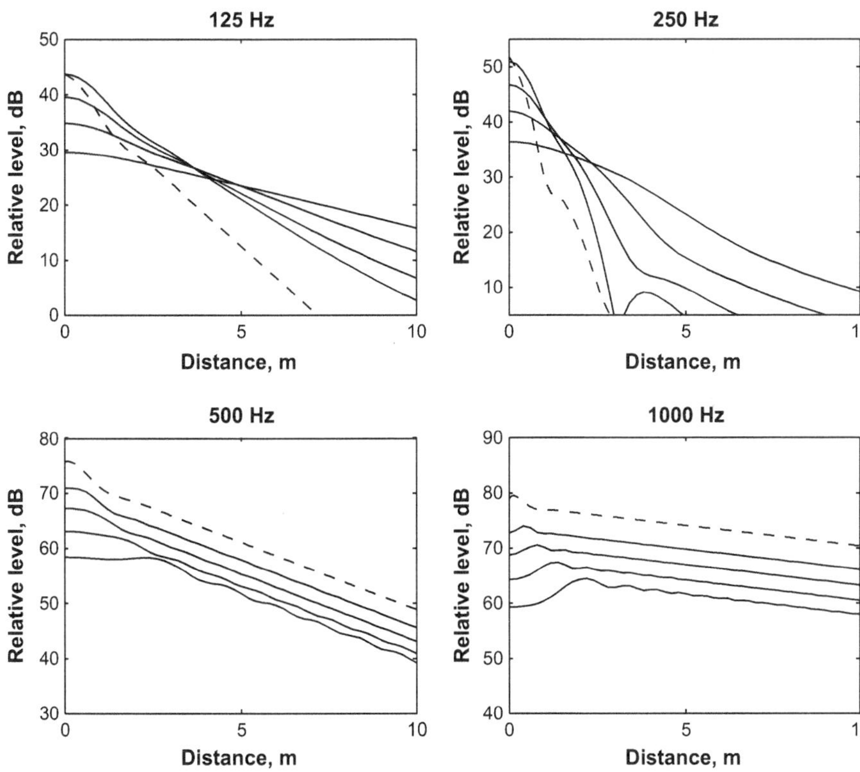

FIGURE 6-19 Sound radiation from a rail versus distance along the track from the forcing position: —, relative sound pressure level at various lateral distances (0.5 m, 1 m, 2 m, and 4 m) from the rail; – – –, relative source strength obtained from rail vibration. Based on vertical rail vibration, parameters as in Table 3.3

reaches a peak of about 20 dB/m and the sound is again radiated only from the vicinity of the forcing point.

At higher frequencies, propagating waves in the rail are transmitted over a significant length of the track. At 500 Hz the decay rate is 2 dB/m, and at 1000 Hz it is only 0.7 dB/m, see Figure 3.30. The radiation from the rail therefore extends over a considerable length, as seen in Figure 6.19. Particularly for 1 kHz, it can be seen that the radiated level is relatively low for small values of x and reaches a peak at some distance away from the forcing point. This distance increases as the receiver is moved further away from the track. This is due to the propagating wave in the rail which radiates sound at an angle to the normal.

To illustrate this further, Figure 6.20 shows the sound intensity calculated at various positions relative to the rail. At 125 Hz the sound radiation can be seen to be approximately omnidirectional relative to the forcing point. At 250 Hz the sound is again radiated only from the vicinity of the forcing point. Due to the interaction between the two near-field waves at this frequency, the radiation is focused in a region around ±45° relative to the plane of the forcing point, at least at these relatively close distances.

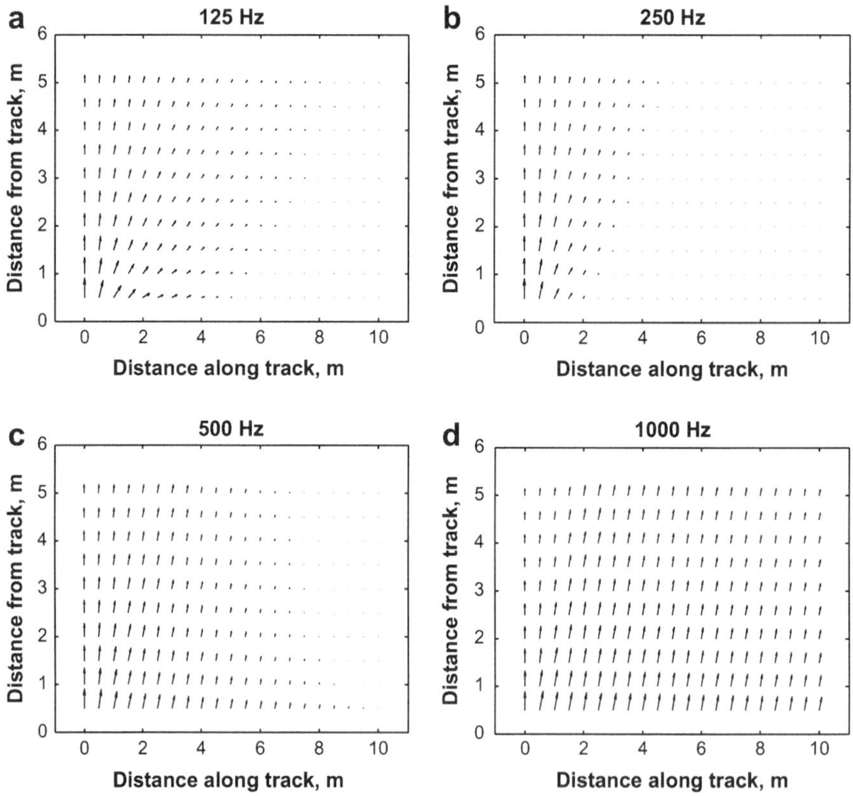

FIGURE 6-20 Sound intensity predicted due to rail vibration at various frequencies. Intensity vectors are shown scaled in decibels with a range of 25 dB

At the higher frequencies the radiation from the rail extends over a considerable length, as already seen in Figure 6.19. Of particular note is the direction of the intensity vectors. Apart from the region close to the forcing point at $x = 0$, the direction of the intensity vector is close to that determined by the wavenumbers in the rail, k_r, and in air, k_{air}. At 1 kHz, $k_r = 4.8$ rad/m and $k_{air} = 18.3$ rad/m, giving an angle to the normal of $\sin^{-1}(k_r/k_{air}) = 15.2°$. This agrees closely with the mean angle over all the points shown (omitting those at $x = 0$), which is 14.8° at 1 kHz, with a standard deviation of 1.9°.

It is clear from these results that, due to the extended nature of the rail as a source, the spatial distribution of the sound field in the direction along the track does not lend itself to the use of a simple directivity factor.

6.4.5 Relationship between decay rate and noise

Returning to the calculation of sound power from the rail, as the rail vibration is not uniform along its length it is necessary to take account of the rate of decay of vibration with distance. The general expression for radiated sound power W from a vibrating structure, equation (6.1), can be written in the following form for an infinite rail:

$$W = \frac{1}{2}\rho_0 c_0 \sigma P \int_{-\infty}^{\infty} |v(x)|^2 dx \tag{6.17}$$

where $v(x) = i\omega u(x)$ is the velocity amplitude at x, $\rho_0 c_0$ is the characteristic acoustic impedance of air and σ is the radiation ratio, which depends on frequency. The factor {1/2} appears as the velocity amplitude should be converted to an rms value. P is a perimeter length of the cross-section; since the velocity in the definition of σ should be that normal to the surface, only that part of the perimeter projected onto a plane perpendicular to the motion v is included here. For lateral motion of a rail this is twice the height, but for vertical motion it also includes the top and bottom widths of the foot and head. For simplicity, in this section a simple beam model for the rail vibration is considered, as described in Section 3.2. From the form of motion in equation (3.17), and making use of symmetry about $x = 0$:

$$W = \rho_0 c_0 \sigma P \frac{|v(0)|^2}{2} \int_0^{\infty} \left| e^{-ikx} - ie^{-kx} \right|^2 dx \tag{6.18}$$

The velocity $v(0)$ at the excitation point is determined by the wheel/rail interaction, as described in Chapter 5. The amplitude of each wave is $v(0)/\sqrt{2}$. If a Timoshenko beam is used instead, the wavenumbers in the two exponential terms will differ slightly. The track dynamics has most effect on W via the integral term. Above the cut-on frequency of propagating waves, the evanescent wave component has only a small contribution to the integral since it decays rapidly with x, and the above expression can be approximated by

$$W \approx \frac{1}{2}\rho_0 c_0 \sigma P |v(0)|^2 \int_0^{\infty} \left| e^{-ikx} \right|^2 dx \tag{6.19}$$

For a complex valued wavenumber, $\tilde{k} = k_r + ik_i$ (with k_i negative), the integral in equation (6.19) reduces to

$$\int_0^\infty \left| e^{-i\tilde{k}x} \right|^2 dx = \int_0^\infty e^{2k_i x} dx = \left[\frac{e^{2k_i x}}{2k_i} \right]_0^\infty = \frac{-1}{2k_i} \tag{6.20}$$

giving

$$W \approx \frac{\rho_0 c_0 \sigma P |v(0)|^2}{-4k_i} \tag{6.21}$$

It was shown in equation (3.14) that the imaginary part of the wavenumber, here denoted k_i, is proportional to the decay rate in dB/m, $\Delta = -8.686k_i$. Thus, taking decibels, the sound power level L_W is given by

$$L_W \approx 10\log_{10}\left(4.343\rho_0 c_0 P \frac{v_{ref}^2}{W_{ref}}\right) + 10\log_{10}\sigma + 10\log_{10}\left(\frac{|v(0)|^2}{2v_{ref}^2}\right) - 10\log_{10}\Delta \tag{6.22}$$

where W_{ref} is the reference value used for the definition of sound power level and v_{ref} is the corresponding reference value for velocity. The first term is a constant. Apart from this, L_W is related to the radiation index, velocity level and the decay rate through the term $-10\log_{10}\Delta$. A doubling of decay rate is equivalent to a 3 dB reduction in radiated noise. Consequently, it is usual, as in Chapter 3, to plot track decay rates on a logarithmic scale. This may appear, at first sight, like taking decibels of a decibel value. However, it is emphasized that the decay rate in dB/m is not a simple decibel value but, as has been seen, is proportional to the imaginary part of the wavenumber, which has units rad/m.

The effect of the approximation in equation (6.19) relative to the full result of equation (6.18) is illustrated in Figure 6.21(a) for an Euler–Bernoulli beam on a single layer support of stiffness 100 MN/m^2, as used in Section 3.2. The dashed line indicates the effect of ignoring the evanescent wave completely, whereas the solid line is the effect of including both waves but adding their powers (each determined according to equation (6.19)) and ignoring any interaction between them. Above the cut-on frequency of the propagating waves (200 Hz) the results are close to 0 dB, that is the approximation is valid and the propagating wave dominates the radiation. Below the cut-on frequency, the sound power level is underestimated by 1.8 dB if both waves are included independently. Thus there is a small effect of ignoring the cross term between the two waves; this interaction is important near the forcing point and hence influences the result for cases with high decay rates. The error increases to 4.7 dB if only a single wave is considered, as the two waves are equally important below the cut-on frequency. It may be noted, however, that below the cut-on frequency the rail radiation is small and sleeper noise will tend to dominate the sound radiation from the track (see Figure 2.14).

Figure 6.21(b) shows equivalent results for a beam on a two-layer foundation, as in Section 3.3. This shows similar trends, except that the cut-on of bending waves occurs at 400 Hz. Propagating waves also occur between 100 and 200 Hz but with a high decay rate.

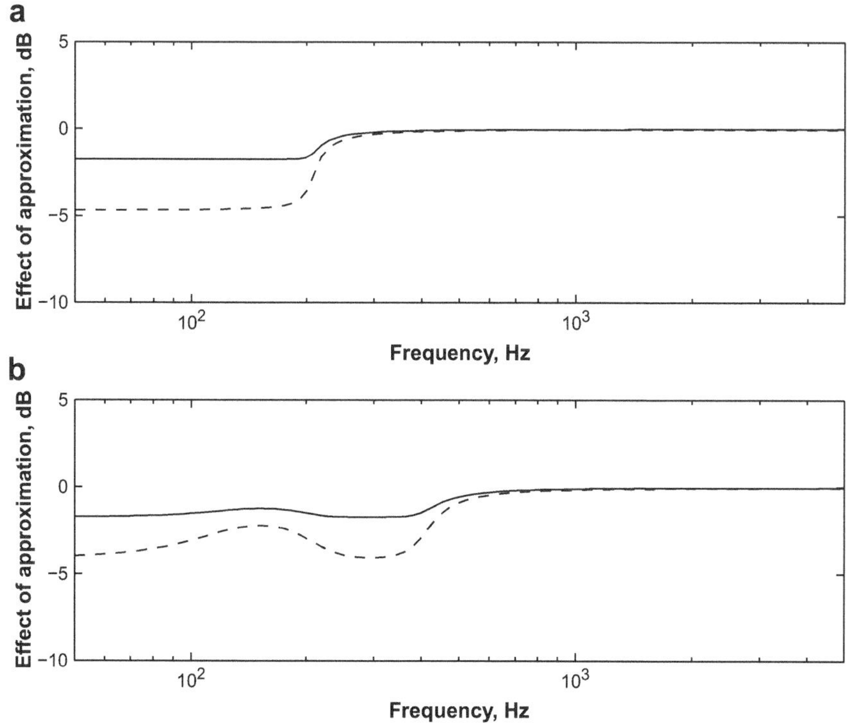

FIGURE 6-21 Effect on sound power estimate of ignoring interaction between two waves in a supported Euler–Bernoulli beam representing a UIC60 rail. —, both waves included but assumed incoherent, – – –, only single (propagating) wave included. (a) Single layer support, parameters as in Table 3.1, (b) two-layer support, parameters as in Table 3.2

6.4.6 Three-dimensional effects on sound power

Although the rail is infinite in length, with a two-dimensional geometry, as has been seen the vibration is not uniform along its length. Nevertheless, the radiation ratio determined using a two-dimensional approach should be satisfactory provided that (i) the structural wavelength in the rail is much greater than the wavelength of sound in air, and (ii) the decay of vibration with distance along the rail is small. If the first condition is not satisfied, acoustic short-circuiting occurs. If the second condition is not satisfied, the rail vibration will be localized close to the wheel/rail contact positions, and will tend to form a point source rather than a line source.

These two requirements have been considered in [6.18] to provide quantitative conditions for the validity of a two-dimensional approach. Calculations were performed using a line array of simple sources, in the same way as described in Section 6.4.4 above. This was done for a range of complex wavenumbers in the rail. Only a single wave was considered in the rail, propagating to $x \geq 0$ and $x \leq 0$. As before, the source strength of the array of sources was chosen according to the rail vibration amplitude and phase at that location.

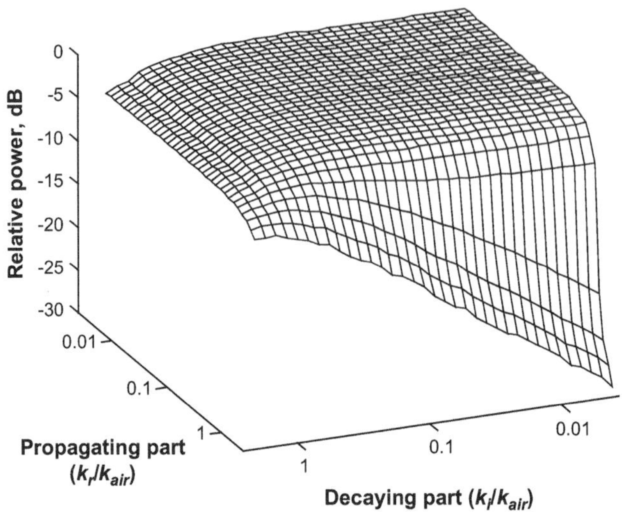

FIGURE 6-22 Variation of sound power W/W_{2D} with rail non-dimensional wavenumber [6.18]

The results were expressed as the ratio between the sound power, W, radiated by the three-dimensional distribution of sources and the result obtained using the two-dimensional approximation, W_{2D}. Figure 6.22 shows this ratio in decibel form as a three-dimensional plot. The horizontal axes are the real and imaginary parts of the rail wavenumber normalized by the acoustic wavenumber, k_{air}. The real part k_r/k_{air} corresponds to propagating waves and the imaginary part $-k_i/k_{air}$ to the decay with distance. It can be seen that the two-dimensional result gives a good approximation to the radiated sound power (ratio close to 0 dB) over much of the range of complex wavenumber, particularly for $k_r \ll k_{air}$ and $-k_i \ll k_{air}$. It was confirmed that the results in this non-dimensional form were independent of frequency. The same results are shown in Figure 6.23 in the form of a contour plot.

When the real part of the wavenumber, k_r, is greater than the acoustic wavenumber, k_{air} (i.e. the wavelength in the rail is smaller than that in air), an acoustic short-circuiting effect occurs and the radiation falls considerably. Conversely, for $k_r \ll k_{air}$ the wavelength of structural waves is large compared with the wavelength of sound and a two-dimensional approach is valid. When the decay rate is high, $-k_i > k_{air}$, the source is localized close to the forcing point and the rail resembles a point source rather than a line source. Since the radiation ratio for a point dipole is proportional to f^4 (Figure 6.3) whereas for the line dipole this is f^3 (Figure 6.5) the sound power additionally falls in proportion to f (and therefore to k_{air}) in this region.

Figure 6.23 shows that sound power will be predicted to within 2 dB of the correct value using a two-dimensional model, provided that the propagating wavenumber, $k_r < 0.65\, k_{air}$ and the decaying part of the wavenumber, $-k_i < k_{air}$. Analysis of the wavenumbers of a rail installed in the track [6.18] showed that for tracks with either stiff or soft rail pads the waves were all in the region enclosed by the 2 dB contour for frequencies above about 250 Hz. These are shown in Figures 6.24 and 6.25. For lower frequencies the waves all have real and imaginary parts that become large compared with k_{air} and a point source model becomes appropriate. No waves appeared in the

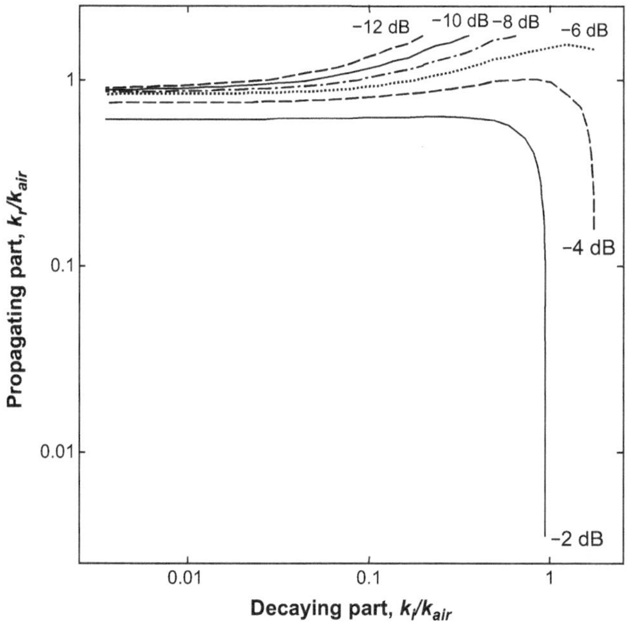

FIGURE 6-23 Contour plot showing variation of sound power W/W_{2D} with decaying and propagating parts of non-dimensional wavenumber [6.18]

short-circuiting region where $k_r > k_{air}$; this is a consequence of the fact that rails have to be stiff in order to support a train and therefore they have high wavespeeds, which turn out to be mostly much greater than the speed of acoustic waves.

6.4.7 Waveguide finite and boundary element methods

An alternative approach can be used to calculate the sound radiation from a rail allowing directly for the three-dimensional effects. For a long structure such as a rail, which has effectively two-dimensional geometry, the waveguide finite element method [6.19, 6.20] can be used to calculate its vibration. This method is based on a two-dimensional finite element discretization of the cross-section. Solutions are found for a sequence of wavenumbers along the rail. The response in the spatial domain can be recovered by using an inverse Fourier transform. This can be directly coupled to a wavenumber boundary element approach [6.21] to calculate the sound radiation from a series of two-dimensional problems at different wavenumbers in the direction along the rail. The radiated power can be determined directly in the wavenumber domain. The sound pressure can also be obtained from an inverse Fourier transform. This approach has been used, for example, to calculate the sound transmission through extruded aluminium panels [6.22].

Such an approach has also been used to calculate the sound radiation from a rail [6.23]. Results for a conventional rail confirmed that the approach described in previous sections gives acceptable results. However, the method was also extended to calculate the radiation from an embedded tram rail, directly accounting for the

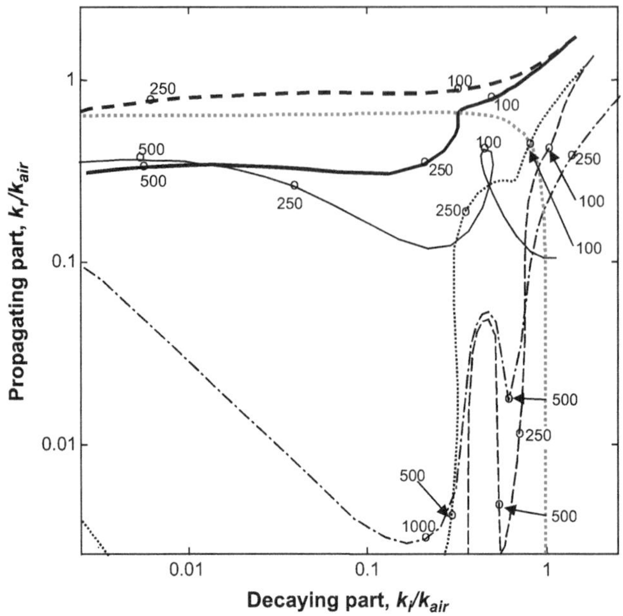

FIGURE 6-24 Propagating versus decaying part of normalized wavenumber calculated for track with soft pads. The grey dotted line is the −2 dB contour from Figure 6.23. Numbers indicate frequency in Hz. ▬, vertical bending wave; · · · ·, vertical nearfield wave; ▬ ▬ ▬, lateral bending wave; ——, torsional wave; – · – ·, web bending wave; – – –, lateral nearfield wave [6.18]

radiation from the upper surface of the embedding material. This method is particularly useful for such situations.

6.4.8 Implications for measuring using a microphone array

An implication of the three-dimensional nature of the sound radiated by a rail is that measurements made using a microphone array can be very misleading in their assessment of the contribution from the rail radiation [6.24]. Microphone arrays have been widely used in order to identify sound sources on moving trains, particularly for aerodynamic sources, see Chapter 8. However, they have also been used to study rolling noise [6.25–6.34]. In such studies it has often been found that the wheel is the dominant source of rolling noise, whereas analysis based on models such as TWINS shows that the rail can be the dominant source in much of the frequency range, as discussed in Chapter 2. By considering the nature of the sound radiation from the rail it is found that there are some fundamental problems in measuring it using a microphone array.

Microphone arrays may be arranged in a single line, which may be horizontal to locate sources along the train or vertical to locate their height. They may also be arranged in T or X shapes, a 'star' shape or a fully two-dimensional arrangement to locate sources in two dimensions. The potential spatial resolution depends on the

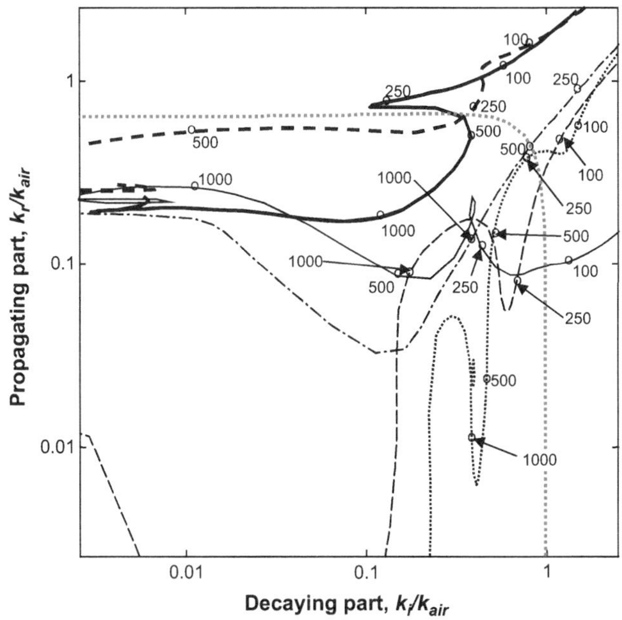

FIGURE 6-25 Propagating versus decaying part of normalized wavenumber calculated for track with stiff pads. The grey dotted line is the −2 dB contour from Figure 6.23. Numbers indicate frequency in Hz. **—**, vertical bending wave; · · · ·, vertical nearfield wave; **– – –**, lateral bending wave; ——, torsional wave; – · – ·, web bending wave; – – –, lateral nearfield wave [6.18]

microphone spacing, along with the acoustic wavelength. More details are given in Chapter 8.

The outputs from each microphone are added together to give the sound arriving from a particular direction. The array can be used with different time delays to determine the radiation in different directions, but in the present application it is normally used with no time delays in order to extract the sound radiated from the sources directly in front of the array. Even in cases where a swept focus is used [6.25] the range of angles considered is generally only a few degrees.

It is not always appreciated that, when a microphone array is used to localize sources, an implicit source model is required. This is usually that the source consists of a distribution of uncorrelated omnidirectional point sources located in a plane at some known distance from the array. This is a quite reasonable assumption for aerodynamic noise but is not correct for the noise radiated by the rail.

The rail consists of an extended, correlated source, not an array of uncorrelated sources. As seen in Section 6.4.4 the radiation occurs in the form of plane waves orientated at a particular angle to the track. This angle was found to be around 15° for radiation from vertical bending waves in Figure 6.20, but can be much larger for lateral bending waves [6.18].

Using a similar modelling approach to that used in Section 6.4.4 above, the result of using a horizontal microphone array to measure the radiation from a rail was simulated in [6.24]. The rail was represented by a series of correlated sources and the sound pressure was calculated at a receiver location at the trackside. An average during a train pass-by was calculated, ignoring the Doppler effect. The pressures were also combined to simulate the output from a one-dimensional microphone array.

Figure 6.26 shows results calculated using the model of [6.24]. This shows the distribution of rail vibration amplitude along the rail, the output from a single microphone and the output from a microphone array for two different frequencies. The microphones are positioned at a distance of 5 m from the rail. The track parameters used here are the same as those used in the results of Section 6.4.4.

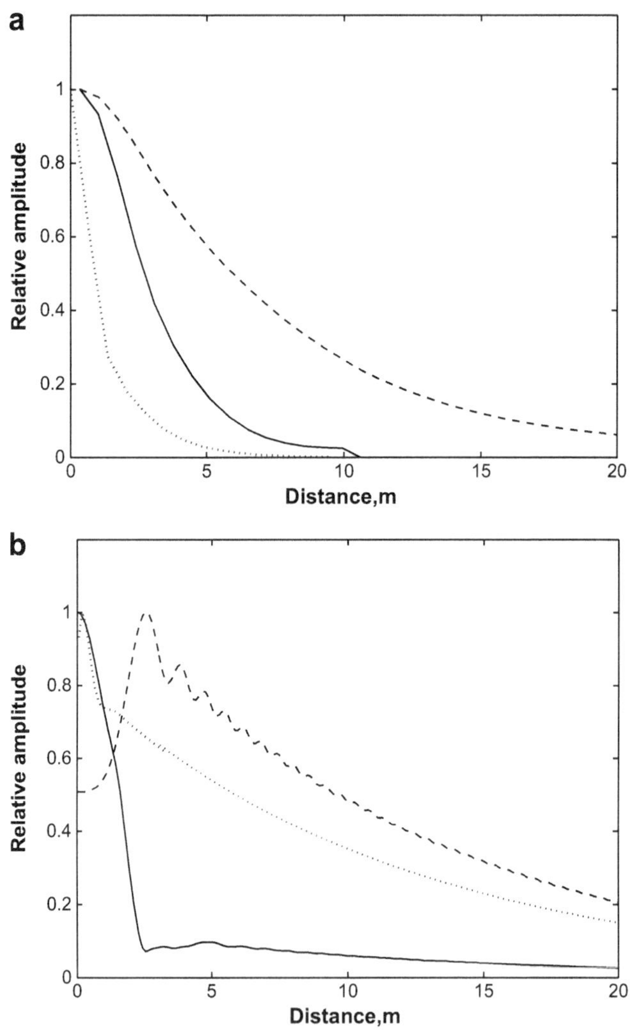

FIGURE 6-26 Simulation of the effect of using a microphone array to measure noise for vertical rail vibration, parameters as in Table 3.3, (a) 125 Hz, (b) 1000 Hz. · · · ·, relative source strength obtained from rail vibration; – – –, magnitude of sound pressure at single microphone versus distance along the track from the forcing position; —, output from microphone array

At 125 Hz the rail vibration is localized close to the forcing point, effectively forming a point source. The pressure at the single microphone decays gradually with distance, according to the radial distance from the forcing point, whereas the result of the microphone array allows the source to be localized more effectively.

The results for 1000 Hz show an extended source on the rail, as in Figure 6.19. The sound pressure results are similar to those seen in Figure 6.19 but the microphone array shows only a local region close to the forcing point. The coherent wavefront radiated at 15° to the normal is effectively suppressed by the array processing. Thus the array only detects the sound radiated by the discontinuity in the bending waves that occurs around the forcing point and effectively forms a point source, rejecting most of the rail radiation.

The overall effect of using a microphone array is summarized in Figure 6.27, from [6.24]. Figure 6.27 shows the difference between the overall sound detected by the microphone array and that measured by a single microphone for vertical and lateral vibration for two different tracks. As free waves propagate in the rail the result from the microphone array can be up to 10 dB too low due to omission of the propagating wave part. A slight overestimation occurs in the region where the decay rates are very high. Figure 6.28 shows the overall spectrum of rail noise and the part that is detected by the microphone array. Clearly, at frequencies above 500 Hz the microphone array tends to underestimate the rail contribution.

Thus it is clear that a horizontal array (or indeed a two-dimensional array) directed normal to the track will not detect a large part of the noise from the rail at high frequencies where free wave propagation occurs in the rail. Differences of up to 10 dB have been found. This may explain why measurements using microphone arrays tend to emphasize the wheel as the dominant source, whereas using theoretical

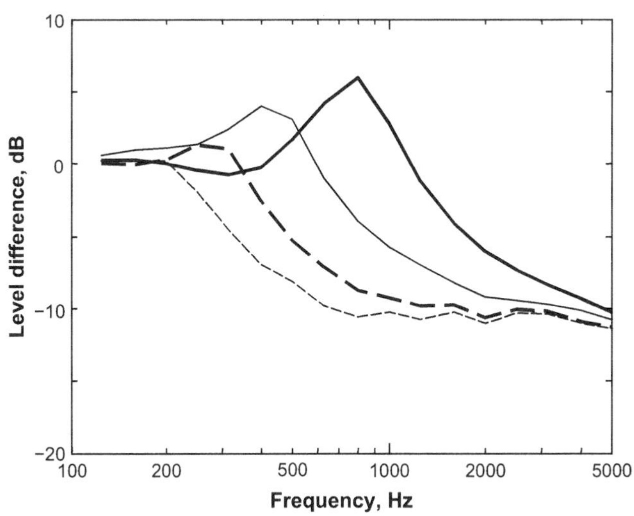

FIGURE 6-27 Overall effect of using microphone array to measure noise from rail vibration. —, vertical vibration, rail pad stiffness 700 MN/m; - - -, lateral vibration, rail pad stiffness 85 MN/m; —, vertical vibration, rail pad stiffness 200 MN/m; – – –, lateral vibration, rail pad stiffness 40 MN/m, from [6.24]

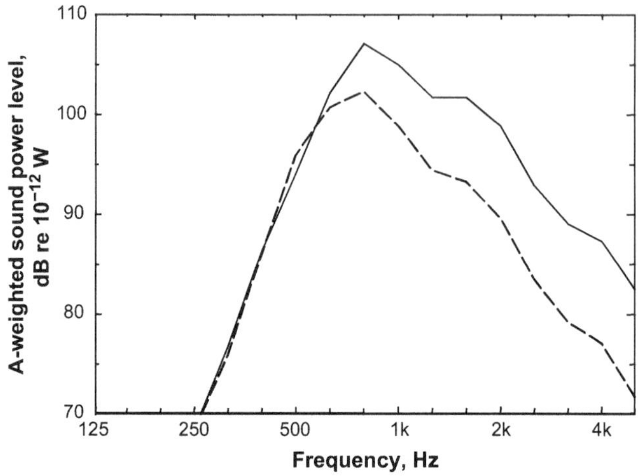

FIGURE 6-28 Effect of using microphone array on estimate of rail component of noise. —, from TWINS prediction (vertical pad stiffness 200 MN/m), - - -, as inferred from microphone array, from [6.24].

models the rail is also found to be an important source in many situations. Although it may be possible to use different focus angles to avoid this problem, the need for an implicit source model remains an issue and great caution is required when microphone arrays are used with extended coherent sources.

6.5 SLEEPER RADIATION

6.5.1 Radiation ratio of a single sleeper

The sound radiation from an individual sleeper can be considered as that from a flat rectangular surface located in a plane baffle. Its lower surface is hidden by the ballast and can therefore be neglected. Sleepers are typically 2.5 m long and 0.2 to 0.25 m wide. The vibration of concrete sleepers has been described in Section 3.7. It was seen that the response of the two ends can be considered to be independent at low frequencies (see Figure 3.58), where the response due to excitation at one rail consists of a pitching motion. The bending wavelengths of flexural modes occurring above 120 Hz are considerably longer than the acoustic wavelength (see Figure 3.54). The critical frequency, according to equation (6.16), is around 65 Hz (assuming $E = 3.75 \times 10^{10}$ N/m^2, $\rho = 2400$ kg/m^3, $\nu = 0.15$ and $h = 0.25$ m). This is lower than the natural frequency of any bending modes of the sleeper. Thus the sleeper vibrates like a rigid piston at low frequencies while the effect of bending modes at higher frequencies can be ignored.

Figure 6.29 shows results obtained using a Rayleigh integral for various dimensions of sleeper in which the sleeper vibration is assumed uniform. These radiation ratios all tend to unity above about 800 Hz. Between 300 and 800 Hz they are independent of the length, and determined only by the width, a, which here is 0.2 m. In this region the radiation ratio is given approximately by

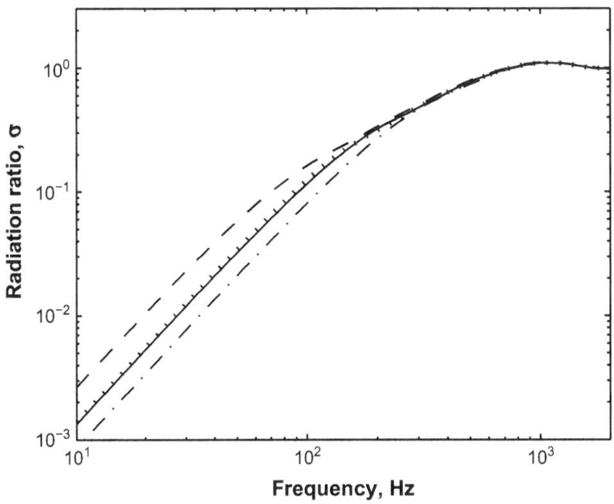

FIGURE 6-29 Radiation ratio of a sleeper set in a rigid baffle: – – –, 2.5 × 0.2 m; —, 1.25 × 0.2 m; – · – ·, 0.8 × 0.2 m; · · · ·, 2.5 × 0.2 m for pitching motion

$$\sigma \approx \frac{2.8af}{c_0} \tag{6.23}$$

At low frequencies the length of the sleeper, L, is also important and the radiation ratio is given approximately by

$$\sigma \approx \frac{2\pi aLf^2}{c_0^2} \tag{6.24}$$

This corresponds to the result in equation (6.5) for a baffled disc if a^2 is replaced by S/π where $S = aL$ is the surface area.

In addition, Figure 6.29 shows results for the 2.5 m long sleeper in which a linear dependence of vibration amplitude is assumed with one rail seat stationary (similar to Figure 3.54(a)). This result is similar to that for uniform motion based on using half the length.

6.5.2 Effect of multiple sleepers excited by the rail

When the rail is excited at a point, the track vibrates over a certain distance and all the sleepers within that distance also vibrate with a relative amplitude and phase that depends on the corresponding rail vibration. Figure 6.30 shows the relative amplitude and phase of several sleepers for the track parameters of Table 3.3. For these parameters, at low frequencies, sleeper 2 at 0.6 m from the excitation point has an amplitude approximately half that of sleeper 1 at 0 m and they are roughly in phase. Subsequent sleepers have much smaller vibration amplitude. Significant phase changes only occur once the bending waves in the rail cut on above 400 Hz, where the sleepers have more similar vibration amplitudes. However, they are also increasingly isolated from the rail at these higher frequencies.

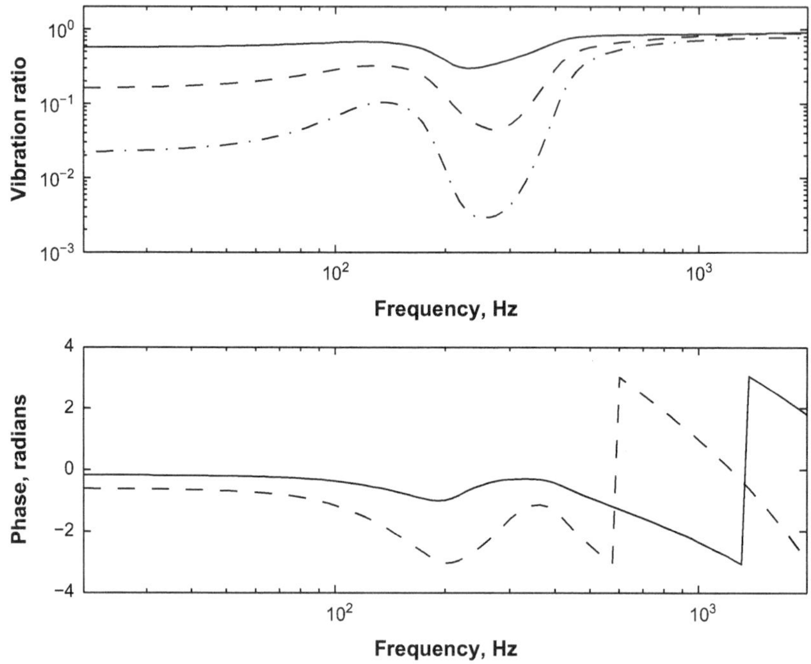

FIGURE 6-30 Vibration of successive sleepers relative to sleeper 1 at 0 m. —, sleeper 2 at 0.6 m; – – –, sleeper 3 at 1.2 m; – · – ·, sleeper 6 at 3.0 m

At low frequencies the distance between adjacent sleepers is small compared with the acoustic wavelength. As a result, it is no longer possible to treat each sleeper as an independent source. Figure 6.31 shows the radiation ratio calculated using the same Rayleigh integral approach as in the previous section but now including three adjacent sleepers. The outer ones are assigned a vibration amplitude of half that of the central one; all are assumed to be vibrating in phase. At low frequency this leads to an amplification of the sound radiation as these three sleepers become a single composite source. Note that the acoustic wavelength becomes equal to 1.2 m, twice the sleeper separation distance, at 286 Hz, above which the radiation ratio drops.

In order to find the increase at low frequencies in the combined radiation ratio relative to the result for a single sleeper, it is necessary to determine the combined squared volume velocity. If the amplitudes are in the ratios α:1:α, the combined squared volume velocity is $(1+2\alpha)^2 v_0^2 S^2$ where S is the surface area of one sleeper and v_0 is the vibration velocity of the central sleeper. The sum of the individual squared volume velocities is $(1+2\alpha^2)v_0^2 S^2$. Thus the ratio of these two is given by $(1+2\alpha)^2/(1+2\alpha^2)$. For $\alpha = 0.5$ this gives a ratio of 2.67 as found in Figure 6.31 below 100 Hz.

Taking the ratio of these two curves gives the amplification caused by the interaction between multiple sleepers. This is shown in Figure 6.32. Compared with these results from the Rayleigh integral, similar results can be found by using a single monopole to represent each sleeper, especially at low frequencies. Two results are also shown in which the source strengths at each frequency are chosen in

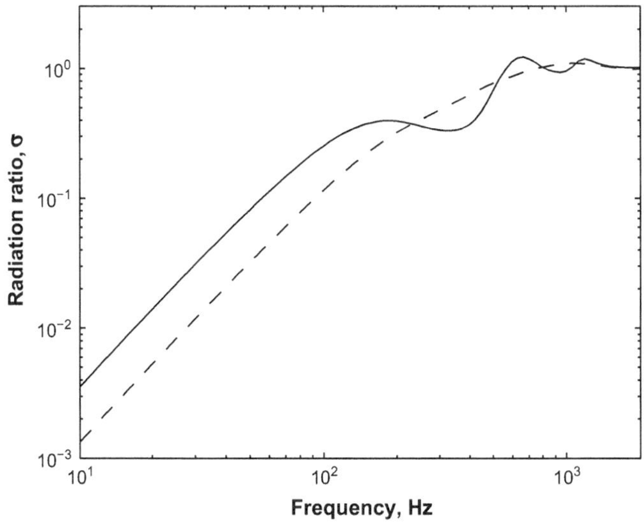

FIGURE 6-31 Radiation ratio of sleepers set in a rigid baffle: – – –, single sleeper 1.25 × 0.2 m; —, three sleepers 1.25 × 0.2 m with vibration amplitudes in the ratio 0.5:1:0.5

accordance with the relative vibration of the various sleepers, as shown in Figure 6.30. This shows that it is sufficient to consider only three sleepers. Although more sleepers fall within the range defined by the acoustic wavelength at low frequency, the vibration is sufficiently localized that these additional sources have

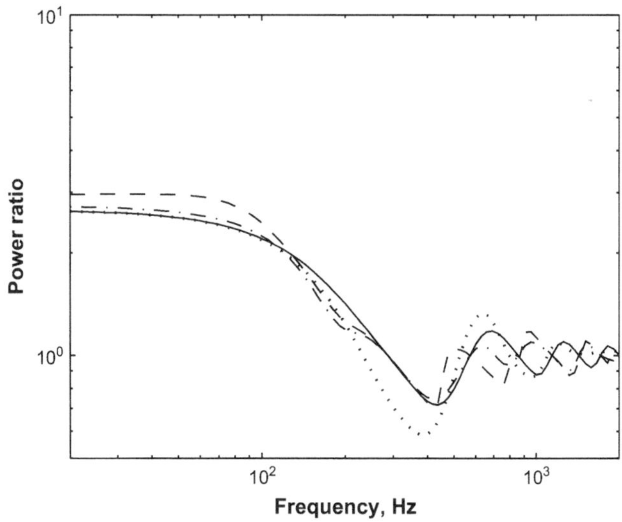

FIGURE 6-32 Increase in sleeper radiation ratio due to multiple sleepers: —, for three sleepers with vibration amplitudes in the ratio 0.5:1:0.5; – · – ·, for three sleepers with amplitudes from track vibration model; – – –, for 21 sleepers with amplitudes from track vibration model; · · · ·, from Rayleigh integral for three sleepers with vibration amplitudes in the ratio 0.5:1:0.5

negligible source strength. At higher frequencies, where more sleepers vibrate, they radiate independently.

It is unclear to what extent the ballast also vibrates and radiates sound. If it does this could increase the sound radiation at low frequencies, although the radiation ratio will not be significantly affected.

6.6 SOUND PRESSURE LEVELS DURING TRAIN PASSAGE

In this section the determination of the sound pressure during a train passage is discussed. Only results for distances that are relatively close to the track are covered, standard measurement positions being located at 7.5 m or 25 m [6.35]. For propagation over larger distances, air absorption and meteorological effects should also be considered, see, e.g., [6.36, 6.37].

6.6.1 Sound pressure during the passage of a series of wheels

For a receiver position at the trackside, the sound pressure level varies with time as the train passes, as indicated in Figure 2.1. Using the results of this chapter for the sound power of wheels, rails and sleepers, and the corresponding directivities, it is possible to predict the evolution of the sound pressure level or the average level during a pass-by. The geometry is shown in Figure 6.33. The train consists of a series of sources moving at speed V in the x direction; the receiver is located at a lateral distance y and possibly at a height h $(= -z)$ relative to the sources.

Consider first a single point monopole, located at (0, 0, 0). A line of receivers is located at a lateral distance y, between $x = -L/2$ and $L/2$. From equation (6.2), the mean-square pressure at a single position $(x, y, 0)$ is given by

$$\overline{p^2}(x, y, 0) = \frac{\rho_0 c_0 W}{4\pi(x^2 + y^2)} \tag{6.25}$$

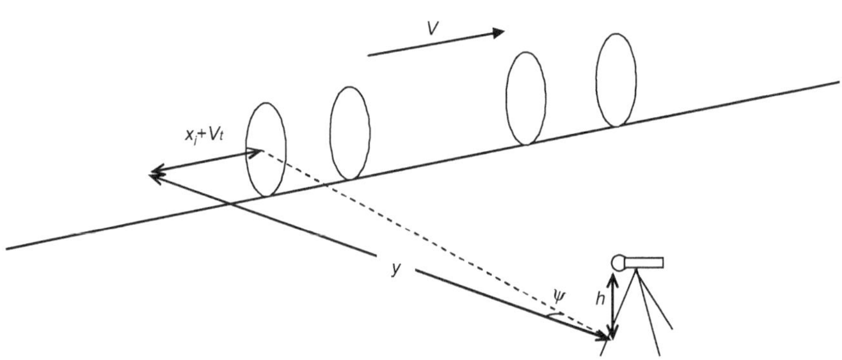

FIGURE 6-33 Motion of a series of wheels past a microphone position

where W is the sound power emitted by the source. The average mean-square pressure along the line of receivers is therefore

$$\langle\overline{p^2}\rangle = \frac{\rho_0 c_0 W}{4\pi}\int_{-L/2}^{L/2}\frac{dx}{(x^2+y^2)} = \frac{\rho_0 c_0 W}{2\pi L y}\tan^{-1}\left(\frac{L}{2y}\right) \tag{6.26}$$

This is equivalent to the average mean-square pressure at a single point y when a monopole source moves past it from $x = -L/2$ to $L/2$ (ignoring the Doppler effect and any time lag due to the propagation of sound).

A similar result is obtained if the single moving source is replaced by an array of incoherent sources located along the line from $x = -L/2$ to $L/2$. If their sound power per unit length is $W' = W/L$ then in the limit $L \to \infty$:

$$\langle\overline{p^2}\rangle = \frac{\rho_0 c_0 W'}{4y} \tag{6.27}$$

For a height h, this equation is modified to

$$\langle\overline{p^2}\rangle = \frac{\rho_0 c_0 W'}{4\sqrt{y^2+h^2}} \tag{6.28}$$

This can be used to predict the average mean-square pressure, provided that $y \ll L$. The sound pressure level reduces with 3 dB per doubling of distance in the same way as for a line monopole, see equation (6.14). However, the present result is a factor of $\pi/2$ (or 2 dB) greater. The reason for this difference is that, for a line monopole, the sound intensity vectors are normal to a cylindrical surface, whereas for an array of incoherent monopoles there are components of intensity that are not perpendicular to this surface and which do not cancel each other out. Thus the net intensity normal to the surface is less than $\overline{p^2}/\rho_0 c_0$.

Next, a point dipole source is considered, emitting a sound power W and orientated with its maximum radiation in the y direction. Introducing the directivity from equation (6.8), the mean-square pressure at position $(x, y, 0)$ is given by

$$\overline{p^2}(x,y,0) = \frac{3\rho_0 c_0 W}{4\pi(x^2+y^2)}\cos^2\psi \tag{6.29}$$

where $\psi = \tan^{-1}(x/y)$. The average mean-square pressure along the line of receivers is therefore

$$\langle\overline{p^2}\rangle = \frac{3\rho_0 c_0 W}{4\pi}\int_{-L/2}^{L/2}\frac{y^2\,dx}{(x^2+y^2)^2} = \frac{3\rho_0 c_0 W}{4\pi L y}(\chi + \sin 2\chi) \tag{6.30}$$

where $\chi = \tan^{-1}(L/2y)$. This is, again, equivalent to the average mean-square pressure at a point y when a dipole source moves past it. As before, a similar result is obtained if the single moving source is replaced by an array of incoherent sources located along the line between $x = -L/2$ and $L/2$. In the limit $L \to \infty$:

$$\langle\overline{p^2}\rangle = \frac{3\rho_0 c_0 W'}{8y} \tag{6.31}$$

where $W' = W/L$ is the power per unit length. This result is also higher than for a line dipole, equations (6.14) and (6.15), in this case by a factor of $3\pi/8$ (or 0.7 dB). At a height h relative to the source this is modified to

$$\langle \overline{p^2} \rangle = \frac{3\rho_0 c_0 W'}{8\sqrt{y^2 + h^2}} \cos^2\phi \tag{6.32}$$

where $\phi = \tan^{-1}(h/y)$ is the angle in the vertical plane.

The mean-square pressure from multiple wheels within a train can be found by adding their contributions as incoherent sources, i.e. adding their mean-square pressures. Figure 6.34 shows the sound pressure level evolution predicted for a series of wheels modelled using a monopole or dipole directivity. As discussed in Section 6.3.2 the directivity of a wheel can be considered as a combination of monopole and dipole components. In these examples each source is allocated a nominal sound power of 1 W. The positions of the sources are shown at the top of each figure; their spacing represents nominal bogie vehicles 20 m long with a bogie wheelbase of 1.8 m. Results are shown for two distances, y, away from the sources. Even for small distances the individual wheels in a bogie cannot be easily separated. Further away pairs of bogies also merge into a single peak. This effect is even more marked for monopole sources than for dipoles.

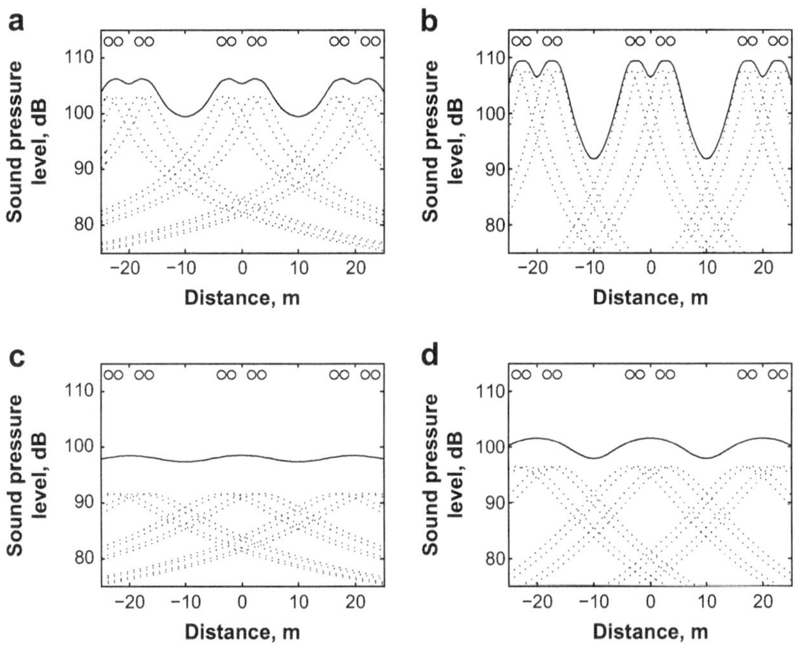

FIGURE 6-34 Evolution of sound pressure level during the passage of a series of sources representing the wheels of a train. (a) Monopoles, distance 2 m; (b) dipoles, distance 2 m; (c) monopoles, distance 7.5 m; (d) dipoles, distance 7.5 m. —, total sound pressure level; ····, contributions from each wheel; ○, location of each wheel

In some situations it is necessary to include the contribution of the sound radiated by both wheels on an axle. This depends on the degree of shielding offered by the underside of the vehicle and is therefore likely to be more significant for freight vehicles than for passenger vehicles, which have more equipment mounted underneath them.

6.6.2 Sound pressure due to the track

A similar model to that described above can be used for the contribution of the sleepers, which can be represented as monopoles moving with each wheel. However, the rail has been shown to be an extended source which does not radiate with a simple directivity and so a different approach is needed.

As a first approximation the average pressure at some distance from the rail can be calculated using the coherent source equations, (6.14) and (6.15), where the sound power per unit length W' is determined using the spatially averaged rail vibration and the appropriate (two-dimensional) radiation ratio.

To estimate the time history, the approach of Section 6.4.4 can be used to account for the effect of propagating waves in the rail. Here, the rail is replaced by an array of point sources, the contributions of which, at the receiver position, are added accounting for their relative phase. Examples were shown in Figure 6.19.

In addition, the rail is excited by multiple wheels. Although the vibration from one wheel/rail contact consists of coherent propagating waves, the contributions from the various wheels in a train should be added incoherently, that is in terms of mean-square responses.

6.6.3 Effects of source motion

In the analysis of the previous two subsections, the effect of source motion has been included only quasi-statically. The sound from a moving source undergoes the well-known Doppler effect, in which the perceived frequency is higher than that emitted as the source approaches and lower as it recedes. The perceived frequency at the receiver location, f_{rec}, can be written as

$$f_{rec} = \left(\frac{c_0}{c_0 + V\sin\psi}\right) f_e \tag{6.33}$$

where f_e is the frequency emitted by the source (in its own frame of reference), V is the train speed and ψ is the angle to the normal at the time of emission, shown in Figure 6.33. This ranges from $-\pi/2$ to $\pi/2$ as the train passes, so that f_{rec} is initially greater than f_e. The total change in frequency is thus

$$\Delta f_{rec} = \left(\frac{c_0}{c_0 - V} - \frac{c_0}{c_0 + V}\right) f_e \approx \left(\frac{2V}{c_0}\right) f_e \tag{6.34}$$

At 160 km/h (44 m/s) this corresponds to a frequency change of $0.26f_e$, slightly more than the bandwidth of a one-third octave band, which is $0.23f_e$. At 300 km/h (83 m/s) the frequency changes from $1.32f_e$ to $0.81f_e$, a change of more than two one-third octave bands.

Another, related, effect of source motion is that the intensity is greater in front of the source than behind. This makes the pass-by time history asymmetrical [6.6], although the effect is small for typical train speeds.

6.6.4 Ground effects

When sound propagates in air above a ground plane it is also reflected by the ground. For a rigid ground this can be represented simply by an image source located below the ground plane, as shown in Figure 6.35. The complex pressure amplitude due to the direct and reflected paths is

$$p = p_0\left(D(\theta_1)\frac{e^{-ikr_1}}{r_1} + D(\theta_2)\frac{e^{-ikr_2}}{r_2}\right) \tag{6.35}$$

where D is the directivity factor of the source, k is the acoustic wavenumber and p_0 is a reference pressure. The above expression can also be written in the form

$$\frac{p}{p_f} = \left(1 + \frac{D(\theta_2)r_1}{D(\theta_1)r_2}e^{-ik(r_2-r_1)}\right) \tag{6.36}$$

where p_f is the free-field pressure, equal to the first term in equation (6.35). The two distances, r_1 and r_2, can be found from simple geometry:

$$r_1 = \sqrt{d^2 + (h_r - h_s)^2};\ r_2 = \sqrt{d^2 + (h_r + h_s)^2} \tag{6.37}$$

where h_s and h_r are the heights of the source and receiver, respectively, and d is the horizontal distance. The phase difference between the direct and reflected sound is $k(r_2 - r_1)$. At low frequencies, where $k(r_2 - r_1) \ll 1$, the two contributions add in phase with one another and the pressure amplitude is approximately doubled. However, at higher frequencies the phase difference becomes important. At some frequencies the pressure from the image source is added to that from the direct source, at others there is destructive interference.

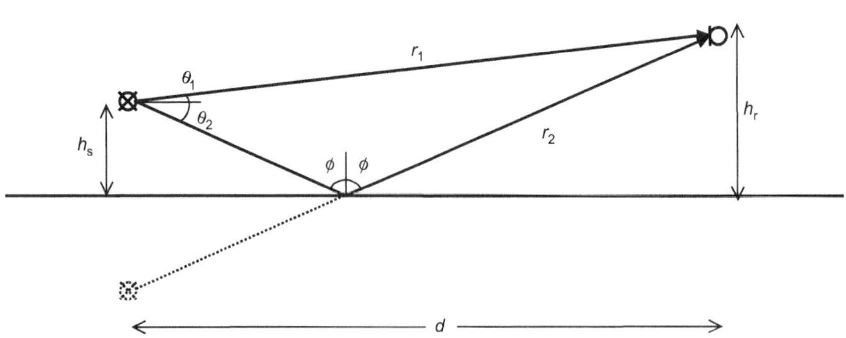

FIGURE 6-35 Location of an image source indicating the path lengths of direct and reflected sound

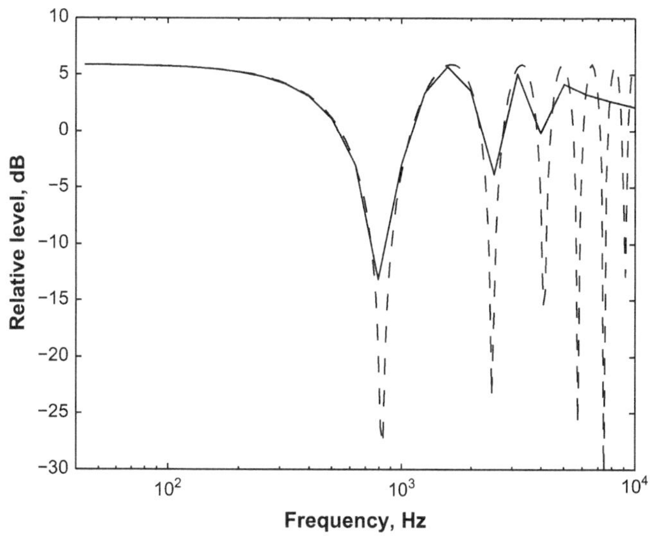

FIGURE 6-36 Effect of rigid ground, monopole source, $h_s = 0.5$ m, $h_r = 1.6$ m, $d = 7.5$ m. – – –, narrow-band result; —, in one-third octave bands

An example of this is shown in Figure 6.36 in the form of the change in sound pressure level relative to the free-field situation. At low frequency the sound pressure level is increased by 6 dB due to the pressure doubling. For this geometry the first interference dip, where $k(r_2 - r_1) = \pi$, occurs at 800 Hz. At higher frequencies there are many peaks and dips but the one-third octave band spectrum tends to an average increase of 3 dB.

When the source moves along a line in front of a receiver location, as discussed in Section 6.6.1, the ground dip changes in frequency as the distance from the source to the receiver varies. Figure 6.37(a) shows how this shifts from 800 Hz to 2 kHz as the source moves from $x = 0$ to 20 m. The average mean-square pressure during the passage of the source therefore has a much less distinct ground dip, even when calculated in narrow frequency bands, as seen Figure 6.37(b).

Where, as is more usual, the ground plane is not fully reflective, it can be described by an amplitude reflection coefficient, R. This is the ratio of the reflected to the incident complex wave amplitudes. For a plane wave incident on a locally reacting ground plane at an angle ϕ to the normal, the reflection coefficient can be expressed in terms of the impedance of the ground:

$$R(\phi) = \frac{(z_n' \cos \phi - 1)}{(z_n' \cos \phi + 1)} \tag{6.38}$$

where $z'_n = z_n/\rho_0 c_0$ with z_n the normal specific acoustic impedance of the ground and $\rho_0 c_0$ the specific impedance of air. Inserting this into equation (6.36) gives

$$\frac{p}{p_f} = \left(1 + R(\phi)\frac{D(\theta_2) r_1}{D(\theta_1) r_2} e^{-ik(r_2 - r_1)}\right) \tag{6.39}$$

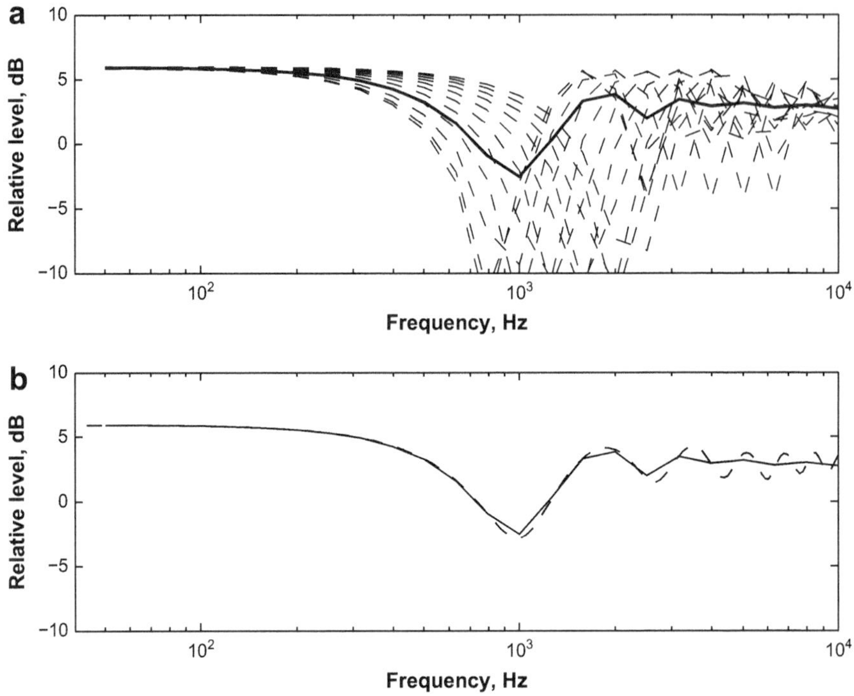

FIGURE 6-37 Effect of rigid ground, monopole source, $h_s = 0.5$ m, $h_r = 1.6$ m, $d = 7.5$ m for passage of source along $x = 0$ to 20 m. (a) In one-third octave bands: – – –, individual results at 1 m intervals; —, average over passage. (b) Average over passage: – – –, narrow-band result; —, in one-third octave bands

If the source spectrum is fairly flat, an average over one-third octave bands can be introduced [6.38]:

$$\left|\frac{p}{p_f}\right|^2 = \left(1 + |R(\phi)|^2 \left(\frac{D(\theta_2)r_1}{D(\theta_1)r_2}\right)^2 + 2|R(\phi)|\frac{D(\theta_2)r_1}{D(\theta_1)r_2}\frac{\sin\left(k(r_2 - r_1)\Delta\right)}{k(r_2 - r_1)\Delta} \times \cos\left(k(r_2 - r_1)\sqrt{1 + \Delta^2} - \beta\right)\right) \tag{6.40}$$

where $\Delta = (f_u - f_l)/2f$ is half the relative bandwidth of the frequency band ($\Delta = 0.1154$ for a one-third octave band) and β is the phase of $R(\phi)$.

The impedance of the ground, z_n, can be estimated using a number of models or measurement methods [6.37]. A commonly used model is due to Delany and Bazley [6.39]. They measured the impedances of a large number of fibrous absorbent materials and derived the following empirical model in terms of their flow resistivity, σ_e:

$$z_n' = 1 + 9.08\left(\frac{1000f}{\sigma_e}\right)^{-0.75} - 11.9i\left(\frac{1000f}{\sigma_e}\right)^{-0.73} \tag{6.41}$$

This form of the equation applies for flow resistivities in Pa.s/m^2 and a time dependence of $e^{i\omega t}$ for consistency with the rest of this book.

This model is attractive as it depends only on a single parameter, the flow resistivity, σ_e. The flow resistivity of a porous material can be measured by passing a steady air flow through a sample of the material. It is defined as the pressure gradient divided by the mean flow velocity (volume flow rate per unit cross-sectional area). For a ground, however, the 'effective flow resistivity' is used, which can be derived by fitting the above model to measured ground impedances. A selection of typical values of the effective flow resistivity σ_e for different grounds is given in Table 6.1, from [6.40]. It may be noted that the original data used to produce the model of equation (6.41) in [6.39] were limited to a range $8000 \leq \sigma_e \leq 4.5 \times 10^4$ Pa.s/m^2 and $|z_n'| \leq 4$, yet the above model is often used for a much larger range of values of σ_e.

Impedances estimated from equation (6.41) are plotted in Figure 6.38 for several values of σ_e. As frequency increases the impedance tends to $z_n = \rho_0 c_0$ ($z'_n = 1$) but at low frequencies it has large real and (negative) imaginary parts. A surface with an impedance similar to that of air will tend to absorb sound effectively; from equation (6.38), $R(\phi) = 0$ when $z_n' = 1/\cos\phi \geq 1$. Conversely, when the impedance is large most of the sound will be reflected.

The effect of these impedances on the sound pressure, for the example given above, is shown in Figure 6.39. The ground dip is shifted to lower frequencies due to phase changes introduced in $R(\phi)$, but for this geometry the shift is only significant for very low values of σ_e. The average sound pressure level at high frequencies reduces from 3 dB above that in free field, for the rigid ground, to 0 dB as the ground becomes more absorptive. The magnitude and phase of the reflection coefficients, $R(\phi)$, for this example are shown in Figure 6.40(a) and (c).

The reflection coefficient is modified if the incident sound field is a spherical wave rather than a plane wave [6.38]:

$$R_{sph} = R(\phi) + (1 - R(\phi))F(w) \tag{6.42}$$

where $F(w)$ is known as the 'boundary loss factor' [6.37]

$$F(w) = 1 - i\sqrt{\pi}\rho e^{-w^2}\operatorname{erfc}(iw) \tag{6.43}$$

TABLE 6.1 EXAMPLES OF EFFECTIVE FLOW RESISTIVITY OF VARIOUS TYPES OF GROUND, FROM [6.40]

Ground surface	Effective flow resistivity, σ_e, Pa.s/m^2
Snow	$1\text{–}5 \times 10^4$
Grass, rough pasture, etc.	$1.5\text{–}3 \times 10^5$
Roadside soil	$3\text{–}8 \times 10^5$
Limestone chips (10 to 25 mm)	$1.5\text{–}4 \times 10^6$
Rain-packed earth	$4\text{–}8 \times 10^6$
Asphalt	$\sim 3 \times 10^7$

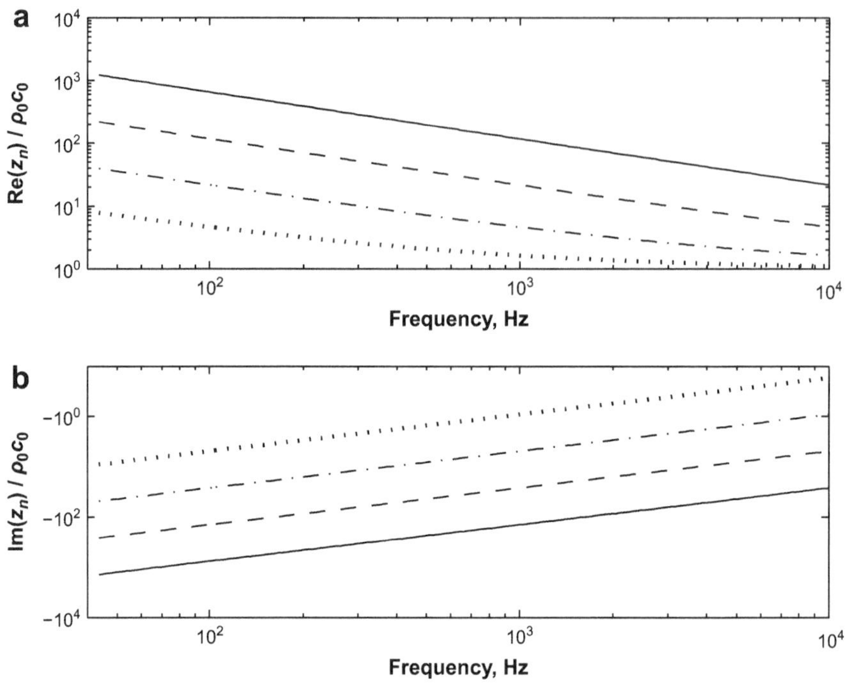

FIGURE 6-38 Ground impedances from equation (6.41). (a) Real part, (b) Imaginary part. —, $\sigma_e = 3 \times 10^7$ Pa.s/m^2; – – –, $\sigma_e = 3 \times 10^6$ Pa.s/m^2; – · – ·, $\sigma_e = 3 \times 10^5$ Pa.s/m^2; · · · ·, $\sigma_e = 3 \times 10^4$ Pa.s/m^2

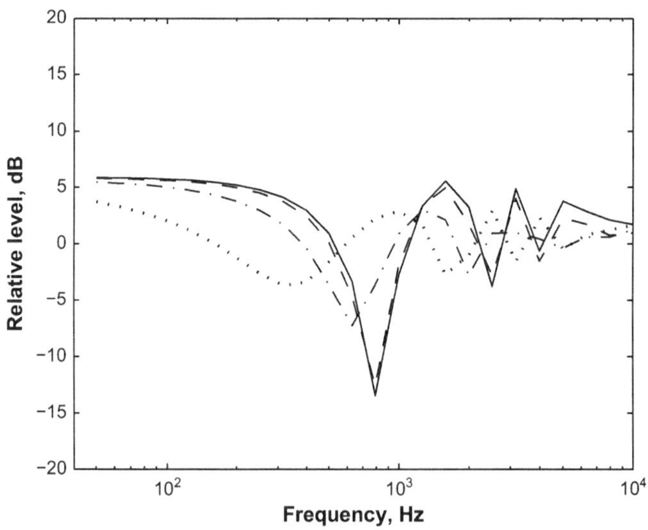

FIGURE 6-39 Effect of ground reflection for various ground impedances, monopole source, $h_s = 0.5$ m, $h_r = 1.6$ m, $d = 7.5$ m. —, $\sigma_e = 3 \times 10^7$ Pa.s/m^2; – – –, $\sigma_e = 3 \times 10^6$ Pa.s/m^2; – · – ·, $\sigma_e = 3 \times 10^5$ Pa.s/m^2; · · · ·, $\sigma_e = 3 \times 10^4$ Pa.s/m^2

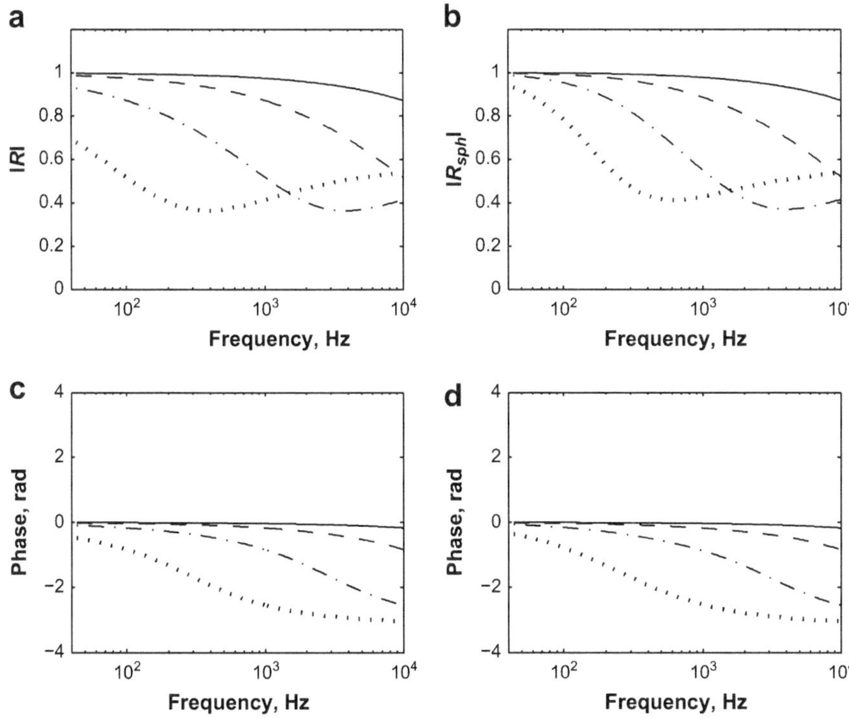

FIGURE 6-40 Amplitude reflection coefficient R for various ground impedances, $h_s = 0.5$ m, $h_r = 1.6$ m, $d = 7.5$ m. —, $\sigma_e = 3 \times 10^7$ Pa.s/m^2; – – –, $\sigma_e = 3 \times 10^6$ Pa.s/m^2; – · – ·, $\sigma_e = 3 \times 10^5$ Pa.s/m^2; · · · ·, $\sigma_e = 3 \times 10^4$ Pa.s/m^2. (a) and (c) Plane wave reflection coefficient, (b) and (d) spherical wave reflection coefficient

erfc denotes the complementary error function and w is the 'numerical distance'

$$w = \sqrt{\frac{-ikr_2}{2}}\left(\cos\phi + \frac{1}{z'_n}\right) \tag{6.44}$$

(again this has been rewritten for a time dependence of $e^{i\omega t}$). This has the effect of ensuring that $R \to 1$ for grazing incidence. An asymptotic expansion of equation (6.43) is given in [6.38]. For the above example, the magnitude and phase of R_{sph} are shown in Figure 6.40(b) and (d). The effect is relatively small here but should be taken into account for propagation over larger distances where the incidence will be nearer to grazing ($\phi \to \pi/2$).

6.7 VALIDATION MEASUREMENTS

The theoretical models described in this and the previous chapters were implemented in a package of computer programs, called TWINS (Track-Wheel Interaction Noise Software [6.17]). Experimental validation of the TWINS model was carried out for ERRI in 1992 [6.41, 6.42]. Three different wheel and track designs were considered, with three or four train speeds in the range 50 to 160 km/h for each combination, giving a total of 25 combinations (one wheel type was unavailable at two of the sites). Simultaneous measurements were performed of wheel and rail vibration and of radiated noise. Measurements of the wheel and rail roughness profiles were also carried out to provide the input to the predictions. As a check, various other parameters were recorded, such as the wheel and rail mobilities, wheel damping, and the decay rates of rail vibration along the track.

In [6.43] an updated version of the model was checked against this original dataset. The validation was also extended by considering a number of quite different wheel and track designs developed in the EU projects Silent Freight and Silent Track and tested in 1999.

In Figure 6.41 the predictions (from roughness to noise) using this updated model are compared with measurements in terms of overall A-weighted level for the 1992 tests. For each of the 25 combinations of wheel, track and train speed that were tested, the total A-weighted noise has been predicted. The measurements (and predictions) were based on an average of three positions microphone close to the track. The results are consistently slightly overpredicted, with a mean of +2.7 dB. The spread of these results is relatively small with standard deviations of only 2.0 dB. The large range of values found here, between 78 and 104 dB, occurs mainly due to the speed range, the different wheel and track types contributing much smaller differences.

Spectral results from these comparisons are summarized in Figure 6.42(a). This shows the difference between predicted and measured noise spectra, averaged

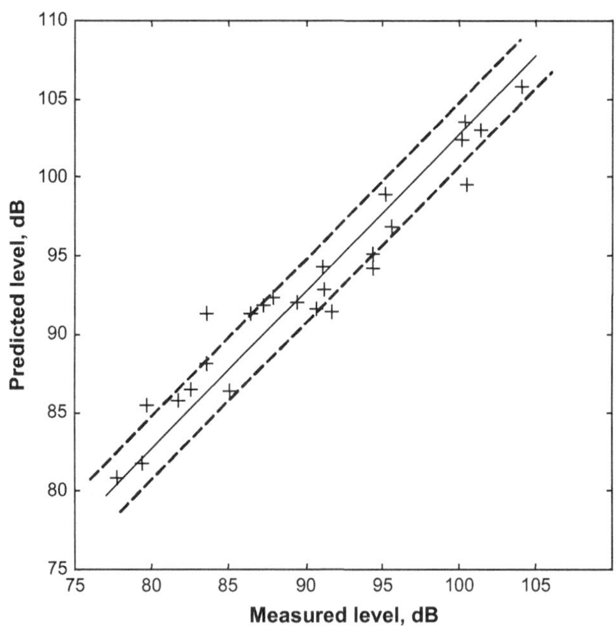

FIGURE 6-41 Predicted noise plotted against measured noise for all 25 cases from 1992. Continuously supported rail model using measured decay rates. —, mean difference; – – –, mean ± one standard deviation; +, actual results [6.43]

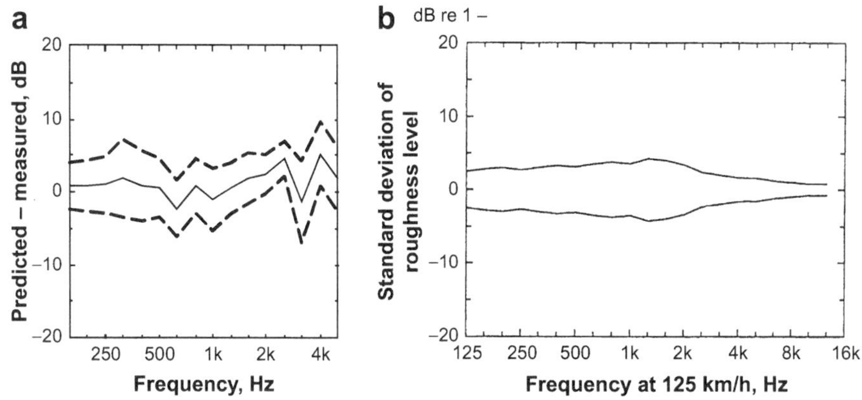

FIGURE 6-42 (a) Predicted noise spectra minus measured noise spectra, averaged over all 25 cases from 1992. Continuously supported rail model using measured decay rates. —, mean difference; – – –, mean ± one standard deviation [6.43]. (b) One standard deviation range of roughness spectra relative to their respective means for data measured on nominally identical wheels and tracks [6.42]

over all 25 wheel/track and speed combinations. While the standard deviation is rather large (3–5 dB), the mean value is very close to 0 dB, indicating a good prediction.

The cause of the large spread in the results can be attributed to the roughness data. For example, variations between nominally similar wheels are shown in Figure 6.42(b) and are of the same order of magnitude as the variations in Figure 6.42(a). Additionally, uncertainties in the position of the contact patch, and hence which roughness measurements should be used, will also have an influence.

In Figure 6.43 the overall predictions are compared with measured noise for the results of the 1999 Silent Freight/Silent Track tests (described in Chapter 7). These results were all measured at 100 km/h so the range of values is smaller than previously, being due mainly to design differences and also partly due to differences in roughness levels. The results here are consistently slightly underpredicted, with a mean of −1.7 dB. The spread of these results is again relatively small with standard deviations of 1.9 dB (note the different scale from Figure 6.41).

The predictions from the 1999 tests could be improved by using predicted track decay rates rather than measurements, suggesting inadequacies in these measurements, and also by including a correction for modal sleeper behaviour [6.43]. Figure 6.44(a) shows spectral results from the 1999 tests including these corrections. A consistent difference remains, particularly at low frequency, which may be due to the omission of wheel roughness measurements (the predictions were based only on rail roughness as the wheel measurements were found to be incorrect).

In Figure 6.44(b) results are shown for noise spectra predicted from measured vibration levels, again compared with the corresponding measured spectra. These results have a considerably smaller standard deviation (1.5–2 dB) and the mean is close to 0 dB over the whole frequency range. This indicates that the radiation part of the model, described in this chapter, is particularly reliable.

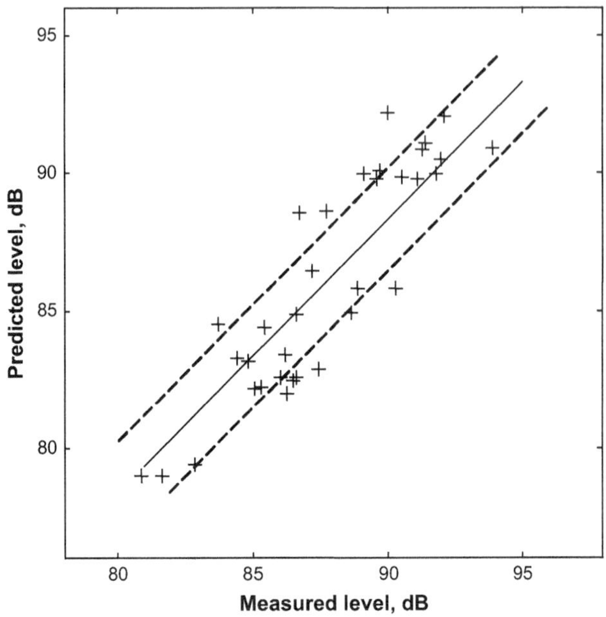

FIGURE 6-43 Predicted noise plotted against measured noise for all 34 cases from 1999. Continuously supported rail model using measured decay rates. —, mean difference; – – –, mean ± one standard deviation; +, actual results [6.43]

The remaining discrepancies between predictions and measurements appear to be due to uncertainties in the roughness input and some inadequacies in the prediction of vibration. Nevertheless, the model is considered to be sufficiently reliable to allow it to be used for studies of noise reduction measures, which will be considered in the next chapter.

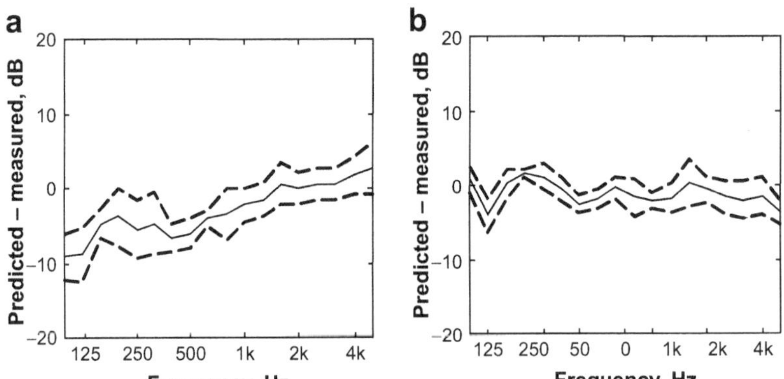

FIGURE 6-44 Predicted noise spectra minus measured noise spectra, averaged over all 34 cases from 1999. (a) Predictions from roughness using continuously supported rail model using measured decay rates and including measured ratio of sleeper vibration. (b) Predictions based on measured vibration. —, mean difference; – – –, mean ± one standard deviation [6.43]

REFERENCES

6.1 F.J. Fahy and P. Gardonio. *Sound and Structural Vibration: Radiation, Transmission and Response*, 2nd edition. Academic Press, Oxford, 2006.

6.2 J.W. Strutt (Lord Rayleigh). *Theory of Sound*, 2nd edition. Dover Publications, New York, 1945. Reissue. See section 278.

6.3 C.E. Wallace. Radiation resistance of a rectangular panel. *Journal of the Acoustical Society of America*, **51**, 946–852, 1972.

6.4 D.A. Bies and C.H. Hansen. *Engineering Noise Control*, 2nd edition. E&FN Spon, London, 1996.

6.5 A.D. Pierce. *Acoustics: An Introduction to its Physical Principles and Applications.* Acoustical Society of America, Woodbury, NY, 1989.

6.6 P.M. Morse and K.U. Ingard. *Theoretical Acoustics.* Princeton University Press, Princeton NJ, 1986.

6.7 P.J. Remington. Wheel/rail rolling noise, I: theoretical analysis. *Journal of the Acoustical Society of America*, **81**, 1805–1823, 1987.

6.8 P.J. Remington. Wheel/rail noise – Part I: Characterization of the wheel/rail dynamic system. *Journal of Sound and Vibration*, **46**, 359–379, 1976.

6.9 E. Schneider, K. Popp, and H. Irretier. Noise generation in railway wheels due to rail-wheel forces. *Journal of Sound and Vibration*, **120**, 227–244, 1988.

6.10 D.J. Thompson. Predictions of acoustic radiation from vibrating wheels and rails. *Journal of Sound and Vibration*, **120**, 275–280, 1988.

6.11 T.W. Wu. *Boundary Element Acoustics: Fundamentals and Computer Codes.* WIT Press, Southampton, 2000.

6.12 U. Fingberg. Ein Modell für das Kurvenquietschen von Schienenfahrzeugen. *VDI Fortschrittberichte*, Reihe 11, Nr. 140, 1990.

6.13 D.J. Thompson and C.J.C. Jones. Sound radiation from a vibrating railway wheel. *Journal of Sound and Vibration*, **253**, 401–419, 2002.

6.14 D.J. Thompson and M.G. Dittrich. Wheel response and radiation – laboratory measurements of five types of wheel and comparisons with theory. *ORE Technical Document DT*248 (*C*163), Utrecht, 1991.

6.15 X. Zhang and H.G. Jonasson. Directivity of railway noise sources. *Journal of Sound and Vibration*, **293**, 995–1006, 2006.

6.16 M.G. Dittrich, W.J. van Vliet, and D.J. Thompson. Characterisation of track response and radiation. Step 3 – response of tracks 4 and 5, TNO report TPD-HAG-RPT-91-0161, 1991.

6.17 D.J. Thompson, M.H.A. Janssens, and F.G. de Beer. TWINS: Track-Wheel Interaction Noise Software, theoretical manual (version 3.0). TNO report HAG-RPT-990211, Delft, 1999.

6.18 D.J. Thompson, C.J.C. Jones, and N. Turner. Investigation into the validity of two-dimensional models for sound radiation from waves in rails. *Journal of the Acoustical Society of America*, **113**, 1965–1974, 2003.

6.19 U. Orrenius and S. Finnveden. Calculation of wave propagation in rib-stiffened plate structures. *Journal of Sound and Vibration*, **198**, 203–224, 1996.

6.20 C-M. Nilsson. Waveguide Finite Elements Applied on a Car Tyre. PhD thesis, Royal Institute of Technology, Aeronautical and Vehicle Engineering MWL, 2004. TRITA-AVE 2004:21, ISBN 91-7283-798-5.

6.21 D. Duhamel. Efficient calculation of the three-dimensional sound pressure field around a noise barrier. *Journal of Sound and Vibration*, **197**, 547–571, 1996.

6.22 C-M. Nilsson, A.N. Thite, C.J.C. Jones, and D.J. Thompson. Estimation of sound transmission through extruded panels using a coupled waveguide Finite Element-Boundary Element method. *Proceedings of 9th International Workshop on Railway Noise*, Feldafing, Germany, September 2007.

6.23 C-M. Nilsson, C.J.C. Jones, D.J. Thompson, and J. Ryne. A waveguide finite element and boundary element approach to calculating the sound radiated by tram rails. Journal of Sound and Vibration, 2008, accepted for publication.

6.24 T. Kitagawa and D.J. Thompson. Comparison of wheel/rail noise radiation on Japanese railways using the TWINS model and microphone array measurements. *Journal of Sound and Vibration*, **293**, 496–509, 2006.

6.25 B. Barsikow, W.F. King III, and E. Pfizenmaier. Wheel/rail noise generated by a high-speed train investigated by a line array of microphones. *Journal of Sound and Vibration*, **118**, 99–122, 1987.

6.26 B. Barsikow. Experiences with various configurations of microphone arrays used to locate sound sources on railway trains operated by the DB AG. *Journal of Sound and Vibration*, **193**, 283–293, 1996.

6.27 G. Hölzl. Low noise goods wagons. *Journal of Sound and Vibration*, **193**, 359–366, 1996.

6.28 J.D. van der Toorn, H. Hendricks, and T.C. van den Dool. Measuring TGV source strength with Syntacan. *Journal of Sound and Vibration*, **193**, 113–121, 1996.

6.29 M.G. Dittrich and M.H.A. Janssens. Improved measurement methods for railway rolling noise. *Journal of Sound and Vibration*, **231**, 595–609, 2000.

6.30 S. Brühl and A. Röder. Acoustic noise source modelling based on microphone array measurements. *Journal of Sound and Vibration*, **231**, 611–617, 2000.

6.31 C. Hanson and B. Barsikow. Noise sources on Amtrak's high speed train. *Proceedings of Inter Noise 2000*, Nice (France), August 2000.

6.32 A. Nordborg, A. Martens, J. Wedemann, and L. Willenbrink. Wheel/rail noise separation with microphone array. *Proceedings of Inter Noise 2001*, The Hague (Netherlands), August 2001.

6.33 K.G. Degen, A. Nordborg, A. Martens, J. Wedemann, L. Willenbrink, and M. Bianchi. Spiral array measurements of high-speed train noise. *Proceedings of Inter Noise 2001*, The Hague (Netherlands), August 2001.

6.34 T. Kitagawa, Y. Zenda, Y. Abe, and Y. Ogata. Sound radiated by vibration of railway wheels. *Proceedings of Inter Noise 2001*, The Hague (Netherlands), August 2001.

6.35 International standard: Railway applications – Acoustics – Measurements of noise emitted by railbound vehicles. International Standards Organization, ISO 3095:2005.

6.36 T.F.W. Embleton. Tutorial on sound propagation outdoors. *Journal of the Acoustical Society of America*, **100**, 31–48, 1996.

6.37 K. Attenborough. Sound propagation in the atmosphere. In: M.J. Crocker (ed.), *Handbook of Noise and Vibration Control*,. John Wiley & Sons, Hoboken, NJ, 2007.

6.38 C.I. Chessell. Propagation of noise along a finite impedance boundary. *Journal of the Acoustical Society of America*, **62**, 825–834, 1977.

6.39 M.E. Delany and E.N. Bazley. Acoustical properties of fibrous absorbent materials. *Applied Acoustics*, **3**, 105–116, 1970.

6.40 T.F.W. Embleton, J.E. Piercy, and G.A. Daigle. Effective flow resistivity of ground surfaces determined by acoustical measurements. *Journal of the Acoustical Society of America*, **74**, 1239–1244, 1983.

6.41 D.J. Thompson, B. Hemsworth, and N. Vincent. Experimental validation of the TWINS prediction program, Part 1: Method. *Journal of Sound and Vibration*, **193**, 123–135, 1996.

6.42 D.J. Thompson, P. Fodiman, and H. Mahé. Experimental validation of the TWINS prediction program, Part 2: Results. *Journal of Sound and Vibration*, **193**, 137–147, 1996.

6.43 C.J.C. Jones and D.J. Thompson. Extended validation of a theoretical model for railway rolling noise using novel wheel and track designs. *Journal of Sound and Vibration*, **267**, 509–522, 2003.

CHAPTER 7

Mitigation Measures for Rolling Noise

7.1 INTRODUCTION

As explained in Chapter 1, there are some important principles that should ideally be followed in order to achieve successful noise control. In particular, the dominant sources must be identified and the parameters that influence them should be understood before attempting to implement noise control measures. This approach has been the motivation for the development of the theoretical models for rolling noise that have been described in Chapters 2–6. The dominant source of railway noise in many situations has been identified as rolling noise. Moreover, using the models it is now possible to quantify the contributions of the various component sources. These will differ from one situation to another, because of which, noise reductions achieved in one case are not universally applicable. As has been seen, both the wheels and the rails are often important sources which contribute significantly to the overall sound level, so that effective noise control usually requires both sources to be reduced. This is illustrated in general terms in the box on next page. If one source is reduced by X dB when there are other sources present, the overall reduction is always less than X dB. Such effects are also made less intuitive by the use of the decibel scale.

Having developed this understanding, in this chapter the next steps in the noise control process are considered in relation to rolling noise: the development and implementation of low noise designs. The theoretical models allow the influence of various design parameters to be investigated. Once the most promising parameter changes have been identified, realistic designs can be produced accounting for practical constraints of the railway environment. Furthermore, the models can be used to interpret the results of experiments and to translate them to other situations.

Although various potential solutions for reducing rolling noise have been proposed over the years, often practical tests have been conducted without first following the above process. When these produce a disappointing result it may be because the solution being tested is not effective, but it may also be because of a failure to account for other source contributions.

In comparing the efficiency of various solutions, it is vital that account is taken of the balance of sources in the initial situation in each case. For example, an effective reduction in wheel noise will have little impact on the overall noise in situations where the track is the dominant source (see box on next page). More critical still is the need to avoid increasing the contribution of one source by measures taken to reduce another. Fortunately, this latter aspect is rarely a major concern for wheel/rail

Noise reduction in the presence of multiple sources

Suppose that there are two sources which contribute sound pressure levels L_1 and L_2 at the receiver location. If these sources can be assumed to be incoherent, which is usually the case, the combined sound level is given by

$$L_{total} = 10\log_{10}(10^{L_1/10} + 10^{L_2/10})$$

If source 1 is now reduced by a certain amount while source 2 remains unchanged, the effect on the total will always be *less* than the change in L_1. This is illustrated in the figure below. At the left of the figure, L_2 is initially 10 dB lower than L_1 and the effect on the total of reducing L_1 is quite large, although still less than the change in L_1 itself. At the right of the figure, L_2 is initially 10 dB greater than L_1 and the effect of reducing L_1 is negligible. If L_1 and L_2 are initially equal (the centre of the graph), a reduction of 10 dB in L_1 leads to a change of only 2.6 dB in the total.

If more than two sources are present, L_2 could be considered to be the sum of all remaining sources.

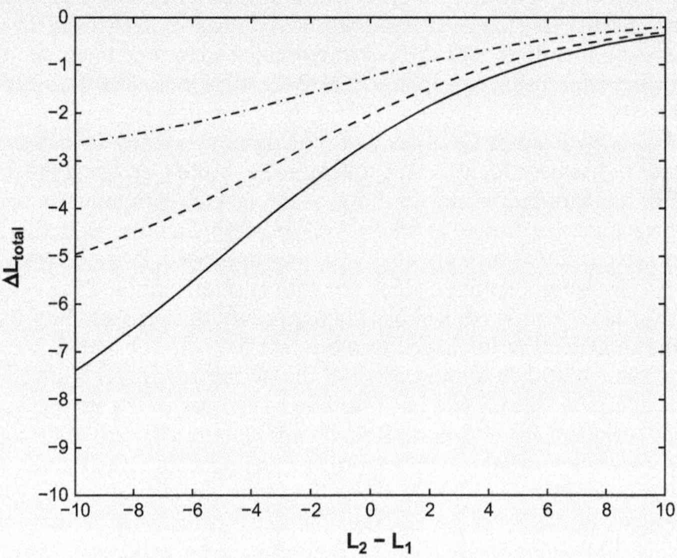

Change in L_{total} due to a reduction of 3 dB (– · – ·), 6 dB (– – –) or 10 dB (—) in L_1.

noise, as changes to the wheel or track mostly have only secondary effects on the other component.

The structure of the theoretical model for rolling noise is shown again in Figure 7.1, from which the various parameters that can be influenced can be identified. These different aspects will be considered in turn in the various sections of this chapter.

First, the potential will be considered of reducing the excitation due to the roughness. This will directly affect all components of noise. Particularly for freight

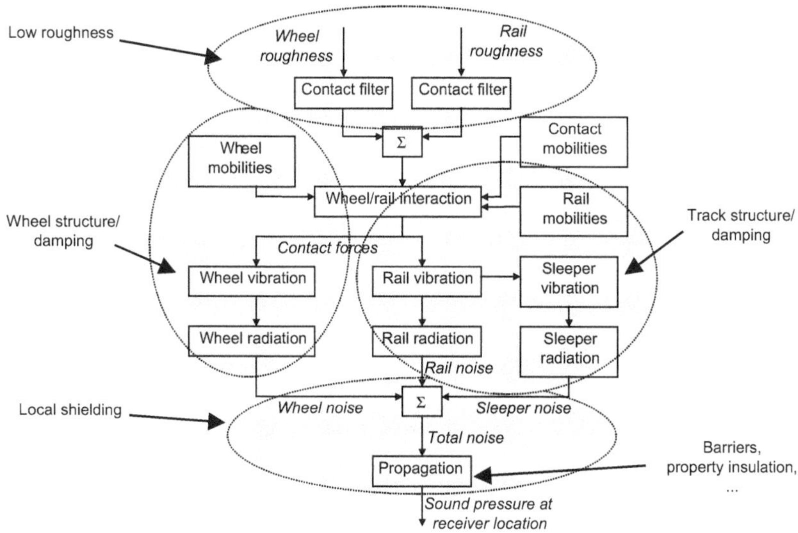

FIGURE 7-1 Schematic view of the model for rolling noise, showing the main potential means of reducing rolling noise

wagons in Europe, there has been considerable scope to reduce noise by this means by changes in the braking system. However, mostly wheel and rail running surfaces are actually very smooth. Once cast-iron brake blocks have been eliminated, further reductions in the excitation will be difficult and it will be necessary to consider the other techniques highlighted.

Consideration will next be given to various measures that can be applied, in turn, to wheels and track in order to reduce their radiated noise for a given roughness input. These include added damping, structural modification and the use of vibration isolation. All of these are quite standard techniques for noise control [7.1], the difficulty being to adapt them to the railway environment.

The discussion continues with a section covering shielding applied close to the source. Finally, traditional noise barriers are discussed briefly, although they are beyond the main scope of this book.

Throughout the chapter, results are drawn from various research projects, particularly the work of the European Rail Research Institute (ERRI). Formerly known as ORE (Office for Research and Experiment), this was the research arm of the International Union of Railways (UIC) until 2004. Their work led to the development and validation of the TWINS model [7.2, 7.3]. This was followed in the 1990s by the OFWHAT project ('Optimized Freight Wheel and Track') [7.4, 7.5] (see box on page 226) and parallel projects on high speed trains and track in France [7.6–7.8] (see box on page 228). Later a number of collaborative projects were supported by the European Union jointly with industry: 'Silent Freight' and 'Silent Track' [7.9–7.12] (see box on page 230) and 'Eurosabot' [7.10, 7.13]. In the Netherlands a national project 'Stiller Treinverkeer' (STV) was carried out in parallel with these [7.14, 7.15] and is described in the box on page 232. More details of each of these projects are also given in [7.16]. In each case theoretical modelling was used to establish low noise designs before implementing them in prototypes for testing.

Optimized Freight Wheels and Track (OFWHAT)

By the early 1990s it was identified that the reduction of noise from freight wagons was the first priority for the railways of Europe. As well as reducing the wheel roughness by changes to the braking system, potential reductions were seen to be possible by modifications to the wheels and the track. The development of prototypes of noise-optimized wheels and tracks was undertaken within the work of the European Rail Research Institute (ERRI). Use was made of concepts identified during the development of the TWINS models. In particular, it was clear that simultaneous action was required on both wheels and track. TWINS was also used as a design tool in developing specifications for the dynamic characteristics of the optimized components and predicting their efficiency in terms of noise reduction if the prototypes were to conform to the specification.

Prototype track development

For lower speeds (60 to 100 km/h) the track contribution was identified from TWINS calculations to be more important than the wheel contribution. A significant part of the potential noise reduction would therefore have to come from optimization of the track. The agreed starting point was a track with a 9 mm thick pad and bi-bloc sleepers, as used particularly in France. Options investigated consisted of stiffening the rail pad, increasing its damping loss factor and adding tuned absorbers to the rail to increase the decay rates (see Figure 7.28(a)).

Prototype wheel development

For the wheel design, the starting point was the standard UIC 920 mm diameter freight wheel. The development of noise-optimized wheels was carried out on the assumption that tread braking would continue to be required. The main concept considered was a wheel-mounted tuned absorber, tuned to two of the most important modes at frequencies of 1720 and 2330 Hz (see Figure 7.19). In addition, an optimized wheel shape was developed with a diameter of 860 mm and a thick web. The tests also included an existing small wheel of diameter 640 mm.

Test results

A test track was built at the Velim test centre (Czech Republic), with the reference track and three modified sections: two variants of optimized pad and one section with rail absorbers. A test train was run with wagons with standard wheels, wheels with absorbers, shape-optimized wheels and small wheels. The tests were carried out at 60 and 100 km/h. The results at 100 km/h are summarized in Table 7.2 (page 273). The largest reduction of 7 dB was obtained with wheels with absorbers on optimized track with absorbers.

In such a research setting it is possible to establish the effects of different measures more precisely than when they are implemented in service. Nevertheless, many of the techniques referred to are now well on the way towards implementation in practice. The work of the IPG programme (Innovation Programme Noise) in the Netherlands, for example, has assisted in demonstrating and evaluating a number of noise reduction technologies *in situ* [7.17–7.20].

The main purpose in this chapter is to highlight the principles behind the various potential noise control techniques rather than to discuss particular products.

Inevitably, this is done by reference to specific results, especially concentrating on results that have been published and those in which the authors have been involved. However, they will be placed strongly in the context of the models in order to provide a basis for understanding their operation.

7.2 REDUCTION OF ROUGHNESS

From Figure 7.1 it is clear that reductions in the combined wheel and rail roughness will lead to reduced noise in direct proportion to the change in roughness. Nevertheless, it is important that account is taken of both wheel and rail roughness. In the same way as for noise components (see box on page 224), if they are initially of similar magnitudes, a reduction by 10 dB of one component of roughness will lead to less than a 3 dB reduction in the total. Worse than this, if effort is spent on reducing the smoother of the two components, the effect on the noise could be negligible.

7.2.1 Effects of braking system

It has been known since the late 1970s that if cast-iron block brakes are replaced by disc brakes, this can lead to reductions in noise levels on good track of up to 10 dB [7.21] (see also Chapter 2). This is related to the presence of corrugations on the running surface of wheels that are braked using cast-iron blocks. These can be seen in the photograph in Figure 7.2(b), whereas the disc-braked wheel in Figure 7.2(a) is smooth. These corrugations on the wheel form the dominant roughness unless the rail is also corrugated.

In the 1990s, Dings and Dittrich [7.22] reported a survey of the roughness of many Dutch wheels and rails and confirmed this conclusion. They showed, moreover, that the roughness of wheels fitted with disc brakes and supplementary cast-iron tread brakes could be greater than those of purely tread-braked wheels. Example results are given in Figure 7.3 in the form of the total roughness level summed over the bands 16 to 250 mm and the overall A-weighted noise level from the

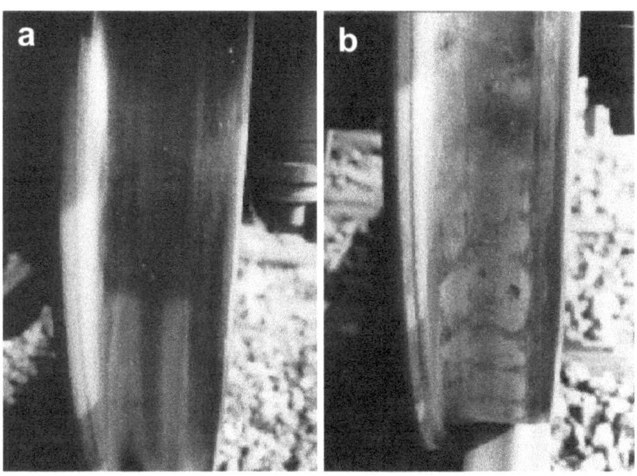

FIGURE 7-2 Running surface of wheels. (a) Disc braked (b) braked by cast-iron blocks (photograph: Pieter Dings)

MONA–RONA–VONA projects

Three linked projects, MONA, RONA and VONA, were carried out in France between 1994 and 1996 coordinated by the SNCF. The purpose was to devise and demonstrate noise control measures for rolling noise of high speed trains (TGV), in parallel with the OFWHAT project, which was aimed at freight traffic. The emphasis was placed on short-to-medium-term solutions that could readily be applied using existing technology. The three projects covered roughness (in particular rail grinding), wheel noise and track noise.

The starting point was defined in terms of the relative contributions of wheel, track and aerodynamic noise to the pass-by noise of a trailer TGV bogie at 300 km/h: aerodynamic noise was found to dominate the spectrum for frequencies up to 1000 Hz, the track in the region 1250–2000 Hz and the wheel above 2000 Hz, although the track also has a significant contribution at these higher frequencies. The aerodynamic noise and rolling noise contribute approximately equally to the overall level at this speed.

MONA – Meulage Optimisé vis à vis des Nuisances Acoustiques
(Grinding optimized with respect to acoustic nuisance)

Investigations were performed into the roughness left by rail grinding. It was identified that, by using a final grinding pass at a lower train speed, the wavelength of the peak due to grinding could be reduced so that it lies at a frequency of less importance to the noise.

An 'acoustic grinding' strategy was also developed for the high speed network, based on monitoring the network every six months using a microphone fitted under the bogie of a track geometry measurement vehicle.

RONA – Roues Optimisées vis à vis des Nuisances Acoustiques
(Wheels optimized with respect to acoustic nuisance)

The purpose of RONA was to design quieter wheels that could be fitted to TGV trailer bogies as a retrofit. These should have no significant weight increase. A wheel diameter of 920 mm was retained throughout. A large number of variants were considered using the TWINS software. Laboratory testing was also carried out. The most promising wheel designs were then chosen for implementation and field testing. The main wheel designs considered were a 'symmetric' profile (Figure 7.12), a composite wheel with an aluminium web and steel tread (Figure 7.21), wheels with tuned absorbers and web screens.

VONA – Voie Optimisés vis à vis des Nuisances Acoustiques
(Track optimized with respect to acoustic nuisance)

The main purpose of the VONA project was to develop low noise track designs using the TWINS model. These were intended for use on high speed lines, although they would be applicable to other lines too. Tests were then carried out on a 100 m long test section before the chosen solutions were tested in running tests in conjunction with RONA. The solutions considered were optimized rail pads and tuned absorbers (Figure 7.28(b)). The fact that aerodynamic noise is the dominant source below 1000 Hz meant that it was not necessary to consider measures to reduce the sleeper noise.

Continued

MONA-RONA-VONA projects—Cont'd

Combined results RONA-VONA

The field tests of the RONA and VONA solutions were carried out simultaneously. The overall reductions obtained by the various combinations are given in Table 7.3. The effect at 150 km/h was considerable, but the effect at 300 km/h was limited to 3 dB by the presence of aerodynamic noise. Nevertheless, by using intermediate measurements and calculations using TWINS the reduction in rolling noise could be quantified. Reductions in rolling noise as high as 8 dB were obtained.

corresponding wheels measured very close to the track. This measure of 'overall' roughness does not take correct account of the weighting of different bands [7.23] (see Section 5.6.4) nor is the rail roughness included, but nevertheless the results show a fairly clear trend.

Disc brakes are now widely used for passenger rolling stock. Modern high speed trains such as the latest generations of TGV and the ICE use disc brakes without supplementary tread brakes. As a result, these trains are no noisier at 300 km/h than traditional (tread-braked) stock at conventional speeds of 140 to 160 km/h (see Figure 2.3).

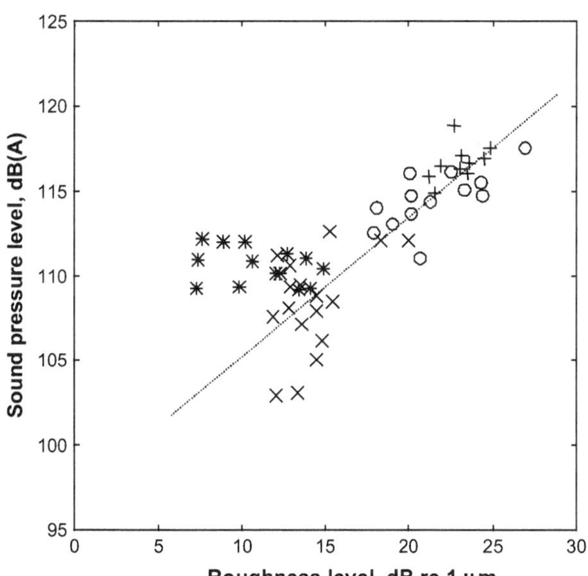

FIGURE 7-3 Noise levels measured close to the track from individual wheels, plotted against total roughness level of these wheels. +, disc and cast-iron block; ○, cast-iron block only; ∗, disc and sinter block; ×, disc only; ····, fitted curve [7.22]

Silent Freight and Silent Track

Two linked projects, Silent Freight and Silent Track, were supported by the EU and coordinated by ERRI. Each included several industrial partners as well as railway and research partners. Silent Freight concentrated on freight vehicles and started in 1996 while Silent Track focused on reducing the noise from the track and commenced a year later. They followed on from the developments in OFWHAT and culminated in a joint measurement exercise in 1999 in which prototypes of the various solutions that had been developed were tested.

Silent Freight

The aim of Silent Freight was to reduce the vehicle component of noise by 10 dB. This consists predominantly of the wheel contribution as it was shown that the bogies and vehicle superstructure had negligible effect on the overall noise (see section 4.7). The starting point was the UIC 920 mm standard freight wheel. The main solutions considered were:

- Optimized wheel shape, including a slight reduction in wheel diameter from 920 to 860 mm and increase in web thickness. Two such designs were developed (Figure 7.15), which were shown to be acceptable from a thermo-mechanical point of view for tread braking using cast-iron brake blocks (the design developed in OFWHAT was not, due to its thicker web). One of these designs was constructed for the field tests.
- Damping treatments – a tuned absorber system which is attached to the inside face of the tyre was developed. A constrained layer treatment was also considered in calculations but not pursued in the field tests.
- Ring dampers – a novel ring damper was developed and implemented in the tests.
- Web shields fitted to wheels to shield the radiation from the axial motion of the wheel web. These consisted of 1 mm steel panels, resiliently mounted from the hub and tyre of the wheel adding 7 kg to the weight of the wheel (Figure 7.25).
- A perforated wheel was developed to reduce the sound radiation from the wheel at low frequencies (Figure 7.26).
- Bogie shrouds (in combination with low barriers close to the rail, see Silent Track). These were developed to be usable in any country in Europe, so that the combined gauging constraints from several countries restricted the allowable envelope. Consequently, a gap of 118 mm remained between the bottom of the shroud and the top of the barrier which severely limited its effectiveness.

Silent Track

The aim of Silent Track was complementary to Silent Freight: to reduce the component of noise from the track by 10 dB. The starting point was a track with UIC60 rails, monobloc concrete sleepers and 10 mm studded rubber pads. The main solutions considered were:

- Optimized rail pad, with increased stiffness. This gave a reduction of track noise of 2 dB.
- Rail damper to increase the decay rate of vibration along the rail. This was a twin mass-spring system attached continuously along the rail on each side (Figure 7.28(c)).
- Modified rail cross-section with narrower foot to reduce sound radiation from motion in the vertical direction (Figure 7.34(d)). This necessitated the development of an alternative rail fastening system which held the rail under the head to prevent excessive rail roll while allowing vertical flexibility. A new sleeper design was also included in this track section but its effect was not quantified separately.
- Low barriers very close to the rail.

Continued

Silent Freight and Silent Track—Cont'd

Combined field tests

The main solutions developed in these two projects were implemented as prototypes in an extensive field test in May 1999. The reference situation was dominated by noise from the track, in part due to the relatively low speed (100 km/h) and the fact that tests were carried out at over 30°C so that the rail pads were rather soft. The track noise was about 8 dB(A) greater than the wheel noise. This meant that the reductions in wheel component could not be measured directly but had to be deduced from a combination of noise and vibration measurements with interpretation from the theoretical model. Combined results are summarized in Table 7.4 on page 274.

The bogie shrouds and barriers were tested separately from the other solutions, except that the track with the rail damper was also present. Table 7.1 on page 268 lists the reductions. Although not all solutions were tested together, it is expected that reductions of up to 10 dB are achievable using a combination of rail damping, wheel treatments, shrouds and barriers.

Alternative brake block materials also offer a solution which can be less costly than disc brakes, discussed in more detail in the next section in relation to freight vehicles. However, although the sinter block materials tested in [7.22] were found to produce very smooth wheel surfaces, the wheels were not as quiet as might be expected, as shown in Figure 7.3. This is thought to be due to hollow wear of the wheel profiles leading to greater noise generation [7.24] and the fact that the wheels were rougher towards the edge of the contact band than in the centre, although the reasons were never fully resolved. It is clear, however, that a roughness measurement on a single line around the wheel perimeter may not give a result that corresponds reliably to the effect of roughness on noise. The sintered material also has the disadvantage that excessive amounts of copper particles are emitted into the environment.

Figure 7.4 shows noise spectra for wheels all of the same type but with different braking systems. These were measured at 1 m from the nearest rail in order to identify the noise from individual wheels. Also shown are the corresponding roughness spectra formed of the average of a number of lines across the expected contact patch. The rail roughness at the site is also shown. This shows again that disc-braked wheels have lower noise and roughness than those with cast-iron brake blocks.

Composite blocks have also been introduced, for example on the power cars of TGVs. Figure 7.5 shows further examples of roughness spectra measured on wheels braked using different braking systems. The wheels with composite brake blocks have very smooth running surfaces. The average spectral levels for composite blocks are up to 15 dB lower than those for cast-iron blocks, the largest differences being between wavelengths of 25 and 50 mm. There are, however, quite large variations between wheels with the same braking system. Measured roughness from 99 Swedish wheels are given by Johansson [7.25] which show the same trends.

Roughness levels, and hence noise, can grow quite quickly after wheels have been reprofiled. Results are shown in Figure 7.6 measured at a monitoring station in the Netherlands [7.26]. Noise levels were measured at 7.5 m from the track at a site which was smooth (similar roughness to the ISO3095 limit) and for train speeds in the range 120 to 135 km/h. With this monitoring system it is possible to associate

ICES-STV project Stiller Treinverkeer

ICES-STV was a multi-partner project that ran in the Netherlands from 1994 to 1999. It was part financed by industry, and as with other projects was focused on demonstrating noise reductions for freight traffic. Low noise concepts were developed using theoretical modelling and were integrated into two container wagons and an experimental slab track.

Tests were carried out in 1998 for the validation and demonstration of the concepts. These were unusual in that, rather than test a whole train, a single freight wagon was detached from the propelling locomotive and allowed to coast over the test section of track. In this way contributions from adjacent vehicles, which may have interfered with the measurement of a quiet wagon, were eliminated. A small correction was required for differences in speed as the coasting wagon gradually slowed down over the various test sections (by 10–15 km/h from an initial speed of 100 km/h over the whole test section). Noise was measured at 7.5 m from the track and vibration of the various track components was measured, as well as of the wheels and on the vehicles.

On the vehicle the most important consideration was to eliminate the cast-iron brake blocks. By doing so the wheel roughness could be reduced and both wheel and track noise components could be reduced considerably. Several alternative braking concepts were considered. A composite brake block, which did not roughen the wheel, was implemented but since this did not deliver sufficient braking force a magnetic brake acting on the rail was added as well. A second vehicle was fitted with drum brakes, which like disc brakes do not act on the wheel running surface, but which were considered to be a cheaper alternative to disc brakes. Unfortunately, this vehicle suffered technical problems in the tests and no results were obtained. A reference vehicle fitted with cast-iron brake blocks was also measured.

Other measures, implemented on the vehicles with novel braking systems, were wheel dampers and 'wheel skirts', a partial enclosure around each wheel with absorptive treatment.

A major achievement was the development of a new small rail profile, which was embedded in a stiff layer on a concrete slab (Figure 7.36). This was continuously supported by the stiff embedding material which provided a considerable damping effect on the wave propagation in the rail, restricting the rail radiation.

The slab track was also tested in configurations with a layer of absorptive material on top of the track slab, with integral absorptive mini-barriers and with both (see Figure 7.36). A standard ballasted track and a slab track with embedded UIC54 rails were also tested for reference. The latter is approximately 2–3 dB noisier than ballasted track.

The results of the demonstration tests, after correction to a constant speed of 100 km/h, can be summarized as follows:

- *Composite brake blocks*: overall reduction of 6 dB.
- *Mini-barriers*: overall reduction of 6 dB.
- *Absorptive panels on track slab*: approximately 2 dB reduction in overall noise (unfortunately some of the material found its way onto the railhead causing an increase in noise for which adjustments had to be made).
- *Wheel damper*: 4 dB reduction in wheel noise.
- *Wheel skirt*: 1 dB reduction in wheel noise (unfortunately most of the absorptive material did not remain in place rendering the skirts ineffective).
- *Slab track with small rail section*: approximately 3 dB reduction in track noise compared with ballasted track (after correction for roughness effects).

Overall, the project achieved its objective of a reduction of at least 10 dB: the combination of measures listed above gave an overall reduction of 11 dB, and once the various disturbing factors had been allowed for the potential reduction was 16 dB.

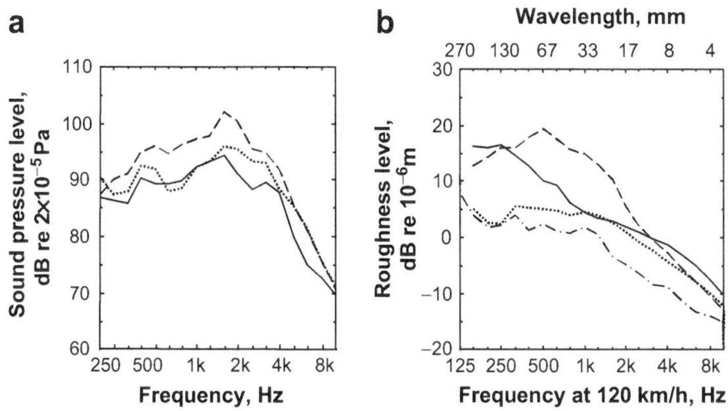

FIGURE 7-4 Noise and roughness spectra of wheels with various types of braking system. (a) Average sound pressure measured at 1 m from the rail and 0.5 m above the top of the railhead, (b) average roughness for wheels of each type of braking system. —, disc brakes only; – – –, cast-iron block brakes; · · · ·, sinter block brakes; – · – ·, rail roughness [7.24]

noise levels with individual vehicles. Results are plotted from two specific ICR passenger coaches, equipped with disc brakes and additional cast-iron or composite (LL) block brakes. The noise level of these vehicles is plotted against time. Although the date of reprofiling (or re-entry into service) was not known precisely, it is

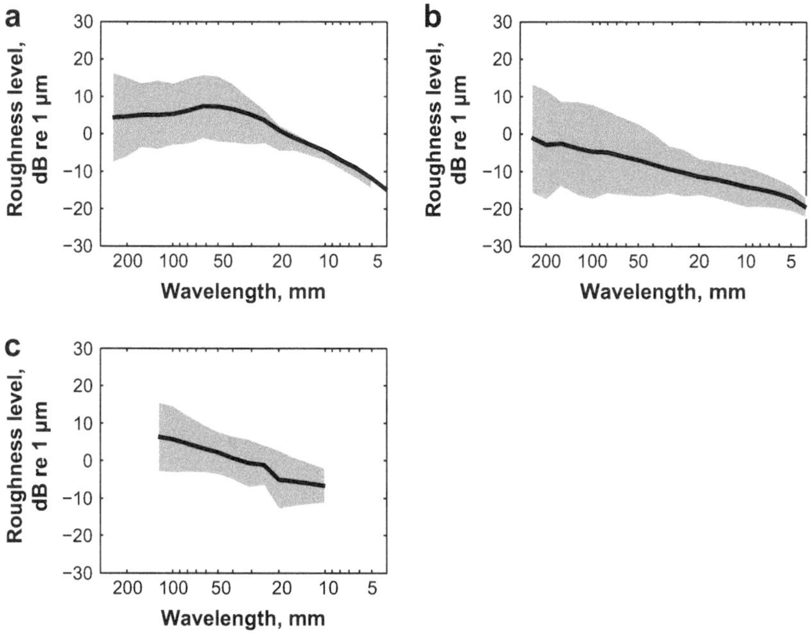

FIGURE 7-5 Typical wheel roughness spectra showing mean and ±2 standard deviations. (a) Cast-iron block brakes, (b) composite block brakes, (c) disc brakes. Source: SNCF

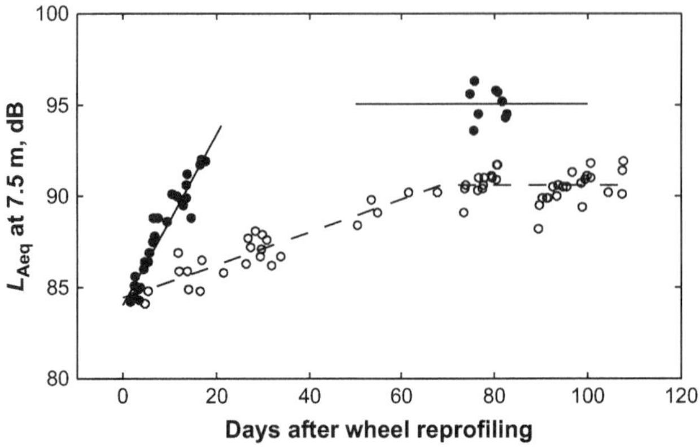

FIGURE 7-6 Development of noise level after reprofiling of wheels. Noise measurements at 7.5 m from the track, train speed 120–135 km/h. ●, —, cast-iron brake blocks; ○, – – –, composite (LL) brake blocks; results provided by Edwin Verheijen, courtesy of ProRail

reasonable to assume that both vehicles commenced with the same noise levels after reprofiling. The straight line trends have been fitted to the data by eye. The actual distances travelled by the vehicles are not available. On average this vehicle type runs about 20 000 km per month, or 700 km per day. However, the vehicles were recorded passing the monitoring station up to three times per day. It is known that they then ran up to 1800 km in a day and made up to 90 station stops in that time. Unfortunately, there are also long periods when the vehicles in question did not run past the monitoring station and it is not known if they ran on another route or not at all.

The noise levels from the vehicle with cast-iron block brakes increased rapidly. The level increased by about 5 dB in the first six days, during which time the vehicle ran about 10 000 km. After about 20 days (approximately 20 000–30 000 km) it reached an average level of about 11 dB above the initial noise level. In contrast, the noise level from the vehicle with LL blocks grew more slowly, taking about two months to reach a constant level, which was around 5 dB lower than that for the vehicle with cast-iron blocks.

7.2.2 Alternative brake blocks for freight vehicles

While passenger rolling stock has increasingly used disc brakes, until recently freight vehicles in Europe have retained cast-iron brake blocks. This is due to the higher cost of disc brakes, the difficulty of running trains with mixed braking systems and most importantly the fact that rules for interoperability of freight vehicles required cast-iron brake blocks. The Eurosabot project [7.13] had the objective to find low cost solutions to reduce the wheel roughness, and hence the noise, of tread-braked freight vehicles. These were intended to be suitable for retrofit application in a fairly short time scale.

The aim of the project was to provide the design principles that could be used to develop a brake block to replace cast-iron blocks with only minimal changes to the brake cylinder, the rigging or the brake pressure (this corresponds to the terminology 'LL-block'). Unfortunately, this was not achieved within the timeframe of the project. One brake block prototype had suitable friction characteristics but required further development to meet the functional requirements and to evaluate performance under operating conditions. One composite block passed the brake and endurance tests but gave only a limited expected noise reduction. Another failed in the endurance tests. Meanwhile, a sinter block passed brake tests and had an expected noise reduction of 3–6 dB(A) but its long-term braking and acoustic performance had yet to be established [7.13]. Moreover, all but the first of these prototypes had different friction characteristics to cast-iron blocks. Consequently, they were not suitable for a simple retrofit and would require modifications to the braking system.

Eurosabot was more successful in developing theoretical models for the wheel roughness generation process [7.27–7.29]. From both rig tests and field tests on cast-iron blocks it was shown that roughness growth occurs due to hot spot contraction, a thermo-elastic instability phenomenon, and a wear regime called 'galling' in which block material is transferred to the wheel running surface. Sinter blocks also show hot spot formation, but the protrusions on the wheel are probably worn away by the more abrasive sinter blocks. They give very smooth wheels, as seen above, but the wheel wear is high. Composite blocks tend to form less severe hot spots and smooth wheel surfaces but they impose high thermal loads on the wheel and their braking performance is less stable. Despite advances in modelling, it was concluded that manufacturing a low noise brake block remains a trial and error process.

Although the Eurosabot project did not achieve its direct objectives, it has facilitated the development and implementation of composite brake blocks which has been pursued further by the UIC Action Plan described in the following section.

The Euro Rolling Silently (ERS) project (2002–2005) continued the development of new LL-type brake blocks for retrofitting to tread-braked freight wagons [7.16]. By the end of this project, three promising prototype blocks were pre-homologated by UIC and were being further tested in revenue service to get information on life-cycle costs.

7.2.3 UIC Action Programme

The development of braking systems for freight vehicles which ensured smooth wheels has been stimulated by the UIC (International Union of Railways) Action Programme [7.30, 7.31]. The main issues were not so much the development of prototype solutions, but rather the adaptation of known solutions from passenger to freight traffic.

Composite brake blocks were already in use for passenger rolling stock and were proven solutions in terms of both general application to the railway system and acoustically, giving a reduction up to 10 dB(A). Distinction is made between K blocks (composite blocks) of which several are available and LL blocks which are composite blocks with a similar friction characteristic to cast iron and could therefore be used as a simple retrofit in existing wagons.

Application of composite blocks to freight vehicles has involved selecting suitable brake blocks, assessing their general braking performance, and developing wheels capable of withstanding higher temperatures due to braking by composite blocks. In parallel, investigations were carried out into other system problems such as ensuring that electrical conductivity was sufficient to operate track circuits, and winter time behaviour in northern European countries. Several types of composite blocks for freight vehicles have been or are being homologated by UIC.

In the IPG project in the Netherlands, 'pilots' were established in which several freight trains were fitted with K or LL blocks. Their performance was monitored over a period of time. The effect on noise of using some of these is affected by other changes; for example, wheel designs were changed and in some cases wheel dampers fitted. Nevertheless, a consistent reduction of 7–9 dB was achieved [7.17]. In addition, passenger vehicles have been fitted with LL blocks, see Figure 7.6.

Figure 7.7 summarizes a number of measured noise levels from freight wagons fitted with K blocks [7.16]. These are plotted against the number of axles per unit length (APL), as the average pass-by noise level should increase roughly according to $10 \log_{10}$ (APL). This is partly accounted for in the TSI for Conventional Rail [7.32] as the noise limits are defined for wagons in three categories of APL: below $0.15 \, m^{-1}$, between $0.15 \, m^{-1}$ and $0.275 \, m^{-1}$, and above $0.275 \, m^{-1}$. Results are shown from different measurements in Switzerland, Germany, Austria and France. Wagons with cast-iron brake blocks can be seen to exceed the limits, whereas those with composite brake blocks mostly do not. The latter can be seen to be mostly 10 dB or more quieter than the wagons with cast-iron blocks. In fact, the limit values for freight wagons were set in the TSI with the intention of ensuring that future wagons would have noise levels corresponding to the use of composite brake blocks.

7.2.4 Rail roughness

The range of roughness levels found on rails is even larger than that seen for wheels. At some sites the rail is smoother than most wheels while at others severe corrugations may form on the rail head (see Figure 2.5). In the presence of short pitch rail corrugations, the noise from both tread-braked and disc-braked stock rises to a level that can be 10 dB higher than the normal level for tread-braked stock [7.21], as

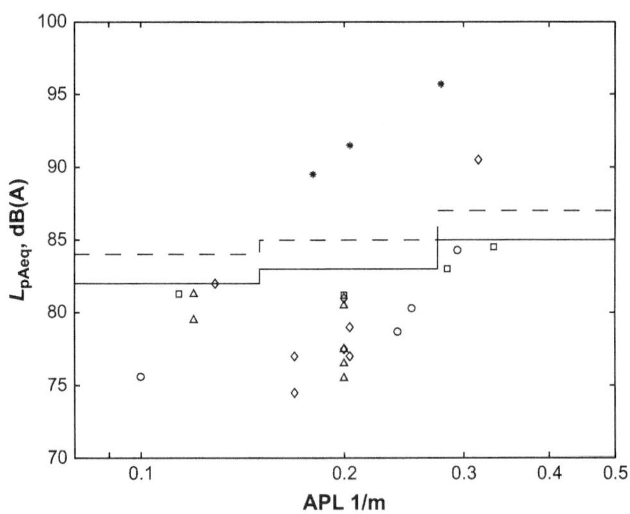

FIGURE 7-7 Equivalent pass-by noise level measured at 80 km/h at 7.5 m from the track for various types of freight wagon in different countries, plotted against number of axles per unit length (APL). ◇, Switzerland (K-blocks); Δ, Austria (K-blocks); ○, France (K-blocks); □, Germany (K-blocks); ∗, cast-iron blocks; —, TSI limit for new rolling stock; – – TSI limit for refurbished rolling stock [7.16]

discussed in Chapter 2. As a result, the phenomenon of rail corrugations is often known as 'roaring rail'. Wavelengths of around 3–8 cm and peak-to-trough amplitudes of up to 0.1 mm (greater for longer wavelengths) are common [7.33]. The normal remedy is to grind the rail, although this is mostly done only for reasons of preventing rail defects and fatigue cracks, not for acoustic reasons. It is not often borne in mind that rail grinding itself is a noisy process, usually carried out at night.

Although the conventional rail grinding process using rotating grinding stones can make the rail considerably smoother, a noticeable tonal peak is often left at a wavelength associated with a resonance of the grinding stones and their drive system. This wavelength depends on the speed of the grinding train but is typically 20 to 30 mm which can be important for noise generation. By using a final grinding pass at a lower train speed, the wavelength of the grinding peak can be reduced so that it lies at a frequency of less importance to the noise generation. Figure 7.8 shows spectra of rail roughness measured after conventional grinding and the modified procedure [7.16]. Alternative grinding processes involving longitudinally oscillating stones have also been developed which avoid this problem.

In Germany, rail grinding is carried out according to acoustic criteria. Sections of 'specially monitored track' may be designated which can be associated with a source term in environmental calculations that is 3 dB lower than normal [7.34]. This has been officially recognized since 1998 and by 2005 covered almost 1000 km of the DB network. The track is monitored twice per year by measuring the sound using a dedicated monitoring coach. 'Acoustic' grinding is used when the level exceeds a particular limit, which is 6 dB above the expected level after grinding. The roughness is also monitored after grinding. A typical grinding interval is about four years, depending on the traffic levels. The 'acoustic' grinding procedure involves planing or milling followed by grinding using oscillating stones at a speed of approximately 1.2 km/h [7.34].

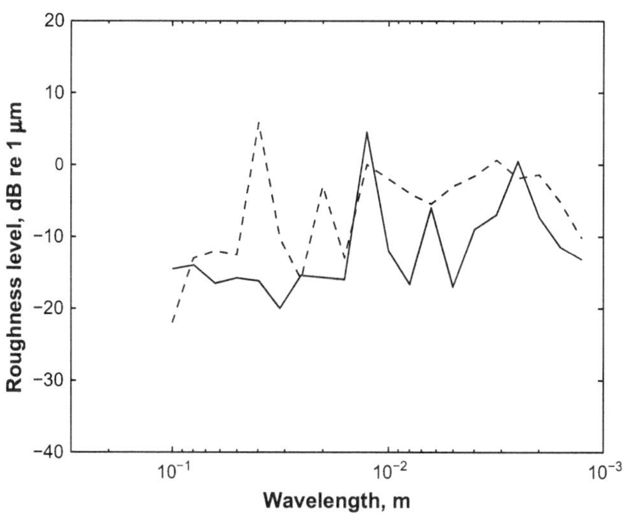

FIGURE 7-8 One-third octave spectra of rail roughness after grinding at two grinding speeds [7.16]. – – –, normal grinding speed; —, reduced speed (2.5 km/h)

7.2.5 Avoiding rail corrugation

There are many factors that affect the development of rail roughness in general and rail corrugation in particular. Although this is too large a problem to tackle adequately within the scope of this book a brief summary is in order.

Various forms of corrugation are known to occur due to different mechanisms [7.33]. It is now generally believed that 'roaring rail' short pitch corrugation on straight track is caused by uneven wear of the rail head, with the wavelength determined by a mechanism associated with anti-resonances of the track vertical dynamics. At an anti-resonance the track appears very stiff to the wheel/rail contact and the contact force can be large for a given roughness input. There is a strong anti-resonance associated with the pinned–pinned mode at around 1 kHz for excitation above a sleeper (see Chapter 3). Additionally, there is an anti-resonance in the region 200–500 Hz associated with the sleeper acting as a vibration absorber. Which of these is involved in corrugation formation depends on train speed. In both cases it is advantageous to use soft rail pads [7.35, 7.36].

In [7.37] it is described how a change in the automatic operation of the Victoria Line of London Underground led to a change in the dominant wavelength of corrugation. This could be directly linked to the frequency of the anti-resonance due to the pinned–pinned mode of the track, the corresponding wavelength changing when the train speed changed.

Although the formation of corrugation has been the subject of research for many years, e.g. [7.38–7.40], the development of broad-band roughness on the rail (and wheel) is less well understood. It is clear that a perfectly smooth surface does not exist, yet the levels of roughness present on rails and wheels are generally very low. In [7.41] a model is presented for the growth of broad-band roughness. This has been extended in [7.42] to consider the effect of rail dampers on roughness growth. Validation of such models is difficult as long time scales are often involved. Roughness growth may vary between different sites in ways that may not necessarily be anticipated beforehand, making controlled tests difficult. The rail steel metallurgy, surface contamination and friction, track dynamics, vehicle dynamic behaviour and the mix of train speeds, axle loads and tractive forces all appear to have an influence. Monitoring all these influences over long periods of time is unrealistic.

7.2.6 Changes to the contact zone

The wheel/rail contact zone can, at least in theory, affect noise production in two ways. First, the Hertzian contact spring introduces resilience between the wheel and rail. At high frequencies this absorbs some of the roughness excitation, as discussed in Section 5.2. If the contact spring can be made softer, this effect can be enhanced and the noise can be reduced, in particular that from the track. This is illustrated in Figure 7.9, which shows the wheel and rail vertical vibration at the contact point for a unit roughness for various values of contact stiffness, calculated using the models from Chapter 5. As the contact stiffness is reduced, the rail vibration is lowered, especially at high frequency. The wheel vibration is also reduced in this region.

The predicted effect of softening the contact spring on the overall noise is shown in Figure 7.10 [7.24]. Unfortunately, it has to be appreciated that such changes are very difficult to achieve. Large changes to the wheel and rail transverse profiles lead to only marginal differences in contact stiffness. Only very radical changes, such as

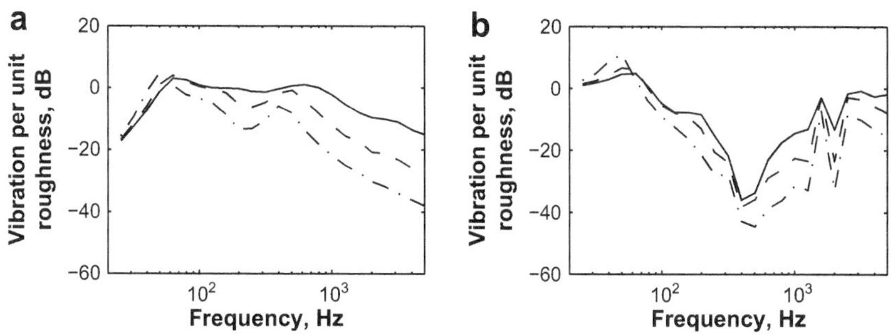

FIGURE 7-9 Vertical vibration at the wheel/rail contact for unit roughness based on vertical interaction only and mobilities of Figure 2.16. (a) Rail, (b) wheel. —, $K_H = 1.14 \times 10^9$ N/m; – – –, $K_H = 3.0 \times 10^8$ N/m; – · – ·, $K_H = 1.0 \times 10^8$ N/m

the use of different materials or a resiliently mounted tread [7.43], are likely to achieve these changes, and they probably have too many associated risks for practical application.

The second potential effect of the contact zone is its influence on the contact filter; a longer contact patch leads to a greater filter effect, as discussed in Section 5.4. Once again, practical changes to wheel and rail transverse profiles will have only a small effect on this. Changes to wheel diameter can be more significant, as discussed in Section 5.4, so that smaller wheels can lead to increased noise from the track (see also Section 7.3.1 below).

A rubber tyre can be seen as an extreme case of these changes to the contact zone. The contact stiffness is greatly reduced and the contact filter effect increased. While

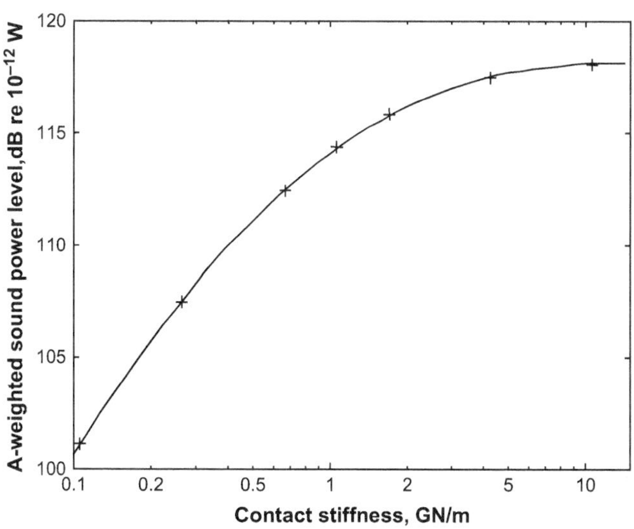

FIGURE 7-10 Effect of change in contact stiffness on overall noise radiation. Usual values are around 1 to 1.5 GN/m [7.24]

this is not the complete answer to the railway noise problem (see box on page 4), rubber-tyred metro systems may have lower rolling noise levels than conventional steel-wheeled systems.

7.3 WHEEL SHAPE AND DAMPING

Returning to Figure 7.1, once the roughness has been minimized within practical constraints the next steps that can be taken require measures to be applied to both the wheel and track. This section focuses on the wheel, so that reductions are mostly quoted in terms of their effect on the wheel component of noise. Various conventional noise control techniques can be applied to the wheel, such as added damping, structural modification (e.g. stiffening) or vibration isolation by introducing resilient layers. Changes to the noise radiation properties by local shielding of parts of the wheel or the use of perforation may also be considered. In most cases these various measures only affect the wheel component of noise, their effect on track noise being mostly less than 1 or 2 dB.

7.3.1 Wheel shape optimization

The cross-sectional shape of the wheel can have a significant influence on the noise radiated. The most important aspects of wheel design for its noise radiation are the shape of the web and the wheel diameter.

As pointed out in Chapter 4, the modes of vibration of the wheel that are principally excited by the roughness are those with large radial components. Most of the sound is then radiated by axial motion of the wheel web, although the radial motion of the tyre is also important. The radial and one-nodal-circle axial modes are strongly coupled together, particularly for a wheel with a curved web, due to the asymmetry of the wheel cross-section and the proximity of their natural frequencies. If this coupling can be reduced, the vertical excitation due to roughness will produce smaller amplitudes of axial motion on the web and consequently reduced noise radiation. A straight web greatly assists this process, although the tyre region must inevitably remain asymmetrical due to the presence of the flange on one side of the running surface.

An early attempt to optimize the wheel shape was carried out by the Technical University of Berlin on the basis of laboratory testing at a scale of 1:4. The most promising design was then tested at full scale by the German Railways (DB) [7.44]. This wheel, which had no additional damping treatment, was almost as quiet as two different damped wheels, also tested in [7.44], which will be referred to in the following section. Compared with the Intercity passenger wheel (Ba92), it had a thicker, straight web and larger radius transitions between web and tyre and between web and hub.

These designs were later assessed using the TWINS model and the results are shown in Figure 7.11 [7.45]. Predicted results are shown for these two wheel designs and for a standard UIC freight wheel (Ba02) for comparison. All three wheels have a diameter of 920 mm and the roughness spectrum is assumed to be the same in each case. The predictions are given in terms of the sound power emitted by a single wheel (omitting track noise). An overall difference of up to 6 dB in the wheel component of sound can be seen, with significant differences particularly at high frequencies. This

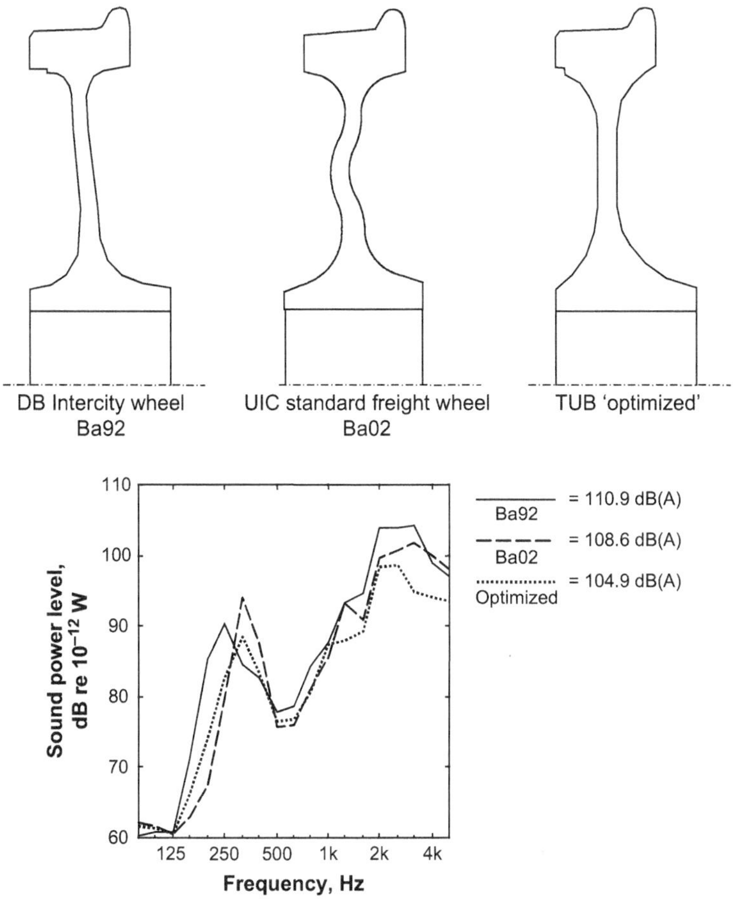

FIGURE 7-11 TWINS predictions of the wheel sound power component from three types of wheel, assuming a train speed of 160 km/h and the same roughness spectrum in each case, typical of a disc-braked wheel [7.45]

Intercity wheel also has a web which is nearly straight, but it is actually particularly noisy; this shows the benefit of assessing a wheel design using the theoretical model. The reason for its poor acoustic performance is that the web is rather thin, the region between the web and tread has very sharp transitions and the tyre is thin. Consequently, this wheel has a series of one-nodal-circle modes in which the web motion is large, which are strongly excited by the vertical displacement input due to roughness [7.46].

In the RONA project (box on page 228) a large number of wheel designs were assessed theoretically. In order to try to compensate for the asymmetry of the tread region, additional mass was added to the inside of the tyre on the opposite side from the flange. In initial tests, three such wheel shapes were compared; these were designed for use in a TGV trailer bogie, [7.47]. In each, the mass of the wheel was increased by about 40 kg compared with the standard design. In tests at 300 km/h the wayside noise above 1600 Hz, indicative of the wheel component, was reduced by 4–5 dB and the overall noise by 2 dB(A). A similar design, known as JR13, was used in later combined tests and is shown in Figure 7.12. These later tests included wheel vibration measurements from which it was shown that the wheel component of noise was reduced by about 3 dB overall [7.16].

As well as changes to the cross-section shape, small wheels are also promising. The main effect of reducing the wheel diameter is to increase the natural frequencies of the wheel. Figure 7.13 shows the natural frequencies of a set of straight-webbed wheels of differing diameters, adapted from [7.48]. The natural frequencies of each set of modes is increased as the diameter is reduced, although those of the one-nodal-circle axial modes show the greatest effect. The upper limit of excitation frequencies is determined by the contact filter (Chapter 5). This has been calculated as the frequency at which the roughness wavelength equals the contact patch length, $2a$, here assuming a wheel load of 50 kN, a rail radius of curvature of 0.3 m and a speed of 160 km/h. This frequency changes only slowly as the wheel diameter is reduced, as shown in Figure 7.13. Consequently the number of modes that are excited is effectively reduced as the wheel size is reduced.

The radiating area is also reduced when the wheel is made smaller but this effect is much more modest than the effect on natural frequencies. For example, changing from 920 to 740 mm diameter gives a reduction in wheel surface area (including the tyre) of 27%, equivalent to a reduction of 1.4 dB.

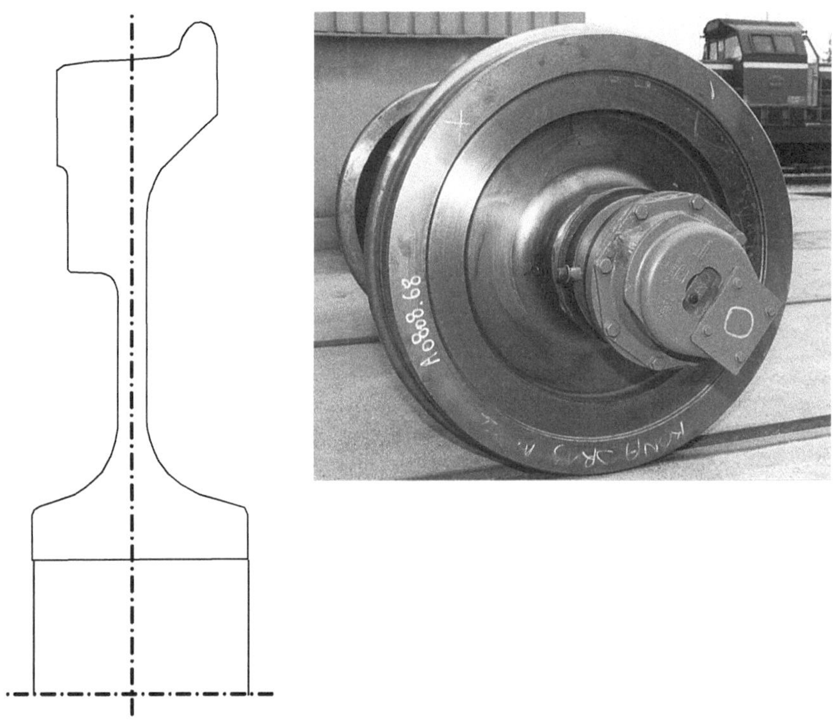

FIGURE 7-12 Experimental optimized wheel shape JR13 for application on TGV [7.16]

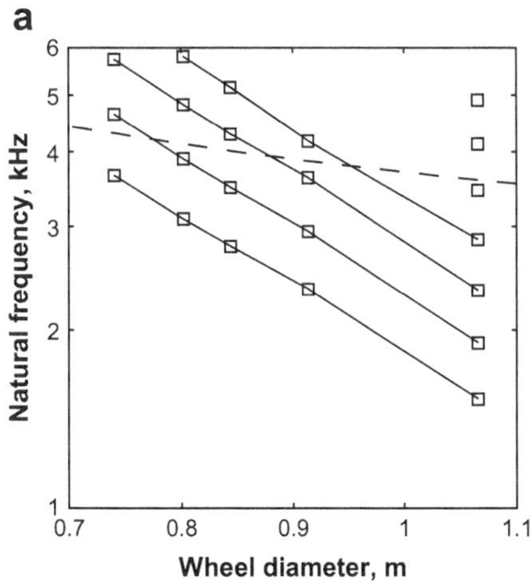

FIGURE 7-13 Effect of wheel diameter on natural frequencies of straight webbed wheels. (a) One-nodal-circle axial modes; (b) radial modes. □, natural frequencies of modes with $n \geq 2$; – – –, 'cut-off' frequency of contact filter (adapted from [7.48])

In an early theoretical study, reported in [7.49], it was shown that the noise from a wheel could be reduced by more than 10 dB by a combination of a straight, thick web and a reduction in the wheel diameter from 920 to 740 mm. This wheel design has not been applied in practice, however.

Within the OFWHAT project (box on page 226), a wheel designed to similar principles but with a diameter of 860 mm was tested. This was predicted to reduce the wheel component of noise by 4 dB but, in the experiments, a reduction of only 1.5 dB was found [7.4, 7.5]. A more radical effect was demonstrated by testing an existing wheel with a diameter 640 mm and very thick web. This was found to reduce the wheel component by as much as 18 dB. However, due to the shorter contact patch, the roughness filtering was lower and subtle changes also occurred in the track response so that the track noise was found to increase by about 2 dB [7.4]. Results of TWINS predictions for this wheel and the standard 920 mm wheel are shown in Figure 7.14, from [7.50].

In practice, although new vehicles can be designed to use small wheels, there are significant difficulties in applying such optimal wheel shapes as retrofit solutions to existing vehicles. Only a small change in wheel diameter can be tolerated within an existing bogie. Moreover, for tread-braked stock, the stresses and deformations caused by thermal loading due to braking have to be kept within strict tolerances. This means that a straight web, as found in the wheels described above from refs [7.44, 7.49], can only be used for disc-braked stock.

Nevertheless, within the Silent Freight project (box on page 230), it was shown that it is still possible to reduce the wheel component of noise significantly while keeping within acceptable limits for thermo-mechanical behaviour. Two such 'optimized' wheel shapes, with a diameter of 860 mm, are shown in Figure 7.15 along with the reference UIC 920 mm freight wheel. One of these designs, Figure 7.15(b), was predicted to produce 3 dB less noise than a standard freight wheel at 100 km/h,

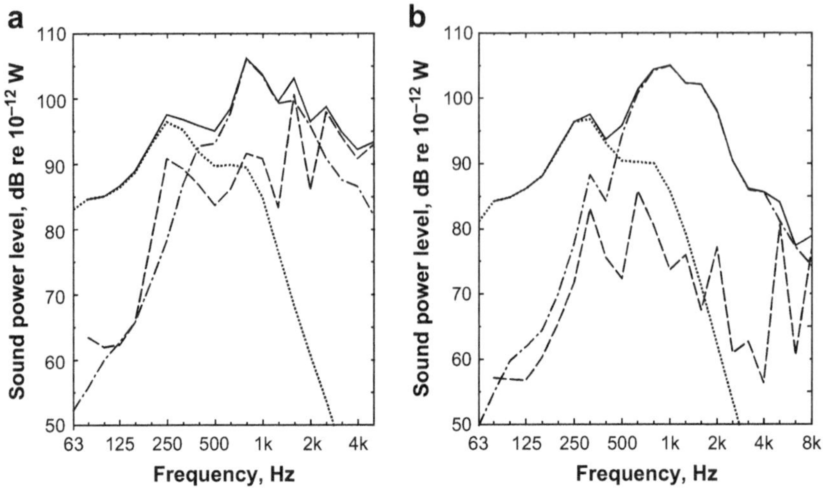

FIGURE 7-14 TWINS predictions of the wheel sound power component from two types of wheel, assuming a train speed of 100 km/h and the same roughness spectrum in each case, typical of a tread-braked wheel. (a) 920 mm diameter standard freight wheel, (b) 640 mm diameter solid web wheel. —, total; – – –, wheel; – · – ·, rail; · · · ·, sleeper, [7.50]

as shown in Figure 7.16 [7.51]. Measurements confirmed this 3 dB reduction [7.9, 7.16]. In combination with constrained layer damping (described in Section 7.3.2 below), the overall benefit compared with the undamped standard wheel was predicted to rise to more than 5 dB. The other design shown (Figure 7.15(c)) was implemented in the tests in combination with wheel web perforation (see Section 7.3.4) but in this form was not found to give significant noise reductions despite encouraging predictions.

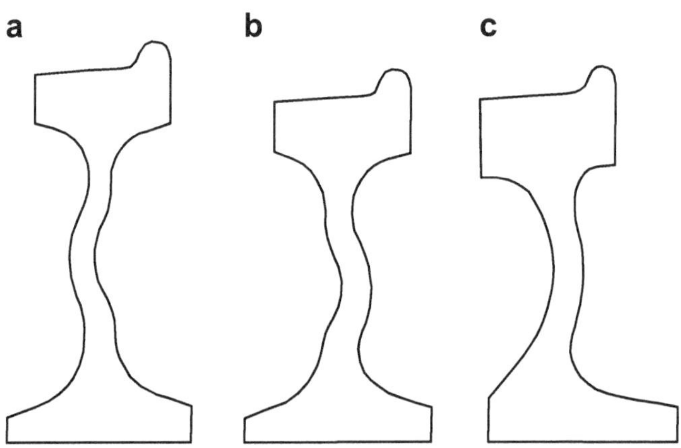

FIGURE 7-15 Wheel cross-sections from Silent Freight project [7.16]. (a) Reference wheel 920 mm diameter, (b) shape-optimized 860 mm diameter wheel, (c) alternative 860 mm diameter wheel design

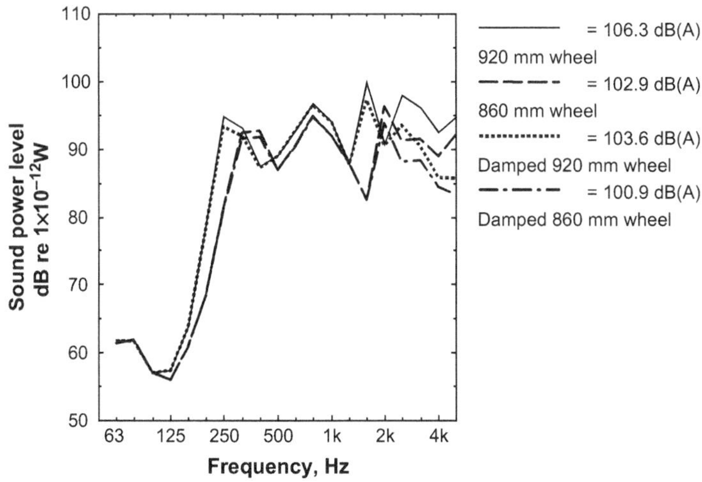

FIGURE 7-16 Predicted sound power from a UIC 920 mm freight wheel at 100 km/h and from the shape-optimized wheel (Figure 7.15(b)) with and without constrained layer damping treatment. The roughness is typical of tread-braked stock [7.51]

The development of these 860 mm shape-optimized wheels suitable for tread-braking demonstrates the conflict between the requirement for a radially stiff wheel, to obtain wheel natural frequencies as high as possible, and the requirement for a wheel with low residual stresses and low permanent deformation after onerous braking cycles, for which a thin flexible web is preferable. For reasons of the safe operation of freight wagons internationally, and due to the introduction of K blocks (see Section 7.2.3) the standards for thermal stresses are being set even more tightly than the draft versions used in the shape optimization within Silent Freight. It seems, therefore, that the scope for an acoustically optimized tread-braked wheel shape will, in future, be very limited. On the other hand, where disc brakes are used, the constraints to wheel shape optimization are much less and there is considerable scope for using low noise wheel designs, including those with a straight web.

7.3.2 Increased wheel damping

The low initial damping of wheels appears at first sight to make them very suitable for the addition of damping treatments. However, it should be recalled from Chapters 4 and 5 that the coupling with the rail increases the initial damping considerably, so that, to control rolling noise, additional damping should exceed this 'rolling damping'. Typically, the damping ratios rise from around 10^{-4} for a free wheel to greater than 10^{-3} for a wheel rolling on a track, see Table 4.2.

Although this chapter concentrates on rolling noise, it will be seen in Chapter 9 that wheel damping treatments are very effective in reducing or eliminating the squealing noise that occurs in sharp curves. The success in treating curve squeal noise, where a small amount of additional damping is often sufficient to give large noise reductions, does not indicate that a damping treatment will necessarily be effective in reducing the wheel component of rolling noise. Laboratory tests on damped wheels are of little value and can be quite misleading unless the resulting damping values are used in a rolling noise prediction that takes account of the true coupling with the track.

Figure 7.17 shows results of increasing the damping of a wheel as predicted using the TWINS model. A UIC 920 mm freight wheel is modelled initially with damping ratios of 10^{-4} in all modes with $n \geq 2$; a higher modal damping is assumed for modes with $n = 0$ and 1, as discussed in Section 4.2.3. In subsequent calculations the minimum damping ratio has been progressively increased.

The effect on the wheel mobility is shown in Figure 7.17(a) in one-third octave bands. This shows the change in mobility level ($20 \log_{10} |Y_w|$) relative to the initial situation. At 1.6 kHz and above, where modes are strongly excited in the radial direction (see Figure 4.3), an increase in damping by a factor of 10 leads to a 10 dB reduction in the one-third octave band level of mobility. The amplitude at a resonance peak is reduced by a factor of 10 (20 dB), see equation (5.29), but the frequency band average is reduced by only 10 dB. For most one-third octave bands at low frequencies, where there are no modes in the band, an increase in the damping has no effect on the wheel mobility. The small effect found in the 400 Hz and 1 kHz bands corresponds to the $n = 2$ and 3 axial modes which appear as narrow peaks in the radial mobility, see Figure 4.3.

The effect on the noise radiated by the wheel is shown in Figure 7.17(b). As expected, there is no reduction in noise at low frequencies, where the damping does not affect the mobility. However, even at high frequencies, due to the 'rolling damping' the noise reduction is much more modest than the reduction in mobility would imply. For example, in the 1.6 kHz band, for a 100-fold increase in the damping the noise reduction is only 6 dB, whereas the mobility is reduced by nearly 20 dB (and the amplitudes of resonance peaks by 40 dB). Increasing the damping has a greater benefit at higher frequency, where the effective damping of the wheel due to interaction with the track is lower (see Section 5.2.4), but it remains much less than found in the mobility.

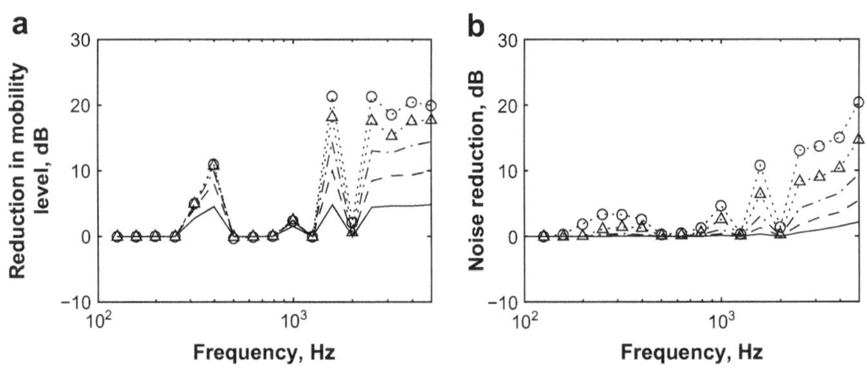

FIGURE 7-17 Effect of increasing the damping of the wheel relative to an initial situation with wheel damping ratios of 10^{-4} in the modes with $n \geq 2$. (a) Reduction in radial mobility level, (b) reduction in noise radiated by the wheel. Damping ratios increased to: —, 3×10^{-4}; – – –, 10^{-3}; – · – ·, 3×10^{-3}; △, 10^{-2}; ○, 3×10^{-2}

Increasing the wheel damping also affects the rail noise. The peaks and troughs seen in the rail vibration in Figure 5.4 occur as a result of the lightly damped wheel modes. A 100-fold increase in the wheel damping can lead to a reduction of 2–3 dB in rail noise in the one-third octave bands above 1.6 kHz, although this frequency region is usually dominated by the wheel component of noise. The overall reduction in the component of noise from the rail is negligible as this is dominated by lower frequencies.

7.3.3 Damping treatments

Turning to specific wheel dampers, a number of different types have been developed and implemented in practice. One means of adding damping to the wheel is to use a constrained layer treatment, see Figure 7.18(a). This consists of a thin layer of highly damped visco-elastic material sandwiched between the wheel and a stiff constraining plate. Such treatments have been used successfully, for example in the UK where they have been applied to the wheels of tread-braked multiple unit trains since the late 1980s to prevent curve squeal (see Chapter 9). For tread-braked wheels, constrained layer damping can only be applied to the wheel web, due to the high temperatures reached in the tread region during prolonged periods of drag braking. Under these circumstances, even on the web, the materials have to be capable of surviving temperatures of over 200°C.

Calculations of the effects of such damping treatments applied to the web of a wheel [7.51] have shown that a practical constrained layer treatment can reduce the wheel component of rolling noise by about 3 dB. The mass added to the wheel by such treatments is limited to a few kilograms. These analyses used parameters corresponding to a realistic choice of damping material and allowed for the temperature and frequency dependence of the material properties. Predicted results are shown in Figure 7.16 for two wheel types. These were based on a maximum thickness of constraining plate of 1 mm steel, as the damping treatment needs to be formed to

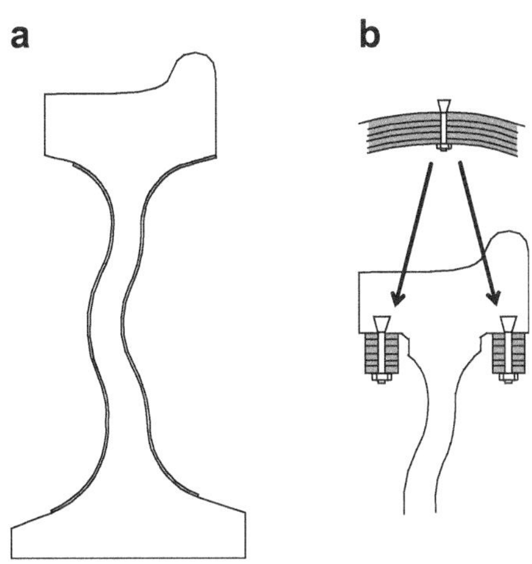

FIGURE 7-18 Examples of wheel damping devices. (a) Constrained layer damping (thin layer of visco-elastic material backed by thin stiff constraining layer), (b) tuned resonance devices [7.45]

fit the shape of the wheel web. To improve on this result a thicker constraining plate would be needed, which would be more appropriate for disc-braked wheels with a straight web.

Field tests in Italy on the ETR500 high speed train have demonstrated that a commercially available constrained layer damping treatment can reduce the overall rolling noise by 4 to 5 dB between 200 and 300 km/h [7.52]. In this case the constraining plate was 1 mm aluminium [7.53]. As with earlier results, this high noise reduction is probably related to the dominance of the wheel noise in the particular situation and is not necessarily universally applicable.

Tuned resonance dampers, such as those shown in Figure 7.18(b), have been used successfully. For example in Germany [7.44] reductions in overall noise of 5–8 dB(A) were reported for train speeds of 200 km/h. The relatively low levels of noise radiated by the ICE high speed train have also been attributed to the use of such dampers. However, such reductions are not universal. As these results were at high speeds and with disc-braked roughness spectra, the wheel can be expected to have a large contribution to the overall spectrum before the wheel damping treatment is applied. Moreover, the starting point in these tests appears to be a wheel design which is noisy for other reasons, as shown in Figure 7.11. Another important point is that, to fit the dampers to the wheel, part of the inner face of the tyre has to be machined away. This first makes the wheels noisier so that tests of the same wheels with and without dampers can be misleading.

Tests of similar dampers fitted on different wheel designs were performed in the Netherlands and France in the 1980s at speeds of 120 km/h. These results were much less impressive. Overall reductions in A-weighted noise level of only 1 to 3 dB were measured [7.54, 7.55]. Could it be that the technology stopped working once it crossed national borders? Part of the answer lies in the differences in wheel designs used in these countries. Nonetheless, spectral results indicated that differences still occurred at frequencies above 1.6 kHz, where the wheel noise is expected to be significant and where reductions of up to 5 dB were found. Thus the wheel component may still have been reduced considerably, but the track component of noise limited the overall benefit in the situations considered. For the full benefit to be seen it would be necessary to test them in combination with a low noise track.

In the OFWHAT project [7.4, 7.5], wheels fitted with simple 'tuned absorbers' were tested in combination with other measures. In principle, such absorbers are mass–spring systems added to a structure to remove energy from particular resonances. By using a system with a high damping loss factor, such an absorber also acts as a source of added damping at frequencies above the mass–spring resonance frequency, where the mass becomes inertially 'grounded'. Two sets of absorbers, tuned to frequencies of 1720 and 2330 Hz, were fitted to a standard UIC 920 mm freight wheel. These consisted of steel discs bolted through holes in the wheel web and supported by rubber pads. They are shown in Figure 7.19. It was found in running measurements that the wheel component of noise was reduced by 4 dB by these absorbers. This was less than predicted using TWINS (6 to 9 dB), the difference being believed to be due to the fact that the absorbers did not conform fully to the performance specification [7.4].

Within the Silent Freight project an alternative tuned absorber system was investigated consisting of nine masses attached by resilient layers to the inner edge of the tyre. This was fitted to the 860 mm optimized wheel of Figure 7.15(b). This, in combination with the change in wheel shape, was found to reduce the wheel noise by

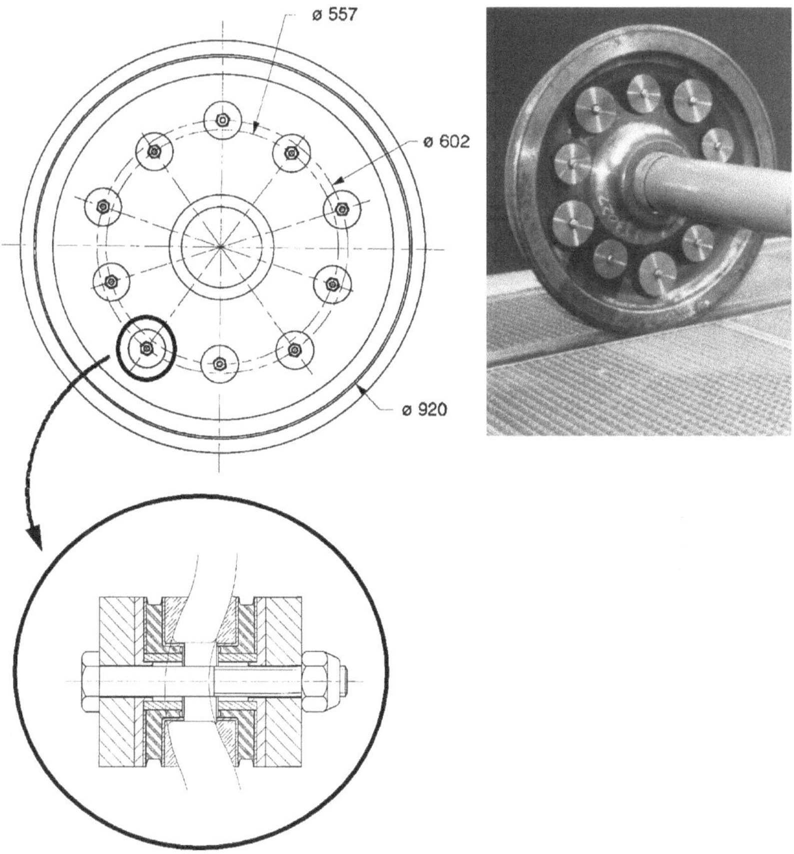

FIGURE 7-19 Tuned absorbers developed in the OFWHAT project fitted to freight wheels [7.16]

at least 7 dB(A) [7.56]. However, such an absorber is not practical for tread-braked wheels due to the high temperatures occurring on the tyre during braking, which would damage the elastomer forming the resilient layer.

In Figure 7.20 the so-called shark fin damper is illustrated. This commercially available design consists of damped sandwich plates mounted from the inside edge of the wheel tyre. Färm [7.57] evaluated the use of this form of damper on the wheels of a new Intercity train and found reductions in A-weighted overall noise of 1 dB at 80 km/h and 3 dB at 200 km/h. The noise above 2 kHz was reduced by 4–5 dB, mainly due to the reduction in wheel noise, but also due to a modest reduction in rail noise. Small reductions were found below 1 kHz but these are less likely to be due to the damper and may therefore be attributable to other differences such as in roughness.

Ring dampers, consisting of a metal ring sprung into a groove, have been used for many years to increase the damping of a wheel to help reduce squeal noise, see Chapter 9. In Silent Freight, a novel ring damper was developed in an attempt to reduce rolling noise [7.58]. In the field tests the results corresponded to a reduction

of about 2 dB in wheel component [7.16]. Again, care is needed if the damped wheel is compared with the same wheel without the rings, as this was modified from the reference wheel by machining part of the inner face of the tyre, creating a noisier reference situation.

In the EU project 'Silence' [7.59] an all-metal wheel damper has been developed, see Figure 7.21. This avoids the use of elastomeric materials which are adversely affected by the high temperatures on a tread-braked wheel. This relies on friction between metal plates to provide damping. Although it is difficult to obtain sufficient damping using such a mechanism, initial results appeared promising.

Tests have been carried out on a TGV fitted with ring dampers on some of the wheels. These consist of a damped metal ring attached in a groove inside the wheel tyre and have also been used for trams. Results indicated that the noise at 300 km/h

FIGURE 7-21 Friction damper developed by Lucchini in Silence project. Photo by courtesy of DB

was reduced by about 2.5 dB(A) overall and by up to 5 dB in the frequency range above 1.6 kHz [7.60].

It has been seen, therefore, that a variety of wheel dampers have been developed. Provided that they increase the damping of the wheel sufficiently to exceed the effective damping of a rolling wheel they can be quite successful in reducing the wheel component of noise, but their effect on the total noise depends on the balance of wheel and track sources in the initial situation.

7.3.4 Multi-material wheels

The use of aluminium for the wheel web would allow it to be made much thicker, and therefore stiffer, without a weight penalty. A steel tread should be retained, however, as this is the preferred material for the contact region. Such a wheel, denoted Alu4, was developed in the RONA project. This had the same mass as the reference TGV wheel, despite having a web thickness of 89 mm, almost four times as thick, as shown in Figure 7.22. This has the effect of raising significantly the natural frequencies of the axial one-nodal-circle modes, although the radial modes are relatively unaffected. The noise reduction was predicted as 4.5 dB [7.16]. Measurements (including vibration measurements) indicated a reduction of up to 6 dB in wheel component. A carbon fibre composite web was also considered initially but not found to be promising.

Tuned absorbers, consisting of cylindrical masses in an elastomeric sleeve, were inserted into holes in the thick web of the Alu4 wheel, as shown in Figure 7.22. Ten were fitted, tuned to two of the wheel's radial natural frequencies (1.9 and 2.6 kHz) with a total increase in the mass of only 15 kg. The total reduction in wheel noise for the Alu4 wheel with absorbers, compared with the reference TGV wheel, was predicted as 12 dB [7.16], this being confirmed by measurements.

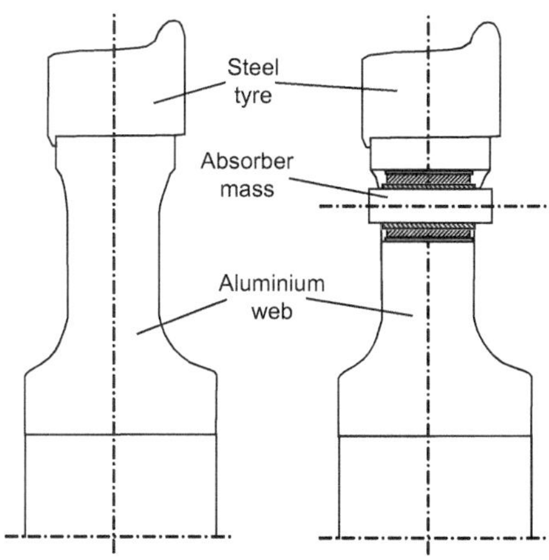

FIGURE 7-22 Wheel with aluminium web (Alu4) including variant fitted with tuned absorbers [7.16]

7.3.5 Resilient wheels

Resilient wheels have a rubber element between the tread and the inner part of the wheel, as shown in Figure 7.23 [7.61]. This has the effect of isolating the wheel web from the tread and also introducing some additional damping to the wheel. In fact, to be effective in isolation the stiffness has to be quite low, whereas to be effective in damping it should be higher. Such resilient wheels are very common in light rail and tramway applications. However, since the tragic accident involving an ICE at Eschede in Germany in 1998 [7.62] due to the failure of a resilient wheel, they have not been considered for use at high speeds.

In [7.61] a resilient wheel for 'light rail' application was analysed theoretically for a range of different values of stiffness of the resilient layer. Figure 7.24 shows the predicted components of noise from the wheel and track for various values of the stiffness of the resilient element. It was found that the noise from the resilient wheel with a notional very stiff material was much less than that from the track, since this wheel had a small radius and straight web. The extreme case of this is where the resilient material is replaced by steel in the model, shown as 'all steel' in Figure 7.24. However, at more typical stiffnesses (moderate/low) of the resilient elements, the noise from the wheel was found to be similar to that of the track.

A similar theoretical study was reported in [7.63] for a wheel intended for freight application. It was shown there, as in Figure 7.24, that by appropriate selection of the stiffness, a resilient wheel can reduce the track component of noise. Reductions of around 2 dB were found and were attributed to a change in the balance of wheel and

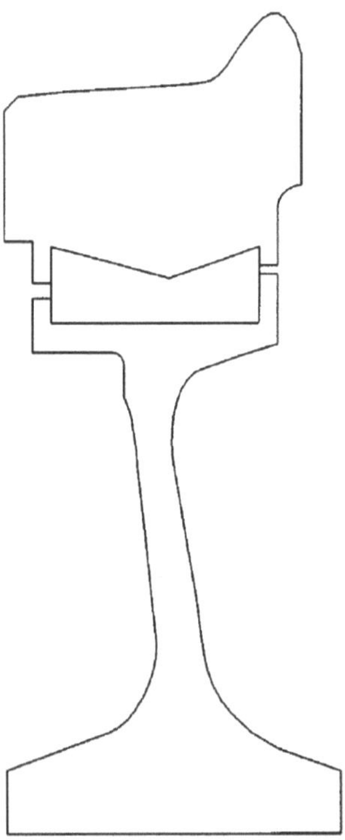

FIGURE 7-23 A typical resilient wheel with 740 mm diameter from a light rail vehicle [7.61]

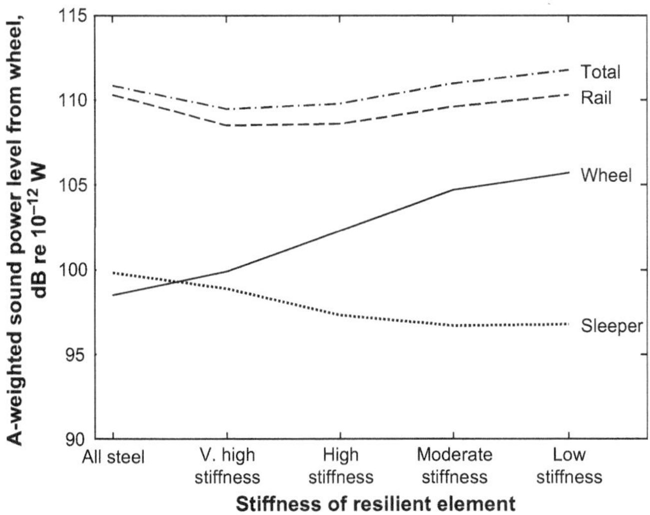

FIGURE 7-24 Overall A-weighted sound power levels from resilient wheels and track for various stiffnesses of resilient element. Resilient wheel based on a 740 mm diameter light rail vehicle wheel, typical ballasted track, train speed 100 km/h, roughness from tread-braked wheels [7.61]

rail vibration amplitudes relative to the roughness, especially between 500 and 1000 Hz. The wheel component was increased, but was still sufficiently small compared with the track component at these frequencies for the rolling noise to be reduced slightly for a train speed of 100 km/h. This has not been applied in practice, however.

7.3.6 Reducing the sound radiation from the wheel

All the above measures are intended to reduce the vibration of the wheel and hence its radiated noise. In addition it is possible to attempt to minimize the sound that is radiated due to a particular vibration level.

One means of reducing the radiation is to mount a shield on the wheel so that sound radiation from the web region is obstructed. It is the web that produces most of the noise from the wheel, particularly for a curved web design, although the tread, particularly its radial motion, should not be neglected. Such a shield must be mounted in such a way that it does not vibrate significantly, as then it would radiate sound itself; thus it must be resiliently mounted and reasonably well damped. It should also be made of a material which gives a high enough acoustic transmission loss while still being flexible. The use of a thin metal plate means that bending waves excited in the panel are short compared with the wavelength of sound in air and therefore have poor sound radiation. Good sealing is also required so that no sound can escape through gaps. In addition the damping of the wheel may also be increased somewhat by the mounting arrangement.

In the RONA project such web screens, consisting of 1 mm thick steel plates, were mounted resiliently and clipped onto the wheel. They were fitted to a wheel which had been optimized mechanically, that is in terms of mass. Its mass was only 335 kg but its initial noise radiation was greater than the reference wheel. The two screens on a wheel each added 15 kg to the mass, but their application on this particular wheel still left the mass below that of the reference wheel. The wheel component of noise was reduced relative to the reference wheel by about 6 dB [7.16].

Wheels with similar shields were constructed and tested within the Silent Freight project [7.9]. These were based on the optimized wheel cross-section of Figure 7.15(b). They are shown in Figure 7.25. Compared with the reference wheel, this was estimated to reduce the wheel component of noise by about 8 dB [7.16], although it was difficult to assess because of the dominance of track noise.

The concept of a perforated wheel was studied in the Silent Freight project [7.64]. The idea was to reduce the radiation efficiency by introducing acoustic short-circuiting between the front and back of the wheel web. The effect depends on the size and spacing of the holes [7.65]. Due to the thickness of the web and practical limitations on the distance between holes and their size, the perforated wheel was shown to be promising at low frequencies, where a 6–9 dB reduction was expected, but no appreciable effect was predicted above about 1 kHz which is where the wheel noise is usually dominant. A perforated wheel was constructed, see Figure 7.26, based on the alternative optimized wheel cross-section shown in Figure 7.15(c). Unfortunately, no noise reduction was found, although in combination with the ring damper a modest reduction of around 2.5 dB was found [7.16].

7.4 TRACK RESPONSE AND RADIATION

In the same way as for the wheel, the noise from the track can be dealt with by using a range of measures such as damping treatments, vibration isolation, structural

a b

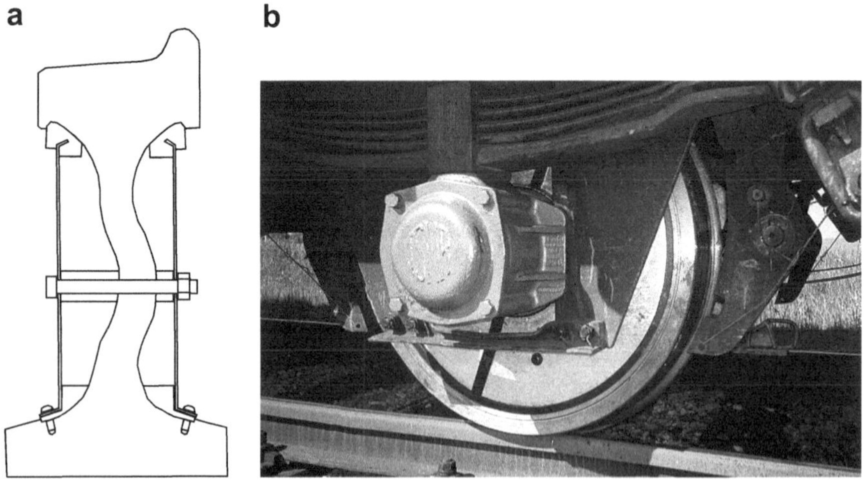

FIGURE 7-25 Shape-optimized wheel fitted with web shields developed in Silent Freight [7.16]. Photo: Rick Jones, used courtesy of UIC

FIGURE 7-26 Silent Freight shape-optimized 860 mm wheel with perforated web. Photo: Rick Jones, used courtesy of UIC

modification or changes to the sound radiation properties. However, the very different dynamic properties of a track compared with a wheel mean that the performance of these measures is rather different. For example, unlike a wheel, the track does not have lightly damped resonances so that the effect of structural modification is much more limited. Damping treatments can still be effective but their effect is seen in terms of increasing the spatial decay rate rather than reducing resonant amplitudes.

7.4.1 Rail pad stiffness

The noise radiated by the track is strongly related to the stiffness of the rail fastening, in particular the rail pad between the rail and the sleeper. As described in Chapter 3, soft pads cause the rail to become decoupled from the sleeper, which minimizes noise from the sleeper but allows the rail to vibrate more freely. Waves can therefore travel over a greater distance and the noise from the rail is increased, as shown schematically in Figure 3.7. Conversely, with stiff pads the contribution from the rail is reduced but that from the sleeper is increased. Measured decay rates in the track for three different pads were shown in Figure 3.10, confirming the dependence on pad stiffness.

The dependence of the various components of noise on the rail pad stiffness is illustrated in Figure 7.27. A compromise can be reached when the sleeper contribution is equal to the combined vertical and lateral rail component [7.66]. For the situation modelled here, this occurs at a pad stiffness of about 2500 MN/m. This is typical of a traditional stiff 5 mm thick pad under load [7.67] (see also Table 3.8) but is very stiff by current standards, and mostly would be unacceptable for new installations. The pad stiffnesses referred to here, and shown in Figure 7.27, are the high frequency, low amplitude tangent stiffnesses, not the static or low frequency, large amplitude secant stiffnesses often used to evaluate rail pads (see also Section 3.8).

In the OFWHAT project alternative rail pads were investigated. Two 'optimum' pads were compared with a fairly soft reference pad. Both were stiffer and one also had a high damping loss factor. A difference of 4 to 5 dB in track component of noise was measured for the first pad, confirming qualitatively the results of Figure 7.27. However, attempts to double the damping loss factor of the pad, which should produce a further 2 dB reduction in track noise, were less successful.

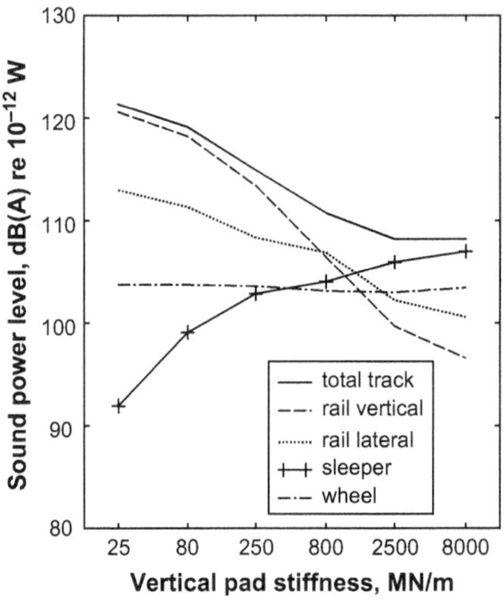

FIGURE 7-27 Example of predicted sound power due to one wheel and the associated track vibration versus high frequency rail pad stiffness. Calculations using TWINS for a standard 920 mm freight wheel at 100 km/h, with a typical tread-braked roughness

Similarly, in the VONA project 'optimized' pads were tested that had a vertical dynamic stiffness of 560 MN/m compared with 250 MN/m for the reference pad (values based on fitting the track model to the measured frequency response function). The damping loss factor was slightly higher (0.26 compared with 0.18, determined in the same basis). The reduction in track component of noise was found to be 3 to 4 dB.

In the Silent Track project a stiffer rail pad was also tested, consisting of a 10 mm thick studded EVA pad. This had a dynamic stiffness measured in laboratory tests of 430 MN/m. It gave a reduction of track noise of 2 dB compared with a soft natural rubber pad measured in the laboratory as having a dynamic stiffness of 56 MN/m. This reduction was less than expected theoretically, but the tests were affected by high temperatures of over 30°C which may have had an influence on the results.

To reduce track forces and damage to sleepers and track components, it has become common practice to use relatively soft rail pads with dynamic stiffnesses in the range 80–400 MN/m. According to the prediction model, a track with soft rail pads should be considerably noisier than a track with stiff rail pads for the same roughness, see Figure 7.27. For example, a track with a pad of stiffness 80 MN/m is predicted to be about 8 dB noisier than one with a stiffness of 800 MN/m. However, experimental evidence suggests that the differences are smaller than this. Part of the reason seems to be that the pad stiffness found by fitting the model to measured track responses and decay rates is higher than that obtained in direct laboratory measurements, particularly for soft pads. This might be related to the influence of the clips.

Moreover, soft studded or grooved rubber pads have a strongly load-dependent stiffness, designed to prevent rail roll-over. The effect of this load dependence on noise has therefore been studied. In [7.68] an analysis is given of the effect of the preload on the point receptance of the track. The preload has a considerable effect on the point receptance, but the decay rate is affected much less. By including the effects of the preloads exerted by additional wheels [7.69], it has been shown that the noise predicted using the unloaded stiffness is about 4 dB too great for the soft pad. The corresponding difference for a stiff pad is only 1 dB. Altogether, therefore, a soft studded rubber pad, which has a stiffness of about 80 MN/m as measured on a single assembly, should give an increase in track noise of only about 3 dB compared with a pad of stiffness 800 MN/m. This agrees more closely with the results of the Silent Track tests discussed above.

Although it is well established that using a stiff rail pad can reduce the track component of noise, soft pads are being used increasingly for non-acoustic reasons. Instead, the decay rates must be increased by means of the application of damping treatments, discussed in Section 7.4.3 below, which allow soft pads to be retained.

7.4.2 Track mobility

The dynamic response of the track, as characterized by its mobility, has an effect on the interaction of the wheel and track. As seen in Chapter 5, the component (wheel, track or contact spring) with the highest vertical mobility at a particular frequency, is induced to vibrate with the amplitude of the roughness. Between about 100 and 1000 Hz it is the track that mostly has a higher mobility than either the wheel or the contact spring, as shown in Figure 2.16. It therefore vibrates with an amplitude close to that of the roughness. Above 1000 Hz the contact spring is more flexible than the rail and, apart from frequencies close to wheel natural frequencies, the contact spring absorbs most of the roughness excitation.

Changing the track mobility thus has no effect at all on its noise radiation in the region where it already has a greater mobility than the wheel or the contact spring. Only if the track mobility can be brought down well below that of these other elements will there be any effect. On the other hand, at high frequencies, where the track mobility drops below that of the contact spring, reducing the rail mobility has a beneficial effect. However, in this region above 1 kHz, the mobility is influenced only by the mass and bending stiffness of the rail itself, not by the support structure. Increasing the size of the rail to reduce its mobility would have a negative effect on the sound radiation (see Section 7.4.5 below). In any case, the rail is not usually the dominant source above 2 kHz.

In the VONA project, a number of alternative rail shapes were considered theoretically. For a constant mass, noise reductions of less than 1 dB were predicted. A rail section with a mass of 74 kg/m, A74, which is similar to the standard UIC 60 profile but with a thicker web, was estimated to give a reduction of about 1.5 dB. However, due to difficulties in obtaining approval, this was not implemented in the field tests.

7.4.3 Rail damping treatments

As seen in Section 7.4.1 above, it is desirable to increase the track decay rates without introducing stiffer rail pads in order to maintain the isolation of the sleeper. If applied on a track with a soft pad, therefore, added rail damping can achieve significant noise reductions.

The problem with adding damping to the rail is that it is already highly damped. The decay rates presented in Figure 3.10, for example, correspond to an equivalent damping loss factor of the rail of around 0.02 at high frequencies, and considerably greater than this at low frequencies. This is many times larger than the damping of the wheel, even allowing for the additional damping present during rolling (a loss factor of 0.02 corresponds to a modal damping ratio of 0.01).

Tuned absorber systems are an attractive option for application to rails. The theory of such a tuned absorber system is presented in [7.70]. The efficiency increases as the active mass increases. The internal damping of the absorber should also be sufficiently high to give a broad bandwidth of operation. In [7.71] it was found that a damping loss factor of at least 0.35 in the absorber was sufficient – if it is any greater than this the added benefit is small. Tuning frequencies should be around 1 kHz where the peak in the rail noise spectrum occurs; the use of multiple tuning frequencies can also broaden the peak in the decay rate. In the present section some practical implementations are discussed.

In the OFWHAT tests [7.4, 7.5] a prototype absorber was tested which was mounted on the rail foot, shown in Figure 7.28(a). This produced a measured reduction in track noise component of 2 dB(A) for a train speed of 100 km/h when mounted on a track with stiff pads. Unfortunately, the combination of absorbers with soft pads was not tested, but in this situation the absorber could be expected to yield a considerably greater benefit.

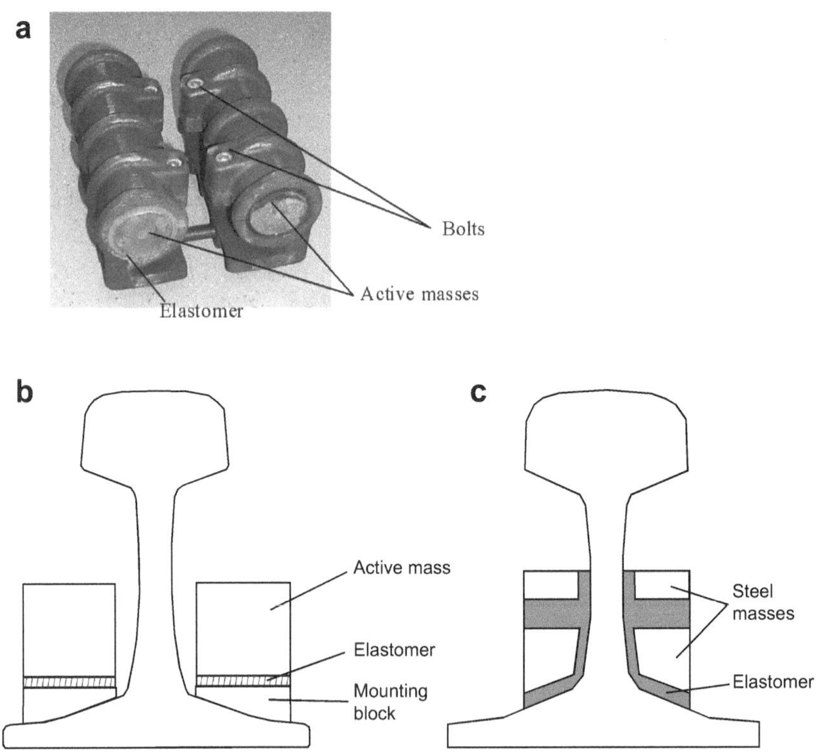

FIGURE 7-28 Rail dampers (a) developed in OFWHAT, (b) developed in VONA, (c) developed in Silent Track [7.16]

Tuned absorbers were also designed in the VONA project. These used rectangular steel blocks of dimensions 200 × 45 × 45 mm to obtain an active mass of 3 kg on each side of the rail in each sleeper span. They were glued to the top of the rail foot via an inclined block to give a vertical mounting surface, see Figure 7.28(b). An initial design involving a clamping arrangement at the edge of the rail foot was found to be less successful due to the flexibility of the rail foot. A total mass of 9 kg was added to each sleeper span. Tuning frequencies of 1000 and 2000 Hz were chosen using different elastomeric elements, which were mounted in alternate sleeper spans. The effect on the track noise was estimated to be 4 dB, and the combined effect of this and stiff pads was 6 dB.

Within the Silent Track project a novel absorber system was developed, shown in Figure 7.28(c). This comprises steel masses embedded in a high damping elastomer forming a two-degree-of-freedom absorber. It is bonded to the rail *in situ* or can be supplied pre-fitted to the rails. Decay rates measured on a 4 m length of free UIC60 rail fitted with this absorber are presented in Figure 7.29. These decay rates are very large at high frequencies but quite small at low frequencies. This is in part a consequence of the design, since the tuning frequencies are set to correspond to the broad peak in the rail noise spectrum between 500 and 2000 Hz. These decay rate values should be added to the decay rates already present in the track to give an overall decay rate for a damped track [7.70]. Comparing these results with Figure 3.10, it can be seen that the absorber would increase the damping of vertical vibration above about 500 Hz for the soft pad (shown in Figure 7.29), and above about 800 Hz for the intermediate pad. Below these frequencies the decay rates are generally so high

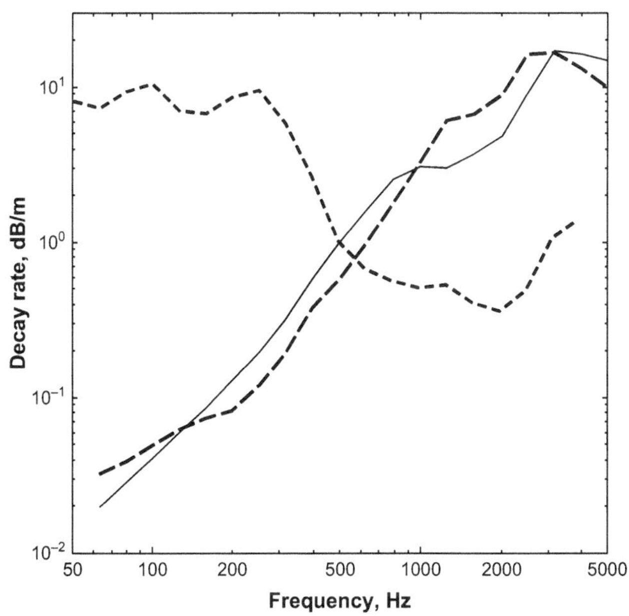

FIGURE 7-29 Measured decay rates on a 4 m length of free rail fitted with the Silent Track tuned absorber system [7.45]. —, vertical vibration; — — —, lateral vibration; - - -, vertical decay rates on track with soft pads from Fig. 3.10.

(around 10 dB/m) that damping treatments are unlikely to have any effect. In any case it is the sleepers rather than the rail that become the dominant source below these frequencies.

For a track with a pad stiffness of 300 MN/m (i.e. equivalent to the intermediate pad of Figure 3.10) this system was predicted to reduce the track component of noise by about 6 dB(A). The reduction will be less for stiffer pads and greater for softer pads. Because the track decay rates of a damped track are largely determined by the damping from the absorbers, the noise level for a track with absorbers is found to be almost independent of pad stiffness; the reductions quoted are therefore more indicative of the starting point than the efficiency of the absorber.

The noise reduction obtained by using this absorber is shown in Figure 7.30 [7.71]. The left-hand graph shows predicted results in terms of the sound power radiated by the track. This shows a benefit for frequencies above about 500 Hz. The right-hand graph shows the measured sound pressure level. In this case the wheel contribution is also present. Even though a noise-reducing wheel has been used (with web shields, see Figure 7.25), the noise above 2 kHz is likely to be contaminated by the wheel contribution. Nevertheless, the measurements agree reasonably well with the predictions. In fact both are consistent with reductions of about 6 dB(A) in the rail component of noise.

Since these early prototype tests, considerable experience has been gathered on the use of various designs of rail damper in service and various commercially available designs have been developed. The acoustic benefit clearly depends on the properties of the track to which they are fitted, in particular the rail pad stiffness.

Extensive tests have been carried out in France on two types of rail damper [7.72], shown in Figure 7.31. Both are designed to be attached to the rail between sleepers

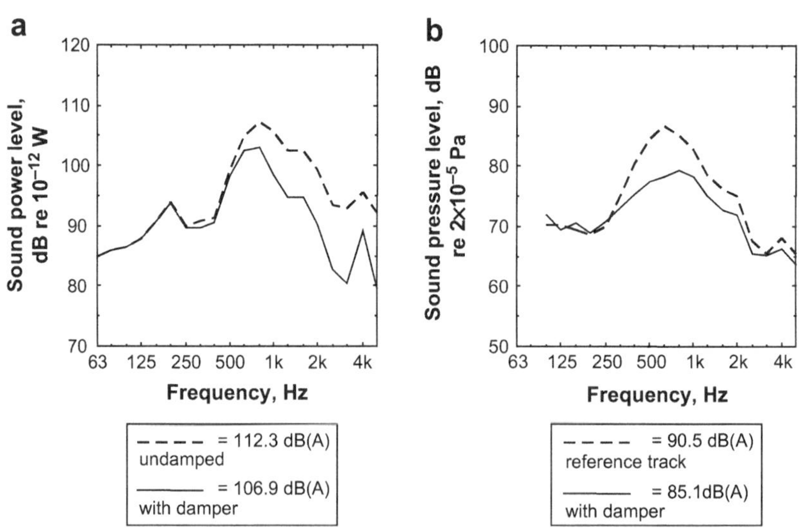

FIGURE 7-30 Effect of Silent Track tuned absorber system. (a) Predicted sound power from the track. (b) Measured total sound pressure from rolling noise at 100 km/h using a vehicle fitted with a noise reducing wheel. Average from three microphones at 3 m from near rail [7.71]

FIGURE 7-31 Two types of rail damper installed in service. (a) Corus, (b) Schrey & Veit

using a clipping mechanism. The one in Figure 7.31(a) was a development of the damper developed in Silent Track. The one in Figure 7.31(b) includes assemblies of plates below the rail as well as mounted against the web. The track in this case had a 9 mm grooved rubber rail pad with a stiffness identified as 380 MN/m. Overall reductions of 2–4 dB were found. In addition, one of these dampers (Figure 7.31(b)) was installed on a high speed line in France. This gave only a small noise reduction, less than 1 dB overall [7.59], as a consequence of the greater contribution of the wheel at high speed and, in addition, the presence of aerodynamic noise.

Tests have also been carried out in Germany with a rail damper similar to that shown in Figure 7.31(a) [7.73]. These were applied on two tracks with pad stiffnesses of 200 and 800 MN/m. The decay rates measured on the track with softer pads, with and without dampers, are shown in Figure 7.32(a). Differences of up to a factor of 10 (equivalent to 10 dB) are found in the region 800 to 2000 Hz.

Noise measurements have shown overall reductions of 3–4 dB for the track with soft pads, depending on the train type. For example, Figure 7.32(b) shows the average noise spectra for freight vehicles; for regional passenger trains the reductions

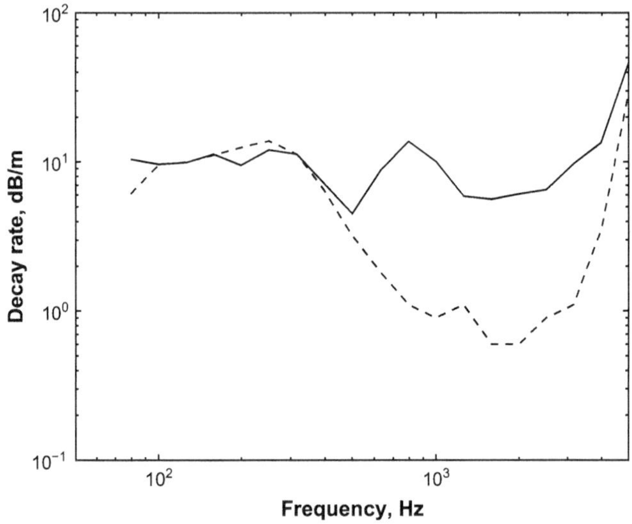

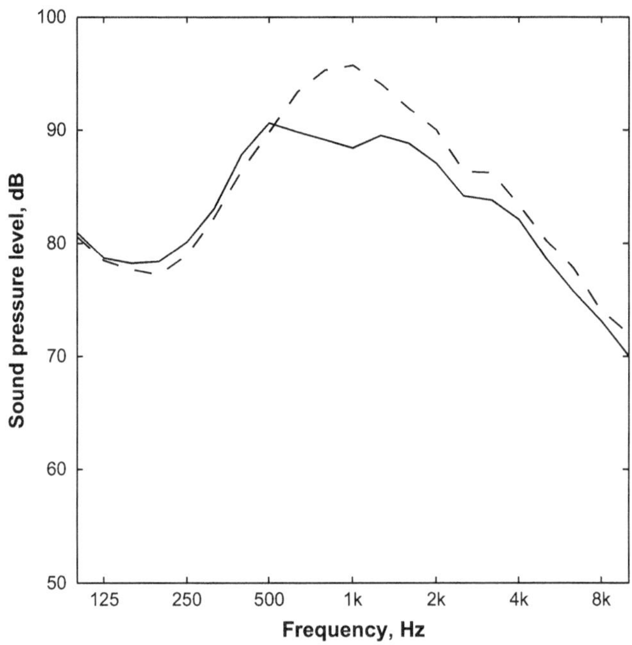

FIGURE 7-32 Measurements of Corus rail dampers on DB track. (a) Vertical track decay rates; (b) average noise from freight trains at 3 m from the track. – – –, reference track; —, damped track

were slightly lower. On the track with stiffer pads the effect was only about 1.5 dB. However, comparisons with theoretical predictions [7.74] confirm that the wheel component of noise is responsible for limiting the benefit, even in the region 500–1000 Hz, and indicate that the rail component of noise may have been reduced by as much as 8–9 dB on the track with soft pads and by 5–6 dB on the stiffer track. Preliminary results also indicate that rail dampers may give an additional benefit by reducing the tendency of the rail roughness to develop [7.42].

In the Dutch IPG programme [7.19, 7.20] a number of rail damper installations have been tested. As the track in the Netherlands is based on a stiffer rail pad than those used in the French or German tests, the benefit was found to be lower, typically 2–3 dB, but this is still considered worthwhile in many situations.

Adding constrained layer damping to the rail, either to the web or under the foot, was studied numerically in the Silent Track project and found to give only a very small benefit. A 1 mm layer of damping material and 2 mm constraining layer was considered, applied on either the foot or the web. Even on a track with a low pad stiffness of 80 MN/m, the track noise was predicted to be reduced by only about 2 dB(A), and these reductions were concentrated at high frequencies. For stiffer pads, the reductions were found to be negligible.

Some commercial systems are available which appear to have the form of a constrained layer treatment. However, these have much thicker layers of both visco-elastic material and constraining plate. Analysis of their performance has led the authors to believe that they operate on the principle of a tuned absorber.

In summary, the use of tuned absorber systems can be very effective in reducing the noise radiated by the rail. These can be installed on the rail in such a way that they do not affect the normal track construction and do not interfere with maintenance operations. To be effective these must add considerable mass to the rails. A removable attachment using clips or bolts is favoured by most administrations over a bonded installation.

7.4.4 Rail shape optimization

In principle, reduction in the size of the rail could lead to significant reductions in its sound radiation. This occurs by affecting the radiating area and, more importantly, the radiation ratio (see equation (6.1)). Figure 7.33 [7.45, 7.75] shows the radiation ratio predicted using a two-dimensional boundary element model for two notional modified rail sections as well as for the standard UIC60 section. These rail sections are shown in Figure 7.34. One is a rail with a reduced height, achieved by shortening the web (Figure 7.34(b)); the other, labelled 'bullhead', has the same height as UIC60 but has a 'head' at the bottom as well as the top, so that it is only 70 mm wide instead of 150 mm (Figure 7.34(c)). The radiation ratio curves for a standard rail peak between 500 and 1000 Hz, above which they tend to unity. At lower frequencies they fall at a rate proportional to the cube of the frequency, as discussed in Section 6.4.1. By reducing the size of the rail cross-section, the frequency at which they reach unity can be raised, as shown in Figure 7.33. Reducing the height of the rail can thus lower the radiation from lateral motion, while reducing the width can minimize the component of sound from vertical motion. The corresponding reductions in surface area are more modest.

In practice, the scope for reducing the dimensions of the rail is limited. Nevertheless, this concept led to the development by British Steel and British Rail Research of a low height rail known as 'hush rail' [7.76]. This was 110 mm high with a mass of 50 kg/m, similar to the rail section shown in Figure 7.34(b). It was tested in practice in 1990, although the tests were inconclusive. For practical reasons the rails were installed on wooden sleepers but this led to high levels of noise from the sleepers

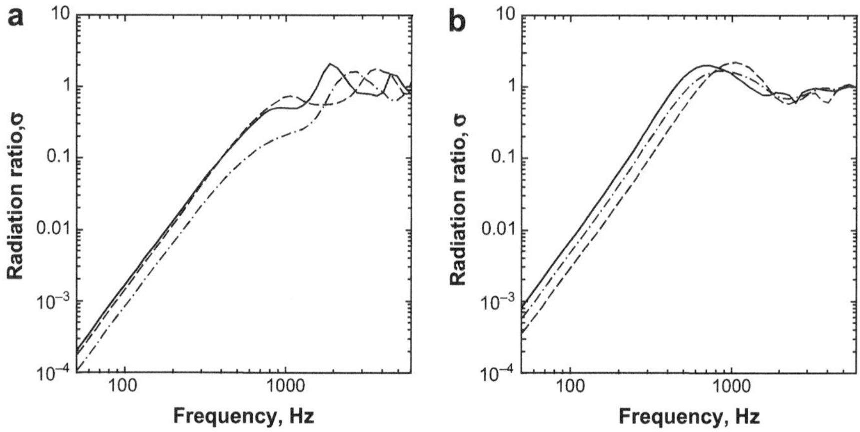

FIGURE 7-33 Radiation ratio of rails of various dimensions vibrating uniformly vertically or laterally, calculated using boundary elements. (a) vertical vibration; (b) lateral vibration. — UIC60 trail; – – –, low height rail; –·–·, 'bullhead' rail. [7.45, 7.75]

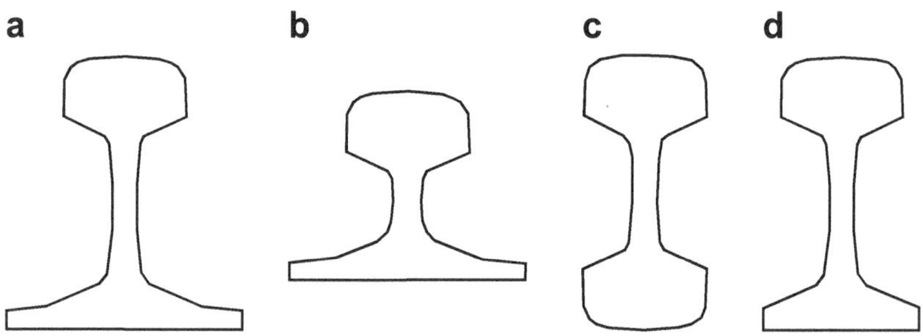

FIGURE 7-34 Rail sections (not to scale). (a) UIC60, (b) low height rail based on UIC60 as used for Figure 7.33, (c) 'bullhead' rail as used for Figure 7.33, (d) narrow foot as tested in Silent Track

which unfortunately masked any effect of the low height rail [7.77]. These rails have been used on slab tracks in tunnel to reduce the structure height but no further noise tests have been conducted.

Where the dominant source of noise from the rail is its vertical vibration, as is the case for low pad stiffnesses (see Figure 7.27), the dimension which should be minimized is the width rather than the height. Gains of up to 4 dB(A) can be expected from a rail with a narrower foot [7.78], although this requires a modified rail support system to prevent excessive rail roll. Such a rail was tested within the Silent Track project [7.9], as shown in Figure 7.34(d). This was found to reduce the rail component of noise by about 3 dB. Although such benefits appear promising, they would require considerable changes to the construction of the railway.

7.4.5 Embedded rail

Rails embedded in an elastomer are often used for tramways to facilitate street running. They also form a relatively soft rail fastening system for which the decay rates are not too low. At first sight, an embedded rail appears to offer a means of reducing the radiating area of the rail to a portion of the railhead, giving benefits in the acoustic radiation similar to those shown for small rail sections in Figure 7.33. However, there are two aspects of the embedded rail which lead to greater sound radiation than might be expected from a small rail section [7.79].

First, the free surface of the elastomer also vibrates with the rail, at some frequencies even amplifying the rail vibration. The surface area is therefore considerably greater than that of the railhead. It is only at higher frequencies that the vibration of the elastomer is smaller than that of the rail. Although the elastomer itself has short wavelengths and high damping, it is driven by the vibration in the rail and so shares its wavelength. As a result, no acoustic short-circuiting effects occur within the elastomer itself, although at some frequencies, where the free surface vibrates out of phase with the rail head, there is potentially some reduction in the radiation.

A second aspect is the fact that by embedding the rail, the dipole characteristic of a free rail is replaced by a monopole source characteristic, in a similar way to mounting a loudspeaker in a box in order to improve its efficiency. This can considerably increase the radiation ratio at low frequencies, similar to the effect seen in Figure 6.16.

7.4.6 Ballastless track forms

Ballastless track, or 'slab track' is growing in popularity in some countries for high speed lines. In such a track, the rail is attached by resilient fasteners to a concrete base, typically 0.5 m thick. This gives a track with potentially lower maintenance costs, since the track vertical and lateral profiles do not need to be maintained, for example by tamping the ballast at regular intervals.

Unfortunately, slab track has a reputation for being rather noisy. Increases in noise level relative to ballasted track of between 2 and 4 dB are typically found. The reason for this is often given as the fact that the acoustic absorption offered by the ballast has been removed. However, this is likely to explain only about 1 dB of the difference. A more important effect lies in the fact that slab track usually incorporates softer rail fasteners. They are intended to substitute an extra compliance in the rail support to compensate for the loss of the vertical compliance normally provided by the ballast layer. As seen in Section 7.4.1 above, soft rail fasteners can lead to higher rail noise as the rail can vibrate over a greater length.

The concrete slab usually has such a high impedance that its vibration is much less than that of the rail, and its noise radiation is thus also negligible. Therefore track models for slab track can be based simply on a single layer of resilience under the rail (see Section 3.2).

Acoustic treatments, such as that shown in Figure 7.35, have been demonstrated on slab tracks in Germany [7.16, 7.80, 7.81]. The full treatment shown in Figure 7.35 gave a noise reduction of about 6 dB. Omitting the raised barriers lowered the benefit to about 4.5 dB. A more limited absorptive treatment, which left the rail exposed, gave a noise reduction of about 3 dB. This was nevertheless sufficient to make the track equivalent to a conventional ballasted track with concrete sleepers [7.81].

Since the concrete slab usually has a high mechanical impedance, the stiffness of the rail support can be increased considerably without the component of noise radiated by the slab becoming significant compared with the rail noise. The use of

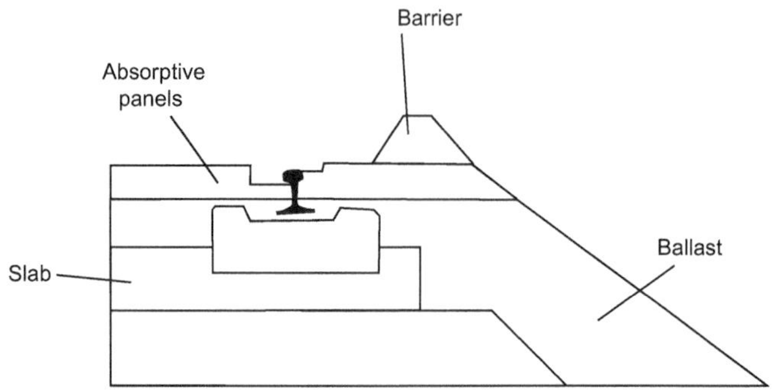

FIGURE 7-35 Absorptive treatment applied to slab track [7.16, 7.80, 7.81]

a stiffer rail support will increase the attenuation of vibration along the rail. In the STV project a track was designed in which the rail was continuously supported by embedding it in a visco-elastic material within a channel in the slab. This made it possible to use a smaller rail section as the track bending stiffness is partly provided by the slab. A small rail profile was developed, referred to as SA42. This rail had a mass per unit length of 42 kg/m but a height of only 80 mm, half the height of the reference UIC54 rail. This was continuously supported by the stiff embedding material, which provided a considerable damping effect on the wave propagation in the rail, restricting the rail noise component. The radiating surface area of the rail and embedding material was also minimized, thus reducing the sound radiation. This was measured as giving approximately 3 dB reduction in track noise compared with ballasted track and 5 dB reduction compared with a slab track fitted with embedded UIC 54 rails. An alternative analysis based on reciprocal vibro-acoustic measurements indicated a noise reduction from the SA42 track of 7 dB [7.15] which was more consistent with theoretical estimates [7.82, 7.83].

The slab track was also tested in configurations with a layer of absorptive material on top of the track slab, with integral absorptive mini-barriers and with both. These are shown in Figure 7.36. The barriers were designed around the Dutch national structure gauge and came to a height of 0.7 m above the top of the rail. They were constructed of concrete elements with a layer of rockwool and an absorptive top layer. These mini-barriers gave a reduction of 6 dB and the absorptive panels on the track slab a reduction of approximately 2 dB in the overall noise [7.15].

Some problems occurred in the tests due to the absorptive material from the track slab becoming detached and finding its way into the wheel/rail contact region. This led to increases in noise for which adjustments had to be made. The prototype track was dismantled after the tests.

7.5 SHIELDING MEASURES

7.5.1 Bogie shrouds and low barriers

The final area of noise control that is possible close to the source, as identified in Figure 7.1, is to reduce the sound transmission to the receiver. This can be done, for

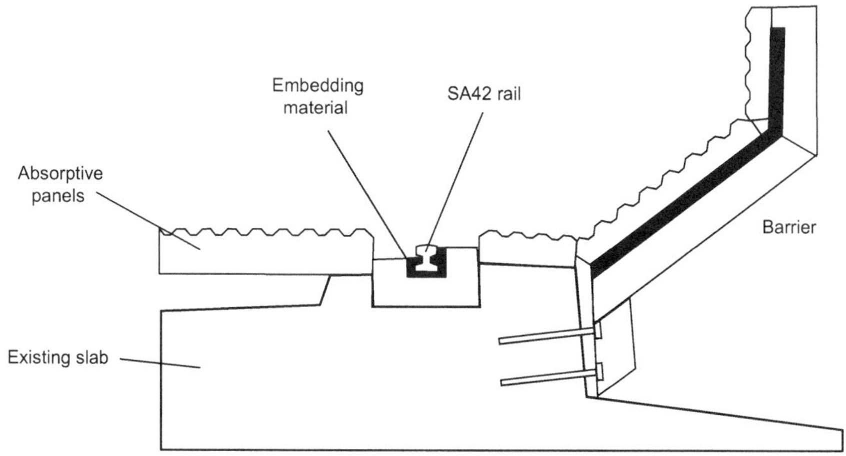

FIGURE 7-36 Slab track with SA42 rail section, absorptive treatment and integral mini-barriers [7.16]

example, by introducing shielding in the form of vehicle-mounted 'shrouds' and track-mounted low barriers. Such a configuration is shown schematically in Figure 7.37.

This concept has been considered in a number of theoretical and practical studies. A successful demonstration project was carried out in the UK in the early 1990s [7.84]. For combinations of low barriers mounted close to the rail and bogie-mounted shrouds reductions of 8 to 10 dB(A) were found. The use of either low barriers or vehicle-mounted shrouds in isolation is expected to lead to much smaller reductions in noise [7.85].

Modelling of the performance of such systems can be achieved using a combination of statistical energy analysis and boundary element predictions [7.85, 7.86]. Not surprisingly, one of the most critical parameters of such systems is the size of any gap between the top of the barrier and the bottom of the shield. The system

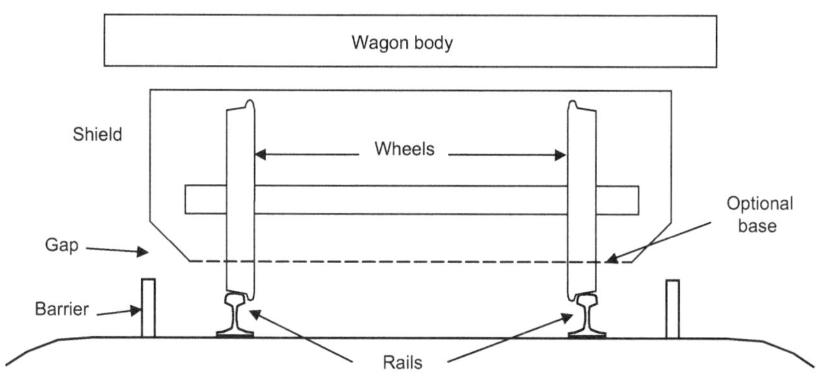

FIGURE 7-37 The principle of bogie shields and low, close barriers. The gap should be as small as possible, and absorbent material should be included on the inside faces of the shield

TABLE 7.1 MEASURED REDUCTIONS IN A-WEIGHTED NOISE LEVEL IN dB FROM BOGIE SHROUDS AND LOW BARRIERS IN SILENT FREIGHT/SILENT TRACK [7.16, 7.87]

	Barriers alone	Shrouds alone	Shrouds and barriers
Reference track	1	1	2
Track with rail dampers	0	2	3

developed in Silent Freight and Silent Track was intended to be usable in any country in Europe, so that the combined gauging constraints from several countries restricted the allowable envelope [7.87]. Consequently, a gap of 118 mm remained between the bottom of the shroud and the top of the barrier which severely limited its effectiveness. Another limiting factor is that, when used on tracks with a low decay rate of vibration, the contribution of the part of the rail not contained within the bogie shield is large enough to limit the overall noise reduction. When used on a track fitted with rail dampers, the benefit of the bogie shield is therefore greater. The results obtained from the shrouds and barriers in Silent Freight/Silent Track are summarized in Table 7.1.

7.5.2 Trackside barriers

The installation of noise barriers alongside railway lines as a means of reducing noise has become quite commonplace in many countries. According to a study carried out by the UIC, quoted in [7.88], about 1000 km of barriers have already been constructed along European railway lines. On the Betuwe line in the Netherlands, which was opened in 2007, there are 160 km of noise barriers on 160 km of route, even though much of it runs parallel to motorways. Although these barriers represent only about 2–3% of the total cost of the project (€4.7bn), if the three tunnels on the route are also included, the proportion spent on 'noise mitigation' rises to around 20%. Control at source is now increasingly seen as potentially more cost effective than secondary measures such as noise barriers [7.88, 7.89]. While this book focuses on control at source, a brief discussion of the effect of noise barriers is in order. For more detailed general coverage of the performance of noise barriers the reader is referred, for example, to [7.90, 7.91, 7.92].

Noise barriers prevent sound from travelling directly from a source to a receiver, but due to diffraction some noise still reaches the 'shadow zone' behind the barrier. Prediction methods are often based on the Fresnel number, N, derived from the theory of diffraction:

$$N = \frac{2}{\lambda}\delta = \frac{2}{\lambda}(A + B - d) \tag{7.1}$$

where λ is the acoustic wavelength and δ is the 'path length difference'. This is the difference in length between the path over the top of the barrier $A + B$, as shown in Figure 7.38, and the direct path in the absence of the barrier, d. Maekawa [7.93] gave a chart to allow the barrier insertion loss to be determined from N, with the horizontal axis adjusted to make this a straight line. An alternative is provided by the Kurze–Anderson formula for the insertion loss IL[7.94]:

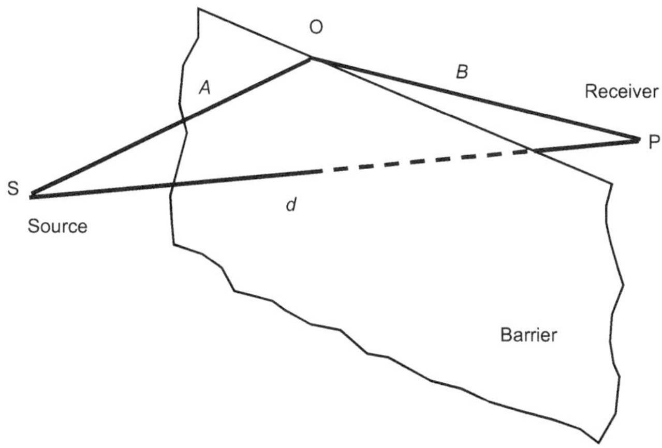

FIGURE 7-38 Geometry of sound diffraction over the top of a barrier

$$\mathrm{IL} = 5 + 20\log_{10}\left(\frac{\sqrt{2\pi N}}{\tanh\sqrt{2\pi N}}\right) \tag{7.2}$$

This is shown in Figure 7.39. A line source (or moving point source) gives up to 5 dB less attenuation than indicated by this formula. This can be estimated by using a modified formula [7.92]

$$\mathrm{IL} = 5 + 15\log_{10}\left(\frac{\sqrt{2\pi N}}{\tanh\sqrt{2\pi N}}\right) \tag{7.3}$$

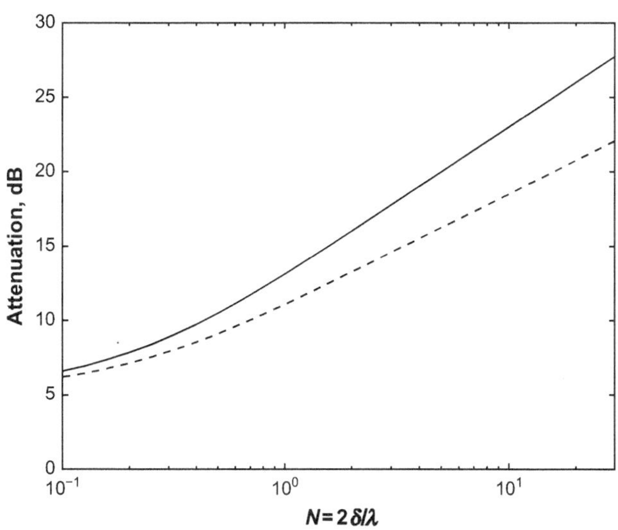

FIGURE 7-39 Attenuation due to noise barrier. —, predicted using equation (7.2) for a point source; – – –, predicted using equation (7.3) for a line source

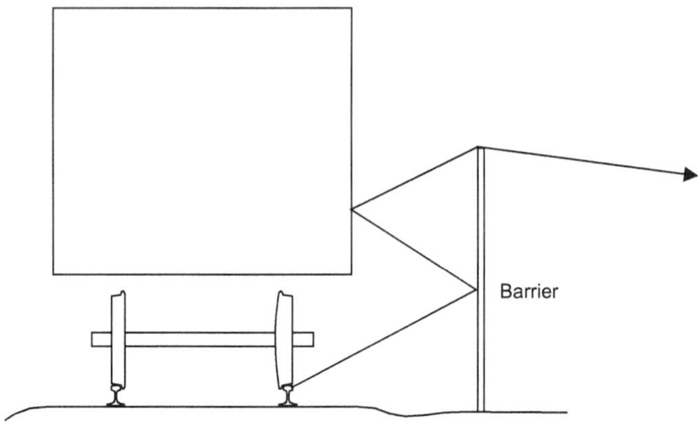

FIGURE 7-40 Reflections between barrier and vehicle body

also shown in Figure 7.39. Maekawa [7.95] showed that this tends to overestimate the insertion loss at small values of N but gives reasonable results at larger values.

In these results the insertion loss (level difference between situations without and with the noise barrier) is frequency dependent due to the presence of λ in N. Low frequencies correspond to small values of N and therefore small insertion losses.

The above formulae apply to a semi-infinite barrier in free space; they do not allow for the effects of ground reflections. A full calculation will take into account image sources to allow for ground reflections on both the source and receiver sides of the barrier. In some situations a barrier can actually remove the effect of the ground dip (see Section 6.6.4) causing a much lower attenuation than given by the simple model.

Multiple reflections between the side of a train and the barrier can also compromise the effectiveness of a barrier. This is shown in Figure 7.40. For this reason absorptive barriers should be used where possible. Hemsworth indicates that the maximum effectiveness of a reflective barrier in reducing railway noise is about 15 dB(A), which requires the barrier to be at least 3 m taller than the source [7.96, 7.97]. For an absorptive barrier the maximum attenuation that can be achieved is increased by about 5 dB.

Most environmental noise calculation schemes and software contain methods to calculate the effectiveness of barriers, often based on the methods given in [7.98]. Boundary element methods can also be used to evaluate the performance of particular designs, see, e.g., [7.91].

7.6 COMBINATIONS OF MEASURES

A number of measures have been described in this chapter, which can reduce one or more of the components of noise from the wheel/track system. Several solutions for reducing wheel noise have been commercially available since the 1980s; successful solutions for track noise have reached the market more recently.

Apart from reductions in surface roughness, none of the technologies for noise control at source discussed in this chapter can achieve reductions of 10 dB or more in overall noise by themselves. Mostly, the effect on overall noise is less than 3 dB. To obtain larger reductions, it is necessary to use a combination of different measures. In assessing combinations of measures it is important to remember that their effects cannot necessarily be simply added. The combined effect may be greater or smaller than the sum of the individual benefits gives an explanation of this.

Clearly, it is important to identify correctly the relative importance of the various sources in the initial situation. As has been seen, wheel and track components are often similar in their contribution to the overall sound. To achieve a reduction of the order of 10 dB, a large reduction in both track noise and wheel noise components is required, usually meaning that careful analysis is required, followed by separate treatments for each source. Some examples are given from the various research projects introduced at the start of this chapter.

The results obtained in the OFWHAT project are summarized in Table 7.2 in terms of the reduction of overall A-weighted noise level at 100 km/h. Here the track noise component in the reference situation was around 4 dB greater than that from the wheel so that the changes are minimal when only the wheel is modified. For the best combination of wheel and track tested, reductions of about 7 dB were achieved.

The overall noise reductions obtained in the combined field tests of the RONA and VONA solutions for TGV are given in Table 7.3. The effect at 150 km/h was considerable, but the overall effect at 300 km/h was limited to 3 dB by the presence of aerodynamic noise. Nevertheless, by using intermediate measurements, and calculations using TWINS, the reduction in rolling noise could be quantified, as shown in Table 7.3. This reduction was found to be as large as 8 dB. At 150 km/h the track was the dominant component so that the modified wheels gave only small reductions, but at 300 km/h the wheel contribution appeared to be initially slightly greater than that of the track.

The main solutions developed in Silent Freight and Silent Track were implemented as prototypes in an extensive field test in May 1999. The reference situation here was dominated by noise from the track, in part due to the relatively low speed (100 km/h) and the fact that tests were carried out at over 30°C so that the rail pads were rather soft. The track noise was found to be about 8 dB greater than the wheel noise in the reference situation. This meant that the reductions in the wheel component of noise could not be measured directly but had to be deduced from a combination of noise and vibration measurements, with interpretation provided by the theoretical model.

Table 7.4 lists the combined effects. From these overall reductions it appears that the web shields and tuned absorbers used together with the optimized wheel both give reductions in wheel component of about 8 dB compared with the reference wheel. In combination with the rail dampers these gave the best overall reduction of 7 dB on the track with conventional rails and 8 dB on the track with modified rails. The noise levels were also recalculated for a tread-braked wheel roughness spectrum and found to give slightly smaller reductions in A-weighted noise level than those obtained for the actual roughness spectrum present in the tests [7.9].

As already explained, these various results cannot be directly applied more generally without first understanding the initial situation and determining the effect on each noise component separately.

Adding the results of different noise control measures

Suppose two noise control measures X and Y each lead to a noise reduction of 5 dB. What is the result of implementing both together? The answer might appear to be a 10 dB reduction, but unfortunately things are rarely that simple.

The 'law of diminishing returns'

In many situations the combined result of applying X and Y is *less* than the sum of their separate effects. For example, in figure (a) below the overall level may consist of a primary source and a secondary one, indicated by the two grey bars, which add together to give the level indicated by the outer histogram. If X or Y each affect only the primary source, the overall reduction may be 5 dB in each case, but the result of combining X and Y will be considerably less than 10 dB, in this example only 7 dB. This secondary source may also represent a flanking path that is unaffected by the measures. Note also that the reduction achieved by X or Y is less than their effect on the primary source (about 8 dB in this example). This is often the reason why noise control in practice gives less benefit than is predicted beforehand.

A similar effect may occur due to the frequency dependence of the measures, as illustrated in figure (b). Here, X and Y are assumed to reduce the dominant frequency components but once one measure has been applied, these frequency components may no longer be dominant. The presence of other frequency bands that are less affected by the measures has a similar effect to flanking paths discussed above.

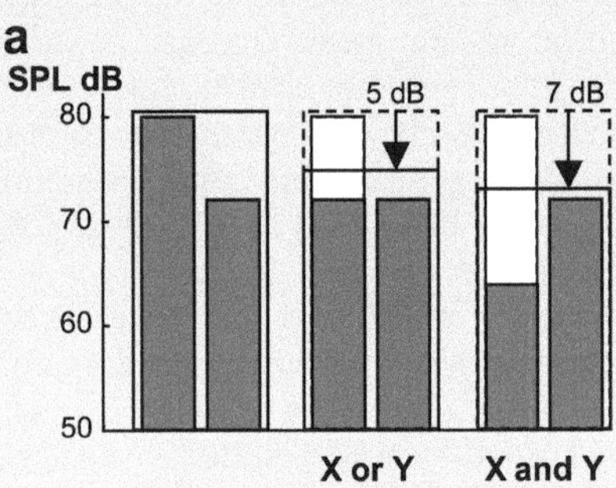

Presence of other sources or flanking paths

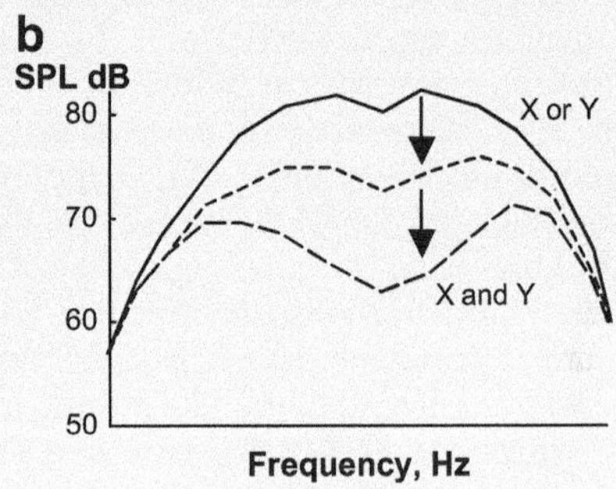

Noise reduction is not the same across the whole frequency range

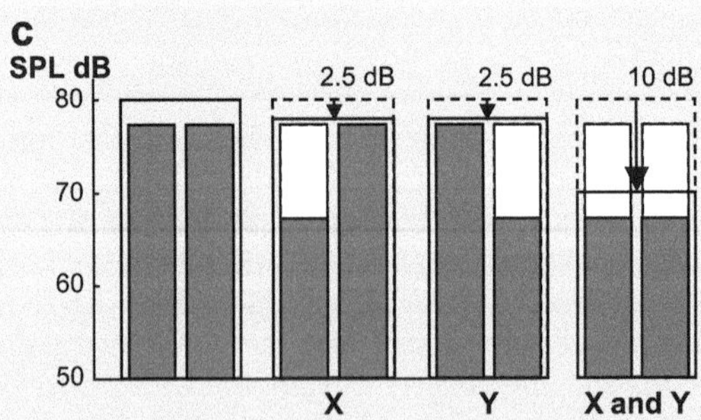

Continued

Adding the results of different noise control measures—Cont'd

'Greater than the sum of the parts...'

In other situations, however, applying X and Y together may actually give a *greater* reduction than applying them individually. This can occur if the two measures are effective on different sources or transmission paths. In the example shown in figure (c), there are initially two equal contributions. X reduces one of them by 10 dB, while Y reduces the other by 10 dB. The overall effect of each measure is therefore only 2.5 dB, but taken together their effect is 10 dB (unless there is also a third source or path of course).

The implementation of K (or LL) brake blocks has been something of a success story: the UIC Action Programme and the implementation of limit values through the TSI has stimulated the introduction of low noise vehicles (Figure 7.7). However, this has merely brought the noise from freight vehicles down to levels similar to current passenger stock. It should not be expected that such large effects can be achieved easily by further lowering of the limit values. Wheel roughness is now at a similar level to good quality rail roughness, so further benefits will be small even for a large effort. Furthermore, it is interesting to note that, although these noise reductions have been achieved by lowering the wheel roughness levels, in terms of the radiating components it is the track vibration which is usually dominant for freight vehicles (see Figure 2.14). As a result it is unlikely that further significant reductions in noise levels from freight vehicles can be achieved by modifications to the vehicles alone such as wheel damping. It will be important to take account of the track radiation by extending the limits to cover infrastructure as well as vehicles.

It is clear that there is no panacea for rolling noise but, by a suitable choice of combinations of measures, significant noise reductions can be achieved. Inevitably, this involves compromise on other requirements such as weight, ease of maintenance and cost. Where this involves action on both the vehicles and the track there are additional complications if these are owned by different parties. Yet, if action is taken only on the vehicle (or only on the track) the effect will probably be disappointing.

TABLE 7-2 OVERALL REDUCTIONS IN A-WEIGHTED NOISE LEVEL IN dB ACHIEVED IN THE OFWHAT PROJECT FOR A SPEED OF 100 KM/H [7.4, 7.5]

	UIC 920 mm wheel	UIC 920 mm wheel with absorbers	860 mm shape-optimized wheel	640 mm wheel
Reference track	0.0	1.0	0.3	0.3
Optimized pads	2.5	4.2	3.0	4.3
High damping optimized pads	3.0	5.0	3.3	5.3
Ditto plus rail absorbers	3.7	7.1	4.2	6.9

TABLE 7-3 REDUCTION IN A-WEIGHTED NOISE LEVEL IN dB AT 150 KM/H MEASURED FOR COMBINATIONS OF WHEEL AND TRACK DESIGNS IN RONA–VONA. RESULTS AT 300 KM/H ARE REDUCTION IN ROLLING NOISE OBTAINED FROM MEASUREMENTS USING CALCULATIONS TO REMOVE AERODYNAMIC COMPONENT [7.16]

	Reference track		Optimized pads		Optimized pads + tuned absorbers	
	150 km/h	300 km/h	150 km/h	300 km/h	150 km/h	300 km/h
Reference wheel	0	0	4.5	1.4	7.2	1.6
JR13 steel wheel, 'symmetric' profile	1.9	0.9	6.2	3.3	8.2	4.7
Alu4, aluminium web, steel tread	1.1	1.9	5.4	4.4	–	5.8
Mechanically optimized wheel with web screens	0.8	2.7	4.4	5.0	7.0	6.2
Alu4 wheel with tuned absorbers	1.7	3.2	6.3	6.1	7.9	8.5

Such aspects require economic or political solutions in order to stimulate the implementation of the technical solutions described here. That is beyond the scope of the present book.

TABLE 7-4 REDUCTION IN OVERALL A-WEIGHTED NOISE LEVEL IN dB RELATIVE TO REFERENCE WHEEL/TRACK FOR VARIOUS COMBINATIONS TESTED IN SILENT FREIGHT/SILENT TRACK [7.11, 7.16]

	Wheel						
Track	Reference	Ring damper	Perforated wheel	Perforated + ring	Shape optimized	Shape opt + absorber	Shape opt + web shields
Reference	0.0	−0.3	−0.4	−0.2	0.8	1.1	1.1
Stiff pads	2.1	1.9	*	2.1	*	2.6	2.7
Rail damper	5.4	4.7	*	5.6	*	6.9	6.7
Stiff pads + damper	4.4	3.1	*	4.3	*	5.6	5.4
New track design	2.7	3.3	1.9	2.2	2.7	4.2	3.9
New track + damper	6.1	6.1	*	5.8	*	7.7	8.0

* not tested

REFERENCES

7.1 F.J. Fahy. Fundamentals of noise and vibration control, Chapter 5 in *Fundamentals of Noise and Vibration*. F.J. Fahy and J.G. Walker, (eds). Spon Press, 1998.

7.2 D.J. Thompson, B. Hemsworth, and N. Vincent. Experimental validation of the TWINS prediction program for rolling noise, Part 1: Description of the model and method. *Journal of Sound and Vibration*, **193**, 123–135, 1996.

7.3 D.J. Thompson, P. Fodiman, and H. Mahé. Experimental validation of the TWINS prediction program for rolling noise, Part 2: Results. *Journal of Sound and Vibration*, **193**, 137–147, 1996.

7.4 C.J.C. Jones and J.W. Edwards. Development of wheels and track components for reduced rolling noise from freight trains. *Proceedings of Internoise. Liverpool*, 1996. 403–408.

7.5 P. Fodiman. Line test validation of low noise railway components. *Proceedings of World Congress on Railway Research*, Colorado, USA, 1996.

7.6 L. Guccia. Synthèse finale des travaux, groupe de travil bruit de roulement MONA-RONA-VONA, SNCF Direction de la Recherche, report RVA/LG/roult/c/970729, 1997.

7.7 P. Fodiman, L. Castel and G. Gaborit. Validation of a TGV-A trailing-car wheel with an acoustically optimised profile. *Proceedings of 11th International Wheelset Congress*, Paris, June 1995.

7.8 L. Guccia, P. Fodiman, P.E. Gautier, N. Vincent and P. Bouvet. High-speed rolling noise design and validation of low noise components. *Proceedings of World Congress on Railway Research*, Florence, Italy, 179–187, 1997.

7.9 B. Hemsworth, P.E. Gautier and R. Jones. Silent Freight and Silent Track projects. *Proceedings of Internoise 2000*, Nice, France 2000.

7.10 B. Hemsworth. *Silent Freight/Silent Track/Eurosabot projects – Summary*, ERRI, Utrecht, The Netherlands, 2001.

7.11 B. Hemsworth and R.R.K. Jones. *Silent Freight project – final report*, ERRI, Utrecht, The Netherlands, 2000.

7.12 B. Hemsworth. *Silent Track project – final report*, ERRI, Utrecht, The Netherlands, 2000.

7.13 P. de Vos and S. van Lier. Noise-related roughness on railway wheels generated by tread braking. *Proceedings of Internoise 2000*, Nice, France, 2000.

7.14 J. Lub. The quiet rail traffic (STV) project. Part 2 – demonstrating practical low noise solutions. *Proceedings of the Joint EAA/ASA/DEGA Meeting on Acoustics*, Berlin, 1999.

7.15 J. Lub. Een moedige speurtocht naar stiller treinverkeer: het project STV. *National Congress on Sound and Vibration, Rotterdam, 1999.*

7.16 D.J. Thompson and P.E. Gautier. A review of research into wheel/rail rolling noise reduction. *Proceedings of the Institution of Mechanical Engineers, Part F, Journal of Rail and Rapid Transit*, **220F**, 385–408, 2006.

7.17 J. Peen, B. Mulder, and W.J.G. van Roij. Whispering trains projects: noise reduction of freight wagons using LL-blocks. *9th International Workshop on Railway Noise, Feldafing, Germany, 4–8, September 2007.*

7.18 A.H.W.M. Kuijpers. Rail roughness monitoring in the Netherlands. *9th International Workshop on Railway Noise, Feldafing, Germany, 4–8, September 2007.*

7.19 P.H. van den Dool. Rail dampers, rail infrastructure gets quiet. *Inter-Noise 2007, Istanbul, Turkey, 28–31, August 2007.*

7.20 E. van Haaren and G.A. van Keulen. New raildampers at the railway link Roosendaal-Vlissingen, tested within the Dutch Innovation Program. *9th International Workshop on Railway Noise, Feldafing, Germany, 4–8, September 2007.*

7.21 B. Hemsworth. Recent developments in wheel/rail noise research. *Journal of Sound and Vibration*, **66**, 297–310, 1979.

7.22 P.C. Dings and M.G. Dittrich. Roughness on Dutch railway wheels and rails. *Journal of Sound and Vibration*, **193**, 103–112, 1996.

7.23 D.J. Thompson. On the relationship between wheel and rail surface roughness and rolling noise. *Journal of Sound and Vibration*, **193**, 149–160, 1996.

7.24 D.J. Thompson and P.J. Remington. The effects of transverse profile on the excitation of wheel/rail noise. *Journal of Sound and Vibration*, **231**, 537–548, 2000.

7.25 A. Johansson. Out-of-round railway wheels – assessment of wheel tread irregularities in train traffic. *Journal of Sound and Vibration*, **293**, 795–806, 2006.

7.26 E. Verheijen. Statistical analysis of railway noise: trackside monitoring of individual trains. *9th International Workshop on Railway Noise.* Feldafing, Germany, 4–8, September 2007.

7.27 T. Vernersson. Thermally induced roughness of railway wheels, Part 1: Brake rig experiments. *Wear*, **236**, 96–105, 1999.

7.28 T. Vernersson. Thermally induced roughness of railway wheels, Part 2: Modelling and field measurements. *Wear*, **236**, 106–116, 1999.

7.29 M. Petersson. Noise-related roughness of railway wheel treads – full-scale testing of brake blocks. *Proceedings of the Institution of Mechanical Engineers, Part F, Journal of Rail and Rapid Transit*, **214F**, 63–77, 2000.

7.30 P. Hübner. The Action Programme of UIC, UIP and CER 'Abatement of Railway Noise Emissions on Goods Trains'. *Proceedings of Internoise 2001*, The Hague, Netherlands, 107–112, 2001.

7.31 P.H. de Vos, M. Bergendorff, M. Brennan, and F. van der Zijpp. Implementing the retrofitting plan for the European rail freight fleet. *Journal of Sound and Vibration*, **293**, 1051–1057, 2006.

7.32 Directive 2001/16/EC of the European Parliament and of the Council of 19 March 2001 on the interoperability of the trans-European conventional rail system. *Official Journal of the European Communities*, **L110**, 1–27, 2001.

7.33 S.L. Grassie and J. Kalousek. Rail corrugation: characteristics, causes and treatments. *Proceedings of the Institution of Mechanical Engineers, Part F, Journal of Rail and Rapid Transit*, **207F**, 57–68, 1993.

7.34 B. Asmussen, H. Onnich, R. Strube, L.M. Greven, S. Schröder, K. Jäger, and K.G. Degen. Status and perspectives of the 'Specially Monitored Track'. *Journal of Sound and Vibration*, **293**, 1070–1077, 2006.

7.35 H. Ilias. The influence of railpad stiffness on wheelset/track interaction and corrugation growth. *Journal of Sound and Vibration*, **227**, 935–948, 1999.

7.36 X. Sheng, D.J. Thompson, C.J.C. Jones, G. Xie, S.D. Iwnicki, P. Allen, and S.S. Hsu. Simulations of roughness growth on railway rails. *Journal of Sound and Vibration*, **293** 819–829, 2006.

7.37 S. Grassie, J. Edwards, and J. Shepherd. Roaring rails – an enigma largely explained. *International Railway Journal*, **47** (7), 31–33, 2007.

7.38 K. Hempelmann. Short pitch corrugation on railway rails – a linear model for prediction. *VDI Fortschrittberichte.* Reihe 12, nr, **231**, 1994.

7.39 A. Igeland. *Dynamic train/track interaction: simulation of railhead corrugation growth under a moving bogie using mathematical models combined with full-scale measurements.* Doctoral thesis, Division of Solid Mechanics, Chalmers University of Technology, Göteborg, Sweden, 1997.

7.40 J.C.O. Nielsen, R. Lundén, A. Johansson, and T. Vernersson. Train-track interaction and mechanisms of irregular wear on wheel and rail surfaces. *Vehicle System Dynamics*, **40**, 3–54, 2003.

7.41 J.C.O. Nielsen. Numerical prediction of rail roughness growth on tangent tracks. *Journal of Sound and Vibration*, **267**, 537–548, 2003.

7.42 B.E. Croft, C.J.C. Jones and D.J. Thompson. Reducing wheel-rail interaction forces and roughness growth by application of rail dampers. *Proceedings of the 9th International Workshop on Railway Noise*, Feldafing, Munich, 4–8 September 2007.

7.43 P.J. Remington. The resiliently treaded wheel – a concept for control of wheel/rail rolling noise. *Proceedings of Internoise*, Avignon, France, 1401-1406, 1988.

7.44 G. Hölzl. A quiet railway by noise optimised wheels (in German). *ZEV+DET Glas. Ann.*, **188** (1), 20–23, 1994.

7.45 C.J.C. Jones and D.J. Thompson. Means of controlling rolling noise at source, Chapter 6 in V.V. Krylov (ed) *Noise and Vibration from High-Speed Trains.* Thomas Telford, London, 2001, pp. 163–183

7.46 D.J Thompson and M.G. Dittrich. Railway rolling noise – vibration response measurements of five types of wheel. TNO report TPD-HAG-RPT-90–0098, August 1990.

7.47 P. Fodiman, L. Castel and G. Gaborit. Validation of a TGV-A trailing-car wheel with an acoustically optimised profile. *Proceedings of 11th International Wheelset Congress*, Paris, 1995.

7.48 D.J. Thompson. Wheel-rail noise generation, Part II: Wheel vibration. *Journal of Sound and Vibration*, **161**, 401–419, 1993.
7.49 P.E. Gautier, N. Vincent, D.J. Thompson and G. Hölzl. Railway wheel optimization. *Proceedings of Inter Noise*, Leuven, 1455–1458, 1993.
7.50 D.J. Thompson and C.J.C. Jones. A study of the use of vehicles with small wheels for determining the component of noise from the track. *Proceedings of IOA Spring Conference*, Salford, March 2002.
7.51 C. Jones, D.J. Thompson, A. Frid and M. Wallentin. Design of a railway wheel with acoustically improved cross-section and constrained layer damping. *Proceedings of Internoise 2000*, Nice, France.
7.52 A. Bracciali and M. Bianchi. Lucchini CRS Syope damped wheels noise qualification. *Proceedings of the International Wheelset Congress*, Italy, 2001.
7.53 S. Cervello, G. Donzella, A. Pola and M. Scepi. Analysis and design of a low-noise railway wheel. *Proceedings of the Institution of Mechanical Engineers, Part F, Journal of Rail and Rapid Transit*, **215F**, 179–192, 2001.
7.54 R. Hemelrijk. The effect of wheel dampers on the sound emission of trains (in Dutch). *Proceedings of the Dutch Acoustical Society (NAG)*, **67**, 27–37, 1983.
7.55 ORE. Reduction of running noise by wheel-mounted devices. *Question C163 Railway Noise, Report RP13, Utrecht, 1989.*
7.56 P. Bouvet, N. Vincent, A. Coblenz and F. Demilly. Rolling noise from freight railway traffic: reduction of wheel radiation by means of tuned absorbers. *Proceedings of Internoise 2000*, Nice, France.
7.57 J. Färm. Evaluation of wheel dampers in an intercity train. *Journal of Sound and Vibration*, **267**, 739–747, 2003.
7.58 I. Lopez, J. Vinolas, J. Busturia and A. Castanares. Railway wheel ring dampers. *Proceedings of Internoise 2000*, Nice, France, 2000.
7.59 W. Behr and S. Cervello. Optimization of a wheel damper for freight wagons using FEM simulation. *9th International Workshop on Railway Noise, Feldafing, Germany, 4–8, September 2007.*
7.60 F. Létourneaux. High speed railway noise: assessment of mitigation measures. *9th International Workshop on Railway Noise, Feldafing, Germany, 4–8, September 2007.*
7.61 C.J.C. Jones and D.J. Thompson. Rolling noise generated by wheels with visco-elastic layers. *Journal of Sound and Vibration*, **231**, 779–790, 2000.
7.62 Anon. DB grounds ICE1 fleet after Eschede disaster. *Railway Gazette International*, **154**, 449, 1998.
7.63 P. Bouvet, N. Vincent, A. Coblentz and F. Demilly. Optimisation of resilient wheel for rolling noise control. *Proceedings of the 6th International Workshop on Railway and Tracked Transit System Noise*, Ile des Embiez, France, 264–273, 1998.
7.64 A. Daneryd, J. Nielsen, E. Lundberg and A. Frid. On vibro-acoustic and mechanical properties of a perforated railway wheel. *Proceedings of the 6th International Workshop on Railway and Tracked Transit System Noise*, Ile des Embiez, France, 305–317, 1998.
7.65 A. Putra and D.J. Thompson. Sound radiation from a perforated unbaffled plate. *International Congress on Acoustics*, Madrid, 3–7, September 2007.
7.66 N. Vincent, P. Bouvet, D.J. Thompson, and P.E. Gautier. Theoretical optimization of track components to reduce rolling noise. *Journal of Sound and Vibration*, **193**, 161–171, 1996.
7.67 D.J. Thompson and J.W. Verheij. The dynamic behaviour of rail fasteners at high frequencies. *Applied Acoustics*, **52** (1), 1–17, 1997.
7.68 T.X. Wu and D.J. Thompson. The effects of local preload on the foundation stiffness and vertical vibration of railway track. *Journal of Sound and Vibration*, **219**, 881–904, 1999.
7.69 D.J. Thompson, C.J.C. Jones, T.X. Wu and G. de France. The influence of the non-linear stiffness behaviour of rail pads on the track component of rolling noise. *Proceedings of the Institution of Mechanical Engineers, Part F, Journal of Rail and Rapid Transit*, **213F**, 233–241, 1999.
7.70 D.J. Thompson. A continuous damped vibration absorber to reduce broad-band wave propagation in beams. *Journal of Sound and Vibration*, **311**, 824–842, 2008.
7.71 D.J. Thompson, C.J.C. Jones, T.P. Waters, and D. Farrington. A tuned damping device for reducing noise from railway track. *Applied Acoustics*, **68**, 43–57, 2007.

7.72 F. Létourneaux, F. Margiocchi, and F. Poisson. Complete assessment of rail absorber performances on an operated track in France. *World Congress on Railway Research*, Montreal, Canada, June 2006.

7.73 B. Asmussen, D. Stiebel, P. Kitson, D. Farrington, and D. Benton. Reducing the noise emission by increasing the damping of the rail: results of a field test. *9th International Workshop on Railway Noise*. Feldafing, Germany, 4–8, September 2007.

7.74 B. Croft and C.J.C. Jones. Modelling the effect of rail dampers on railway rolling noise and the rate of railhead roughness growth. Silence Subproject G2 Technical Report, ISVR Contract Report CR07/17, December 2007.

7.75 D.J. Thompson and C.J.C. Jones. A review of the modelling of wheel/rail noise generation. *Journal of Sound and Vibration*, **231** (3), 519–536, 2000.

7.76 European Patent: Improvements in railways. British Steel plc and British Railways Board, filed 21.09.89.

7.77 C.J.C. Jones. Reduction of noise and ground vibration from freight trains. *I.Mech.E. International Conference on Better Journey Times – Better Business, Birmingham, 87–97, paper C514/082/96, 1996.*

7.78 N. Vincent. Rolling noise control at source – state of the art survey. *Journal of Sound and Vibration*, **231**, 865–876, 2000.

7.79 C-M. Nilsson, C.J.C. Jones, D.J. Thompson, and J. Ryne. A waveguide finite element and boundary element approach to calculating the sound radiated by tram rails. *Submitted for publication.*

7.80 R.J. Diehl, R. Nowack, and G. Hölzl. Solutions for acoustical problems with ballastless track. *Journal of Sound and Vibration*, **231**, 899–906, 2000.

7.81 G. Hauck, W. Weissenberger, J. Scheuren, and E. Lange. Untersuchungen zur Verringerung der Schallabstrahlung von 'Feste Fahrbahnen' durch absorbierende Beläge. *Eisenbahntechnische Rundschau*, **44**, 559–565, 1995.

7.82 M.H.A. Janssens. Low noise slab-track design: acoustic development and final tests. *Proceedings of the Sixth International Congress on Sound and Vibration*, Lyngby, Denmark, 36, 2643–2652, 1999.

7.83 S. van Lier. The vibro-acoustic modelling of slab track with embedded rails. *Journal of Sound and Vibration*, **231**, 805–817, 2000.

7.84 R.R.K. Jones. Bogie shrouds and low barriers could significantly reduce wheel/rail noise. *Railway Gazette International* 459–462, July 1994.

7.85 C.J.C. Jones, D.J. Thompson, and T.P. Waters. Application of numerical models to a system of train- and track-mounted acoustic shields. *International Journal of Acoustics and Vibration*, **6** (4), 185–192, 2001.

7.86 C.J.C. Jones, A.E.J. Hardy, R.R.K. Jones, and A. Wang. Bogie shrouds and low trackside barriers for the control of railway vehicle rolling noise. *Journal of Sound and Vibration*, **193**, 427–431, 1996.

7.87 R. Jones, M. Beier, R.J. Diehl, C. Jones, M. Maderboeck, C. Middleton and J. Verheij. Vehicle-mounted shields and low trackside barriers for railway noise control in a European context. *Proceedings of Internoise 2000*, Nice, France, 2000.

7.88 J. Oertli. The case for retrofitting freight wagons. *9th International Workshop on Railway Noise*, Feldafing, Germany, 4–8, September 2007.

7.89 J. Oertli. The STAIRRS project, Work package 1: A cost-effectiveness analysis of railway noise reduction on a European scale. *Journal of Sound and Vibration*, **267**, 431–437, 2003.

7.90 J.P. Arenas. Use of barriers, Chapter 58 in *Handbook of Noise and Vibration Control*. M.J. Crocker, (ed.). John Wiley & Sons, Hoboken NJ, 2007.

7.91 K. Horoshenkov, Y.W. Lam, and K. Attenborough. Noise attenuation provided by road and rail barriers, earth berms, buildings and vegetation, Chapter 122 in *Handbook of Noise and Vibration Control*. M.J. Crocker, (ed.). John Wiley & Sons, Hoboken NJ, 2007.

7.92 G.S. Anderson and U.J. Kurze. Outdoor sound propagation, Chapter 5 in *Noise and Vibration Control Engineering*. L.L. Beranek and I.L. Ver, (eds). John Wiley & Sons, New York, 1992.

7.93 Z. Maekawa. Noise reduction by screens. *Applied Acoustics*, **1**, 157–173, 1968.

7.94 U.J. Kurze and G.S. Anderson. Sound attenuation by barriers. *Applied Acoustics*, **4**, 35–53, 1971.

7.95 Z. Maekawa. Environmental sound propagation. *Proceedings of the 8th International Congress on Acoustics*, London, UK, 1974.

7.96 B. Hemsworth. Rail system environmental noise prediction, assessment and control, Chapter 121 in *Handbook of Noise and Vibration Control*. M.J. Crocker, (ed.). John Wiley & Sons, Hoboken NJ, 2007.

7.97 B. Hemsworth. Noise barriers for fast passenger trains. *Proceedings of Internoise 77*, Zürich, Switzerland, B465–470, 1977.

7.98 ISO 9613–2:1996, Acoustics – attenuation of sound during propagation outdoors – Part 2: General method of calculation. International Standards Organization, 1996.

CHAPTER 8

*Aerodynamic Noise**

8.1 INTRODUCTION

Aerodynamic noise is a phenomenon associated with high speed operation. The pursuit of high speeds has been a preoccupation of railways since they first opened. In the Rainhill Trials of 1829, prior to the opening of the Liverpool and Manchester Railway, Stephenson's *Rocket* achieved a speed of 46.8 km/h. By the end of the nineteenth century the companies of the British west and east coast routes were competing to reach Aberdeen first from London. In the famous 'Race to the North' of the 1890s they averaged over 100 km/h. In the 1930s several streamlined steam locomotive types were introduced, with the LNER's *Mallard* in the UK achieving the world speed record for steam traction of 201 km/h in 1938, which still stands.

The Shinkansen 'bullet' trains were introduced in Japan in 1964, running initially at 210 km/h, with progressive increases in operational speed since then. This heralded the start of modern high speed operation, based on fixed formation multiple unit trains. High speed diesel trains began running in Britain in 1976 with a maximum speed of 200 km/h. Trains à Grande Vitesse (TGV) commenced operating at 270 km/h in France in 1981, while the ICE has been running in normal service in Germany since 1991. Such high speed trains now operate at 300 km/h or more in France, Germany, Italy, Spain, Japan, South Korea, Taiwan, Belgium and Britain with other countries due to join this shortly and speeds of 350 km/h a likely prospect in the next few years. In 2007 the SNCF achieved a world speed record of 574.8 km/h with a TGV. High speed intercity passenger trains, Figure 8.1, can compete effectively with air travel.

As speeds increase, however, noise inevitably also increases. This means that when the construction of new high speed lines is proposed, this often leads to opposition on the grounds of noise, even more than for conventional lines. It has been seen in earlier chapters that rolling noise levels increase at a rate of $30 \log_{10} V$ (with V the speed). The noise from aerodynamic sources increases

* This chapter has been written in collaboration with Pierre-Etienne Gautier

FIGURE 8-1 Various examples of high speed trains. Courtesy of SNCF, used with permission

more rapidly with speed, with levels increasing at typically around $60 \log_{10} V$, and consequently they become dominant above a certain speed, referred to as the transition speed.

Figure 8.2 shows an example of the dependence of noise level on train speed. In this graph, the overall noise at 25 m from a TGV-A[1] is presented as a function of speed [8.1]. These values are the sound level maxima from the front and rear power cars and the average level of the trailer vehicles between the two maxima. A change in slope can be seen at around 300 km/h, although this transition will be more gradual than indicated by the two regression lines. If the trends of rolling noise and aerodynamic noise are straight lines that cross at 300 km/h, the total level will be 3 dB higher than either line at this speed, as the two sources are independent and add incoherently. Conventional rolling stock has a lower transition speed, since it is less well streamlined, but this is still normally beyond the range of operational speeds.

For a high speed train, aerodynamic sources are important for both interior and exterior noise, although the discussion here focuses mainly on exterior noise. Interior noise is considered in Chapter 14. Extensive research into railway aerodynamic noise has been carried out since the early 1990s, particularly in the framework of the German–French cooperation, Deufrako [8.2, 8.3], as well as in Japan.

Aerodynamic sources of noise differ from the others described in this book, which are due to vibrating solid surfaces (see Chapter 6). The fact that the sound is

[1] Note that the results shown in Figure 8.2 were obtained in 1989/90 during trials at very high speed. The rolling stock used for the tests were TGV-A trainsets with disc brakes on the trailer vehicles and supplementary tread brakes with cast-iron blocks on the power cars. More recently, all TGV vehicles have been fitted with composite brake blocks on the powered axles, including the initial PSE sets which have also had the tread brakes removed from the trailer vehicles.

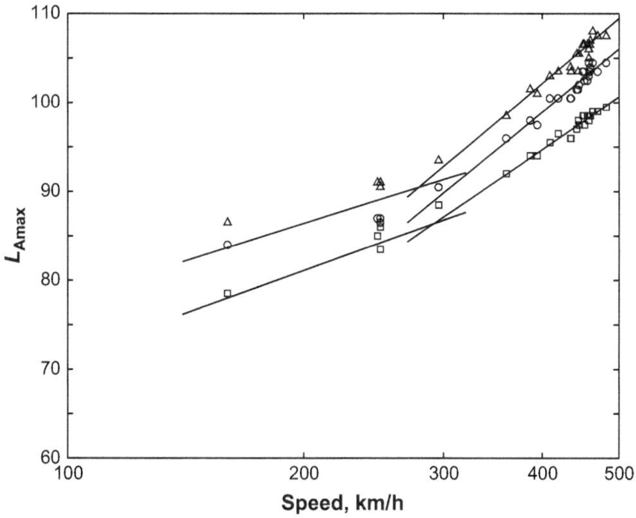

FIGURE 8-2 Maximum sound level of TGV-A at 25 m as a function of speed (data from [8.1]). △, leading power cars; ○, rear power cars; □, trailer vehicles

generated within the acoustic medium, through which it then propagates, increases the complexity of both measurement and analysis. Moreover, the proportion of the flow energy that is actually converted into sound is very small. A short section follows in which the basic principles of aerodynamic noise are summarized, although a detailed treatment is beyond the present scope. Following this, the experimental and theoretical tools commonly used to investigate the phenomena are described and an overview is given of the various sources of aerodynamic noise present on a train. Finally, some recent developments aimed at reducing aerodynamic noise are presented.

8.2 BASIC PRINCIPLES

Theoretical studies of aerodynamic noise have focused extensively on the problems of noise from aircraft, and particularly the jet engine [8.4]. By comparison with a jet engine, a train has a relatively low flow velocity but quite large dimensions. More recently, with the significant reductions that have taken place in jet engine noise, work on aircraft noise has also focused on airframe noise, particularly during landing. For this, the speeds are lower (250–400 km/h) and the effects of discontinuities such as flaps and landing gear are important, making the problem quite similar to that of high speed trains. Aerodynamic noise is also an issue for road vehicles, particularly fast cars, but is then of most significance for interior noise.

Aerodynamic noise is caused by the flow of air over the train as it travels at high speed. This flow is very complex and includes a turbulent boundary layer around the train. For a flow over a flat surface, the flow speed at the surface reduces to zero due to friction whereas at a sufficient distance from the surface it is equal to the free velocity U. (For a moving train the same principles apply in terms of the velocities

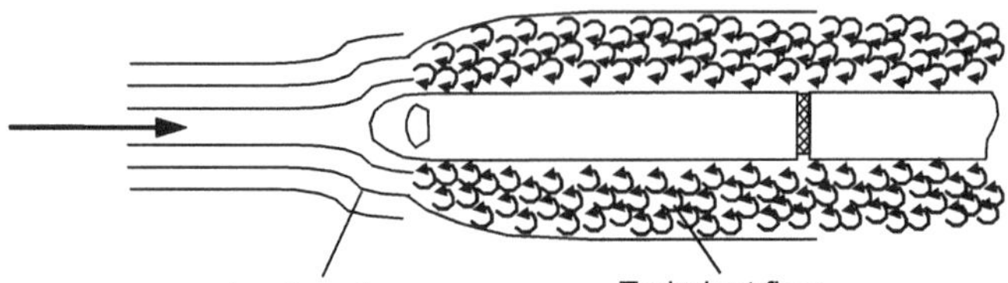

FIGURE 8-3 Illustration of the formation of a turbulent boundary layer around a train. The flow is shown relative to the train

relative to the train.) Between these two extremes a boundary layer is formed. Initially, near the leading edge this boundary layer is *laminar*, in which parallel particles move at different speeds, but due to viscosity this laminar flow quickly forms into a *turbulent boundary layer* as shown schematically in Figure 8.3. Many experiments have shown that, in general, the boundary layer becomes turbulent at a Reynolds number of about 2×10^6 (see box below). Applying this simple formula for a train speed of 300 km/h ($U = 83$ m/s), this corresponds to a distance of only 0.4 m.

Measurements of the boundary layer thickness on a high speed train were carried out by Crespi *et al.* [8.5] using a laser Doppler velocimeter mounted on board a high speed train. The boundary layer thickness is defined as the average distance from the

Non-dimensional quantities in aerodynamics

A number of non-dimensional quantities are used in aerodynamics and aeroacoustics. Often aerodynamic phenomena are found to be equivalent if these non-dimensional quantities are preserved, enabling the use of scale model tests, for example.

Mach number: non-dimensional velocity. This is the flow velocity U expressed as a proportion of the speed of sound in the fluid, c_0:

$$M = U/c_0$$

In air, $c_0 = 343$ m/s at 20°C. A high speed train travelling at 300 km/h (83 m/s) thus has a Mach number $M = 0.24$.

Reynolds number: represents the ratio of inertial forces to viscous forces in the fluid

$$\text{Re} = \frac{UL\rho_0}{\mu}$$

where U is the flow velocity far from any solid objects, L is a typical length of the body disturbing the flow, ρ_0 is the density of the fluid and μ is the viscosity. For air, the ratio $\rho_0/\mu \approx 6 \times 10^4$.

Strouhal number: non-dimensional frequency:

$$\text{St} = \frac{fD}{U}$$

where f is the frequency. For example, for flow over a cylinder of diameter D, vortices are shed at a frequency corresponding to St ≈ 0.2, giving rise to Aeolian tones at this frequency. (This occurs for Reynolds numbers above about 3×10^5.)

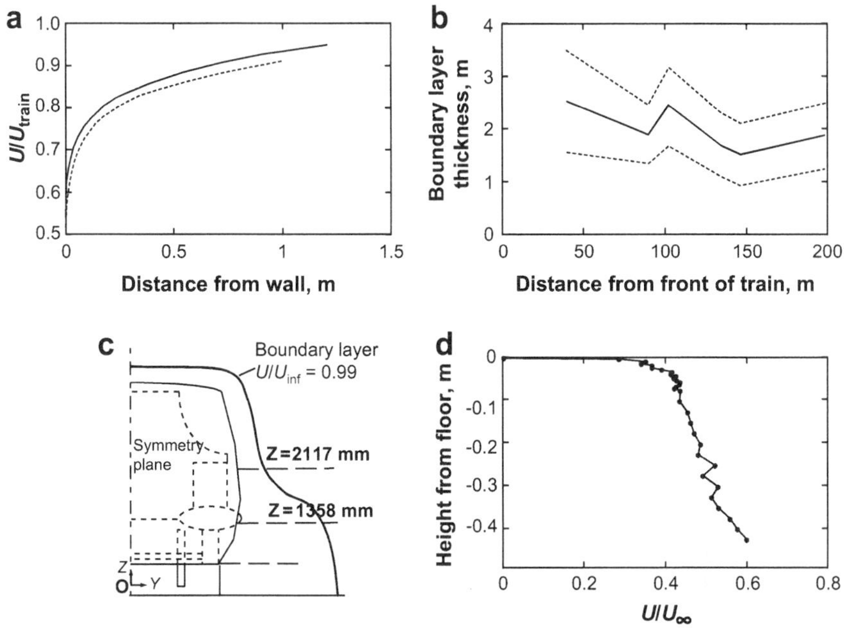

FIGURE 8-4 Measurements of boundary layer thickness on a TGV. (a) Velocity profile of the mean streamwise component; —, 300 km/h; - - -, 220 km/h; (b) boundary layer thickness at various locations along the train at 220 km/h; —, mean; - - -, error band [8.5]; (c) boundary layer shape in the inter-coach region; (d) mean velocity profile under the coach (from [8.6])

train surface at which the flow velocity is 99% of the free flow velocity. These measurements showed that the boundary layer has a relatively constant thickness along the train of about 2 m as shown in Figure 8.4, although it increases rapidly towards the lower region of the vehicle due to ground friction and the blocking effect of the bogie. Measurements were also taken under the coach [8.6], shown in Figure 8.4(d).

More recently detailed measurements and calculations have been carried out on a 1:15 scale model of a TGV-R in a wind tunnel [8.7]. These have confirmed that the boundary layer does not grow much in width along the length of the train, but that turbulence around the bogies is particularly important in generating the boundary layer in the lower region.

It is not immediately obvious, however, how aerodynamic fluctuations such as this lead to noise. The origins of aeroacoustics can be traced to Lighthill's influential work [8.8]. He showed that there is an exact mathematical analogy between the sound radiated by a turbulent flow and that radiated by a set of equivalent *quadrupole* sources set in a non-moving medium.

While a monopole source represents fluctuations in the fluid *mass* within a given volume, a dipole represents a fluctuating *force* acting on the fluid [8.9]. A dipole can also be modelled as two monopoles pulsating out of phase with each other, separated by a small distance (see also Section 6.2). Dipoles are less efficient radiators of sound than monopoles due to inherent cancellation within the source region and they are increasingly less efficient as frequency reduces. This is because they have no net source strength, but the source contributions from the two monopoles arrive at a receiver location at slightly different times. At low frequency the time difference is small compared with the period of oscillation and their contributions largely cancel.

Similarly, quadrupoles can be represented as two dipoles acting out of phase with each other, thus giving no net force but a fluctuating *stress*. They are much less efficient radiators than monopoles or dipoles. Consequently, it is seen that the sound produced by a turbulent flow is only a very small fraction of the flow energy. Dimensionally, the sound power produced by a free turbulent flow is proportional to [8.8]

$$W_{rad} \propto \frac{\rho_0 U^8 l^2}{c_0^5} \tag{8.1}$$

where l is the width of the flow, U is the flow velocity, c_0 is the speed of sound and ρ_0 is the fluid density. The strong speed dependence (sound level increasing with $80 \log_{10} U$) means that this source of noise increases in importance at high speeds. The efficiency of sound generation (sound power divided by the power of the fluid flow) is proportional to M^5, where M is the Mach number, see box on page 284. The constant of proportionality is of the order of 10^{-4} [8.8].

However, for exterior railway noise it is not this free turbulence that is the main source of aerodynamic noise. The dominant sources are more typically dipole-type sources. There are a number of types of such sources:

- Flow over solid cylindrical objects, for example parts of the pantograph (current collector) and other roof-mounted equipment, or handrails. This gives rise to vortex shedding and tonal noise (see box on page 284).
- Flow over discontinuities, for example the inter-coach gaps or bogie region, which can produce broad-band noise due to flow separation and reattachment.
- Flow over cavities or louvers, for example the inter-coach gaps or pantograph recesses, which can produce tonal noise due to interaction between the flow and acoustic resonances of the cavity. The typical flow in the pantograph recess is shown in Figure 8.5, derived from [8.10].

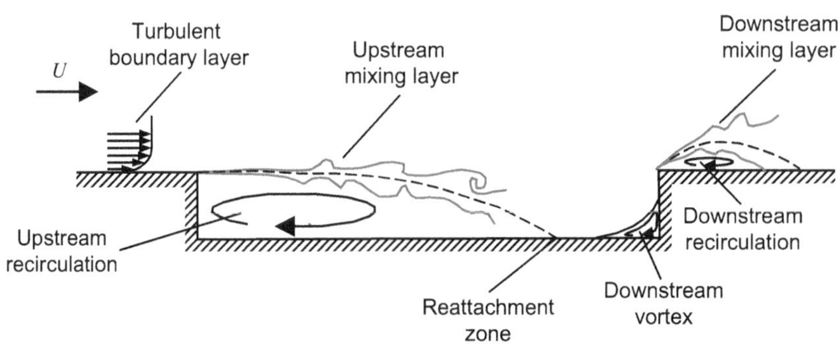

FIGURE 8-5 Features of flow over a cavity which is long compared with its depth such as the pantograph recess (from [8.10])

- Turbulent boundary layer over curved or rough surfaces which can produce dipole rather than quadrupole noise, particularly where flow separation occurs, for example at the nose and rear of the train.

Dipole aerodynamic sources increase in sound power according to the sixth power of flow speed, U^6; hence, the sound level in decibels increases at a rate of $60 \log_{10} U$. However, as the frequency content also changes with increasing speed, shifting from low towards higher frequencies, the A-weighted sound pressure level may increase at a slightly higher rate than this.

The most important sources of aerodynamic noise on a high speed train vary from one train to another (see also Section 8.5.1) but are usually:

- the bogies, particularly the leading bogie in a train (see Section 8.5.4);
- the pantograph(s), their recesses in the roof and any other roof-mounted equipment such as insulators (see Sections 8.5.2 and 8.5.3);
- the nose of the train (see Section 8.5.5);
- the gaps between coaches (see Section 8.5.4);
- ventilation grilles;
- protruberances such as door handles and steps;
- resonant cavities (see Section 8.5.6);
- the turbulent boundary layer on the train surface (this has a much lower intensity but is distributed over a larger area).

Of these, the sources located towards the top of the train are particularly important when noise barriers are present, as a typical barrier of 2 m height will shield the noise sources from the lower part of the train, including the rolling noise

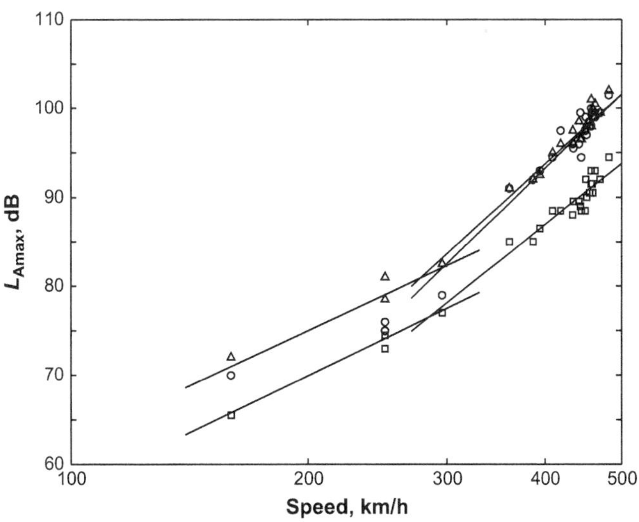

FIGURE 8-6 Maximum sound level of TGV-A at 25 m as a function of speed in the presence of a 2 m high barrier (data from [8.1]). △, leading power cars; ○, rear power cars; □, trailer vehicles

from the wheels, but will leave the upper sources exposed. Figure 8.6 [8.1] shows the dependence of sound level on speed for a train behind a 2 m high barrier located at 9 m from the track centre. The microphone was located at rail height, the track being on an embankment. Comparing these results with those in Figure 8.2 without a barrier, this shows that the transition speed between rolling noise and aerodynamic noise is reduced to between 270 and 290 km/h. While this barrier gave a reduction of 14 dB at a speed of 160 km/h, the effect was only about 8 dB at 350 km/h. Moreover, for a receiver height of 2 m above rail height, the barrier effect at high speed was reduced to only about 3 dB [8.1].

Apart from tonal contributions, most aerodynamic noise is dominated by low frequency components below about 500 Hz. Figure 8.7 shows a typical set of noise spectra at 25 m from a high speed train. At 200 km/h there is a clear peak at around 2–3 kHz caused by wheel noise, and as speed increases this peak increases in both level and frequency. However, the aerodynamic noise at lower frequencies rises more rapidly with increasing speed, so that at 350 km/h the spectrum is almost flat. For frequency bands up to 630 Hz the speed exponent of these data is mostly between 5 and 6 (i.e. a dependence of 50–60 $\log_{10} V$) whereas around the peak at 2–3 kHz it is about 3. Above 3 kHz the speed exponent is again larger due to the shift in the contact filter effect towards higher frequencies (see Chapter 5).

From results such as those shown in Figures 8.2 and 8.6 it is commonly assumed that the overall noise can be decomposed into two components: rolling noise with a speed dependence of about $30 \log_{10} V$ and aerodynamic noise with a speed dependence of around $60 \log_{10} V$. Various different empirically based analysis techniques have been used to derive these curves. However, the results can differ

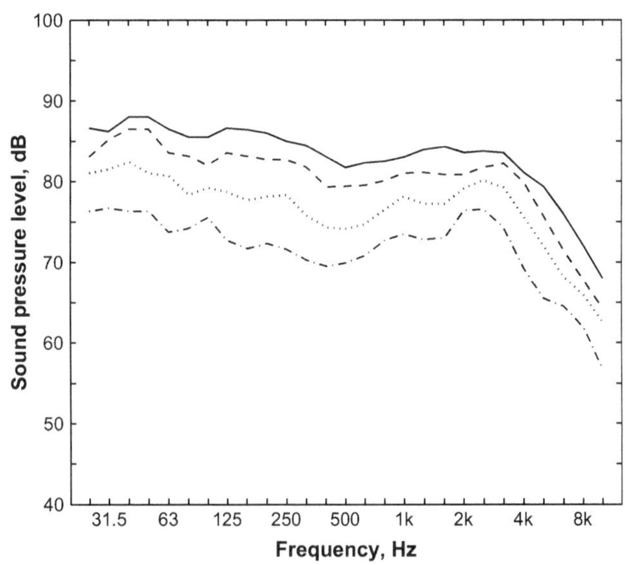

FIGURE 8-7 Typical spectra of noise at 25 m from a TGV-Duplex train running at different speeds; – · – ·, 200 km/h; · · · ·, 250 km/h; – – –, 300 km/h; —, 350 km/h (data from [8.11])

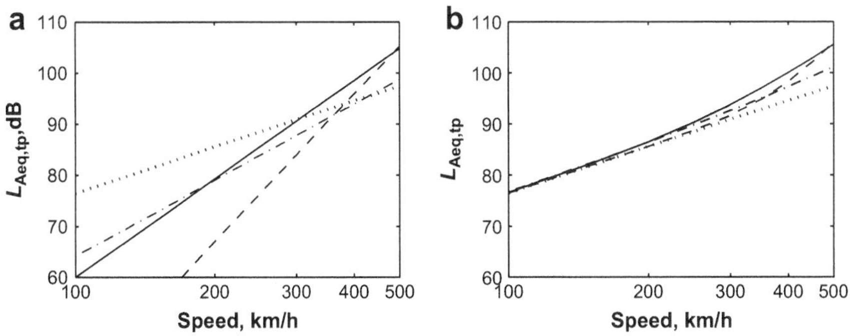

FIGURE 8-8 (a) Various estimates of aerodynamic noise of a TGV; (b) effect on total noise in combination with 30 $\log_{10} V$ curve estimate of rolling noise. —, equation (8.2); – – –, equation (8.3); – · – ·, MAT2S model; · · · ·, rolling noise (30 $\log_{10} V$)

widely, as illustrated in Figure 8.8, which compares the results of several different estimates of aerodynamic noise. These are as follows:

1. Empirical relation given in [8.2]. This is based on (a) measurements on ICE/V at speeds between 200 and 406 km/h [8.12] giving an estimate for vortex shedding noise and (b) an empirical formula derived from measurements of airframe noise on 'clean' aircraft, that is without deployment of landing gear and flaps, for speeds below 400 km/h [8.13, 8.14]. From these two results a range was identified in [8.2], shown in Figure 8.9 which has a mid-point given approximately by

$$L_{\text{Aeq,tp(aero)}} = 64 \log_{10}(V/100) + 60 \tag{8.2}$$

(In [8.2] the terminology $L_{\text{Aeq,tp}}$ was not used but this is what was meant). The contribution of turbulent boundary layer noise is also discussed in [8.2], and is also shown in Figure 8.9. However, for a train of length 200 m as used here, it can be seen that this would be insignificant for the overall A-weighted level, even at 500 km/h.

2. Results of a so-called 'mixed method' are also given in [8.2]. A regression curve for the speed range 99 to 203 km/h is used to represent the rolling noise, extrapolated to higher speeds. The aerodynamic noise curve is derived from the difference between measured results and the rolling noise curve at higher speeds and is given approximately by

$$L_{\text{Aeq,tp(aero)}} = 96 \log_{10}(V/200) + 67 \tag{8.3}$$

However, it should be noted that using the difference between two sound levels that are relatively close to each other in order to extract the level of a small component is liable to lead to considerable errors.

3. Semi-empirical source models are contained in the software MAT2S developed in Deufrako K2 [8.3] to represent various aerodynamic sources. Each of these models has a different speed dependence, but the overall result shown in

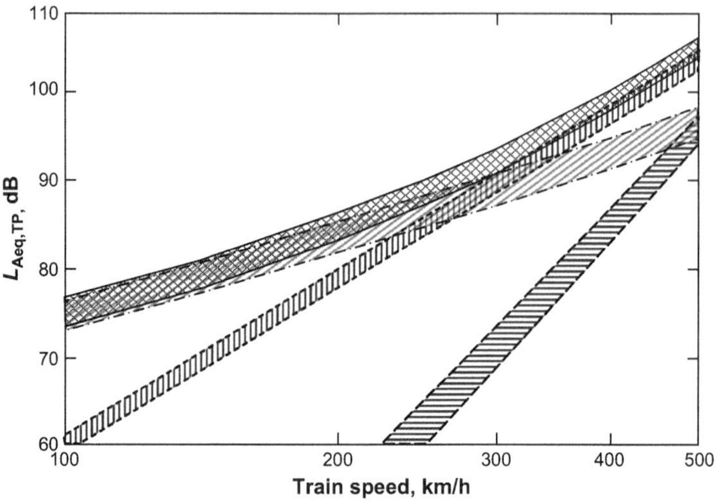

FIGURE 8-9 Estimates of vortex shedding and turbulent boundary layer noise for a train of length 300 m and height 3.8 m. [cross-hatched], total; [horizontal hatched], turbulent boundary layer noise; [vertical hatched], vortex shedding noise; [diagonal hatched], rolling noise [8.2]

Figure 8.8 can be seen to have a lower speed dependence (around $50 \log_{10} V$) than the other two approaches.

Although these results differ widely, it can be seen from Figure 8.8 that large differences in aerodynamic noise estimates give rise to only small differences in total noise, particularly below 350 km/h. Comparing these curves with measured data is therefore inconclusive concerning the contribution of aerodynamic noise. It should also not necessarily be assumed that the rolling noise follows exactly a dependence of $30 \log_{10} V$. Due to the widely differing slopes of the estimates of aerodynamic noise, these span a range of over 12 dB at 200 km/h although they differ by at most 4.5 dB at 350 km/h. The 'transition speeds' deduced from these different estimates also vary between 300 km/h and 430 km/h.

In Figure 8.10 the overall measured noise from TGV-A, TGV-Duplex and Thalys trains is plotted as a function of speed. These results are taken from a number of sources, including an estimate of an overall $L_{Aeq,tp}$ level derived from the data already presented in Figure 8.2. These results show that the overall noise level can be represented quite well by the $30 \log_{10} V$ slope up to around 300 km/h. The increased slope due to aerodynamic noise is evident, particularly above 400 km/h. The measured levels reach about 3 dB above the $30 \log_{10} V$ line at about 370 km/h, which could therefore be considered as the transition speed.

8.3 EXPERIMENTAL TECHNIQUES

Measurements cannot readily be made in the source region, that is within the flow, as there are considerable turbulent pressure fluctuations that would be detected by a microphone but which are not directly associated with sound radiation to the

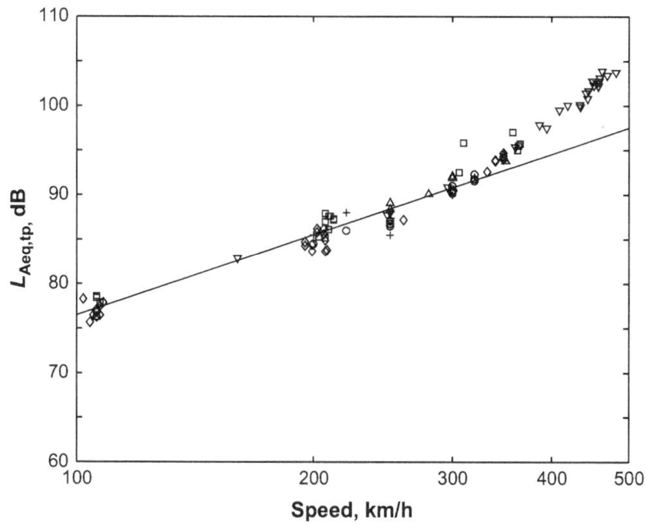

FIGURE 8-10 $L_{Aeq,tp}$ measured at 25 m from TGV-A (▽, data from [8.1] re-evaluated to give $L_{Aeq,tp}$, and □, ◇, from [8.2]), TGV-Duplex (○, △, from [8.15, 8.16]) and Thalys (+, from [8.15, 8.16]). Straight line has a slope of $30 \log_{10} V$ to represent rolling noise

far field. This may be considered as the 'near field' of the equivalent quadrupole sources, which is much stronger than the radiated field. Moreover, placing a microphone in the turbulent flow will disturb the flow and induce additional (dipole) noise radiation unless care is taken in the design of windshields, etc. [8.17].

In Figure 8.6 it was shown that some resolution of the vertical source distribution can be obtained by using barriers of different heights. However, the scope for using this technique to quantify aerodynamic noise is limited as (i) the barrier must not be so close to the train that it disturbs the turbulent boundary layer (i.e. at least 2 m away from the train), which makes the barrier less effective and reduces the resolution; and (ii) even sources that are located out of the line of sight from the microphone have a contribution due to diffraction. Therefore specialized techniques such as microphone arrays [8.17] have been widely applied to the study of aerodynamic noise from trains.

8.3.1 Microphone arrays

The simplest form of microphone array consists of a number of microphones arranged in a line, which record the noise from a passing train simultaneously, see Figure 8.11. By combining the signals from the microphones with a suitable time delay, either during the measurement or by post-processing, the sound from a particular direction can be obtained (so-called beam-forming [8.18, 8.19], see box on page 294). This can be associated with particular source regions, provided that various assumptions are made. In particular, it is usually assumed that the sound sources are located in a plane corresponding to the side of the train, that all sources are incoherent and that they are 'stationary' in the statistical sense. As shown in [8.20], although these assumptions are reasonable for aerodynamic noise they are

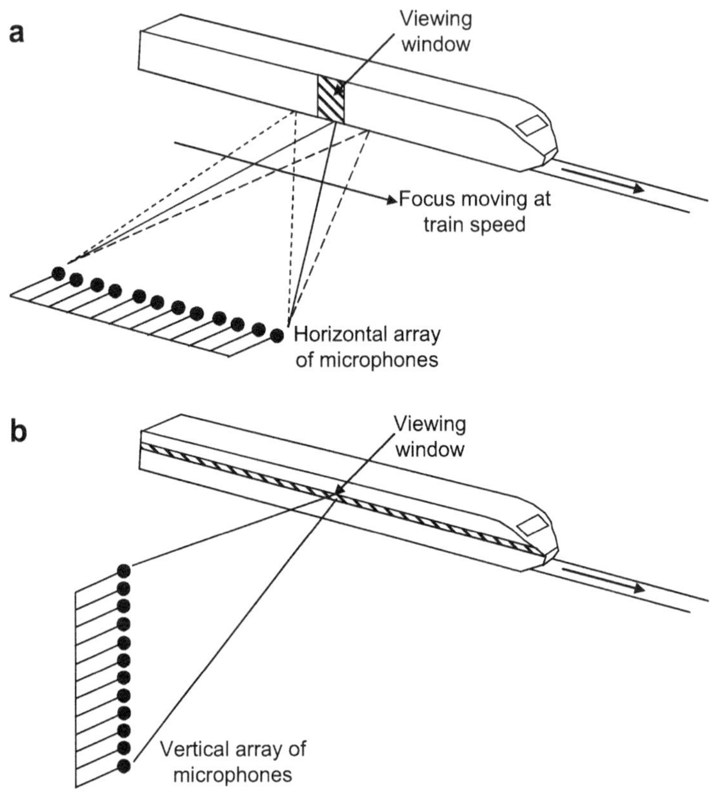

FIGURE 8-11 Principles of a microphone array. (a) Horizontal array with swept focus, (b) vertical array

not appropriate for rolling noise, particularly as the rail is a distributed coherent line source producing a directional source field (see Chapter 6).

By arranging the array vertically, the height of the various sources can be found, while by arranging it horizontally their position along the train can be determined [8.21]. More complex two-dimensional array configurations have been used to locate sources in both directions simultaneously. Two-dimensional arrays have been used based on a +-shape [8.22], X-shape [8.23], T-shape [8.24], star shape [8.16, 8.25], see Figure 8.12, or a spiral [8.26], see Figure 8.13. In [8.27] both +-shape and full planar arrays were used. Apart from the array configuration, the main difference between the various implementations is in the signal processing used: either direct time-domain beam-forming is used or an indirect spatial correlation approach [8.24]. Swept focus techniques have been developed to localize the source on a moving train [8.28]. De-dopplerization should also be applied to remove the Doppler shift due to the moving source [8.28]. A particular problem with measurements on high speed trains is the short time window available from which to determine source locations. This limits the frequency resolution that can be achieved.

It should also be noted that the turbulent boundary layer, especially close to the ground surface, may modify the sound propagation and therefore affect the

FIGURE 8-12 Example of a star-shaped microphone array consisting of 29 microphones [8.16]

directivity of the sources. The flow is known to be important on sound propagation in wind tunnel measurements where convection effects are observed. However, in practice, for a moving train the boundary layer is probably too thin for diffraction effects to be significant. This has been demonstrated by using a known source mounted on a train. Another aspect is that ground reflection has the effect of introducing image sources below the ground which have to be accounted for as well as the direct sources.

In general, the spatial resolution of a microphone array is a function of the frequency, the length of the array, L, and the distance to the source, D (see box on page 294). The main lobe has an *angular* width of approximately $\pm\lambda/L$; the *spatial* resolution is therefore proportional to D, although D should not be less than L. The number of microphones is restricted by the number of channels of simultaneous measurement available. To avoid aliasing problems the microphone spacing should be less than half the acoustic wavelength. Particular microphone spacings are therefore chosen according to the frequency ranges under study, and different configurations may have to be used for different frequency ranges. However, ultimately the resolution is limited to about half the wavelength of sound even where an ideal microphone spacing is used, which means that the spatial resolution at low frequency is inevitably limited.

An example result from the star-shaped microphone array of Figure 8.12 is shown in Figure 8.14 [8.16]. Note that this array is used in three different configurations (different microphone spacings) to cover the whole frequency range. In the particular frequency band shown, 1250 Hz, a strong source can be seen at the leading bogie and others on the leading power car such as fans above the rear bogie. Over

Principles of beamforming using a microphone array

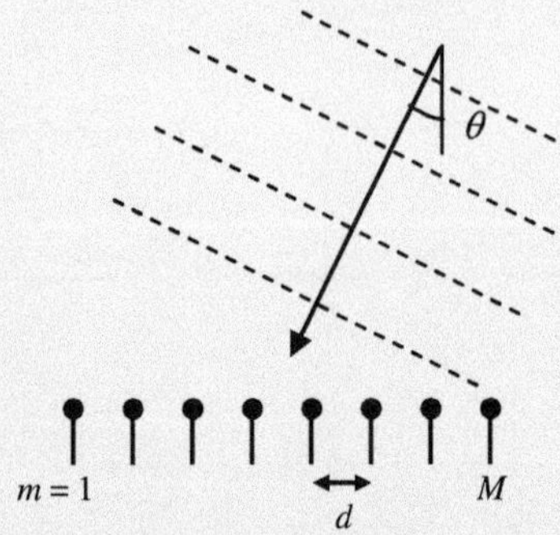

Consider a linear array, as shown in the figure, consisting of M microphones which are equally spaced with separation d. $L = Md$ is the length of the array. If a plane wave is incident at angle θ to the normal, sound will arrive first at microphone M and then at each successive microphone after some delay. Thus the pressure at microphone m will take the form

$$p_m = p_0(t + \tau_m) \tag{1}$$

where $p_0(t)$ is the pressure signal at microphone 1 and τ_m is the appropriate time shift given by

$$\tau_m = \frac{(m-1)d}{c_0}\sin\theta \tag{2}$$

in which c_0 is the speed of sound. The signals from the array are now combined, using suitable weighting (or shading) coefficients w_m and time delays Δ_m, in an attempt to identify the direction of the incident sound. To focus the array in a direction given by an angle ϕ, the output of the array is calculated as

$$a(t,\phi) = \sum_{m=1}^{M} w_m p_m(t - \Delta_m) \tag{3}$$

where the time delays Δ_m for ϕ are given by

$$\Delta_m = \frac{(m-1)d}{c_0}\sin\phi \tag{4}$$

When $\phi = \theta$, the sound at the various microphones is added in phase and the array output $a(t,\phi)$ has a maximum. Simulated examples are shown opposite for an array of 16 microphones as the direction of focus ϕ is varied. In each case a main lobe coincides with the actual angle of incidence, θ. A series of side lobes also occur at a lower amplitude. From (a) the width of the main lobe can be seen to reduce as frequency increases. Identical results are found if frequency is fixed and L is varied. The angular resolution (width of the main lobe) is approximately $\pm\lambda/L$, where λ is the acoustic wavelength [8.18]. In (b) results are shown for different incident angles θ. In (c) it can be noted that at high frequency (or large d) spatial aliasing occurs in which the array cannot distinguish between multiple incident angles. This can in principle occur when $\lambda < 2d$, although in the example shown there is no aliasing at 2 kHz ($\lambda = 1.7d$) – aliasing does occur at this frequency for larger values of θ.

The use of different weighting coefficients is illustrated in (d). The Hanning weighting is proportional to $w_m = 1 - \cos(2\pi m/(M+1))$ (normalized so that the sum of w_m equals 1) whereas in the

Continued

Principles of beamforming using a microphone array—Cont'd

uniform weighting used in the other examples $w_m = 1/M$. The Hanning weighting leads to a broader main lobe but suppresses the side bands. Microphone arrays can also be designed to work with spherical incident fields, where the focus is at a (known) finite distance [8.19].

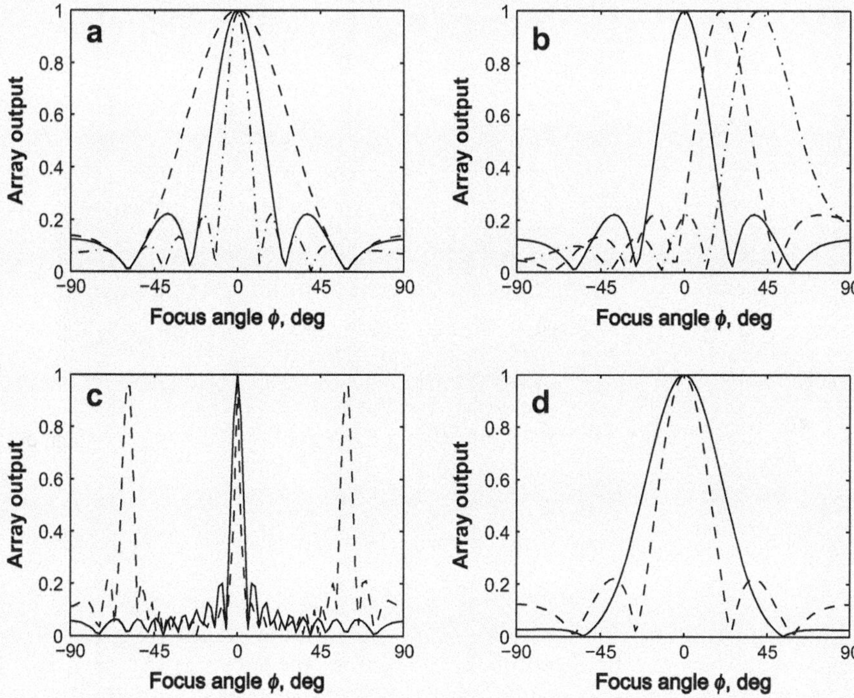

Simulated results of microphone array, 16 microphones at 0.1 m spacing, unless otherwise stated incident sound at $\theta = 0°$ and 500 Hz. (a) Effect of varying frequency: – – –, 250 Hz ($\lambda/L = 0.86$); —, 500 Hz ($\lambda/L = 0.43$); – · – ·, 1000 Hz ($\lambda/L = 0.21$). (b) Effect of varying incident angle: —, $\theta = 0°$; – – –, $\theta = 20°$; – · – ·, $\theta = 40°$. (c) Aliasing effect at high frequency: —, 2000 Hz ($\lambda/d = 1.7$); – – –, 4000 Hz ($\lambda/d = 0.86$). (d) Effect of weighting coefficients: —, uniform weighting; – – –, Hanning weighting

most of the frequency range 250–1250 Hz the leading bogie is found to be the main source from the front power car. In the 500 Hz band, however, the lower windscreen area is found to be the dominant source. This is believed to be due to a large metal water deflector located below the windscreen. The lower part of the cab door also contributes to the noise in the 500 and 630 Hz bands. At higher frequencies, above 2 kHz, the wheels appear to be the main source.

On the trailer cars, the inter-coach gaps can be seen to contribute, although the bogies are mostly more important. On the rear power car, not shown, the pantograph, insulators and cavity together form the most important source over the frequency range 200–1250 Hz.

FIGURE 8-13 Example of a spiral array consisting of 90 microphones (from [8.25, 8.26])

An important question is how to extract source power levels from array measurements. These can be determined readily if (i) the source is compact, (ii) its location is known and (iii) a source model is available, in particular for its directivity. The location can be partially determined from the array output but more reliable results are obtained if *a priori* knowledge of the location can be used. Source models are always required, although these are not often explicitly defined. Instead, an

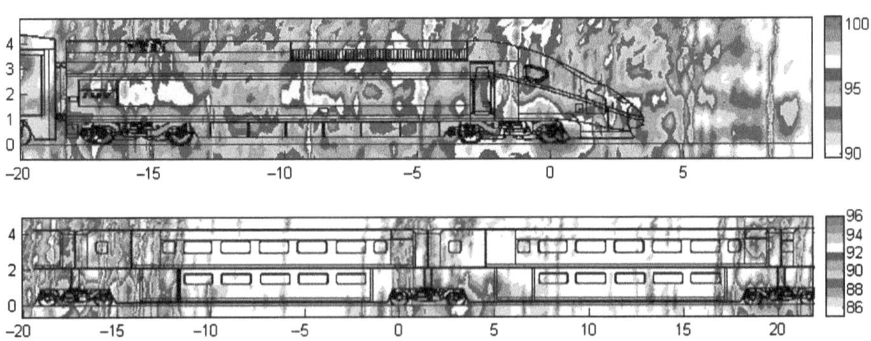

FIGURE 8-14 Noise sources maps of front power car and second and third middle coaches of a TGV Duplex at 300 km/h in the 1250 Hz one-third octave band [8.16]

implicit assumption is made, such as that the source comprises an array of uncorrelated monopole sources.

Various attempts have been made over the years to extract source levels on trains from microphone array measurements. In [8.3, 8.27, 8.29] source strengths were determined by integrating array levels over regions that were defined by the analyst and using an implicit model for the source. In [8.22] an attempt was made, again using an implicit source model consisting of a distribution of sources over the train surface. As the source position was not specified, this approach suffered from the fact that unphysical ghost sources could be wrongly found due to side lobes, which could even be in a region where there is no train.

The turbulent boundary layer on the surface of the train is a distributed source which has a low source level but a large area. It is therefore difficult to measure it reliably due to contamination by side lobes from other sources or by background noise. Moreover, at low frequency the turbulent boundary layer is a correlated source over quite a large length. Sources which are correlated over an extended region are not compatible with the usual implicit source model, as already discussed for the rail [8.20] (see Chapter 6) and require special treatment.

It is possible, in principle, to determine the directivity of sound sources from microphone array measurements as well as their location, provided that they are 'stationary'. This has been demonstrated on a loudspeaker source [8.30] but it has not been applied more generally in array measurements from trains due to the short time window available.

8.3.2 Acoustic mirror

An alternative to the microphone array is an 'acoustic mirror' [8.31], which consists of a parabolic reflector and single or multiple microphones located at its focus. This allows sound at particular locations to be quantified. It is particularly useful in a laboratory situation such as a wind tunnel but has also been used in field tests. Example results are shown in Figure 8.15, which show strong sources at the bogies, especially the leading bogie which is attributed to aerodynamic noise [8.31]. The snowplough region produces considerable noise at higher frequencies, as do the pantographs. The cab door and the wiper are also identified as sources.

8.3.3 Wind tunnels

Wind tunnels are used extensively to test the aerodynamic performance of aircraft and cars. However, to be useful for testing the aeroacoustic behaviour, the wind tunnel must have a low background noise and an anechoic environment. The Maibara wind tunnel of RTRI in Japan is one such facility dedicated to railway noise research [8.32], see Figure 8.16. This has a maximum wind speed of 400 km/h, a cross-section of 3×2.5 m and a length of 8 m. It can be used for testing components, such as pantographs, at full scale. However, to test a whole train, for example for studying nose shape or treatments in the bogie region, it is necessary to use scale models, typically at scale 1:5 or 1:7. Similar facilities are also available in other countries, for example France and Germany, developed initially for aerospace or automotive applications.

Some wind tunnel facilities also include a moving ground plane which gives the correct shear flow between the bottom of the vehicle and the ground. Tests with and

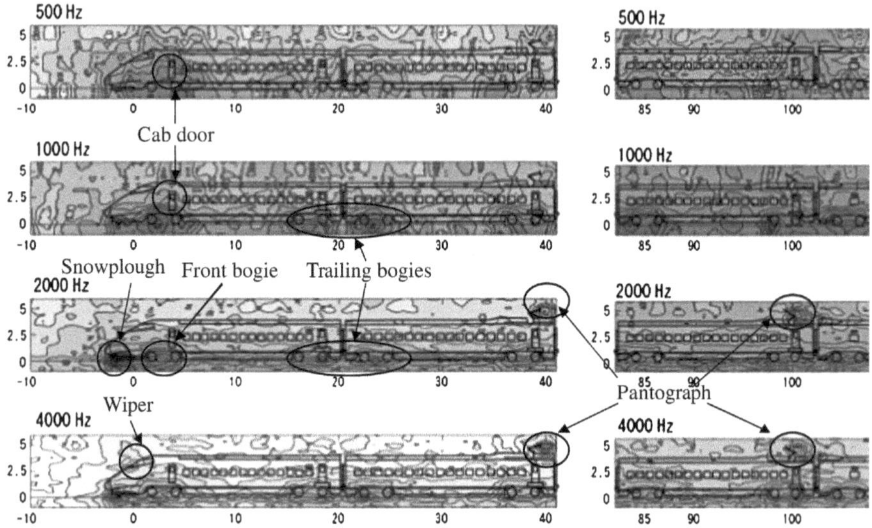

FIGURE 8-15 Noise source distribution on Shinkansen train at 270 km/h obtained by measurement with acoustic mirror (taken from [8.31])

without a moving ground plane [8.33] have shown that a wind tunnel with a fixed ground can still be used as a good approximation of the flow beneath a train provided that the distance from the bottom of the train to the ground is doubled (although this does not easily allow for the effects of the bogie region).

FIGURE 8-16 Maibara low noise wind tunnel. Photo courtesy of RTRI

Microphone arrays or an acoustic mirror can be readily used in a wind tunnel to localize sources of noise. Parietal flow visualization can also be used and local flow velocity can be measured using a hot wire anemometer, e.g. [8.10].

8.3.4 Direct source measurements

Microphone array techniques can be complemented by on-board measurements to give a better characterization of the physical phenomena of aerodynamic sources and even to derive source models. In [8.34] a causality, or coherent output power (COP), technique was presented and used to characterize aerodynamic sources in the bogie and inter-coach spacing areas of a TGV. A number of microphones were located near the sources and anti-turbulence sensors (Neise probes), intended to filter turbulence and to record mainly acoustic waves, were located downstream of the source region.

Using long time averages, the correlation was calculated between the signal recorded by the microphones in the source regions and that received by an anti-turbulence sensor. From this it was possible to determine if the source radiates sound, and to estimate its spectrum. In particular, tonal peaks can be associated with particular source regions. The main problem of this method is the choice of a good sensor which is representative of the source, and to locate it near the source without introducing new noise sources.

8.4 NUMERICAL TECHNIQUES

In recent years major advances have been made in the ability to calculate fluid flows using computational fluid dynamics (CFD) and these can be used to optimize a vehicle for aerodynamic performance (e.g. aerodynamic drag). Nevertheless, there is a large step between such calculations and predictions of aerodynamic noise (computational aeroacoustics, CAA). In general, most applications of CAA have been limited to special cases of limited geometry and practical application to industrial geometry is less common. Nevertheless, there are various numerical approaches that can be taken in order to estimate aerodynamic noise [8.35], some of which have been used to study railway applications.

Direct numerical simulations (DNS) involve simultaneous solution of all scales of turbulence. Although this may seem the ideal method, the range of length scales involved in the geometry of the problem (up to tens of metres) and the turbulence itself (several orders of magnitude smaller) is so large that this is usually only applied to idealized model problems in two dimensions.

One practical way of dealing with modelling turbulence is the widely used Reynolds-Averaged Navier–Stokes (RANS) CFD formulation. This conventional CFD method involves separating the flow into a time-averaged part, which is solved numerically at each grid point and a fluctuating part, which is solved by an approximate model such as a $k - \varepsilon$ model (where k represents the turbulent energy density and ε the turbulent dissipation rate). This is not strictly computational aeroacoustics, but can be considered as an acoustic interpretation of aerodynamic (CFD) calculations [8.35]. In one application [8.6], RANS calculations were carried out with commercial CFD software on the inter-coach gap of a TGV, including the complex bogie geometry. This allowed the turbulence-producing areas to be identified and the different sources to be classified according to their extent and level in terms of turbulent energy. Despite its limitations, this approach can allow a preliminary classification of different solutions and selection of the most promising ones, which can be tested later experimentally. Results of this example are discussed in Section 8.5.4.

In a study of the aerodynamic drag of a high speed train, Paradot [8.7] used a $k - \varepsilon$ RANS model for a five car 1:15 scale model of a TGV-R and compared the results with wind tunnel tests, finding good agreement.

Another approach [8.35] involves separate calculation of aerodynamic and acoustic fields, in which Lighthill's theory for aeroacoustic calculation [8.8] is used for simple geometries. Acoustic source models can be built from CFD data using specific theories for jets, cavities, wakes, etc. For example, in [8.36], a simple cavity with the same aspect ratio as a pantograph cavity was studied. An intrinsic limitation of this method for industrial applications is that each case must be modelled analytically to obtain the source terms.

Large eddy simulations (LES) involve solving the fluid dynamic equations for large-scale eddies in the flow while representing eddies that are small compared with the grid size by an approximate model. This involves more computational effort than RANS but less than DNS. Some examples are given in [8.35] for simple geometries, in particular for steps and pairs of facing steps.

Lattice Boltzmann methods are also being considered. These are based on a statistical mechanics technique which offers some advantages. They have been found to be promising for modelling turbulent source regions around discontinuities for little extra cost compared with LES. Recent applications have included car wing mirrors and it seems practical to extend this to train source regions such as those around bogies.

8.5 REDUCTION OF AERODYNAMIC NOISE

8.5.1 Source contributions

As with other sources of noise, to achieve significant reductions it is important to know the dominant sources first. The main sources contributions from ICE (later called ICE-1) and TGV-A trains as well as the TR07 maglev vehicle were quantified in [8.2, 8.29] using microphone array measurements. The source quantifications were based on the levels measured at 5 m from the surface using cross-shaped arrays. It was assumed that all the energy generated within the main lobe of the array could be associated with the particular source and that side lobe effects could be ignored. Measurements were taken at a range of speeds between 100 and 280 km/h for the ICE and at 100, 200 and 350 km/h for the TGV-A. The results are summarized in Table 8.1 in terms of the sound pressure level at 200 km/h and the speed dependence obtained by fitting a curve of the form $10n \log_{10} V$. The speed exponent n is also shown for each source. These were implemented in the ProHV model [8.2]. It is noted in [8.29] that the turbulent boundary layer noise is expected theoretically to have a speed exponent between 5.5 and 8.5 but that the measurements at low speeds are particularly prone to errors so that, even if the measured results are correct at high speed, the speed exponent (4.3 to 4.6) is probably too low.

In [8.3] these source levels were used to derive $L_{eq,Tp}$ levels for the pass-by of a TGV-A with eight trailer vehicles at 200 and 300 km/h. The results are listed in

TABLE 8-1 **REFERENCE LEVELS FROM PROHV MODEL DERIVED FROM MICROPHONE ARRAY MEASUREMENTS FOR TGV-A AND ICE [8.2, 8.29]. A-WEIGHTED SOUND PRESSURE LEVELS AT 5 M FOR 200 KM/H AND SPEED EXPONENTS (*N*)**

	TGV-A		ICE	
	L_{200} (dB)	*n*	L_{200} (dB)	*n*
Wheel, trailer	90.6	2.9	88.0	2.9
Wheel, front power car	91.5	3.2	91.5	3.3
Wheel, rear power car	90.3	3.0	90.0	2.7
Pantograph, front power car	–	–	94.2	4.9
Pantograph, rear power car	91.8	5.7	92.4	4.0
Cooling fan, front power car	91.6	4.7	90.5	3.6
Cooling fan, rear power car	88.8	4.6	88.0	2.8
Area near front window	91.9	5.1	88.2	2.9
Intercoach spacing	86.6	4.2	86.7	4.7
Bogie	78.7	6.1	78.3	4.3
TBL (/m^2)	70.4	4.3	67.6	4.6

Table 8.2 from which it can be seen that the aerodynamic and rolling noise contributions are estimated to be equal at 300 km/h. The main sources are those grouped as 'compact sources', notably the bogie and inter-coach spacing and the pantograph. The turbulent boundary layer contribution is estimated to be only 3.7 dB less than the total aerodynamic noise at this speed but is almost certainly overestimated at 200 km/h.

More recent measurements have been made in [8.3], including the COP measurements referred to in Section 8.3.4. The rolling noise levels were estimated using TWINS calculations but these were also corrected to correspond more closely to the measurements (due to a higher roughness level in the measurements). These source levels were incorporated into the MAT2S software which allows the pass-by time histories and spectra to be estimated [8.3]. Typical results are given in Table 8.3 for two speeds. Of the various compact sources referred to in Table 8.2, the bogie region appears to be the most important. The pantograph, although often thought of as

TABLE 8-2 **PREDICTIONS FROM PROHV MODEL FOR TGV-A IN TERMS OF $L_{AEQ,TP}$ (IN dB) AT 25 M FROM THE TRACK [8.3]**

	200 km/h	300 km/h
Rolling noise	86.1	91.5
Cooling fan noise	74.2	82.7
TBL	80.0	87.7
Compact sources	79.9	89.0
Total aerodynamic sources	82.9	91.4
Total wayside noise	88.0	94.7

TABLE 8-3 **PREDICTIONS FROM MAT2S FOR MODERN HIGH SPEED TRAIN, $L_{AEQ,TP}$ IN DB AT 25 M**

	200 km/h	300 km/h
Rolling noise	85.6	90.3
Pantograph	56.7	69.2
Bogies	78.1	86.6
Other sources	71.6	81.0
Total aerodynamic sources	79.0	87.7
Total wayside noise	86.5	92.2

a dominant source, has a relatively small contribution to the overall train noise level unless noise barriers are present.

Turbulent boundary layer noise is omitted from this model. From the estimates in Tables 8.1 and 8.2 it may be significant at speeds of 300 km/h and above, although the speed exponent should be higher and the levels at lower speeds therefore lower. On the other hand, the result in Figure 8.9 indicated that it was negligible.

8.5.2 Pantograph noise

Particularly in Japan, where much of the Shinkansen route runs on elevated structures fitted with noise barriers, pantograph noise has been a major issue, leading to considerable efforts to reduce it. Figure 8.17 shows examples of the sound level from Shinkansen trains measured using a microphone array, showing clearly the peaks due to the pantographs.

Although some broad-band noise is produced, the noise generated by the pantograph is mainly tonal. This is caused by vortex shedding around the various cylinders of which it consists, including the collector strip and the horns. These can be a significant source of tonal noise. The physical phenomena are now quite well understood. For example, a typical cylinder of diameter 3 cm in a flow at 300 km/h will generate a tone at about 550 Hz (see box on page 284) Moreover, wind tunnel tests

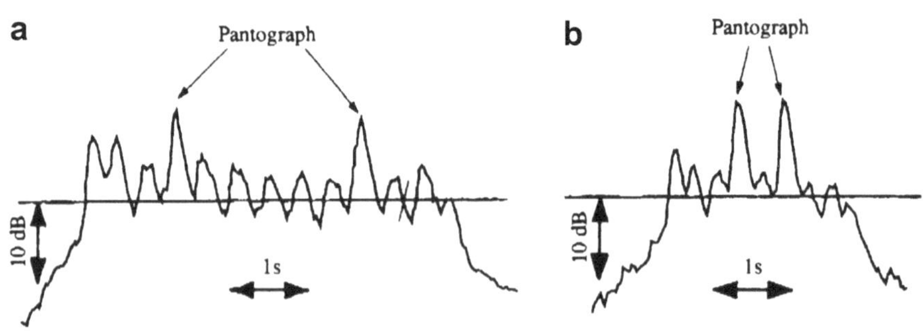

FIGURE 8-17 Time history of A-weighted sound pressure level from Shinkansen measured with a microphone array at 25 m from the track: (a) S-1 train, 235 km/h, (b) S-2 train, 312 km/h [8.33]

have shown that rectangular profiles produce around 6 to 10 dB more noise than the equivalent circular profile, whereas elliptical profiles offer potential reductions of up to 10 dB relative to circular profiles [8.3]. Another technique to reduce the noise from cylinders is to roughen their surface by wrapping cord around them or adding ribs.

A standard DSA 350 SEK pantograph, as fitted to German ICE-1 and ICE-2 trains, was modified in a series of wind tunnel tests to optimize cylinder shapes, as well as to test the principle of using ribs to break up coherent vortex shedding [8.3, 8.35]. In the wind tunnel the total reduction in A-weighted sound level was nearly 5 dB but in field tests the modifications were more limited and the overall noise reduction was not as great. To achieve greater noise reductions new pantograph concepts must be considered.

In Japan, early Shinkansen trains had multiple pantographs. The number of pantographs fitted to more recent trains has been reduced and these are fitted with pantograph 'covers', an example of which is shown in Figure 8.18. While these cannot completely shield pantograph noise, their leading edge has the effect of reducing the velocity of the separating flow and their sides act as noise barriers for the sources on the roof of the train such as the insulators and lower part of the pantograph. However, these pantograph covers themselves generate a certain amount of aerodynamic noise. This is particularly noticeable where smaller vehicle profiles have been introduced as the pantograph cover itself becomes taller.

A new 'low-noise' pantograph that can be used without these 'covers' has been developed by JR East [8.37, 8.38], see Figure 8.19. Based on wind tunnel experiments, initially at reduced scale and then at full scale, this contained a number of features,

FIGURE 8-18 Example of a pantograph cover from Series 700 Shinkansen. Photo courtesy of T. Kitagawa

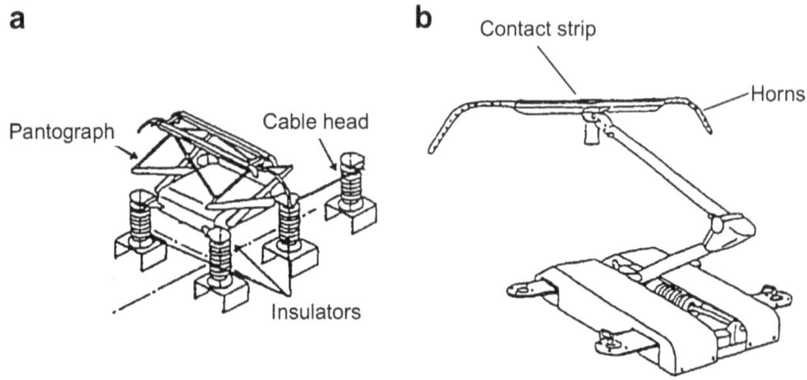

FIGURE 8-19 (a) Conventional pantograph fitted to Shinkansen train, (b) prototype low noise pantograph (from [8.37])

some of which can be seen in Figure 8.19. A large single tube supports the contact strip, with link mechanisms contained inside the tube, thus reducing the number of sources of sound. The relatively large diameter ensures that tonal noise occurs at lower frequencies. The dual contact strip was replaced by a single strip, which had to be designed carefully to achieve aerodynamic stability. Low noise horns were developed with slits in the direction of the air flow. On the underside of the collector shoe, a rough surface was introduced to shorten the correlation length of the vortices generated (this is clearly not possible on the upper surface). In addition, modified insulators were developed that are much longer in the flow direction to reduce the noise they generate and to reduce the number required to support the pantograph from five to two. For application in tests on a train a number of compromises had to be taken but the design was still promising [8.38].

In [8.39] a single arm pantograph fitted to the Series 700 Shinkansen, introduced on JR Central and JR West, is compared with the earlier diamond shape on the Series 300 Shinkansen. In addition a more streamlined pantograph cover was implemented. The effect on noise from this source was found to be about 4 dB(A).

In order to overcome problems of aerodynamic stability of novel pantograph heads, a new actively controlled single arm pantograph (ASP) has been developed in a collaboration between DB and JR East [8.40, 8.41]. This is shown in Figure 8.20. Acoustic development has involved a combination of theoretical modelling and wind tunnel testing. Analysis of results showed that the acoustic radiation is strongly influenced by the pantograph head, the insulators and the interaction between the frame and the lower arm when running with the knee in the upstream position [8.41]. A wing-shaped pantograph head would give a maximum noise reduction but the large aerodynamic forces generated would be hard to control. Results for the ASP are compared with the standard and modified DSA 350 SEK pantograph in Figure 8.21. These are for the quieter 'knee downstream' configuration. A variant with two contact strips (as required for operation on DB) gave reductions of 9 dB(A) compared with the reference case, while a variant with only one contact strip was a further 2 dB(A) quieter. When mounted on novel Japanese insulators the reductions were 11 and 14 dB(A), respectively [8.41].

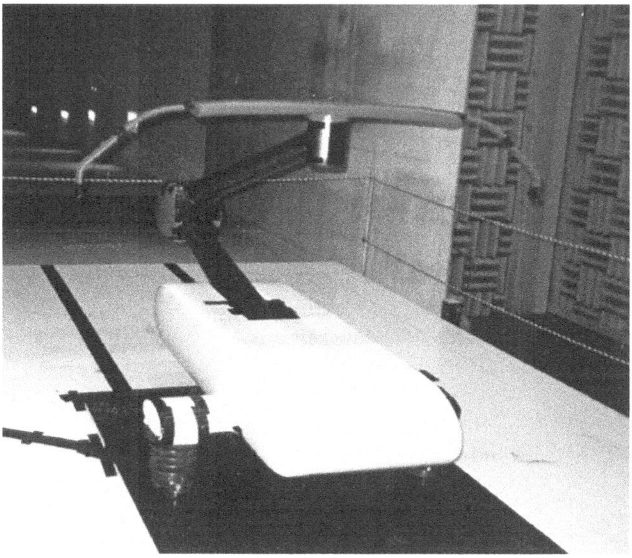

FIGURE 8-20 Prototype actively controlled single arm pantograph ASP (from [8.41])

8.5.3 Pantograph recess

On the TGV, for example, the pantographs are mounted in a recess in the roof which means that when retracted the pantograph is less exposed to the flow. This can be important for trains intended to run under several electrical systems where different pantographs may be required. However, the flow noise induced by the cavity is

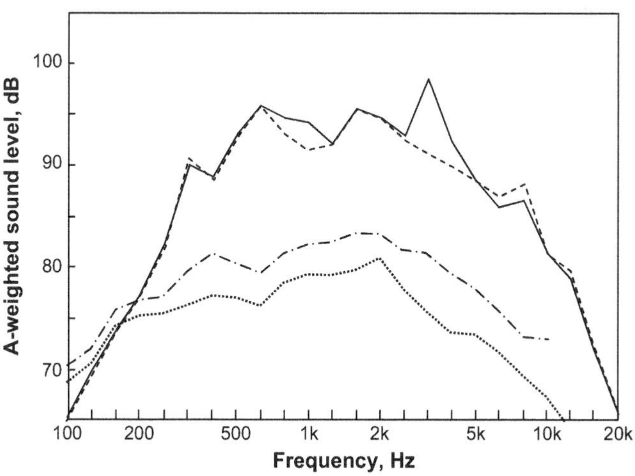

FIGURE 8-21 Pantograph noise radiation measured in a wind tunnel at 330 km/h. —, standard DSA 350 SEK pantograph; - - -, DSA 350 SEK with modifications; – · – ·, ASP with two contact strips; · · · ·, ASP with one contact strip (redrawn from [8.41])

found to be of a similar level to that from the pantograph itself. Extensive studies have been carried out into the flow in such a cavity region [8.3, 8.10, 8.36], see Figure 8.5. Noise reduction would require the flow to be prevented from going inside the recess by the use of spoilers, but no clear solutions have been developed (other than mounting the pantograph directly on the roof).

8.5.4 Bogie and inter-coach spacing

The bogies are complex structures, with little or no streamlining, located in a region of considerable turbulence. It is therefore not surprising that they form significant aerodynamic noise sources, especially the leading bogie of the train, as indicated in Figures 8.14 and 8.15. The air flow through the bogie region is very important for the cooling of brakes, motors and wheels and the bogie by its nature requires considerable freedom to move relative to the vehicle body, all of which makes noise reduction by the use of enclosures very difficult.

The French TGV and its derivatives use articulated Jakobs bogies between the trailer vehicles, in contrast to the more conventional arrangement of other high speed trains such as ICE and Shinkansen. The inter-coach region is shown in Figure 8.22(a) and the turbulence producing areas are shown schematically in Figure 8.22(b). The yaw dampers, which protrude from the side of the bogie, are a particularly important source.

Identification of bogie aerodynamic sources was carried out by SNCF through wayside measurements with the microphone array illustrated in Figure 8.12 [8.3]. Further characterization of these sources was investigated with on-board measurements using the COP technique [8.34]. As mentioned in Section 8.4, CFD techniques using the RANS approach have been used to classify different solutions for the bogie area of a TGV [8.6, 8.42]. The mesh used for the bogie is shown in Figure 8.23. Since only part of the train was modelled, an iteration technique was used to obtain the correct flow into the region of the bogie.

The most successful solutions were then tested in a wind tunnel [8.43] and encouraging results were obtained. Figure 8.24, based on [8.43, 8.44], compares the spectra between the initial configuration and the best combination of solutions obtained for the bogie region between the front power car and the first trailer vehicle.

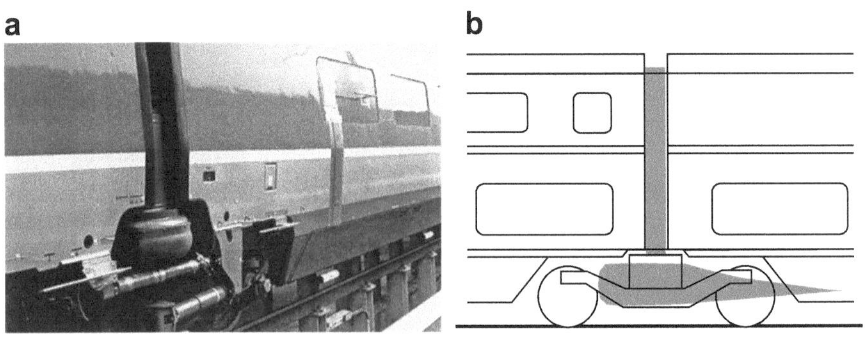

FIGURE 8-22 (a) Inter-coach region of a TGV, showing location of Neisse probes used in [8.34]; (b) turbulence producing areas of TGV inter-coach region

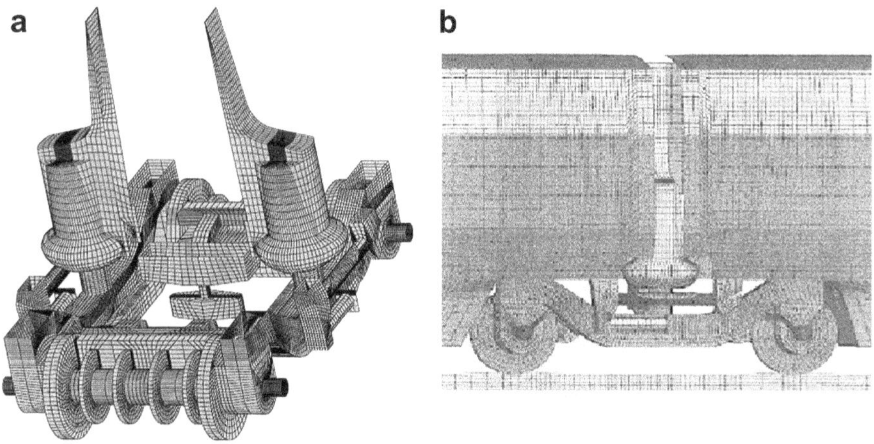

FIGURE 8-23 CFD mesh of bogie surfaces [8.6] and inter-coach region [8.42]

The optimized solution comprised:

- fairings on the bogie (NB these were flush with the body side, which would not be practical on existing bogies);
- completely closed inter-coach gap;
- wider profile of lower part of car body.

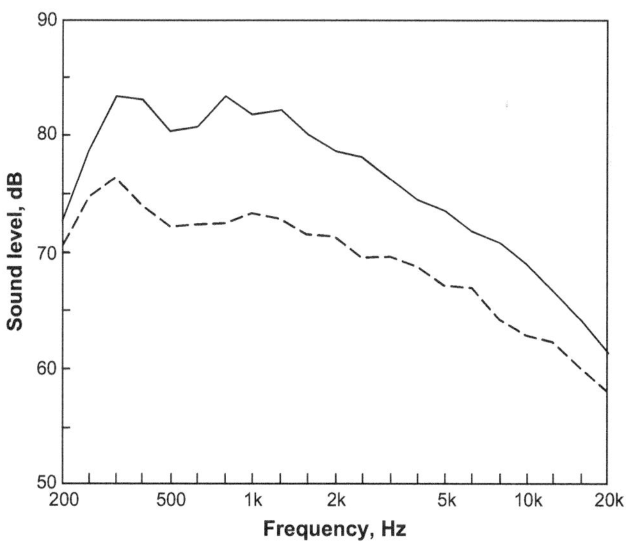

FIGURE 8-24 Spectrum in the bogie area of a 1:7 scale model of a TGV obtained in wind tunnel, comparison of the initial configuration with a good combination of solutions. —, original configuration; – – –, optimized configuration (from [8.43]). NB at full scale the corresponding frequencies are 1/7 of those shown

Together this enabled a reduction of 3 to 10 dB to be achieved across the spectrum, giving a total reduction of almost 8 dB(A) in the wind tunnel. Similar results were found for the bogie region between two trailer cars. For an operational high speed train it is estimated that a substantial reduction of 5 dB(A) could be achieved taking account of other possible sources [8.43], although these results would have to be confirmed with field measurements.

Shielding by bogie fairings is considered to be the most efficient solution for bogie aerodynamic noise due to the complexity of the source region. In [8.45] studies are described of the use of bogie fairings to reduce the drag of high speed trains as well as aerodynamic noise. For use on trains with conventional external wheelset bearings, clearance issues mean that the fairings cannot be mounted flush with the body sides but must be bellied outwards, which leads to some additional aerodynamic noise. From wind tunnel tests it was concluded that drag can nevertheless be reduced by about 10%. Field tests on an ETR500 train were carried out in 2000 [8.46]. This had both standard wheels and wheels fitted with constrained layer damping (see Section 7.3.3). Single microphone measurements at 25 m indicated a noise reduction of around 2 to 3 dB(A) at 300 km/h for the standard wheels when fitted with the fairings. Without the fairings, the damped wheels were 2 to 4 dB(A) quieter than the standard wheels, suggesting that the initial situation was dominated by rolling noise more than aerodynamic noise. For these damped wheels the fairings made only a small difference of about 1 dB(A).

Similar conclusions were reached using microphone array measurements [8.47] which suggest that fairings reduce noise from the bogie region by 2 dB(A) for standard wheels but have almost no effect for the damped wheels. However, other measurements from the same series of tests, at 7.5 m from the track, indicate noise reductions due to fairings for standard wheels of 2 to 4 dB(A) and up to 3 dB(A) reduction due to fairings for the damped wheels [8.48].

In [8.39] attention to detail in the inter-coach region is shown to lead to reductions of 3 dB(A) between Series 300 and 700 Shinkansen. In particular, it is shown to be important to ensure that any misalignments of adjacent vehicles do not lead to the edges of baffles mounted on vehicle ends protruding into the flow. A cover was also mounted at the lower part of the inter-coach region to prevent sound from being transmitted upwards. Roof-mounted insulators for the high voltage cable were also replaced with a streamlined cable connection.

In [8.49] wind tunnel tests on a 1/8 scale model showed that rounding the corners of the vehicle ends (a radius of 200 mm at full scale was used) could be more effective in reducing noise from the inter-coach gap than adding baffles to the coach ends.

The larger loading gauge available for the Shinkansen network, compared with European high speed trains, means that there is more scope for fitting bogie fairings. These have been fitted to several recent experimental trains, for example the Fastech 360 trains of JR East [8.50]. These trains also featured a metallic covering for the inter-coach gap.

Although fairings are known to be effective for reducing aerodynamic noise from the bogie region, as well as rolling noise from the wheels (Section 7.5.1), there are significant practical difficulties in implementing them due to limited clearance, the need to maintain ventilation and the need for access for maintenance.

8.5.5 Nose and body-shell design

Developments of the shape of the nose of the train (see Figure 8.1) have generally been made in order to improve aerodynamic performance, which does not necessarily mean improved acoustic performance. Very long streamlined nose shapes have been used for the latest Shinkansen trains in Japan, mainly in order to reduce the severity of 'micro-pressure waves'. These are shock waves generated in long tunnels with ballastless track and result in a loud impulsive sound at the opposite end of the tunnel [8.51]. However, a long streamlined nose may also have some advantages for aerodynamic noise.

In [8.39] the shape of the Series 700 Shinkansen is compared with that of the earlier Series 300, showing how it is close to the ideal of a constant rate of change of cross-section. In addition, improvements were made to reduce the roughness of the surfaces particularly for the side doors where improved sealing and retractable covers for the handrails were implemented. The gap between the front skirt and the leading bogie was also reduced, see Figure 8.25, and an underfloor cover was introduced under the nose. The leading bogie was also covered by fairings. Overall these changes led to a reduction of about 2 dB(A) for the nose region, although the relative contributions of the various changes are not discussed in [8.39].

Wind tunnel tests on 1/7 scale models in France [8.43, 8.44] also showed a small effect of the nose shape. Figure 8.26 shows results which indicate that the difference

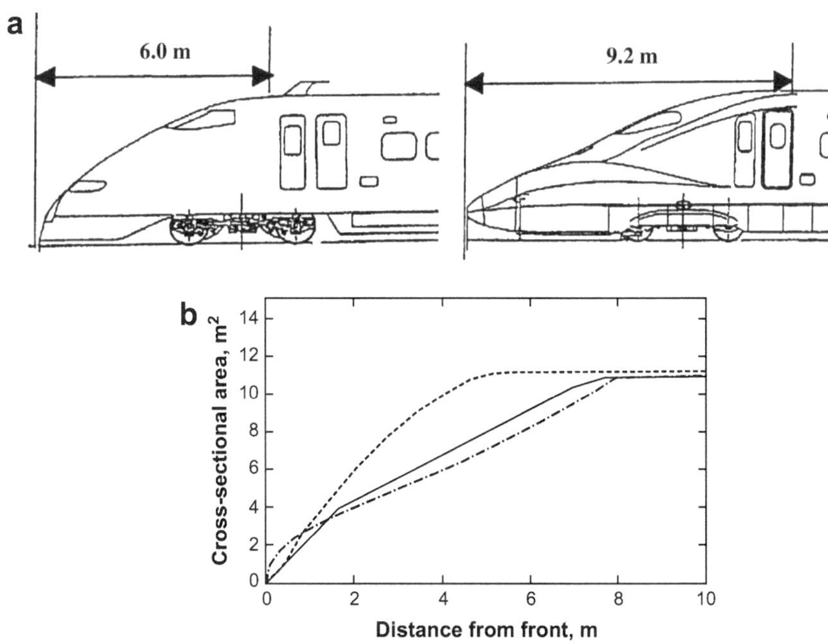

FIGURE 8-25 (a) Cross-section of series 300 (left) and 700 (right) Shinkansen front end designs (courtesy of JR East, used with permission), (b) cross-sectional areas: - - -, series 300; —, series 700; – · – ·, optimum (from [8.39])

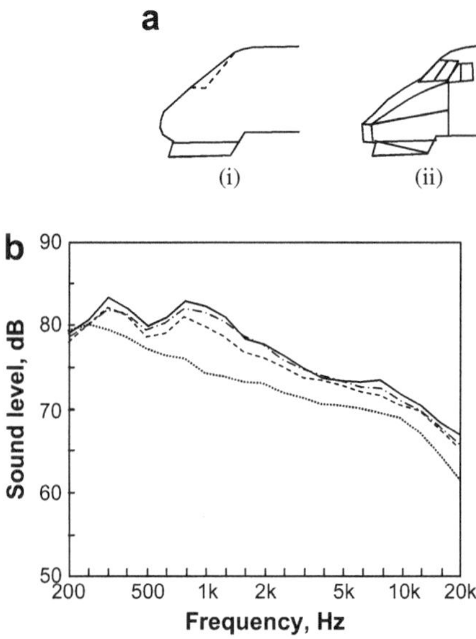

FIGURE 8-26 Results of testing different nose designs at 1/7 scale in wind tunnel (a) nose shapes; (b) noise spectra for different nose designs: —, nose (i); – · – ·, nose (ii); - - -, nose (iii); · · · ·, closed underside; (c) noise spectra for nose (i) with modified front bogie region —, standard bogie; - - -, profiled underside; – · – ·, profiled underside and bogie fairings (redrawn from [8.44]). NB at full scale the corresponding frequencies are 1/7 of those shown

between three nose designs was around 1–2 dB, the long nose being quietest, whereas by closing up the gap under the train model completely the noise level was reduced by more than 5 dB. These results suggest that the improvements to the flow around the leading bogie are the most important aspects of design of the front of the train. Moreover, it was found that by shape modifications to the underside of the front of the power car (widening the underfloor section) to modify the flow in the bogie area, and adding shaped fairings around the leading bogie that deflect the flow away from the leading wheels, the reductions shown in Figure 8.26(c) were obtained. It was concluded that when the bogie region is optimized the long nose (iii) and nose (i) (the TGV-Duplex) are equivalent [8.44].

Other details of design are clearly important, such as smooth body-shells with flush-mounted windows. The cab door and the wiper were noted as important sources in Figure 8.15 [8.31]. In [8.3] the wiper of the ICE/V was shown to be a detectable aerodynamic source but has a level of at least 8 dB less than the front bogie region, the latter also having a larger area. The water deflector below the windscreen on the TGV-Duplex has been noted as a source at 500 Hz in Section 8.3.1.

From measurements on the ETR500 at 290 km/h using a microphone array [8.47] it was observed that cavities for the steps below the driver's door form an important source on the rear power car.

For regions where fan noise is involved, for example louvers, it is not possible directly from array measurements to separate the noise due to flow over the louver irregularity from the fan noise. To do this, frequency analysis may be used if the fan

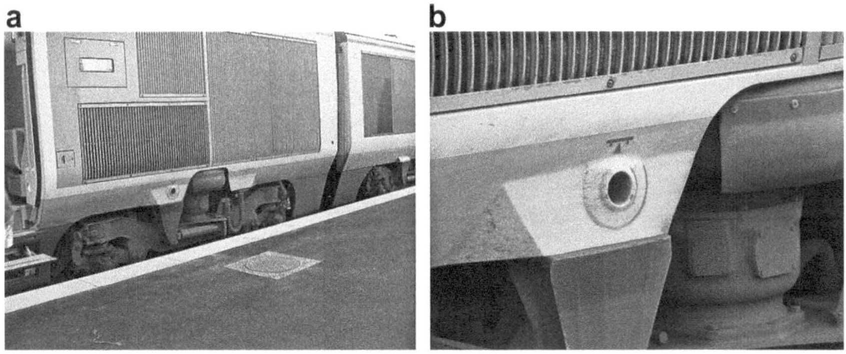

FIGURE 8-27 Jacking holes on Eurostar

speed is known. Alternatively, measurements at standstill can be used to identify the fan noise separately.

8.5.6 Example of acoustic resonators: jacking points on Eurostar

A particular example of excitation by air flow over an acoustic resonator was found on the Eurostar (British/French/Belgian high speed train). Although similar to the TGV design, the jacking holes for lifting the train, see Figure 8.27, are closed at the rear whereas on the conventional TGV designs they are open. These holes have a diameter of about 75 mm and a depth of about 0.3 m. A strong acoustic resonance occurs at about 300 Hz when air flows over the opening – this corresponds to a quarter wavelength in the depth of the holes. Figure 8.28 shows a time–frequency plot of the noise from a Eurostar at 300 km/h in which the peaks at about 300 Hz can be clearly seen at the end of each vehicle (they shift in frequency due to the Doppler effect). Simply by placing tape over the holes, the level in the 315 Hz one-third octave band was reduced by almost 10 dB and the overall level by more than 1 dB(A) in a test.

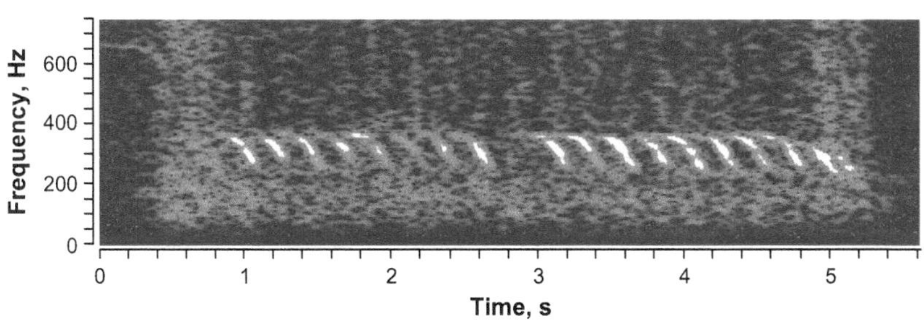

FIGURE 8-28 Time–frequency plot of pass-by noise from Eurostar (dynamic range 40 dB)

8.6 CONCLUDING REMARKS

Following the introduction of high speed trains, aerodynamic noise from trains has been an area of research interest for a considerable time. For current high speed rolling stock it becomes important compared with rolling noise for speeds above about 350 km/h. Array measurements have been used successfully to identify the main aeroacoustic sources. However, correct identification of these sources is indirect, involving either measurements at different speeds or the subtraction of contributions from other known sources (e.g. rolling noise) and quantification of the source strengths remains difficult. Moreover, regression lines which work in terms of the overall contribution of aerodynamic noise are a gross simplification; in practice there are several sources which contribute to the overall level, each of which may have a different characteristic and speed exponent.

Aeroacoustic modelling of trains has not yet been used beyond the research environment and is mostly limited to very simple configurations. Various modelling techniques have been investigated such as a conventional Reynolds-Averaged Navier–Stokes (RANS) model to identify areas of turbulence, specific models for particular geometries, and large eddy simulations (LES) for flow over steps. Lattice Boltzmann methods are also being considered.

Significant progress has been made in reducing aerodynamic noise from pantographs, which is particularly important in situations where noise barriers shield sources from the lower part of the train. Although it is clear how to reduce aerodynamic noise from the bogie and inter-coach region, particularly using fairings, practical difficulties often prevent their implementation.

Finally, it should be noted that aerodynamic noise inside vehicles also depends on the excitation of the carriage walls and roof by the convected turbulent flow, and is discussed separately in Chapter 14.

REFERENCES

8.1 B. Mauclaire. Noise generated by high speed trains. New information acquired by SNCF in the field of acoustics owing to the high speed test programme. *Proceedings Inter Noise*, **90**, 371–374, 1990.

8.2 German–French Cooperation Deufrako Annex K, 1994. Noise sources from high speed guided transport, final report.

8.3 German–French Cooperation Deufrako Annex K2, 1999. Noise sources from high speed guided transport, final report.

8.4 J. Astley *et al.* Predicting and reducing aircraft noise. 14th International Congress on Sound and Vibration. Cairns, Australia, July 2007.

8.5 P. Crespi, R. Grégoire and P. Vinson. Laser Doppler velocimetry measurements and boundary layer survey on-board the TGV high speed train. *Proceedings of World Congress on Railway Research*, Paris, 1994.

8.6 E. Le Dévéhat, C. Talotte, and A. Dugardin. Aerodynamic noise reduction using a CFD approach based on the prediction of the turbulence production area. MIRA International Conference on Vehicle Aerodynamics. Coventry, UK, 1996.

8.7 N. Paradot. *Etudes numerique et experimentale de la resistance à l'avancement d'un train à grande vitesse.* PhD thesis, University of Poitiers, 2001.

8.8 M.J. Lighthill. On sound generated aerodynamically, I. General theory. *Proc. Royal Society, London*, A211, 564, 1952.

8.9 F.J. Fahy. Fundamentals of noise and vibration control, Chapter 5 in *Fundamentals of Noise and Vibration.* In: F.J. Fahy and J.G. Walker (eds.). E&FN Spon, 1998.

8.10 C. Noger, J.C. Patrat, J. Peube, and J.L. Peube. Aeroacoustical study of the TGV pantograph recess. *Journal of Sound and Vibration*, **231**, 563–575, 2000.
8.11 P. Vinson and P. Pinconnat. TGV-Duplex à 350 km/h, bruit rayonné dans l'environnement au passage en champ libre, derrière Merlon et Écran. SNCF report AEF-D R03030/02D-127, 2003.
8.12 B. Barsikow. Schallabstrahlung spurgebundener Hochgeschwindigkeitsfahrzeuge bis 500 km/h. *Proceedings DAGA* '89, 607–610, 1989.
8.13 W.F. King III. On the role of aerodynamically generated sound in determining noise levels from high-speed trains. *Journal of Sound and Vibration*, **54**, 361–378, 1977.
8.14 W.F. King III and D. Bechert. On the sources of wayside noise generated by high-speed trains. *Journal of Sound and Vibration*, **66**, 311–332, 1979.
8.15 P. Fodiman. *The NOEMIE project. Project no. 2002/EU/1663, Final report*. AEIF, 2005.
8.16 C. Mellet, F. Létourneaux, F. Poisson, and C. Talotte. High speed train noise emission: latest investigation of the aerodynamic/rolling noise contribution. *Journal of Sound and Vibration*, **293**, 535–546, 2006.
8.17 T.J. Mueller (ed.). *Aeroacoustic Measurements*. Springer, Berlin, 2002.
8.18 P. Nelson. Source identification and location, Chapter 3 in *Advanced Applications of Acoustics, Noise and Vibration*,. In: F.J. Fahy and J.G. Walker (eds.). Spon Press, London, 2004.
8.19 J.J. Christensen and J. Hald, Beamforming. *Brüel & Kjaer Technical Review* No. 1, 2004.
8.20 T. Kitagawa and D.J. Thompson. Comparison of wheel/rail noise radiation on Japanese railways using the TWINS model and microphone array measurements. *Journal of Sound and Vibration*, **293**, 496–509, 2006.
8.21 B. Barsikow, W.F. King III, and E. Pfizenmaier. Wheel/rail noise generated by a high speed train investigated by a line array of microphones. *Journal of Sound and Vibration*, **118**, 99–122, 1987.
8.22 S. Brühl and A. Röder. Acoustic noise source modelling based on microphone array measurements. *Journal of Sound and Vibration*, **231**, 611–617, 2000.
8.23 B. Barsikow. Experiences with various configurations of microphone arrays used to locate sound sources on railway trains operated by the DB AG. *Journal of Sound and Vibration*, **193**, 283–293, 1996.
8.24 M.M. Boone, N. Kinneging, and T. Van den Dool. Two-dimensional noise source imaging with a T-shaped microphone cross array. *Journal of the Acoustical Society of America*, **108**, 2884–2890, 2000.
8.25 C. Talotte. Aerodynamic noise, a critical survey. *Journal of Sound and Vibration*, **231**, 549–562, 2000.
8.26 A. Nordborg, J. Wedemann, and L. Willenbrink. Optimum array microphone configuration. *Proceedings Inter Noise*, 2000 Nice, France, 2000.
8.27 J.F. Hamet, M-A. Pallas and K.P. Schmitz. Deufrako-1: microphone array techniques used to locate acoustic sources on ICE, TGV-A and Transrapid 07. *Proceedings of Inter Noise 94*, Yokohama, Japan, 187–192, 1994.
8.28 B. Barsikow and W.F. King III. On removing the Doppler frequency shift from array measurements of railway noise. *Journal of Sound and Vibration*, **120**, 190–196, 1988.
8.29 G. Hölzl, P. Fodiman, K.P. Schmitz, M-A. Pallas, and B. Barsikow. Deufrako-2: localized sound sources on the high-speed vehicles ICE, TGV-A and TR 07. *Proceedings of Inter Noise 94*, Yokohama, Japan, 193–198, 1994.
8.30 F. Poisson, J.C. Valière and O. Coste. Directivity pattern measurement of moving acoustic sources. *Proceedings of 8th IEEE Signal Processing Workshop on Statistical Signal and Array Processing*, Corfu, Greece, 90–93, 1996.
8.31 K. Nagakura. Localization of aerodynamic noise sources of Shinkansen trains. *Journal of Sound and Vibration*, **293**, 547–565, 2006.
8.32 T. Maeda and Y. Kondo. RTRI's Large-scale low-noise wind tunnel and wind tunnel tests. *Quarterly Report of Railway Technical Research Institute*, **42**, 65–70, 2001.
8.33 T. Kitagawa and K. Nagakura. Aerodynamic noise generated by Shinkansen cars. *Journal of Sound and Vibration*, **231**, 913–924, 2000.
8.34 N. Frémion, N. Vincent, M. Jacob, G. Robert, A. Louisot, and S. Guerrand. Aerodynamic noise radiated by the intercoach spacing and the bogie of a high-speed train. *Journal of Sound and Vibration*, **231**, 577–593, 2000.

8.35 C. Talotte, P.E. Gautier, D.J. Thompson, and C. Hanson. Identification, modelling and reduction potential of railway noise sources: a critical survey. *Journal of Sound and Vibration*, **267**, 447–468, 2003.

8.36 M. Jacob, V. Gradoz, A. Louisot, D. Juvé, and S. Guerrand. Comparison of sound radiated by shallow cavities and backward steps. AIAA Congress. Seattle, USA, 1999.

8.37 S. Nakagawa. Low noise pantographs and insulators. *Proceedings of World Congress on Railway Research 1997*, Firenze, Italy, 485–490, 1997.

8.38 T. Hariyama, Y. Sasaki, T. Ichigi, and S. Ono. Low noise pantographs and insulators (vol. 2). *Proceedings of World Congress on Railway Research 1999*, Tokyo, Japan, 1999.

8.39 A. Torii and J. Ito. Development of the series 700 Shinkansen train-set, improvement of noise level. *Proceedings of World Congress on Railway Research 1999*, Tokyo, Japan, 1999.

8.40 K. Althammer, W. Baldauf, and T. Lölgen. Considerations for high performance pantographs. *Proceedings of World Congress on Railway Research 1999*, Tokyo, Japan, 1999.

8.41 T. Lölgen. Wind tunnel noise measurements on full-scale pantograph models. *Proceedings Joint ASA/EAA Meeting*, Berlin, Germany, 1999.

8.42 L. Guccia and P.E. Gautier. Aeroacoustic research applied to TGV. *Proceedings of World Congress on Railway Research 1996*, Colorado Springs, USA, 1996.

8.43 O. Grosjean, A. Fages, J.M. Tessier, C. Bertrand, L. Guccia, and S. Guerrand. Réduction du bruit aérodynamique émis par un TGV roulant à plus de 300 km/h. *Proceedings of World Congress on Railway Research* 1997 Firenze, Italy, 1997.

8.44 F. Paulin and O. Grosjean. Etude et optimization aéroacoustique de la partie basse d'un TGV en soufflerie anéchoïque, rapport finale d'essai. Report on ATREBAT project, 1997.

8.45 B. Schulte-Werning, G. Matschke, A. Willaime, A. Malfatti, G. Mancini, and M. Pecorini. High-speed trains with bogie fairings: European research into reducing aerodynamic drag noise. *Proceedings of World Congress on Railway Research 1999*, Tokyo, Japan, 1999.

8.46 L. Gori, G. Pugi, M.T. Cambini and A.G. Violi. Acoustic characterization of high speed train ETR500. *Proceedings of World Congress on Railway Research 2001*, Munich, paper 138, 2001.

8.47 K.G. Degen, A. Nordborg, A. Martens, J. Wedemann, L. Willenbrink, and M. Bianchi. Spiral array measurements of high-speed train noise. *Proceedings Inter Noise 2001*, The Hague, The Netherlands, 2001.

8.48 A. Bracciali and F. Picciolo. Experimental analysis of wheel noise emission as a function of the contact point location. *Journal of Sound and Vibration*, **267**, 469–483, 2003.

8.49 N. Yamazaki and T. Takaishi. *Wind tunnel tests on reduction of aeroacoustic noise from car gaps and bogie sections.* Quarterly Report of RTRI. Railway Technical Research Institute, Tokyo, Japan, 2007. 48, 229–235.

8.50 T. Endo. Fastech 360 prototypes probe ultra high speed territory. *Railway Gazette International*, **161**, 693–697, 2005.

8.51 T. Maeda. Micropressure waves radiating from a Shinkansen tunnel portal, Chapter 7 in *Noise and Vibration from High-Speed Trains*. In: V.V. Krylov (ed.). Thomas Telford, London, 2001.

CHAPTER 9

Curve Squeal Noise

9.1 INTRODUCTION

Curve squeal is one of the loudest and most disturbing noise sources from railways, metros and tramways. It is a strongly tonal noise (i.e. dominated by a single frequency) occurring in sharp curves. Its likelihood, although not necessarily its peak sound level, increases as the curve radius is reduced. Rudd [9.1] gives a rule of thumb that no squeal will occur when the curve radius $R > 100b$ (with b the bogie wheelbase). Thus, for example, a bogie with wheelbase 2.5 m may squeal on curves with $R < 250$ m whereas a 10 m wheelbase two-axle wagon may be expected to squeal on curves up to 1000 m radius. In practice, squeal mostly does not occur in curves of radius greater than about 500 m, and is only sporadic for curves down to about 200 m radius, whereas for curves of 200 m and below it is common.

Such tight curves occur mostly in built-up areas, for mainline railways, tramways and metro rail networks. Its tonal nature means that it is more annoying than a broad-band noise of the same level. Squeal noise therefore causes considerable environmental noise disturbance in the vicinity of curves as well as for rail passengers in vehicles and on stations. A survey for the International Union of Railways (UIC) [9.2, 9.3] found that curve squeal is an important source of disturbance affecting about 7% of railway customers in Europe. It was estimated that in highly populated areas, curve squeal is likely to affect an average of about 1000 inhabitants within 250 m of a squealing curve.

Curve squeal has been found to originate from the self-excited vibration of a railway wheel. The wheel is excited into vibration at frequencies corresponding to its axial normal modes, the main tones occurring at frequencies between about 250 Hz and 5 kHz. Other related forms of curving noise include a low frequency 'graunching' at switches and crossings, possibly due to flange rubbing, and 'juddering' thought to be caused by unstable dynamic behaviour of the vehicle. However, these are not discussed further here. Higher frequency 'flanging' noise is also important.

Time histories of curve squeal show that the noise is not necessarily constant while the vehicle negotiates the curve. Squeal may sometimes occur only for isolated periods and the amplitude of the noise may vary. Moreover, the dominant frequency can change from one wheel mode to another [9.4].

Practical solutions have included adding wheel damping treatment or applying lubrication. However, it is desirable to have a fundamental understanding of the causes of squeal so that, if possible, it can be avoided by appropriate vehicle and/or track design.

Rudd [9.1] identifies three possible excitation mechanisms, each due to stick–slip behaviour in the contact region:

- lateral creepage at the contact between the wheel tread and the top of the rail head;
- wheel flange rubbing on the rail gauge face;
- longitudinal creepage at the contact on the wheel tread due to differential slip.

Longitudinal differential slip arises when the conicity of the wheel is not sufficient to compensate for the difference between the distances that the inner and outer wheel must roll in a section of curved track. The outer wheel has further to travel and cannot rotate fast enough to make up for this added distance, causing it to slip forwards longitudinally. Rudd discounts differential slip as a mechanism for squeal noise, however, partly through experimental findings and also from the hypothesis that the excitation forces do not generate noise as they are within the plane of the wheel (this is not entirely correct due to coupling in the modeshapes). Moreover, independent wheels eliminate differential slip but do not necessarily eliminate squeal. Generally, the literature since Rudd has given little attention to differential (longitudinal) slip as a mechanism for squeal. However, Grassie and Kalousek [9.5] identify it as a possible source of rail corrugation on curves. This is associated with an 'axle wind-up' mode of the wheelset with a frequency in the range 50–100 Hz; the corresponding corrugation wavelengths are around 100–200 mm. Thus, although longitudinal differential slip does occur and may generate low frequency stick–slip, it is not considered relevant to squeal.

Of the two remaining mechanisms listed by Rudd, unsteady lateral creepage is thought to be the main cause of squeal noise, particularly for the leading inner wheel of a bogie. Observations indicate that the highest squeal noise amplitude is usually generated by the leading inner wheel of a four-wheeled bogie or two-axle vehicle. The fundamental frequency of such squeal noise corresponds to a natural frequency of the wheel.

Contact between the wheel flange and rail gauge face may also be relevant to squeal. This occurs at the leading outer wheel (and possibly the trailing inner wheel) in sharp curves. However, flange contact has generally been found to reduce the likelihood of stick–slip squeal due to lateral slip. For example, Remington [9.6] concluded from laboratory experiments that flange contact reduces the level of squeal noise. Nevertheless, flange contact appears to generate a different form of noise referred to as 'flanging noise' or 'flange squeal' [9.7]. Compared with squeal due to lateral slip, this has a considerably higher fundamental frequency, may have a lower level and is often more intermittent in nature. It may also be much more broad-band in nature. Nonetheless, it can also be a source of considerable annoyance.

9.2 CURVING BEHAVIOUR

In order to start understanding curve squeal it is necessary to consider the curving behaviour of a railway vehicle. This is covered in considerable detail in [9.8], see also [9.9]. A railway wheelset consists of two wheels rigidly connected by a single axle. The transverse profiles of wheels are inclined with a conicity which provides for

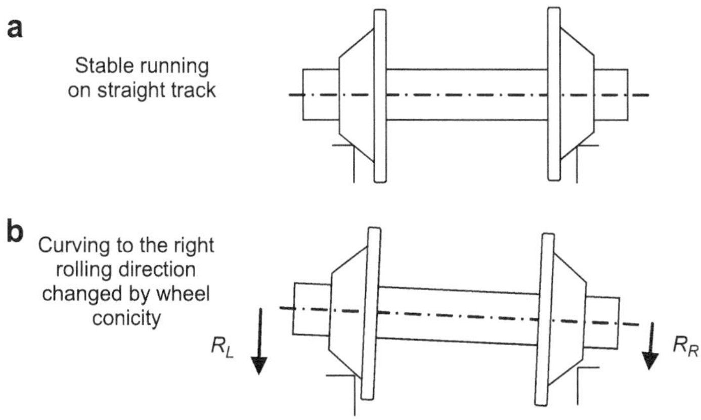

FIGURE 9-1 Schemati\c view of the wheelset showing steering by wheel conicity. R_L and R_R are the rolling radii of the two wheels

natural steering, as shown schematically in Figure 9.1. In a curve, the wheelset moves outwards, partly due to centrifugal forces and also due to the steering mechanism. Consequently, the outer wheel has a larger rolling radius (R_L) than the inner wheel (R_R), allowing it to traverse the longer path around the outer rail. On shallow curves this can account for the difference in the distances that the inner and outer wheels must travel, allowing curving without flange contact. On sharper curves, however, flange contact becomes necessary to provide sufficient restraining force.

A single wheelset such as this does not run stably on the track, but tends to experience kinematic oscillation. In practice, the wheelsets are located within a frame, either a bogie or directly in the vehicle, and are restrained by a large yaw stiffness. This ensures stability of the vehicle, but also causes the curving behaviour to differ from the ideal behaviour [9.8].

If two wheelsets are mounted in a *rigid* bogie, as shown in plan view in Figure 9.2, the axles are constrained to remain parallel to each other and so they cannot both align with the curve radius. The result is a high angle of attack between the wheel and rail in sharp curves, causing high flange forces, wear and increased risk of derailment due to flange climb. In practice, the yaw stiffness of the wheelset within the bogie is optimized to achieve good stability, while retaining suitable curving performance, by allowing the wheelset to align with the curve to some extent [9.8].

FIGURE 9-2 Schematic plan view of a rigid bogie in a curve

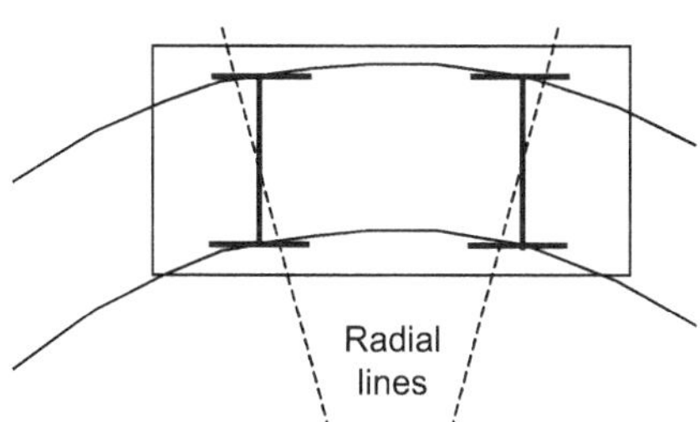

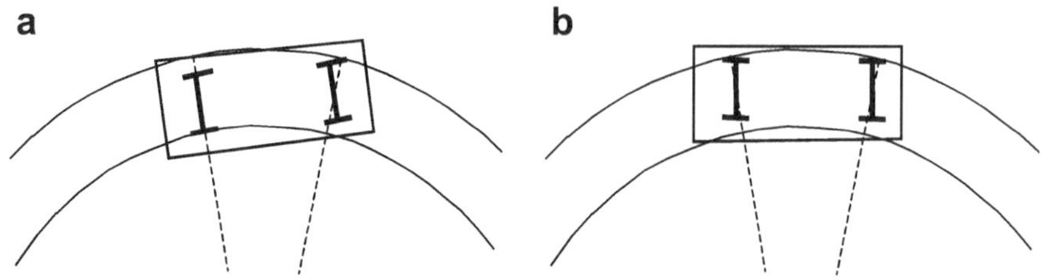

FIGURE 9-3 Schematic plan view of a bogie in a curve. (a) Low speed/small radius, (b) high speed/large radius

The attitude of a bogie in a curve depends on many factors, but particularly the train speed, the curve radius, the wheel and rail transverse profiles and the cant (or inclination) of the track. At low speeds, and in tight radius curves, the front wheelset tends to run in flange contact with the outer rail and the rear wheelset in flange contact with the inner rail, see Figure 9.3(a). As speed increases (or curve radius increases) the rear wheelset moves outwards, see Figure 9.3(b). The angle of attack between the front wheelset and the rail is thus greatest for sharper curves and low speed.

As the leading wheelset has a high angle of attack (yaw angle of the wheelset relative to the rail) this wheelset tries to roll straight ahead but instead is constrained to roll around the curve by the flange. This causes a relative lateral velocity to occur between the wheel and rail, as shown in Figure 9.4. The relative velocity normalized by the rolling velocity is termed the creepage (see Section 5.3). This produces a creep force (friction force) acting in the opposite direction to the relative motion. Typical force magnitudes acting on the four wheels of a bogie are shown schematically in Figure 9.5.

The wheel/rail contact point on the leading inner wheel of a bogie is located towards the field side of the tread. This wheel experiences a high lateral creepage, as shown in Figure 9.6. The leading outer wheel is in flange contact, with the resultant horizontal force acting inwards to ensure that the wheelset remains on the track. This force is a combination of the resolved normal load and an outward-acting creep force. Longitudinal and spin creep forces also act, as shown in Figures 9.5 and 9.6.

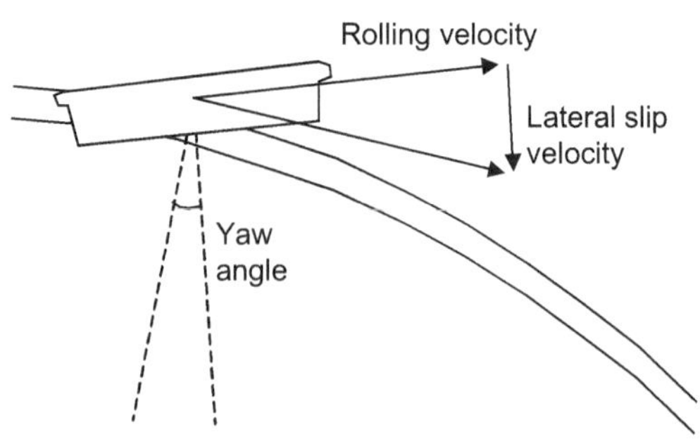

FIGURE 9-4 Generation of lateral creepage by a non-zero yaw angle

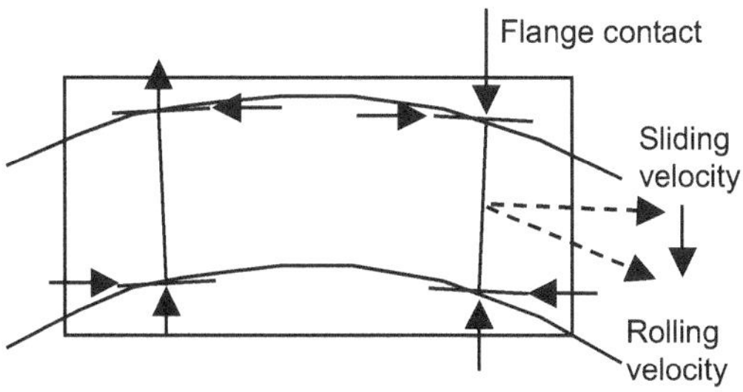

FIGURE 9-5 Schematic view of forces acting on wheels of a bogie in a curve

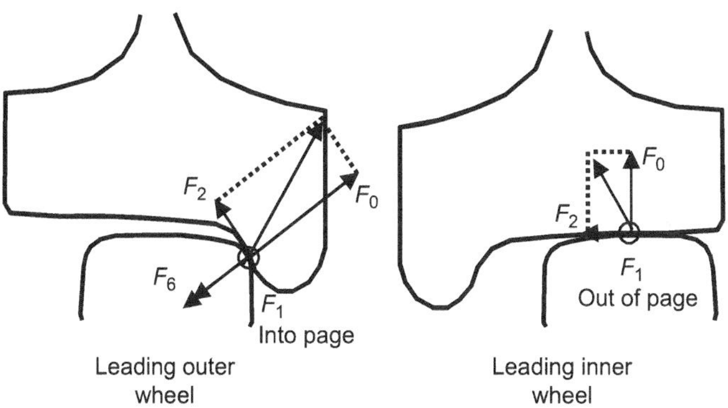

FIGURE 9-6 Schematic view of forces acting on front wheels of a bogie in a curve. F_0 is the normal load, F_2 is the lateral creep force, F_1 is the longitudinal creep force and F_6 is the spin moment

9.3 CREEP FORCES

Creep forces are essentially friction forces and have been described in Section 5.3. The saturation of creep force at high values of creepage is shown in Figure 5.12.

However, it is usually recognized that 'dynamic' or 'sliding' friction coefficients are smaller than 'static' ones ($\mu_d < \mu_s$). Usually, the friction coefficient depends on the sliding velocity, decreasing as the velocity increases. Thus, as creepage increases beyond the saturation point, the creep force once more reduces in amplitude, as shown schematically in Figure 9.7. It is this falling characteristic of the creep force at high creepage that is believed to be the main reason for the unstable dynamic behaviour leading to squeal noise, although there is some evidence that squeal can occur even if the creep force does not fall [9.10].

Vertical and lateral forces have been measured on a reduced-scale twin-disc rig [9.11, 9.12] and are plotted in Figure 9.8 as the ratio of lateral to vertical force (referred to here as the adhesion coefficient, α). They are plotted against the jaw angle in degrees. When expressed in radians this is equivalent to the lateral creepage, i.e. 0.4 degrees is equivalent to 7 mrad or a lateral creepage of 0.007.

Fingberg [9.13] and de Beer *et al.* [9.14, 9.15] use a model by Kraft [9.16] for the dependence of the sliding friction coefficient μ_d on the sliding velocity v:

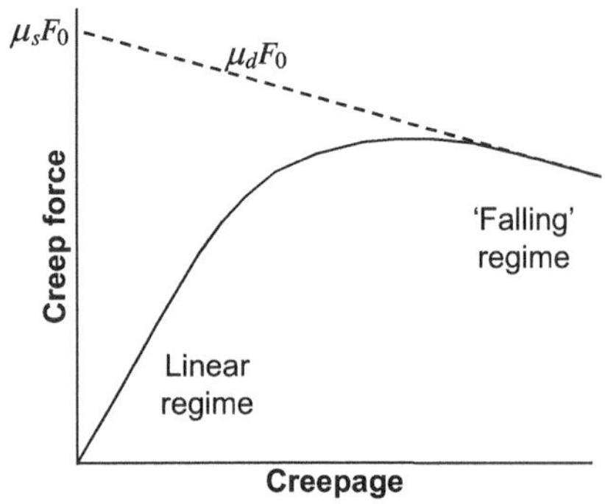

FIGURE 9-7 A typical creep force–creepage relationship for a velocity-dependent friction coefficient

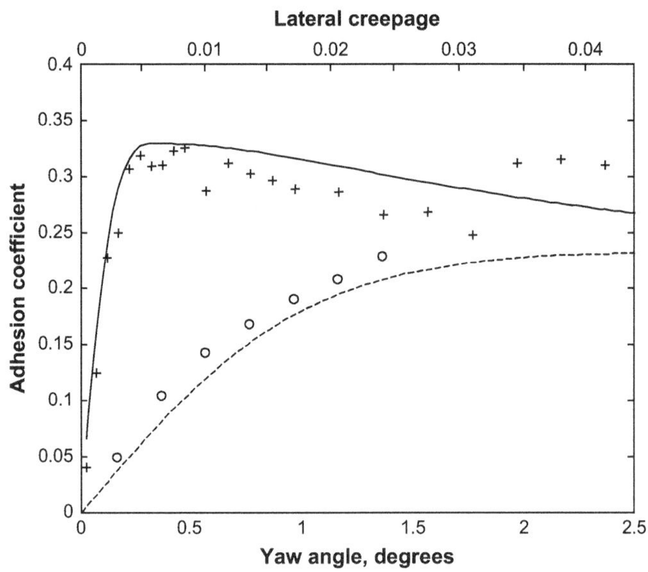

FIGURE 9-8 Lateral adhesion coefficient and lateral creepage in the stable contact position. +, measured, —, predicted, for no longitudinal creepage; o, measured, – – –, predicted, with $\gamma_1 = 0.02$. μ_s taken as 0.33 (from [9.12])

$$\mu_d(v) = \mu_s \left\{ 1 - 0.5e^{-0.138/|v|} - 0.5e^{-6.9/|v|} \right\} \tag{9.1}$$

where μ_s is the static coefficient of friction, and the sliding velocity v can be expressed as $v = \gamma V$ where γ is the creepage and V is the rolling velocity, although note that it is strictly a function of sliding velocity rather than creepage. This formula was derived theoretically but was adapted to fit experimental results. Figure 9.9, taken from [9.12], compares the results of this with various other models from the literature.

One of these is a model given by Rudd [9.1, 9.6]. This represents the adhesion coefficient α (creep force divided by normal load) as:

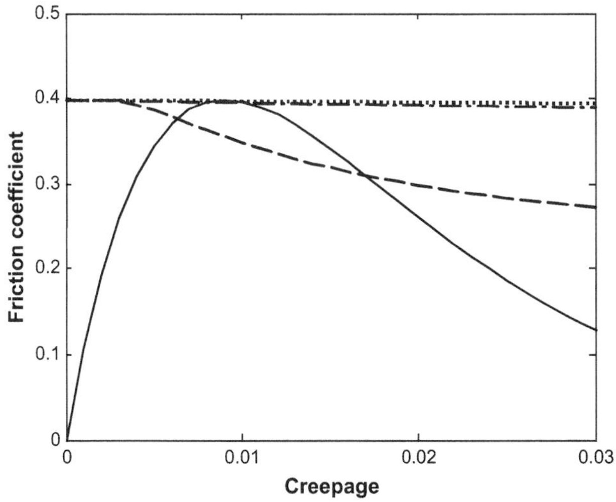

FIGURE 9-9 Comparison of friction or adhesion coefficients from different models of a wheel on a rail for a rolling velocity of 10 m/s. —, adhesion coefficient according to Rudd; – – –, sliding friction according to Kraft; · · · ·, sliding friction according to Poiré/Bochet; – · – ·, as used by Périard, according to Galton. In each case $\mu_s = 0.4$ (for Rudd's model $\mu_0 = 0.4$) (from [9.12])

$$\alpha(\gamma) = \mu_0 \frac{\gamma}{\gamma_0} \exp\left(1 - \frac{\gamma}{\gamma_0}\right) \tag{9.2}$$

where μ_0 and γ_0 are constants (chosen as 0.4 and 0.009, respectively) and γ is the creepage. Note that μ_0 is the maximum value of the curve, not the static friction coefficient. This model includes an approximation to the saturation of creepage as well as the reduction in friction at larger sliding velocities. The model was chosen by Rudd 'to get a feel for' the excitation of squeal [9.1] but comparisons with other models suggest that the negative slope is too steep. The other two models in Figure 9.9 are those considered by Périard [9.17]. They are empirical friction laws derived from measurements on sliding railway wheels at large sliding velocities and show only small reductions of the friction coefficient within the sliding speed range of interest for squeal.

Equation (9.1) can be seen to consist of two falling exponential terms, which can be expressed as:

$$\mu_d(v) = \mu_s\left\{\left(1 - e^{-0.138/|v|}\right) + \left(1 - e^{-6.9/|v|}\right)\right\}/2 \tag{9.3}$$

The first term introduces a falling characteristic at low sliding velocity; it reduces to half its amplitude at $v = 0.2$ m/s. The second term is only effective at large sliding velocity, reducing to half its amplitude at 10 m/s. For the prediction of squeal, only the first term is of relevance. However, the rate of decay in this term was set by Kraft arbitrarily as 50 times that in the second term. In [9.18, 9.19] a more general approach is used based on a single falling exponential, which can be written as

$$\mu_d(v) = \mu_s\left\{1 - \lambda e^{-\kappa/|v|}\right\} \tag{9.4}$$

where λ determines the ratio between the static friction coefficient μ_s and the limit at large sliding velocities, $\mu_d \to \mu_s(1-\lambda)$ and κ determines the rate of decay. From measurements on a 1/5 scale roller rig at about 1.2 m/s and a normal load of 1 kN, κ was found to be approximately 0.05 in [9.12] (for $\lambda = 0.5$). However, it is much more difficult to obtain measurements for a full-scale wheel, so it is unclear if the same values also apply to this case.

9.4 MODELS FOR FRICTIONAL EXCITATION

Curve squeal is a non-linear phenomenon involving effects of wheel and track dynamics and contact mechanics. Two main modelling approaches are possible: linearization of the contact equations and solution in the frequency domain or the solution of non-linear equations in the time domain.

The first approach allows details of the (linear) wheel and track dynamic systems to be included easily. However, it is only possible to determine whether or not the system is unstable, and at which frequency.

For the latter approach it is usually necessary to simplify the wheel and rail dynamics considerably to enable a solution to be obtained. However, it does allow the amplitudes of the vibration and hence noise to be obtained. This approach is illustrated first in the following section using a simple mass–spring system. This may be thought of as representing the dynamic behaviour of a single mode of the wheel.

9.4.1 Stick–slip motion due to difference in static and dynamic friction

The excitation of the wheel in squeal can be considered as a form of 'stick–slip' mechanism in the contact region. Stick–slip occurs in many situations where frictional excitation is present, for example a violin bow acting on the string. As an illustration of 'stick–slip', consider a solid body of mass M resting on a belt moving with speed v_0, as shown in Figure 9.10. It is held against the belt by a normal load N. A spring of stiffness K restrains the mass. The 'static' coefficient of friction is taken as μ_s and the 'dynamic' one as a constant value μ_d, which is assumed to be smaller than μ_s. Note that the velocity v_0 in this model corresponds to the quasi-static lateral sliding velocity of the wheel (i.e. $\gamma_2 V$), not the rolling velocity.

The static friction force at the contact between the body and the belt, $F \leq \mu_s N$, initially causes the body to move with the belt, as long as the friction force F is sufficient to balance the reaction force exerted by the spring, Ku, where u is the displacement. This is referred to as the 'stick' phase of the mechanism. The motion of the mass satisfies

$$u = v_0 t \tag{9.5}$$

As the spring becomes extended, the friction force is eventually overcome by the opposing force of the spring, when $Ku = \mu_s N$. Then the body slips relative to the belt. The motion of the body on the belt is resisted by a dynamic friction force $\mu_d N$ which is

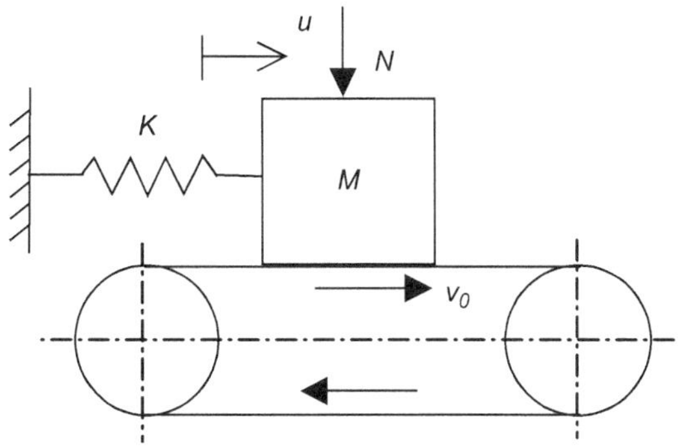

FIGURE 9-10 Schematic view of a constrained mass resting on a moving belt

less than the maximum static friction force. The equation of motion of the mass becomes

$$M\ddot{u} + Ku = \mu_d N \tag{9.6}$$

which has a solution in which the mass exhibits simple harmonic motion

$$u = A\sin(\omega_0(t - \tau)) + \mu_d N/K \tag{9.7}$$

where the amplitude A and lag τ are determined by the initial conditions (at the end of the stick phase). $\omega_0 = \sqrt{K/M}$ is the natural frequency of the mass on the spring.

The contacting surfaces will continue to slip until the relative velocity $(\dot{u} - v_0)$ returns to zero. The contacting surfaces then regain adhesion and the system returns to the 'stick' phase. Alternating stick and slip phases can become periodic.

An example of the displacement, velocity and friction force is shown in Figure 9.11. The amplitudes have been normalized: the displacement is normalized by v_0/ω_0, the velocity by the belt velocity v_0, and the friction force by the normal load N. The motion in the slip phase can be seen to be sinusoidal while that in the stick phase has a constant velocity.

To aid interpretation of these results, a non-dimensional factor β can be defined as

$$\beta = (\mu_s - \mu_d)\frac{N}{v_0 M \omega_0} \tag{9.8}$$

The form of the stick–slip motion depends on the value of this parameter. The situation shown in Figure 9.11 corresponds to a small value of β. Here the stick phase is short and the slip phase has a length almost equal to the natural period of oscillation of the mass on the spring. Thus the stick–slip motion has a similar (slightly lower) frequency to the natural frequency of the oscillator. This corresponds to a low friction difference, high belt velocity or high natural frequency (for a given mass). Since the motion is nearly sinusoidal, higher harmonics of the fundamental frequency are limited in amplitude.

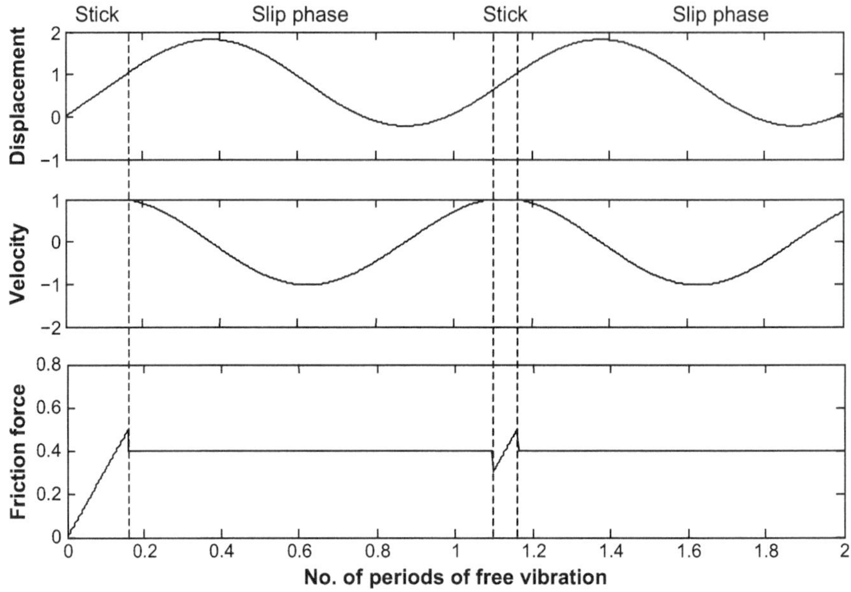

FIGURE 9-11 Normalized displacement, velocity and force characteristics of a simplified stick–slip mechanism: $\mu_s = 0.5$, $\mu_d = 0.4$, $\beta = 0.2$

Figure 9.12 shows a case where β is larger. Here the stick phase is much longer and the slip phase is approximately half a natural period. Thus the overall period is greater than the natural period and the stick–slip motion will have a lower frequency than the natural frequency of the oscillator. This corresponds to a high friction difference, low belt velocity or low natural frequency (for a given mass). The overall dependence of the period on the parameter β is shown in Figure 9.13.

The amplitude of the stick–slip motion is given by

$$A = \frac{v_0}{\omega_0}\sqrt{1 + \beta^2} \tag{9.9}$$

For small β it is approximately equal to v_0/ω_0, i.e. the velocity amplitude equals the steady-state sliding velocity, whereas for large β it is approximately equal to $(\mu_d - \mu_d)N/K$.

It is found that adding damping to the above situation (a damper in parallel with the spring) does not reduce the amplitude appreciably unless the damping is increased above a certain value when it eliminates the stick–slip. This occurs if the damping is sufficient to prevent the velocity returning to V at the end of the slip phase. The minimum damping ratio required to prevent stick–slip calculated using this model is shown in Figure 9.14. An approximate expression for this is found to be

$$\zeta \cong \frac{\beta^2}{4\pi} \tag{9.10}$$

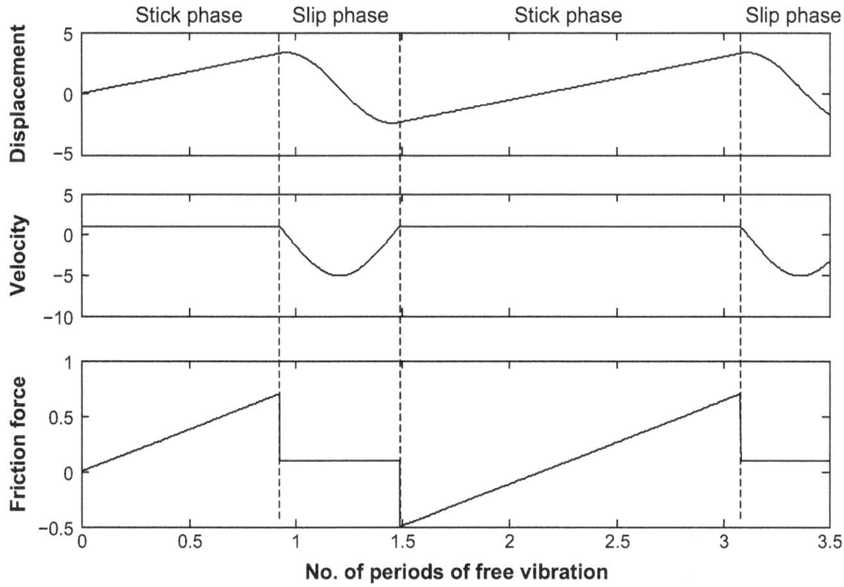

FIGURE 9-12 Normalized displacement, velocity and force characteristics of a simplified stick–slip mechanism: $\mu_s = 0.7$, $\mu_d = 0.1$, $\beta = 5$

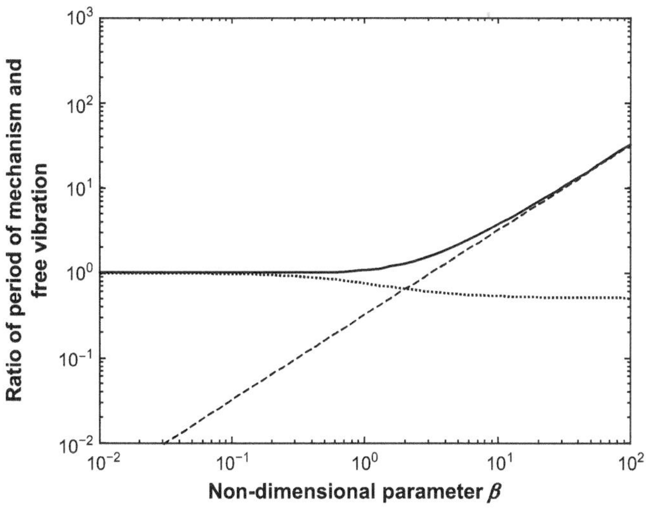

FIGURE 9-13 Ratio of the duration of the mechanism and the period of free vibration. —, period of stick-slip; – – –, length of stick phase; · · · ·, length of slip phase

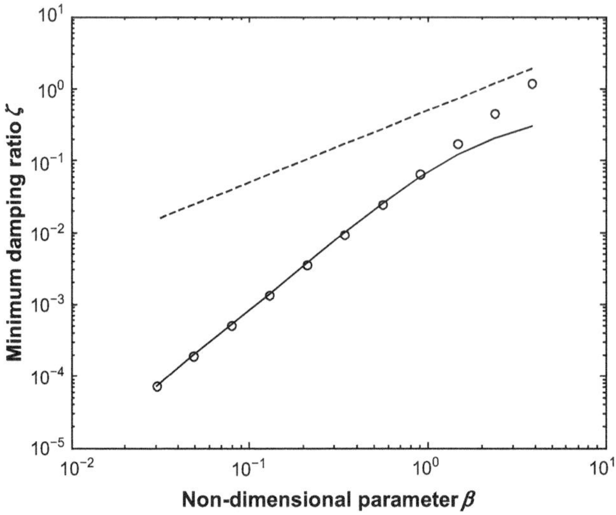

FIGURE 9-14 Minimum damping ratio required to prevent instability. —, numerical solution of damped stick–slip model; ○, approximation for low values of β; – – –, derived from Rudd's expression for the damping loss factor

Results estimated from Rudd's expression (see below) are also shown. This comparison is made by setting

$$\left|\frac{d\alpha}{d\gamma}\right| \cong \frac{\mu_s - \mu_d}{\gamma} \tag{9.11}$$

where α is the adhesion coefficient, which leads to

$$\zeta > \frac{\beta}{2} \tag{9.12}$$

It can be seen that this produces a much larger estimate of the damping required to overcome squeal than the above stick–slip model for small values of β.

This simplified analysis of stick–slip motion demonstrates

- the formation of a *limit cycle*; using different initial conditions leads eventually to the same stable periodic solution;
- the period of oscillation is not necessarily the same as the free period of oscillation of the system;
- as the periodic motion is not sinusoidal it will also exhibit higher harmonics of the fundamental frequency;
- damping can eliminate the stick–slip motion if it is greater than a certain minimum value; otherwise it has negligible effect on the amplitude of the limit cycle.

During curving, the wheel/rail contact can develop periodic motion similar to this stick–slip behaviour. However, it differs from the above since during the 'stick' phase the wheel is in rolling contact with the rail so that some creepage (relative motion) still exists. Moreover, the lateral mobility of a railway wheel exhibits many lightly damped resonances and curve squeal noise corresponds to the stick–slip response of one of these modes.

Substituting parameters for a railway wheel during typical curving, in which the mass–spring system represents a single mode of vibration of the wheel, gives values of β between about 0.1 and 1. This suggests that the stick–slip frequency will be close to the natural frequency of the wheel. This agrees with observations which show that the squeal frequency often corresponds closely to a wheel natural frequency.

9.4.2 Equivalent negative damper

The approach of studying the behaviour in the frequency domain, by linearizing the frictional behaviour, was first adopted by Rudd [9.1]. He described the squeal mechanism in terms of impedances. The change in sign of the slope of the creep force curve at high creepages, see Figure 9.9, is indicated by a 'negative real impedance' (negative damping).

Consider a case of a steady-state lateral creepage γ_2 with a small vibration velocity v_2 superimposed. For the steady-state case the creep force is

$$F_2 = F_0\alpha(\gamma_2) \tag{9.13}$$

where α is the adhesion coefficient, representing the creep curve (Figure 9.9) and F_0 is the normal load. For the case with superimposed vibration, the lateral force has a dynamic component f_2 and the creepage a dynamic component v_2/V. Thus:

$$\begin{aligned} F_2 + f_2 &= F_0\alpha(\gamma_2 + v_2/V) \\ &= F_0\,\alpha(\gamma_2) + F_0\frac{d\alpha}{d\gamma_2}\frac{v_2}{V} \end{aligned} \tag{9.14}$$

Hence the dynamic component of lateral force:

$$f_2 = \left\{\frac{F_0}{V}\frac{d\alpha}{d\gamma_2}\right\}v_2 \tag{9.15}$$

This is equivalent to a 'damper' with rate proportional to the slope of the creep curve. If the curve has negative slope, the damper has a negative rate. If this has a larger negative value than the (positive) damping of the wheel mode, self-excited oscillation (instability) can occur.

When the wheel response is unstable, the creep force and creepage oscillate about a point on the falling part of the creep curve with an increasing amplitude. Non-linearity in the adhesion coefficient prevents this amplitude from growing indefinitely and leads to oscillation with a stable amplitude (limit cycle).

9.5 MODELS FOR SQUEAL

9.5.1 Introduction

Theoretical models for curve squeal have been developed by various authors. As indicated by Rudd [9.1], instability of the lateral friction force is the most likely cause of squeal. Many of the subsequent models that appear in the literature adopt parts of Rudd's approach to the theoretical modelling of squeal.

Fingberg [9.13] and Périard [9.17] have extended this basic model by including better models of the wheel dynamics, the friction characteristic and the sound radiation from the wheel. Time-domain calculations allow the squeal magnitude to be predicted as well as the likelihood of squeal to be determined.

Heckl *et al.* present a theoretical investigation into the vibration response of an annular disc that is excited by an oscillating force, acting at the disc circumference, normal to the plane of the disc [9.20]. Heckl takes a numerical approach in the time domain. The disc response is described by its impulse response function (Green's function). The purpose of this work was to gain an understanding of the vibration response and modal behaviour of the disc under conditions where it generates squeal noise due to lateral creepage. The model does not allow for the geometric parameters of an actual wheel. The friction characteristic was assumed as a simplified bi-linear behaviour. In [9.21] Heckl presents a frequency-domain method using the modal parameters of a wheel to assess which modes of a friction-driven wheel will be unstable, indicated by their growth rate.

De Beer *et al.* [9.15] extended these models, based on excitation by unstable lateral creepage, to include feedback through the vertical force as well as through the lateral velocity. Their model consists primarily of a frequency-domain approach, used to determine instability and to predict which mode is most likely to be excited. This model has been extended further to allow for an arbitrary contact angle and to include lateral, longitudinal and spin creepage [9.19, 9.22]. This should allow it to be applied to flange squeal as well as squeal due to lateral creepage but validation of this aspect has been limited.

Chiello *et al.* [9.23] present a detailed model based on a time-domain approach. Huang [9.19] compares a time-domain approach with the frequency-domain stability analysis.

9.5.2 Linearized frequency-domain approach

The basic frequency domain model of de Beer *et al.* [9.15, 9.22] is presented here. The lateral contact force (tangential to the contact plane) can be written as $F_2 + f_2(t)$ where F_2 is the quasi-static force and f_2 is the time-varying oscillatory force. This can be written as the product of the normal force $F_0 + f_3(t)$, which again has a quasi-static part and a time-varying part, and the adhesion coefficient in the lateral direction, $\alpha_2(\gamma_2 + v_2(t)/V)$, with γ_2 the quasi-steady creepage and v_2 the time-varying relative velocity (due principally to wheel vibration). The adhesion coefficient also depends on the normal load. Thus:

$$\begin{aligned} f_2(t) &= \alpha_2\left(\gamma_2 + \frac{v_2(t)}{V}, F_0 + f_3\right)\left[F_0 + f_3(t)\right] - \alpha_2(\gamma_2, F_0)F_0 \\ &\approx \alpha_2(\gamma_2)f_3(t) + F_0\frac{\partial\alpha_2}{\partial f_3}f_3(t) + \frac{F_0}{V}\frac{\partial\alpha_2}{\partial\gamma_2}v_2(t) \end{aligned} \quad (9.16)$$

where this has been linearized by assuming that time-varying quantities are small and by ignoring terms of second order in small quantities.

The above expression relates the time-varying components of the forces to the time-varying relative velocity due to the rolling contact mechanics. In addition, the forces acting on the wheel, the track and the contact spring induce vibration. Assuming harmonic motion at frequency ω, these can be derived according to their mobilities:

$$-v_{w,2} = Y_{w,22}f_2 + Y_{w,23}f_3 \tag{9.17}$$

$$v_{r,2} = Y_{r,22}f_2 + Y_{r,23}f_3 \tag{9.18}$$

$$v_{c,2} = Y_{c,22}f_2 + Y_{c,23}f_3 \tag{9.19}$$

where v_w and v_r are the wheel and rail velocities which are positive for vibration in the positive coordinate directions and v_c is the contact deflection velocity which is positive for a compression of the contact spring (see Figure 5.10). $Y_{w,ij}$ are the wheel mobilities, $Y_{r,ij}$ are the rail mobilities and $Y_{c,ij}$ are the contact spring mobilities. Indices 2 and 3 indicate lateral and vertical translations. Unlike Chapter 5, the *contact* velocities and mobilities in equation (9.19) relate only to the contact spring and omit the effect of creepage, which is included here separately. The velocities at the contact must obey the continuity relations

$$v_2 = v_{w,2} - v_{r,2} - v_{c,2} \tag{9.20}$$

$$0 = v_3 = v_{w,3} - v_{r,3} - v_{c,3} \tag{9.21}$$

where v_2 is the sliding velocity which is positive for a positive creepage (i.e. the wheel moves in the positive coordinate direction). This can be used with equations (9.17–9.19) to write

$$-v_2 = Y_{22}f_2 + Y_{23}f_3 \tag{9.22}$$

where $Y_{ij} = Y_{w,ij} + Y_{r,ij} + Y_{c,ij}$ is the sum of wheel, rail and contact spring mobilities, and

$$0 = v_3 = Y_{32}f_2 + Y_{33}f_3 \qquad \text{or} \quad f_3 = -\frac{1}{Y_{33}}Y_{32}f_2 \tag{9.23}$$

This allows f_3 to be eliminated from equations (9.22) and (9.16):

$$v_2 = -\left(Y_{22} - \frac{Y_{23}Y_{32}}{Y_{33}}\right)f_2 = Gf_2 \tag{9.24}$$

and

$$f_2 = \frac{F_0}{V}\frac{\partial \alpha_2}{\partial \gamma_2}v_2 - \left(\alpha_2 + F_0\frac{\partial \alpha_2}{\partial f_3}\right)\frac{Y_{32}}{Y_{33}}f_2 = H_1 v_2 + H_2 f_2 \tag{9.25}$$

Equation (9.24) is identical to the model given in [9.15] apart from some simplifications there allowing the rail or contact mobilities to be neglected in some situations. The transfer function G contains information relating to the wheel and rail dynamics and the contact stiffness. The terms H_1 and H_2 in equation (9.25) are due to the creepage behaviour. H_1 is the usual feedback term between lateral velocity and lateral force, which depends on the slope of Figure 9.9, see equation (9.15). H_2 contains a term that is given in [9.15] as a relation between vertical and lateral forces and is proportional to α_2, and another term that is given in [9.22]. However, this latter term is only relevant in the unsaturated regime.

Equations (9.24) and (9.25) form a feedback loop with two branches, as shown in Figure 9.15. This has an open loop transfer function given by $GH_1 + H_2$. As is conventional in control theory, the stability of this loop can be tested by plotting the open loop transfer function on the Nyquist plane (i.e. plotting the imaginary part against the real part). If this encircles the point $+1$, the system is unstable and the frequency at which it crosses the real axis with the largest positive value is the frequency at which it will tend to become unstable. Note that the feedback here is positive; if it is negative, as is more usual, the Nyquist point is -1.

9.5.3 Discussion and results

It may be noted that G has the form $-(Y_{22} - Y_{23}Y_{32}/Y_{33})$. At a wheel resonance, the wheel mobility will usually be the largest contributor to Y_{ij}. Moreover, the wheel mobility will be dominated by a single mode and can be approximated by (see Chapter 4)

$$Y_{w,ij} \approx \frac{\psi_{ni}\psi_{nj}}{2m_n\zeta_n\omega_n} \tag{9.26}$$

where ψ_{ni} is the mode shape of the nth mode for direction i, m_n is the modal mass, ζ_n is the modal damping ratio and ω_n is the excitation frequency which is equal to the resonance frequency. It can then be seen that the terms of the form $Y_{w,22} - Y_{w,23}Y_{w,32}/Y_{w,33}$ are equal to zero. This is only the case for the terms of G, however, if $Y_{w,33}$ is large compared with $Y_{r,33}$ and $Y_{c,33}$. Consequently, the modes which are excited in squeal can be expected to have a small component of their mode

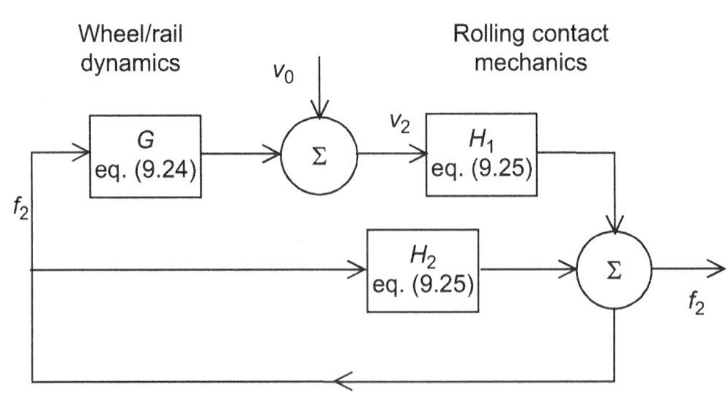

FIGURE 9-15 Feedback loop based on equations (9.24) and (9.25)

shape normal to the surface (so that Y_{33} is *not* dominated by the wheel), while the tangential component is large (so Y_{22} *is* dominated by the wheel).

In [9.19, 9.22] a multi-dimensional version of the above model is presented. This includes the longitudinal and spin creepages as well as that in the lateral direction. Again, this model consists of a feedback loop, but it has a vector/matrix form. However, unlike the previous section, there is no single parameter that can be used to test the feedback gain. Instead, the open loop gain is formed by a 3 × 3 matrix, $[\mathbf{T}] = [\mathbf{G}][\mathbf{H}_1] + [\mathbf{H}_2]$.

The criterion for stability of such a multiple input/multiple output system is that the determinant of the matrix $[\mathbf{I} - \mathbf{T}]$ should not enclose the point 0 on the Nyquist plane. If it does enclose this point, the loop of the locus that crosses the real axis furthest from the origin on the negative axis corresponds to the frequency with the greatest instability. This criterion can be further refined by finding the eigenvalues of the matrix $[\mathbf{T}]$ and plotting these on the Nyquist plane. The eigenvalues that encircle the point +1 are unstable.

Some example results are presented in this section from [9.22] for a UIC standard freight wheel (920 mm diameter). The calculated modes of vibration of this wheel were shown in Figure 4.3.

The wheel mobilities were calculated from a modal summation using natural frequencies and modeshapes calculated by a finite element (FE) model and modal damping obtained from measured data. Note that the natural frequencies differ slightly from those listed in Figure 4.3 as a more refined FE mesh was used. The lateral track mobility was calculated using a multiple beam model from [9.24]. Figure 9.16 compares the magnitude of the lateral point mobilities of the wheel,

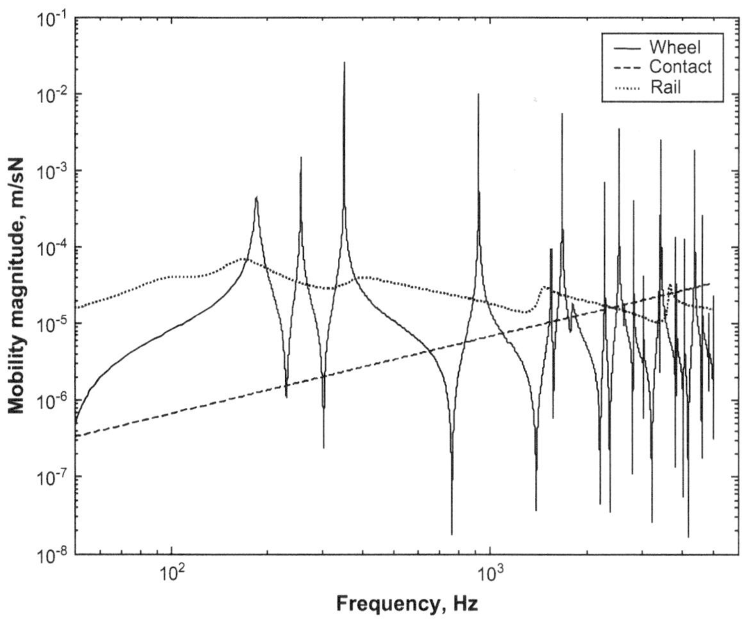

FIGURE 9-16 Lateral point mobilities of the wheel, contact and rail

contact stiffness and the rail at the wheel/rail interface. The dominant peaks are the fundamental axial wheel modes with 0 nodal circles, which correspond to the first row of Figure 4.3.

Stability results have been calculated using steady creepages found previously from vehicle dynamics simulations. Adhesion coefficients and their derivatives were then obtained using the CONTACT software [9.25]. The results are presented here for the case of the leading inner wheel of the bogie where the wheel tyre makes contact on the head of the rail.

The solution of the open loop transfer functions (equations (9.24) and (9.25)) produces a loop gain. However, for the multi-dimensional model the matrix equation produces three complex eigenvalues. The eigenvalue with the largest real part at each frequency from 50 to 10 000 Hz is plotted as a Nyquist plot in Figure 9.17. There are several unstable modes, as a number of loops encircle the point +1. These correspond closely to axial modes of the wheel with 0 nodal circles and n nodal diameters. The real part of open loop transfer function at the frequency at which the curve crosses the real axis is referred to as the loop gain.

By plotting the loop gain against frequency, the unstable wheel modes can be identified, as shown in Figure 9.18. The highest peak at 350 Hz corresponds with the largest contour in the Nyquist plot and is the wheel mode with 2 nodal diameters and 0 nodal circles, i.e. mode (2,0). This is the mode most likely to squeal under these conditions.

The effect on stability due to contact position on the wheel tyre can be investigated by tracking the loop gain of each unstable mode over a range of lateral

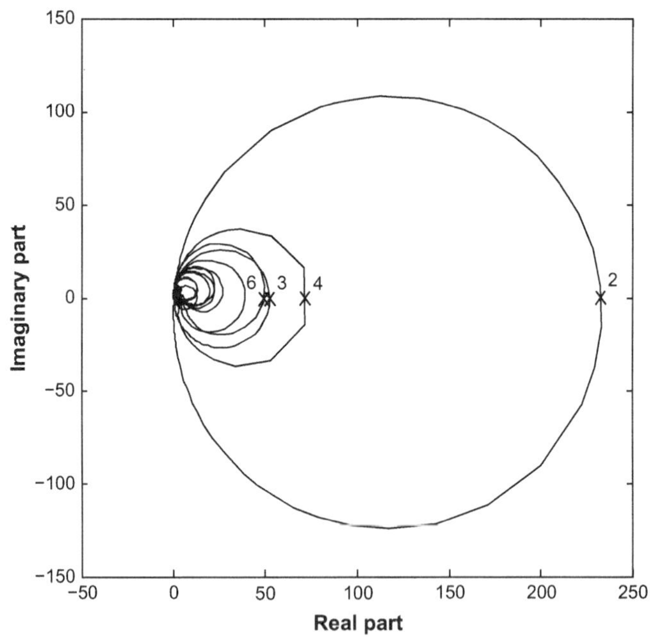

FIGURE 9-17 Nyquist plot of open loop transfer function with +10 mm offset on the head of the rail and 0 mm on the wheel tread. The positions of crossing the real axis and the number of nodal diameters in the four most unstable axial modes are labelled

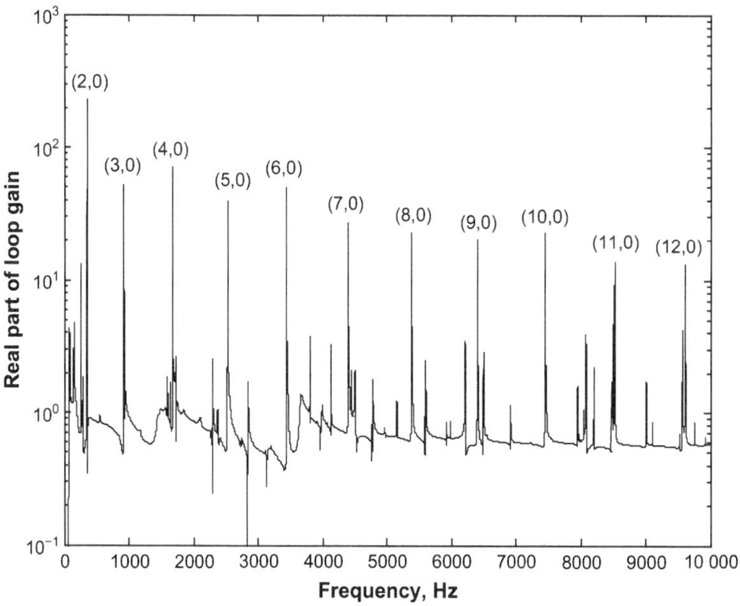

FIGURE 9-18 Spectrum of real part of the loop gain (eigenvalue with the largest real part) for the inner wheel with a contact position at 0 mm

offsets across the tread of the wheel. This is shown in Figure 9.19. The (2,0) mode has the largest loop gain over much of the range of contact positions. The response of this wheel appears especially prone to squeal when there is an offset of between -5 and $+15$ mm on the wheel tread. In other words, squeal is most likely to occur if contact on the wheel tread has shifted across to the field side of the tread. This behaviour is seen on the leading inner wheel of a bogie negotiating curved track, see Figure 9.6. If, on the other hand, the contact position on the wheel is fixed at the nominal running point and the contact on the head of the rail is offset incrementally by up to 20 mm either side of the centreline of the rail, the maximum loop gain does not change significantly.

Results for contact at the wheel flange indicate that predominantly radial modes of the wheel (with $n \geq 2$) are most likely to be excited here [9.22]. These modes are plotted in the second row of Figure 4.3 and have natural frequencies of about 2 kHz and above. These modes have a large component of vibration in the radial direction at the flange. Several modes are found to have similar loop gains. This may account for the higher frequency content of flange squeal and its more multi-tonal character.

In [9.19] more extensive results are given using both the frequency-domain and time-domain approaches. Comparison of results from these two models suggests that the frequency-domain model gives a correct assessment of the instability provided that the creepage is large. However, if the creepage is small, so that the creep force is close to the turning point of the adhesion coefficient, the frequency-domain model predicts instability based on the local slope of the adhesion coefficient whereas the time-domain model predicts that no squeal occurs.

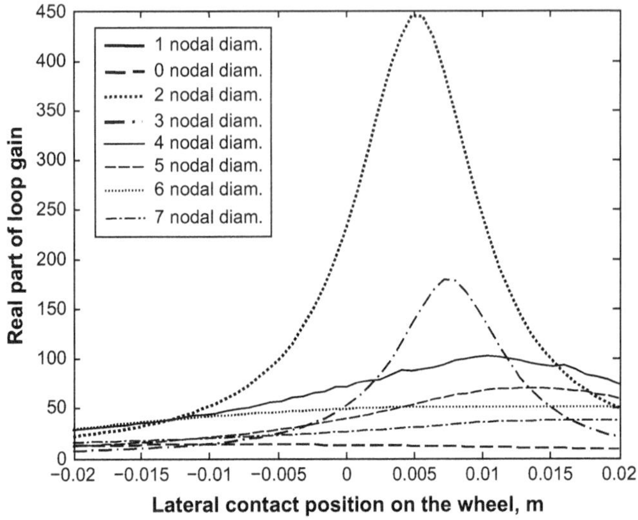

FIGURE 9-19 Real part of the loop gain (eigenvalue with the largest real part) for the inner wheel at various contact positions on the wheel

9.5.4 Summary

It has been shown that squeal is a problem of instability, or self-excited oscillations, which is similar to 'stick–slip'. The resulting wheel vibration radiates noise as squeal. The rail is also excited by the same force and so it will also vibrate at the squeal frequency. However, the rail has a much lower mobility at the wheel resonance frequency (see Figure 9.16) so its response to the force is much smaller. It is therefore responsible for only a small proportion of the noise.

Squeal is usually excited most at the front inner wheel of a bogie. This occurs as the front wheelset usually has the larger angle of attack, the front *outer* wheel is usually in flange contact (flange excitation is a separate phenomenon) while the front inner wheel has a lower normal load so creep saturates more easily than on the outer wheel. 'Axle wind-up' can also occur if the rolling radius difference is insufficient. This gives a low frequency stick–slip in the longitudinal direction but does not excite squeal.

The cause of squeal is the large yaw angle of the front wheelset relative to the rail. This can be between $\psi = b/2R$ and $\psi = b/R$ where b is the wheelbase and R is the curve radius. Taking an intermediate value of $\psi = 0.7\, b/R$, and noting that the creepage $\gamma_2 = \psi$ for steady-state curving and that the creep curve falls for typically $\gamma_2 > 0.007$ (see Figure 9.8) this corresponds to Rudd's rule of thumb of $R/b < 100$ for squeal to occur.

To predict the amplitude of squeal, a time-domain model would be required. The analysis of stick–slip of a mass on a belt has shown that the vibration velocity amplitude is approximately equal to the belt velocity. This corresponds to the creepage multiplied by the rolling velocity. Thus it may be expected that the squeal amplitude will increase with train speed, see also [9.4]. On the other hand, the curving behaviour, and therefore the lateral creepage, is also dependent on the speed (see Figure 9.4) so that the occurrence of squeal may actually reduce as speed is increased in some situations.

9.6 MITIGATION MEASURES FOR CURVE SQUEAL NOISE

In discussing solutions for curve squeal it is of little value to quote decibel reductions. A level reduction says more about the severity of squeal in the starting conditions than about the general effectiveness of a solution. The nature of the instability is such that effective measures should *eliminate* the squeal rather than reduce it. Screening, such as barriers or vehicle mounted shrouds, is therefore not suitable for reducing squeal noise.

It should also be borne in mind that curve squeal tests are very difficult to reproduce due to a high sensitivity to parameters such as temperature, humidity, train speed, track geometry and wheel and rail wear.

Many possible treatments for squeal noise have been developed over the years with varying success. The purpose of this section is to describe the main types of treatment, identifying where possible the mechanism by which they operate as well as advantages and disadvantages. The choice of a treatment for a particular situation will require other considerations to be taken into account. Comprehensive surveys of treatments are given in [9.26, 9.27 and 9.28].

9.6.1 Improved curving behaviour

For new vehicles effective solutions can be sought in the design of the vehicles for curving. In particular, steerable axles can be used in order to reduce the creepages during curving. A system of mechanical linkages was used on a tram in Rotterdam [9.27] and found to reduce the peak level by about 18 dB and the average level by about 7 dB. Unfortunately, this type of design is often in conflict with the requirements for stability at higher speed. On these particular vehicles the mechanism was isolated after some years use.

Other steering mechanisms include cross-bracing between the axles of a bogie [9.8], see Figure 9.20. Such 'radially steered wheelsets' are fitted, for example, to the

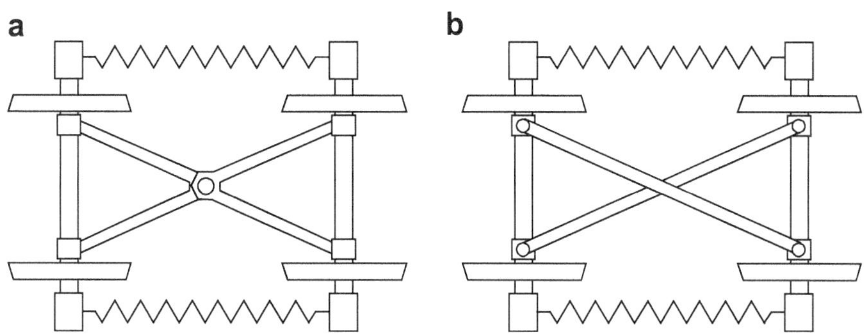

FIGURE 9-20 Examples of bogies with steerable wheelsets. (a) Pivoted arrangement, (b) cross-braced bogie

X2000 in Sweden and ICN in Switzerland to reduce track forces and allow higher curving speeds [9.28] and this can be expected to reduce squeal.

9.6.2 Asymmetrical rail profiles

Asymmetrical rail head profiles are described in [9.29]. These were introduced to reduce wear, and it is reported that they extend the lifetime of the rail from about five to about eight years [9.28]. However, it has also been suggested that they would avoid curve squeal. The purpose of these profiles is to ensure that the correct rolling radius difference is generated while two-point contact on the outer wheel involving the flange is avoided. Although applied at several locations in Austria, observations suggest that no significant reduction of squeal occurs [9.28].

Modified profiles may also be considered to take advantage of the sensitivity of the wheel to the lateral position of the contact point. Wheels have been shown to be more likely to squeal if excited further towards the field side than if excited near the flange [9.15]. However, this principle has not yet been tried in practice.

Gauge narrowing has been used to reduce squeal [9.26, 9.28]. This also has the effect of moving the contact on the inside wheel nearer the flange. In the Netherlands a test with a 7 mm reduction in gauge on a tram curve resulted in a reduction of 22 dB [9.28].

9.6.3 Lubrication and friction modifiers

Known solutions for curve squeal include lubrication using either grease or water. These work by reducing the coefficient of friction, and thereby also the difference between static and sliding friction coefficients. If lubricants are used, it must be ensured that they do not lead to loss of adhesion as this could compromise safety. Grease reduces the coefficient of friction to below 0.2 and is therefore normally only applied to the rail gauge corner or wheel flange. Although this may not be the primary cause of squeal noise, it is possible that it reduces the occurrence of squeal by modifying the curving behaviour, i.e. reducing the angle of attack.

Although lubrication can be applied manually, effective treatments are available based on track-mounted systems that apply grease as necessary on curves using wiping bars. Vehicle-mounted systems are also used. Such systems are normally used to reduce wear rather than to eliminate squeal, but noise reduction is a useful side effect.

Curve squeal is known to occur much less during wet weather. Water mist sprays have therefore also been used effectively to eliminate squeal. At one location at Schaffhausen in Switzerland this technique has been used for over 30 years [9.28]. These have the advantage that they do not pollute the ground. They also have only a temporary effect on adhesion levels. The main disadvantage is the need for measures to deal with periods of frost: either heating or the need to drain the system.

Friction modifiers act by reducing or eliminating the falling friction characteristic without reducing the level of friction too much [9.30, 9.31]. In some cases the friction modifier actually gives a friction characteristic that continues to rise with increasing creepage. An example of friction characteristics measured on a 1/5 scale wheelset in lateral creepage is given in Figure 9.21 [9.32]. These were carried out at a rolling speed of 10 km/h (2.8 m/s). Results from the model of equation (9.4) are shown for $\lambda = 0.5$ and $\kappa = 0.05$ which can be seen to fit the measurements for dry

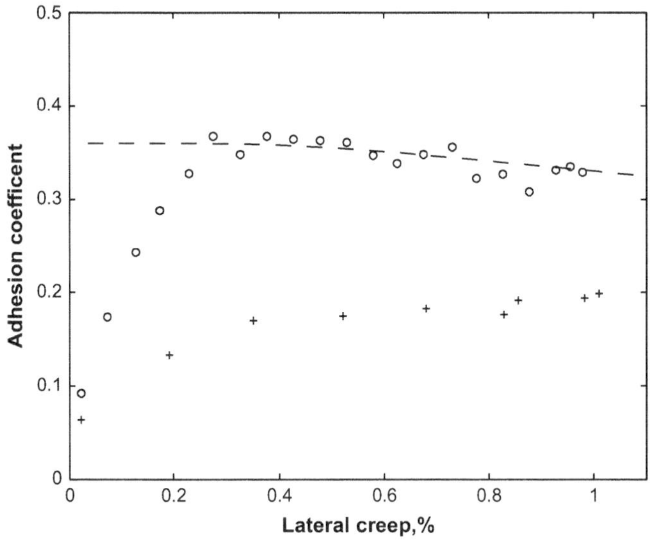

FIGURE 9-21 Measurements of adhesion coefficient in lateral creep measured on 1/5 scale rig. o, clean contact condition; +, after application of HPF friction modifier [9.32]; – – –, model of equation (9.4)

contact. The friction modifier gives a reduced friction characteristic which continues to increase with increasing sliding velocity.

Friction modifiers are available as solid sticks. These can be applied on the vehicle to rub on the wheel tread and/or flange. More recently, they have become available as water-based liquids [9.30] which can be applied either to the track at the entrance to a curve [9.30], see Figure 9.22, or from the vehicle using a spray system [9.33]. They

FIGURE 9-22 Wiping bars used to apply friction modifier in a curve. Photo: Kelsan, used by permission

have been shown to be very effective in eliminating squeal and can be applied to the top of the railhead without compromising traction or braking as the friction coefficient is around 0.35 [9.30].

9.6.4 Wheel damping

Wheel damping treatments are also known to reduce the occurrence of squeal and are commonly used. As seen above, squeal occurs due to the instability of the vibration in a particular wheel resonance. A small increase in the level of damping can often be sufficient to eliminate squeal and produce large noise reductions. The noise reduction is therefore not simply proportional to the level of damping introduced.

Many different means of introducing damping to the wheel have been implemented, as described in Section 7.3. Tuned absorbers consisting of a series of tongues with damping material between them are effective in eliminating or reducing squeal. Other wheel dampers include the 'shark fin' absorber.

Constrained layer damping treatments have also been used effectively to increase the damping of wheels and thereby eliminate squeal. These consist of a thin layer of visco-elastic material, with a stiff backing layer, that are attached to the wheel web. In order to facilitate forming the layer to the shape of the wheel web, particularly on curved webs, the backing layer is often aluminium with a thickness of only 1 to 2 mm. Nevertheless, this is sufficient to add considerable damping to the wheel [9.34]. In [9.26] reductions of 5 to 15 dB are mentioned for constrained layer damping treatments.

Resilient wheels, in which the wheel tyre is mounted on the web via rubber blocks, have been used in tram systems for many years. These introduce a certain amount of additional damping but do not eliminate squeal in all situations.

In addition, ring dampers have been used as a simple means of increasing the damping of a wheel [9.35, 9.36]. These consist of a metal ring that sits in a groove in the inner edge of the wheel tyre. It is believed that friction between the ring and the groove leads to energy dissipation. Although the damping of the wheel is not increased as much as with some of the other treatments discussed above, ring dampers are effective in reducing squeal in some situations. In [9.26] reductions of 5 to 10 dB are mentioned.

9.6.5 Rail damping

Treatments that can be applied locally at sites where squeal occurs will often be more attractive than measures that must be applied to whole vehicle fleets. Applying damping to the rail therefore appears to be a useful method, if it is successful. Indeed, some installations have appeared successful. However, it is unclear how it works, since the rail mobility is not strongly resonant and is hardly affected by added damping treatments. The noise from the rail is undoubtedly reduced, but this is not a major contributor to the overall level.

9.7 CASE STUDY: UK SPRINTER FLEET

In the mid-1980s new Class 150 Sprinter diesel multiple units (DMUs) were introduced onto a wide variety of routes, replacing 1950s designs. A severe squeal noise problem occurred on some curving routes, with high amplitude sustained tones in the region 2 to 5 kHz. Particular problems arose on the Llandudno to Blaenau Ffestiniog line in North Wales. The 14 km section from Betws-y-Coed to Ffestiniog tunnel has many reverse curves of radius 160 to 240 m with no transition curves. The track was fitted with check rails (restraining rails on the inside of the 'low' rail of the curves).

The squeal noise generated was apparently audible at distances up to 12 km away; levels inside the vehicle (with the window open) reached up to 100 dB(A), while close to the wheels noise levels of 130–145 dB were measured. These levels were up to 20 dB greater than usually found in the literature for curve squeal.

An initial investigation by BR Research in 1987 determined that the tones generated corresponded not to the zero-nodal-circle modes (usual in curve squeal) but to radial modes of the wheel. It was concluded that these were excited by contact between the back of the wheel flange and the check rails at the leading inner wheel. The modes of this type of wheel are shown in Figure 9.23. The radial modes have a large radial component of motion at the back of the flange and a small radial component at the tread near the field side. They can therefore be readily excited by the check rail while not being restrained or damped by the contact with the running rail.

The problem is thought to have occurred due to a combination of a stiffer yaw suspension introduced for stability (as this was an all-purpose vehicle with a higher maximum speed than the old stock), a wheel shape giving modes which were easily excited by the check rail and the use of monobloc wheels with low damping.

Track-based or vehicle-based lubrication was not considered suitable due to potential pollution of the local river, which the route follows closely. The removal of

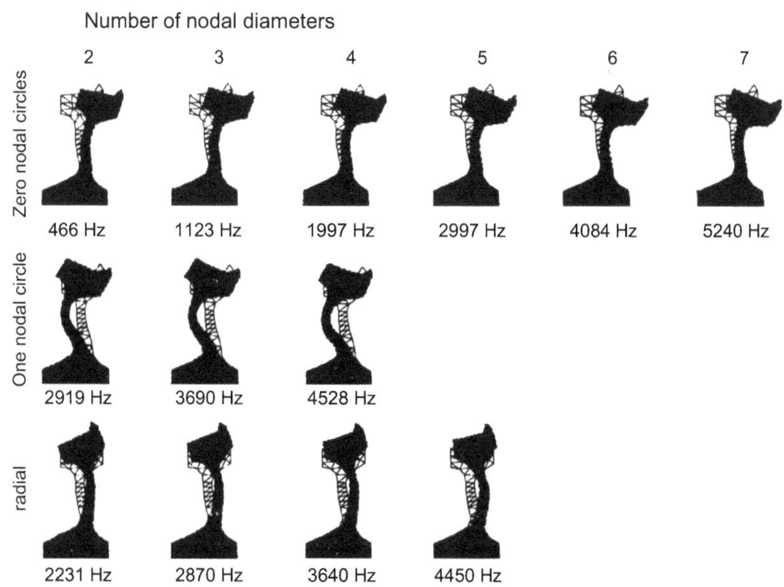

FIGURE 9-23 Modes of vibration of Class 150 wheel

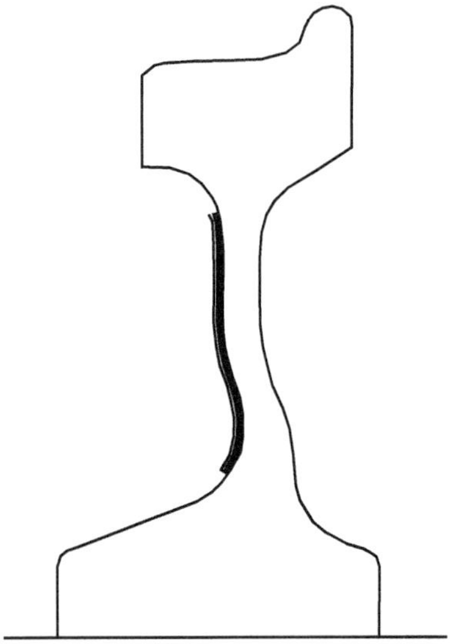

FIGURE 9-24 Constrained layer damping as applied to Class 150 DMU wheel

the check rail was not allowed for safety reasons as at this time the line also carried nuclear flask traffic. Modification of the bogies was not possible as these new vehicles were used on diagrams across the network rather than being captive to the branch line.

The chosen solution comprised a constrained layer damping treatment applied to the wheels, shown in Figure 9.24. This was shown to work in laboratory tests and subsequently in field tests in 1988. The constrained layer damping treatment was eventually fitted to the whole Class 15x DMU vehicle fleet. This treatment is effective in eliminating curve squeal, although with hindsight it could have been better optimized in order to reduce rolling noise as well.

REFERENCES

9.1 M.J. Rudd. Wheel/rail noise – Part II: Wheel squeal. *Journal of Sound and Vibration*, **46**, 381–394, 1976.

9.2 B. Müller. Curve squeal WP2 Inventory of the extent of the problem. Union Internationale des Chemins de Fer, internal report prepared by SBB, 2003.

9.3 B. Müller and J. Oertli. Combatting curve squeal: monitoring existing applications. *Journal of Sound and Vibration*, **293**, 728–734, 2006.

9.4 N. Vincent, J.R. Koch, H. Chollet, and J.Y. Guerder. Curve squeal of urban rolling stock – Part 1: State of the art and field measurements. *Journal of Sound and Vibration*, **293**, 691–700, 2006.

9.5 S.L. Grassie and J. Kalousek. Rail corrugation: characteristics, causes and treatments. *Proceedings of the Institution of Mechanical Engineers*, **207F**, 57–68, 1993.

9.6 P.J. Remington. Wheel/rail squeal and impact noise: what do we know? What don't we know? Where do we go from here? *Journal of Sound and Vibration*, **116**, 339–353, 1987.

9.7 D.T. Eadie, M. Santoro, and J. Kalousek. Railway noise and the effect of top of rail liquid friction modifiers: changes in sound and vibration spectral distruibutions in curves. *Wear*, **258**, 1148–1155, 2005.

9.8 A.H. Wickens. *Fundamentals of Rail Vehicle Dynamics, Guidance and Stability*. Swets & Zeitlinger, Lisse, 2003.

9.9 J.A. Elkins and R.J. Gostling. A general quasi-static curving theory for railway vehicles. *Proceedings of 5th IAVSD Symposium on the dynamics of vehicles on roads and on railway tracks*, 388–406, 1977.

9.10 J.R. Koch, N. Vincent, H. Chollet, and O. Chiello. Curve squeal of urban rolling stock – Part 2: Parametric study on a 1/4 scale test rig. *Journal of Sound and Vibration*, **293**, 701–709, 2006.

9.11 P.P. Kooijman, W.J. van Vliet, M.H.A. Janssens, and F.G. de Beer. Curve squeal of railbound vehicles (Part 2): set-up for measurement of creepage dependent friction coefficient. *Proceedings of Internoise 2000*, Nice, France, vol. 3, 1564–1567, 2000.

9.12 A.D. Monk-Steel, D.J. Thompson, F.G. de Beer, and M.H.A. Janssens. An investigation into the influence of longitudinal creepage on railway squeal noise due to lateral creepage. *Journal of Sound and Vibration*, **293**, 766–776, 2006.

9.13 U. Fingberg. A model of wheel-rail squealing noise. *Journal of Sound and Vibration*, **143**, 365–377, 1990.

9.14 F.G. de Beer, M.H.A. Janssens, P.P. Kooijman, and W.J. van Vliet. Curve squeal of railbound vehicles (Part 1): frequency domain calculation model. *Proceedings of Internoise 2000*, Nice, France, vol. 3, 1560–1563, 2000.

9.15 F.G. de Beer, M.H.A. Janssens, and P.P. Kooijman. Squeal of rail bound vehicles influenced by lateral contact position. *Journal of Sound and Vibration*, **267**, 497–507, 2003.

9.16 K. Kraft. Der Enfluss der Fahrgeschwindigkeit auf den Haftwert zwischen Rad und Schiene. *Archiv für Eisenbahntechnik*, **22**, 58–78, 1967.

9.17 F. Périard. Wheel-rail noise generation: curve squealing by trams. PhD thesis, Technische Universiteit Delft, 1998.

9.18 Z.Y. Huang, D.J. Thompson, and C.J.C. Jones. Squeal prediction for a bogied vehicle in a curve. *Proceedings of 9th International Workshop on Railway Noise*, Feldafing, Germany, September 2007.

9.19 Z. Huang. Theoretical modelling of railway curve squeal. PhD thesis, University of Southampton, 2007.

9.20 M.A. Heckl and I.D. Abrahams. Curve squeal of train wheels, Part 1: Mathematical model for its generation. *Journal of Sound and Vibration*, **229** (3), 669–693, 2000.

9.21 M.A. Heckl. Curve squeal of train wheels, Part 2: Which wheel modes are prone to squeal? *Journal of Sound and Vibration*, **229** (3), 695–707, 2000.

9.22 A.D. Monk-Steel and D.J. Thompson. Models for railway curve squeal noise. *Proceedings of Eighth International Conference on Recent Advances in Structural Dynamics*, Southampton, 13–16 July 2003.

9.23 O. Chiello, J.B. Ayasse, N. Vincent, and J.R. Koch. Curve squeal of urban rolling stock – Part 3: Theoretical model. *Journal of Sound and Vibration*, **293**, 710–727, 2006.

9.24 T.X. Wu and D.J. Thompson. Analysis of lateral vibration behavior of railway track at high frequencies using a continuously supported multiple beam model. *Journal of the Acoustical Society of America*, **106**, 1369–1376, 1999.

9.25 J.J. Kalker. *Three Dimensional Elastic Bodies in Rolling Contact*. Kluwer Academic Publishers, Dordrecht, 1990.

9.26 J.T. Nelson. Wheel/rail noise control manual. TCRP Report 23, 1997.

9.27 F. Krüger. *Schall- und Erschutterungsschutz im Schienenverkehr*. Expert Verlag, Renningen, 2001.

9.28 B. Müller, E. Jansen, and F.G. de Beer. Curve squeal WP3 tool box of existing measures. Union Internationale des Chemins de Fer, report prepared by SBB and TNO, 2003.

9.29 E. Kopp. Fünf Jahre Erfahrungen mit asymmetrisch geschliffenen Schienenprofilen. *Eisenbahn Technische Rundschau*, **40**, 665, 1991.

9.30 D.T. Eadie, M. Santoro, and W. Powell. Local control of noise and vibration with Keltrack friction modifier and Protector trackside application: an integrated solution. *Journal of Sound and Vibration*, **267**, 761–772, 2003.

9.31 D.T. Eadie and M. Santoro. Top-of-rail friction control for curve noise mitigation and corrugation rate reduction. *Journal of Sound and Vibration*, **293**, 747–757, 2006.

9.32 A. Matsumoto, Y. Sato, H. Ono, Y. Wang, M. Yamamoto, M. Tanimoto, and Y. Oka. Creep force characteristics between rail and wheel on scaled model. *Wear*, **253**, 199–203, 2002.

9.33 Y. Suda *et al.* Development of onboard friction control. *Wear*, **258**, 1109–1114, 2005.

9.34 C.J.C. Jones and D.J. Thompson. Rolling noise generated by wheels with visco-elastic layers. *Journal of Sound and Vibration*, **231**, 779–790, 2000.

9.35 P. Wetta and F. Demilly. *Reduction of wheel squeal noise generated on curves or during braking*. 11th International Wheelset Congress, Paris, 1995. 301–306

9.36 J.F. Brunel, P. Dufrénoy, M. Naït, J.L. Muñoz, and F. Demilly. Transient models for curve squeal noise. *Journal of Sound and Vibration*, **293**, 758–765, 2006.

CHAPTER 10

Impact Noise

10.1 INTRODUCTION

As everyone knows, trains go 'clickety-clack'. In fact this familiar noise, caused by impacts at rail joints, is increasingly a thing of the past as continuously welded rail has largely eliminated bolted rail joints. Nevertheless, impact loads at the wheel/rail interface still remain an important source of noise.

Rail joints have not been completely eliminated, as jointed track remains in use on many secondary and freight routes. There are also significant discontinuities in the rail running surface at switches and crossings. Even on continuously welded rail, the welds between two lengths of rail are often not smooth and can cause impacts as the wheel rolls over them. Discontinuities still occur at expansion joints and at insulated joints, used to separate track sections electrically. Finally, wheel defects are another source of impacts. These include wheel flats, caused by wheel sliding, and other out-of-roundness [10.1].

In each case, the excitation mechanism is similar to rolling noise: a vertical discontinuity in the running surface causes a relative motion of the wheel and rail, leading to vibration and noise. The most apparent difference from rolling noise is the discrete nature of the event and therefore the modified character of the noise. In addition, and perhaps more importantly for the modelling, these discrete events may have large amplitudes. Since the contact spring between the wheel and rail is non-linear, a linear approximation, such as was made in Chapter 5, is no longer valid for large deflections. In extreme cases, momentary loss of contact can occur and the contact force tends to zero.

Although an early analysis of impact noise is given in [10.2], based on simple analytical models and scale model experiments, much less effort has been focused on this problem than on rolling noise. Impact loading has been studied more widely from the point of view of potential track damage [10.3–10.7] but then the frequency range of interest is more limited.

As non-linearities have to be taken to into account, the frequency-domain approach adopted in the rolling noise modelling can no longer be used and it is necessary to use a time-domain approach. Such time-domain models, incorporating the non-linearities in the contact zone, have been used, for example, by Clark *et al.* [10.4, 10.5] and Nielsen and Igeland [10.6]. In [10.8] a finite element model of a turnout is used to calculate wheel impacts at a crossing. These models contain large numbers of degrees of freedom to represent the track. Nevertheless, they are mostly limited to a maximum frequency of around 1–2 kHz. While this is sufficient for the purposes of calculating the effects of impact loading, a range up to around 5 kHz is required for acoustic analysis. This would require models of the track with even more degrees of freedom. Moreover, the contribution of wheel modes of vibration at high frequency would need to be included instead of using a lumped mass model.

The models described in this chapter, based on [10.9, 10.10, 10.11], adopt a hybrid approach. In this, simplified models of the wheel and rail are used first in a time-stepping model in order to determine the effects of the non-linearities at the contact zone. Although these models use only a small number of degrees of freedom they are representative of the response at the driving point for a wide frequency range. On the other hand, they do not allow the calculation of the track response at other positions along the rail. The results from this model are then used to determine an equivalent roughness, which can be used to determine the sound radiation using a more complete, but linear, wheel and track model in the frequency domain. It is important that roughness, and not the contact force, is used to transfer the loading from one model to the other. If the contact force were used this would lead to excessive predictions of wheel vibration, not accounting correctly for the damping effect of the wheel/rail interaction (see Section 5.2.5).

10.2 THE EFFECT OF NON-LINEARITIES ON ROLLING NOISE

10.2.1 Wheel/rail interaction

Before studying impact loading, it is useful to consider the effects of non-linearities on the excitation of rolling noise. For this, a simple model of the wheel and rail is used with a 'moving irregularity' approach.

The Hertzian contact spring has a non-linear behaviour due the geometry of the wheel/rail contact, as discussed in Section 5.3. Similar to equation (5.5) but now in terms of displacement rather than velocity, the deflection (compression) of the contact spring can be written as

$$u_c = u_w - u_r + r \tag{10.1}$$

where u_r is the rail displacement, u_w is the wheel displacement (both positive downwards) and r is the roughness forming a relative displacement input between the wheel and rail (positive for an asperity). As the compression is increased, the contact area increases in size and the spring becomes stiffer; as the compression is reduced, the stiffness reduces until total unloading can occur. Assuming a Hertzian contact model, the contact force F is given by

$$F = \begin{cases} C_H u_c^{3/2}, & u_c > 0 \\ 0, & u_c \leq 0 \end{cases} \tag{10.2}$$

where C_H is the Hertzian constant, which depends on the radii of curvature of the surfaces and their material properties. From equation (5.45), C_H can be given as

$$C_H = \frac{2E'\sqrt{R_0}}{3}\left(\frac{2}{\xi}\right)^{3/2} \tag{10.3}$$

where $E' = E/(1 - \nu^2)$ is the plane strain elastic modulus, R_0 depends on the radii of curvature of the two bodies and is given by equation (5.40) and ξ is given by equation (5.46) and listed in Table 5.1. If the relative motion in the contact spring u_c varies in a small range, the above equation can be linearized, as described in Section 5.3. Figure 10.1 shows the non-linear behaviour of equation (10.2) for typical contact geometry and linear approximations for two nominal preloads, 25 and 50 kN. At a given relative deflection the difference between the linear and non-linear models can be seen to be larger for a lower preload.

If, instead, the linear approximation to the contact stiffness of equation (5.47) is used, the wheel/rail interaction model can be written in linear form, $F = K_H u_c$, which also allows the use of a frequency-domain approach.

10.2.2 Simple wheel and track models

In order to examine the effects of the linear approximation, a simple model has been used in which the wheel is represented by a mass [10.9] or a mass and spring [10.12], as shown in Figure 4.12. The point mobility predicted using this model was shown in Figure 4.11. Although the details of the wheel resonances at high frequencies are not included, the simple mass–spring model gives a reasonable representation of the wheel mobility up to 1 kHz, and represents the average behaviour above this.

To represent the track, a model with a minimum number of degrees of freedom is used. Only the point response at the excitation point is required, not the response over the length of the rail. The track can therefore be replaced by an equivalent state space system. This has a frequency response given by

$$H(s) = \frac{b_1 s^3 + b_2 s^2 + b_3 s + b_4}{s^4 + a_1 s^3 + a_2 s^2 + a_3 s + a_4} \tag{10.4}$$

where a_i and b_j are constants and $s = i\omega$. $H(i\omega)$ is the point receptance of the equivalent system. The coefficients a_i and b_j are chosen to ensure that $H(i\omega)$ fits

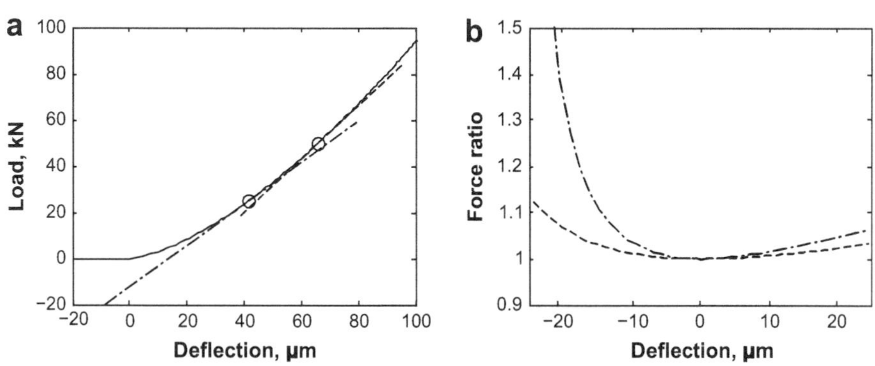

FIGURE 10-1 (a) Non-linear force–deflection relation (—) showing linear approximations at two preloads, (b) ratios of the non-linear contact force to the equivalent linear one. – · – ·, static load $F_0 = 25$ kN; - - - -, static load $F_0 = 50$ kN [10.9]

TABLE 10-1 **COEFFICIENTS FOR USE IN EQUATIONS (10.4, 10.5)**

a_1	1.77×10^3	b_1	3.28×10^{-6}
a_2	1.26×10^7	b_2	1.87×10^{-2}
a_3	7.87×10^9	b_3	23.6
a_4	3.93×10^{12}	b_4	3.97×10^4

closely to the receptance of a track, modelled using a Timoshenko beam on a continuous spring–mass–spring foundation, see Chapter 3. Table 10.1 lists values of the coefficients a_i and b_i for a track with UIC60 rails, a pad of stiffness 583 MN/m^2 per unit length, loss factor 0.25 and a sleeper mass of 270 kg/m [10.9].

This system can also be written in the time domain as

$$\left(\partial_t^4 + a_1\partial_t^3 + a_2\partial_t^2 + a_3\partial_t + a_4\right)u_r(t) = \left(b_1\partial_t^3 + b_2\partial_t^2 + b_3\partial_t + b_4\right)F(t) \tag{10.5}$$

where ∂_t represents differential operator d/dt. $u_r(t)$ and $F(t)$ are the rail displacement and wheel/rail interaction force, respectively. This can be expressed as a series of first-order differential equations in the form [10.13]

$$\begin{Bmatrix} \dot{x}_1 \\ \dot{x}_2 \\ \dot{x}_3 \\ \dot{x}_4 \end{Bmatrix} = \begin{pmatrix} 0 & 1 & 0 & 0 \\ 0 & 0 & 1 & 0 \\ 0 & 0 & 0 & 1 \\ -a_4 & -a_3 & -a_2 & -a_1 \end{pmatrix} \begin{Bmatrix} x_1 \\ x_2 \\ x_3 \\ x_4 \end{Bmatrix} + \begin{Bmatrix} c_1 \\ c_2 \\ c_3 \\ c_4 \end{Bmatrix} F(t) \tag{10.6}$$

where $x_1 = u_r$ and

$$c_1 = b_1;\; c_i = b_i - \sum_{j=1}^{i-1} a_{i-j}c_j \text{ for } i = 2,\,3,\,4 \tag{10.7}$$

If the wheel is represented by a mass, its displacement $u_w = x_5$ satisfies

$$\begin{Bmatrix} \dot{x}_5 \\ \dot{x}_6 \end{Bmatrix} = \begin{pmatrix} 0 & 1 \\ 0 & 0 \end{pmatrix} \begin{Bmatrix} x_5 \\ x_6 \end{Bmatrix} + \begin{Bmatrix} 0 \\ F_0/M_w \end{Bmatrix} + \begin{Bmatrix} 0 \\ -1/M_w \end{Bmatrix} F(t) \tag{10.8}$$

where F_0 is the nominal preload applied to the wheel. This can be extended for the mass–spring model of the wheel (Figure 4.12), see [10.11]. For numerical convenience an additional small mass is then attached on the underside of the spring.

The system of equations (10.2), (10.6) and (10.8) can be solved in a time-stepping procedure, for example using a Runge Kutta method. Note that the wheel and track are represented by linear models, only the contact spring containing non-linearity. The results from this model are compared with those based on using a linear contact spring in place of the non-linear one in equation (10.2).

10.2.3 Results

In [10.9, 10.12] calculation results are presented for a range of broad-band roughness spectra. Figure 10.2 shows four notional spectra of wheel/rail roughness

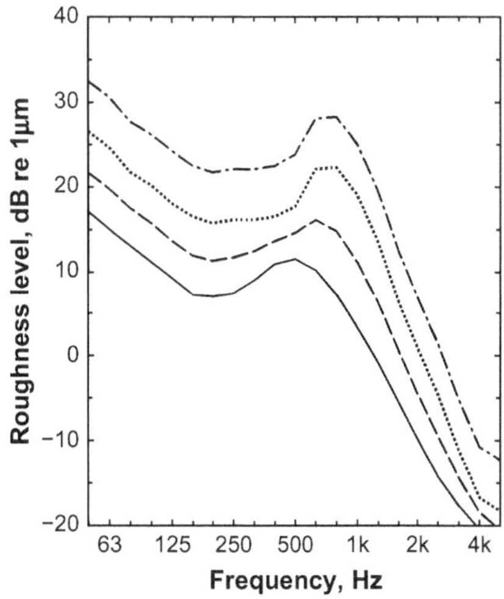

FIGURE 10-2 Assumed one-third octave band roughness spectra. —, cast-iron block-braked wheel on good track at 100 km/h; – – –, intermediate roughness; ····, corrugated rail roughness at 140 km/h; – · – ·, extreme roughness [10.9]

that have been used for the calculations. The solid line represents the roughness of cast-iron block-braked wheels on good quality track. The dotted curve corresponds to a corrugated track. An 'intermediate' roughness is shown mid-way between the first two curves, and an 'extreme' roughness is used which has twice the amplitude of the corrugated track. All these spectra include the effects of the contact filter, which reduces the effectiveness of short wavelength roughness. Summing these spectra for the frequency bands 250 Hz and above, the r.m.s. amplitudes of the roughness, r_{rms}, are 7.6, 13.2, 25 and 50 μm, respectively.

Roughness time histories were generated from the one-third octave band roughness spectra by first creating equivalent narrow-band spectra with random phase at each frequency and then applying an inverse Fourier transform. These time histories $r(t)$ were used as inputs to the wheel/rail system. From this calculation a time history of the wheel/rail interaction force was obtained. This was then transformed into the frequency domain and converted to a one-third octave band force spectrum for ease of presentation. In Figure 10.3, the results of this process using the non-linear model are compared with results from the equivalent linear model.

These results, for the corrugated track roughness, are for seven different values of static wheel load, F_0. Typically, a passenger vehicle has a wheel load of around 50 kN, whereas that for a freight vehicle can vary between about 25 kN, when empty, and over 100 kN, when laden. Wheel loads less than 25 kN are also included in the calculations to illustrate further the dependence on load.

The difference between the linear and non-linear contact force spectra in one-third octave bands is shown in Figure 10.4 for the four levels of roughness. In each case, results are shown for seven values of wheel load, F_0. It is found that the linear and non-linear results are similar for low roughness amplitudes and large wheel loads, giving differences close to 0 dB, but the differences increase for large roughness and/or small loads.

For the tread-braked wheel roughness, Figure 10.4(a), the two models agree very closely for all wheel loads considered. For the moderate roughness, Figure 10.4(b),

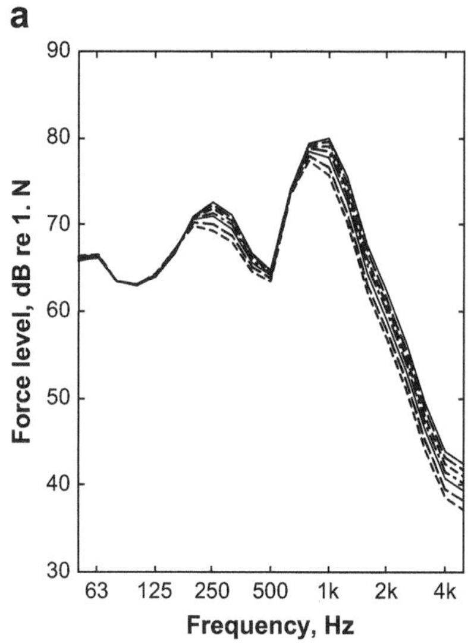

FIGURE 10-3 One-third octave band spectra of the wheel/rail contact force for corrugated track roughness, (a) from the linear model, (b) from the non-linear model. From bottom to top, static load $F_0 = 10$, 15, 25, 35, 50, 70, 100 kN [10.12]

deviations are found for the two lowest wheel loads, although these are lower than would occur in practice. For the corrugated track roughness, Figure 10.4(c), the differences are small for wheel loads of 50 kN and above, but significant for loads less than that. For the extreme roughness, Figure 10.4(d), the differences are significant for all loads except 100 kN.

As the wheel load is reduced, the static deflection of the contact spring becomes smaller, and loss of contact becomes more likely. In Figure 10.5(a), the average absolute difference between the results of non-linear and linear models, averaged over all frequency bands 50 to 5000 Hz, is plotted against r_{rms}/u_0, the ratio of the r.m.s. amplitude of the roughness to the static deflection of the contact spring. For all four roughness spectra considered, when this ratio is less than about 1/3 the differences are small, whereas when it is greater than about 1/2 the differences increase considerably.

In Figure 10.5(b), the percentage of the time over which loss of contact occurs is shown for the same cases, again plotted against the ratio r_{rms}/u_0. This shows a similar shape, although it increases from a slightly higher value of r_{rms}/u_0. This indicates that non-linear behaviour is significant when loss of contact is occurring and just before it starts to occur, but is not significant otherwise, even though the dynamic contact force may reach peak values of two or three times the static value before loss of contact occurs.

10.2.4 Discussion

In [10.9] it was shown that the non-linear effects are significant at frequencies where (i) the dynamic contact spring deflection u_c is a large fraction of the roughness r and (ii) the roughness itself is large. Non-linear effects can be neglected for frequencies below 100 Hz, even though the roughness inputs at these frequencies may be quite large, as the contact spring deflection u_c is only a small fraction of the

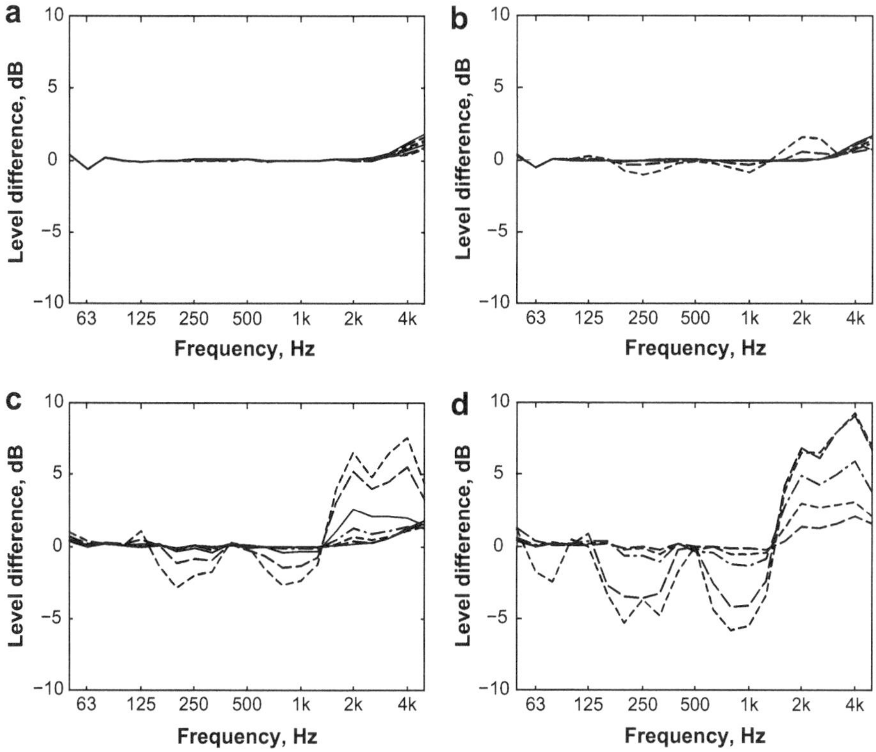

FIGURE 10-4 Difference in one-third octave band spectra of the wheel/rail contact force from the non-linear model compared to the linear model. From largest deviations to smallest, static load $F_0 = 10$, 15, 25, 35, 50, 70, 100 kN. (a) Moderate roughness, (b) intermediate roughness, (c) corrugated rail roughness, (d) extreme roughness [10.12]

roughness input at these frequencies. Larger effects were found for excitation around 200 and 900 Hz. These frequencies correspond to peaks in the contact force for a given roughness input (compare Figure 5.3). Differences can be seen at these frequencies in Figures 10.4(c) and (d). For frequencies above 1 kHz, although $u_c \approx r$ (since $|Y_c| > |Y_w + Y_r|$ for most frequencies above 1 kHz), the roughness amplitude is generally considerably smaller than at lower frequencies. The differences seen at these higher frequencies in Figure 10.4 are due to harmonics of the large excitation around 1 kHz on the corrugated rail, introduced by the non-linear response.

It can be concluded, therefore, that the non-linear contact spring between the wheel and the rail has only a small effect on the wheel/rail response for normal levels of surface roughness. The effects of non-linearity increase as the roughness level increases or the normal load reduces. Significant differences between the results of linear and non-linear models occur when the r.m.s. roughness amplitude (above 250 Hz) exceeds about one-third of the static deflection of the contact spring. This condition corresponds closely to the onset of loss of contact, so that it is concluded that loss of contact is more important to the breakdown of the linear model than the non-linear spring stiffness. The effect on A-weighted noise levels is found to be

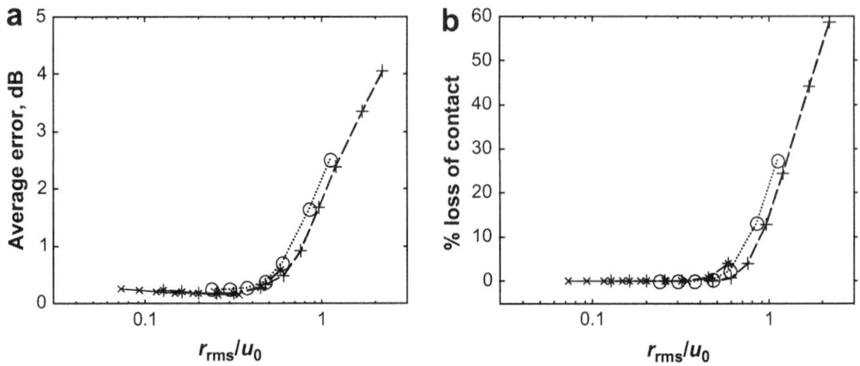

FIGURE 10-5 (a) Average difference in one-third octave band spectra of the wheel/rail contact force from the non-linear model compared to the linear model plotted against ratio of r.m.s. roughness amplitude to static contact deflection. (b) Percentage of time that contact is fully unloaded. ×, moderate roughness; ∗, intermediate roughness; ○, corrugated rail roughness; +, extreme roughness [10.12]

negligible for practical situations [10.12]. However, larger amplitude excitations of the wheel/rail system occur at discrete discontinuities such as wheel flats and rail joints. These are considered in the next sections.

10.3 IMPACT NOISE DUE TO WHEEL FLATS

10.3.1 Contact force

A wheel flat is an area of the wheel tread that has been worn flat, as shown schematically in Figure 10.6(a). This usually occurs due to locking of the brakes under poor adhesion conditions at the wheel/rail contact, for example due to leaves on the railhead during the autumn. Wheels with flats produce high levels of noise, as well as impact loading of the track, which can lead to damage of track components. Most railway administrations have an inspection limit in terms of the length of a wheel flat above which the wheel must be removed from service. Such limits are

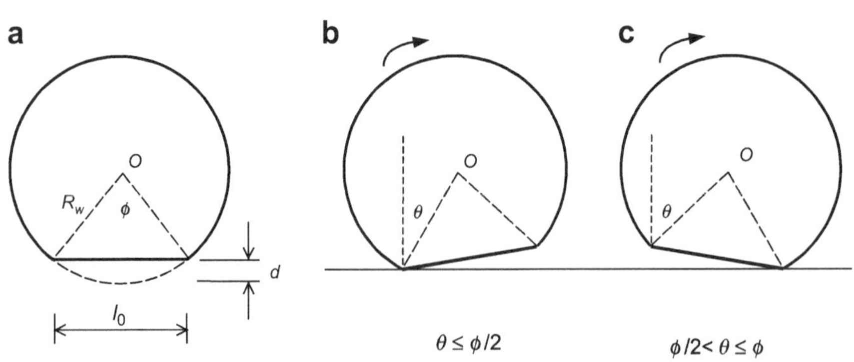

FIGURE 10-6 Rolling of a wheel with an idealized flat [10.10]

typically in the range 40 to 60 mm [10.1]. Typical flats can therefore be up to around 50 mm long, in extreme cases up to 100 mm. After their initial formation, flats rapidly become 'worn', i.e. rounded at their ends due to the high load concentration at the corners. A worn flat of a given depth is longer than the corresponding 'new' flat and is often represented by a cosine function [10.4].

Wheel flats introduce a relative displacement input to the wheel/rail system in the same way as roughness. Roughness has the same effect whether it is located on the wheel or rail surface. In the same way, an indentation of the rail can be envisaged that is equivalent to a wheel flat. The profile shape can be seen to correspond to a circular arc dip in the railhead, as shown in Figure 10.7. The length, l_0, and depth, d, are related by

$$d = \frac{l_0^2}{8R_w} \tag{10.9}$$

where R_w is the wheel radius. However, due to the geometry of the wheel and rail surfaces, the actual displacement input is modified by the wheel curvature in the same way as for roughness, see Section 5.6.2. For the idealized flat shown in Figure 10.6(a), the wheel first pivots downwards on the front corner of the flat, Figure 10.6(b), then pivots upwards again on the rear corner, Figure 10.6(c) [10.10]. The quasi-static motion of the wheel centre is shown in Figure 10.7. This can be considered as the resulting relative displacement input experienced by the wheel/rail system. In [10.10] it is shown that the displacement input due to a worn wheel flat can be represented by a curve of a similar shape to that in Figure 10.7 but elongated.

The response of the wheel and rail can be calculated using the same simple model as described in the previous section. In [10.10] the wheel is represented by a mass and a spring (see Figure 4.12). Figure 10.8 shows examples of the calculated response to a large wheel flat: a new flat of depth 2 mm (length 86 mm). The contact force has a static value of 100 kN.

When the indentation (relative displacement input due to the wheel flat) appears between the wheel and rail, the wheel starts to fall and the rail rises. Since the wheel and rail cannot immediately follow the indentation due to their inertia, the contact force is reduced. Full unloading first occurs at a train speed of 30 km/h, Figure 10.8(a).

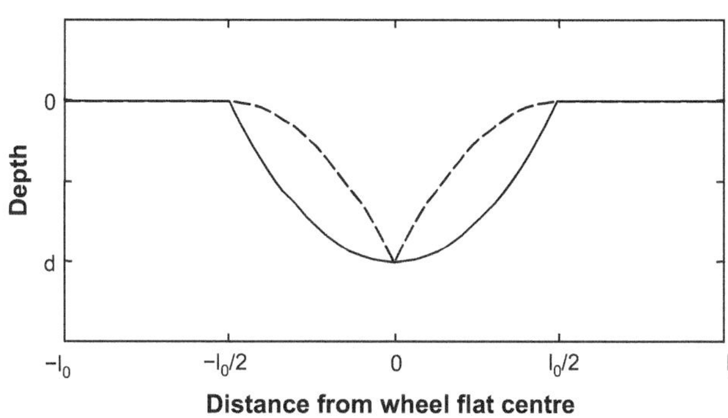

FIGURE 10-7 Wheel flat geometry for new flat of length l_0 and depth d. —, profile shape; – – –, after geometric filtering [10.10]

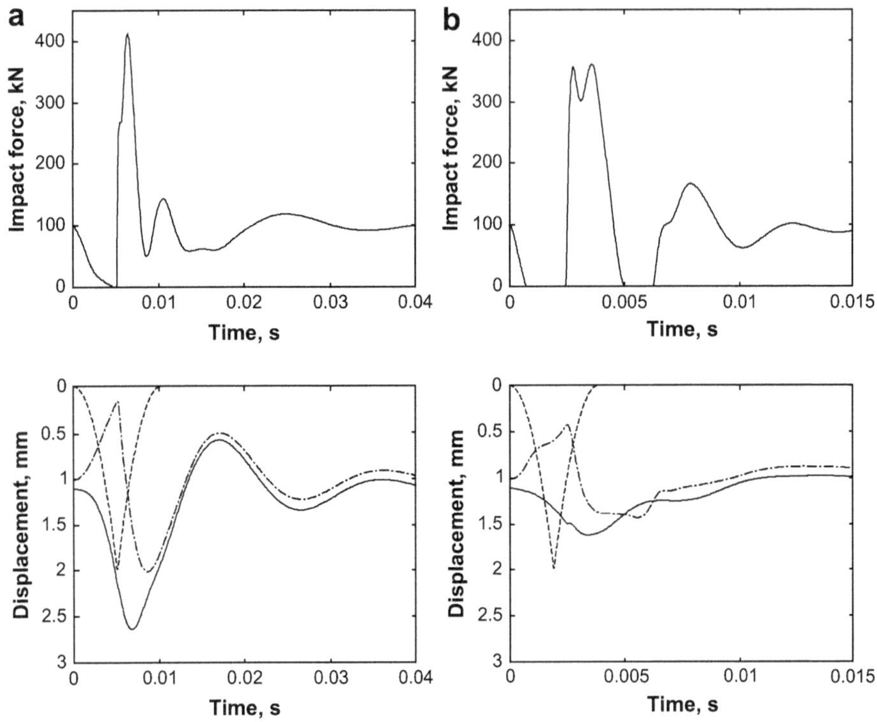

FIGURE 10-8 Wheel/rail interaction force (upper) and displacements of wheel and rail (lower) due to 2 mm newly formed wheel flat (length 86 mm). (a) At train speed 30 km/h, (b) at 80 km/h. —, wheel displacement; – · – ·, rail displacement; - - -, relative displacement excitation [10.10]

In the second half of the flat, as the wheel pivots about the trailing edge of the flat, the relative displacement input reduces again. However, the wheel and rail inertia cause them to continue to move towards each other. The contact force therefore increases rapidly until it reaches its peak. This can be considered as the 'impact'. The peak force in this example is about four times the static load. As the speed increases, contact is lost for longer distances during the unloading phase. At 80 km/h, Figure 10.8(b), a second loss of contact can be seen to occur as the wheel bounces off the rail after the initial impact. The second impact is much smaller than the first one.

Calculation results for a rounded flat of the same depth but overall length 121 mm indicate that the speed at which loss of contact first occurs increases from 30 to about 50 km/h [10.10]. The maximum contact force is plotted against train speed in Figure 10.9 for several cases. The maximum contact force is similar for new and rounded flats of a given depth, but occurs at different speeds.

10.3.2 Noise prediction

The above process is sufficient to estimate the contact forces, including the effects of contact non-linearity. The next stage is to use these results to estimate the radiated noise. For this, the response of the wheel and rail are required, incorporating details

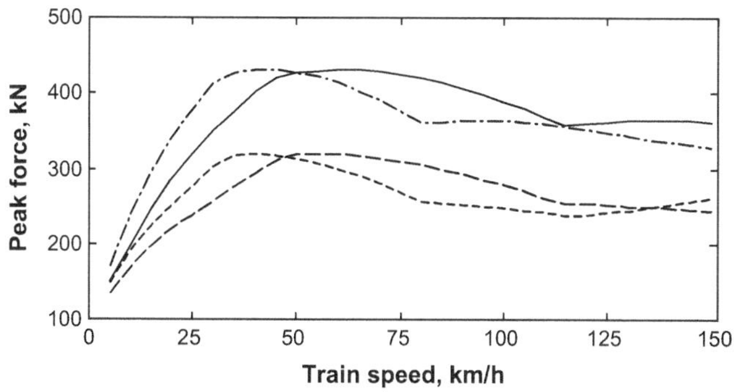

FIGURE 10-9 Peak impact force predicted from different wheel flats. —, due to 2 mm rounded flat (length 121 mm); – – –, 1 mm rounded flat (length 86 mm); – · – ·, 2 mm newly formed flat (length 86 mm); - - -, 1 mm newly formed flat (length 61 mm) [10.10]

of the wheel modes of vibration and the propagation of vibration along the rail. Such effects are included in the conventional rolling noise model.

Unfortunately, it is not possible to use the contact force spectrum obtained from the impact model and apply it directly within a detailed wheel/rail model in the frequency domain, because the interaction force is very sensitive to details of the wheel and track dynamics, as explained in Section 5.2.5. To do so would lead to excessive predictions of wheel vibration.

A hybrid approach has therefore been developed whereby an equivalent roughness spectrum is derived [10.10]. This is defined such that the contact force spectrum obtained using a *linear* model excited by the equivalent roughness spectrum is identical to that obtained from the wheel flat using the above *non-linear* model. At this stage the wheel and track are represented by the same simple elements described above in both cases. The equivalent roughness spectrum can then be used as the input to a more detailed linear frequency-domain model, such as the TWINS model, to predict the noise due to the impact. This procedure has been validated in [10.10] using a simplified wheel model containing a single flexible mode. In addition, in [10.14] radial modes of the wheel were included directly in the non-linear model for an example case. There it was found that the corrections applied in the hybrid approach led to the same result as the direct calculation of interaction force.

Figure 10.10(a) shows results from the hybrid approach. This shows the spectra of sound power due to one wheel and the associated track vibration for a 2 mm new wheel flat at different speeds and for a wheel load of 100 kN. Results correspond to the average over one wheel revolution. Figure 10.10(b) shows, for comparison, corresponding results for roughness excitation due to the moderate roughness considered in Section 10.2 (tread-braked wheel roughness). In both cases the wheel is represented in the frequency-domain model by its full modal basis in the frequency range up to 6 kHz, determined from a finite element model. The track is modelled by a Timoshenko beam, continuously supported on layers of damped springs and mass.

As the speed increases, the noise at frequencies above about 200–400 Hz increases in both cases. The increase in rolling noise with increasing speed is greater than that due to the wheel flat. For the wheel flats that are considered here, which are large compared with typical inspection limits, the noise generated exceeds that due to the tread-braked wheel roughness at all speeds and in

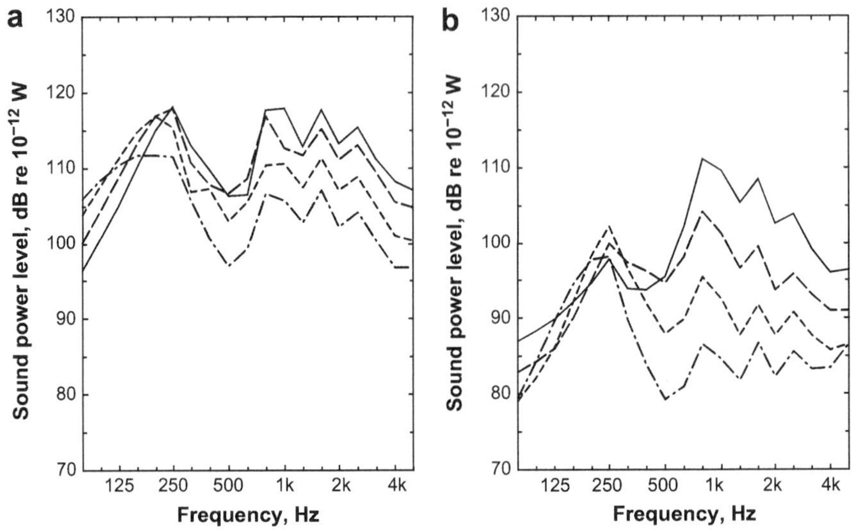

FIGURE 10-10 Sound power level due to wheel and track. (a) 2 mm new wheel flat (length 86 mm), (b) rolling noise from moderate roughness. – · – ·, 30 km/h; - - -, 50 km/h; – – –, 80 km/h; —, 120 km/h [10.10]

all frequency bands. However, the noise due to roughness increases more rapidly with speed so that at sufficiently higher speeds it can be expected to dominate. For corrugated track, the noise due to roughness would exceed that due to the wheel flat at 120 km/h.

Figure 10.11 shows a summary of the variation of the overall A-weighted sound power level with train speed. The predicted noise level due to conventional roughness excitation increases at a rate of approximately $30 \log_{10} V$, where V is the train speed, whereas the noise due to flats increases at an average of around $20 \log_{10} V$ once loss of contact occurs. For example, loss of contact was found to occur for the newly formed 2 mm flat at speeds above 30 km/h and for the rounded 2 mm flat above 50 km/h. This variation with speed indicates that the radiated sound due to wheel flats continues to increase with increasing speed, even though loss of contact is occurring, contrary to [10.2]. The noise from a new flat is greater than from a rounded flat of the same depth, even though the rounded flat is longer. The difference between the sound produced by flats of depths 1 and 2 mm is 3 dB on average but varies between 1 and 5 dB depending on the speed.

Impact noise from wheel flats is also found to increase as the wheel load increases [10.10]. The increase in noise between a load of 50 and 100 kN is about 3 dB. In contrast, the rolling noise due to roughness is relatively insensitive to wheel load (see box on page 151), decreasing slightly with increased load.

10.4 IMPACT NOISE DUE TO RAIL JOINTS

In a similar way to wheel flats, rail joints provide discrete inputs to the wheel/rail system that induce quite large contact force variations. Rail joints can be

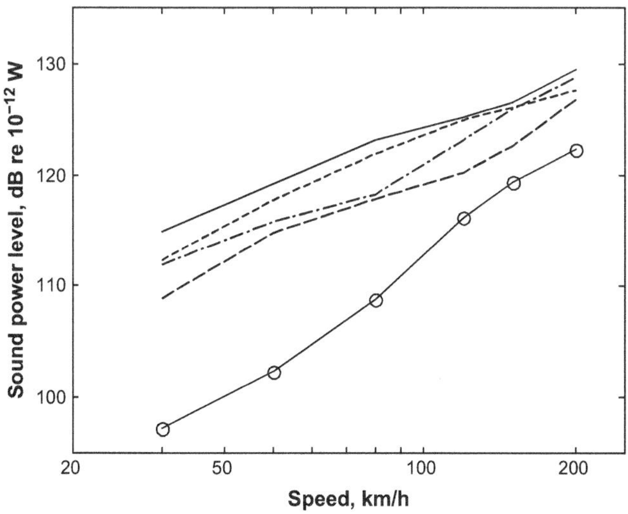

FIGURE 10-11 Sound power radiated by one wheel and the associated track vibration. - - -, 2 mm rounded flat (length 121 mm); – – –, 1 mm rounded flat (length 86 mm); —, 2 mm new flat (length 86 mm); – · – ·, 1 mm new flat (length 61 mm); o—o, rolling noise due to roughness (moderate roughness) [10.10]

characterized by a gap width and a step height (either up or down), as shown in Figure 10.12. Moreover, the rail often dips down to a joint on both sides. Such dips are also present at welds, and are usually characterized in terms of the angle at the joint [10.3].

A similar approach to that described above has been used to study the noise generated at rail joints [10.11, 10.15]. To allow for the reduced bending stiffness at a fish-plated joint a model of a pin-jointed beam was adopted [10.11]. The sound radiation was calculated using the same hybrid method as for the wheel flats, except that a correction had to be introduced to allow for the pin-jointed model of the rail receptance. It was found, for realistic parameter values, that the gap width is insignificant compared with the step height and dip angle and can be neglected.

Results are shown in Figure 10.13(a) for undipped rail joints in the form of the total A-weighted sound power emitted by the wheel and rail during 1/8 s. The results for a step-down joint are found to be virtually independent of the step height (only results for one value are shown) and also change very little with train speed. The same is found to be the case for the peak contact force [10.11]. However, for step-up

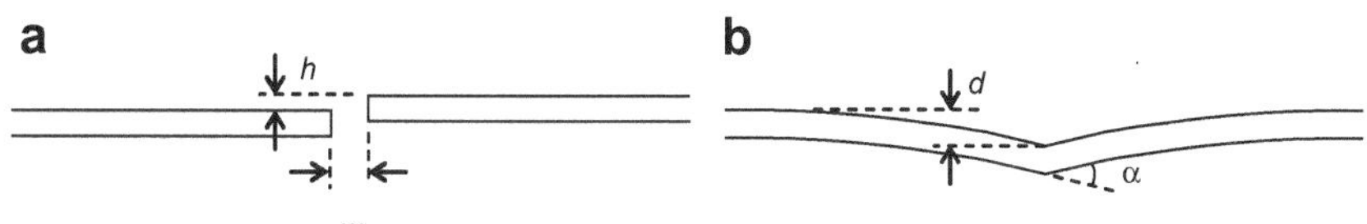

FIGURE 10-12 Parameters describing a rail joint: (a) step height, h; gap width, w; (b) dip, d; dip angle, α

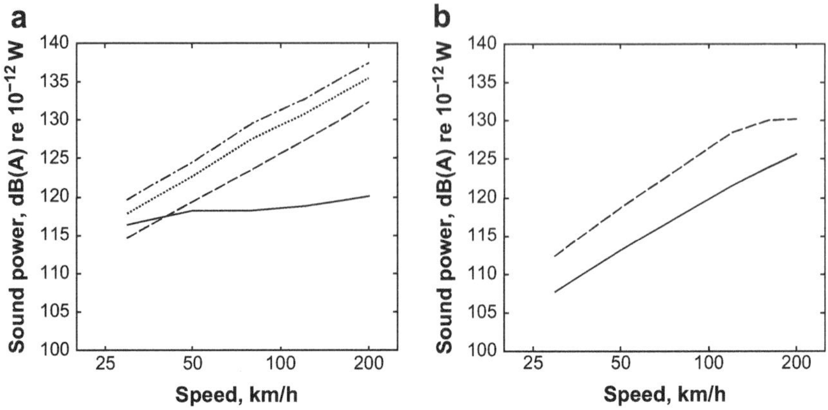

FIGURE 10-13 A-weighted sound radiated by one wheel and the associated track vibration during 0.125 second due to a wheel passing over (a) flat rail joints: – – –, 1 mm step-up; · · · ·, 2 mm step-up; – · – ·, 3 mm step-up; —, 2 mm step-down. (b) Dipped rail joints with no height difference: —, 5 mm dip; – – –, 10 mm dip (all with 7 mm gap) [10.14]

joints both the peak contact force and the sound power level increase with step height and with train speed. The sound power level from a single joint has a speed dependence of around 20 $\log_{10} V$.

In Figure 10.13(b) results are given for dipped joints with no height difference. Here a dip of 5 or 10 mm is considered as a quadratic function over a length of 0.5 m either side of the joint. A dip of 5 mm corresponds to a joint angle of 0.04 radians which is large although not untypical, a dip of 10 mm corresponds to 0.08 radians which is severe. The 10 mm dip produces a similar noise level to a 1 mm step-up undipped joint, although for speeds above 120 km/h the noise level from the dipped joint becomes independent of train speed.

Figure 10.14 shows the predicted noise for joints with both dipped rails and steps. The noise radiation generally increases with speed, regardless of whether loss of contact occurs. For the 5 mm dip, the noise level increases by 8 dB when the step height increases from 0 to 2 mm. For the step-down joints, the noise level is higher than without a step, although at higher speeds the dip has more effect than the step. The results for the 10 mm dip are similar for both step-up and step-down joints indicating the dominance of the dip in this case.

These results may be compared with the rolling noise results given in Figure 10.12 and are found to be 5–20 dB greater. However, this comparison is based on the sound within 1/8 s indicating that the joint can be heard above the rolling noise. To evaluate their effect on the average noise level, the time base used for the joint noise should be adjusted to the average time between joints. Moreover, since the time between rail joints decreases as train speed increases, it is also found that the average noise level from joints increases at a rate of about 30 $\log_{10} V$, similar to rolling noise. Consequently, rolling noise due to the moderate roughness considered above is found to be similar to the average noise due to 5 mm dipped joints with no height difference [10.11]. With a height difference of 2 mm the average noise predicted from the joints increases to almost 10 dB greater than the rolling noise.

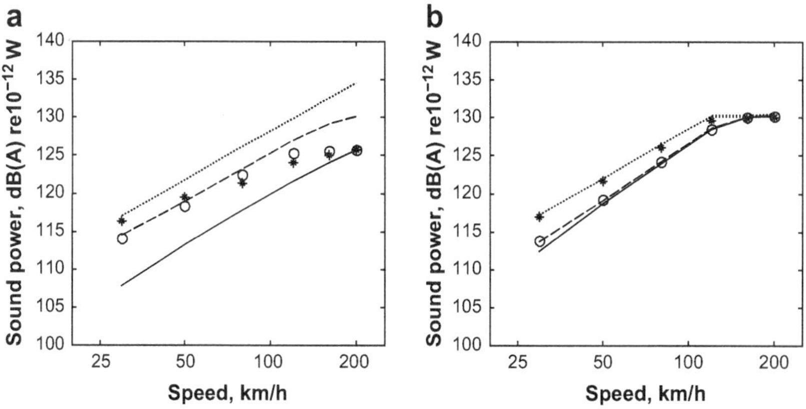

FIGURE 10-14 A-weighted sound power radiated by one wheel and the associated track vibration during 0.125 second due to a wheel passing over different rail joints with 7 mm gap and 5 or 10 mm dip. (a) For 5 mm dip, (b) for 10 mm dip. ····, 2 mm step-up; – – –, 1 mm step-up; —, no height difference; $*$, 2 mm step-down; $\circ$, 1 mm step-down [10.14]

10.5 DISCUSSION

The modelling described in this chapter has been based on interaction in only the vertical direction. Inclusion of the lateral coordinate as well has been considered in [10.16]. For this, the track had to be represented by a similar model to that given by equation (10.3) for both the lateral and cross receptances using higher order polynomials. The wheel and rail are coupled in the lateral direction by a creep force which is sensitive to the normal load, so that if loss of contact occurs this force also becomes zero. The vertical interaction is found to be largely unaffected by the addition of lateral coupling, although the lateral response is only correctly predicted when the lateral coupling is included.

A 1/5 scale model rig has been used for experimental validation of the results of impact noise generation. Tests have been carried out using a simulated wheel flat (in the rail head), dipped joints and stepped joints as well as a relatively smooth rail for a range of preloads and wheel speeds [10.17] and found to confirm the results of the modelling.

In order to control impact noise, the first priority must be to eliminate discontinuities where possible. This reduces the damage to track and wheels as well as reducing noise. The use of good quality continuously welded rail is to be recommended where possible; care is also needed in straightening the rail at welds to avoid dipped joints. Effective wheel-slide protection equipment should be used to minimize the occurrence of wheel flats.

Monitoring systems can be used to detect wheel flats or high impact loads at rail joints or welds. Such condition monitoring is increasingly used for effective management of assets and, although noise control is not usually their primary purpose, it is an added benefit.

Finally, however, it is recognized that some discontinuities in the rail surface are inevitable, for example at switches and crossings, expansion joints and insulated rail joints. Since the excitation mechanism is similar to that for rolling noise, the same mitigation measures can be used as apply to rolling noise. These have been described in detail in Chapter 7.

REFERENCES

10.1 J.C.O. Nielsen and A. Johansson. Out-of-round railway wheels – a literature survey. *Proceedings of the Institution of Mechanical Engineers, Part F: Journal of Rail and Rapid Transit*, **214F**, 79–91, 2000.

10.2 I.L. Vér, C.S. Ventres, and M.M. Myles. Wheel/rail noise, Part II: Impact noise generation by wheel and rail discontinuities. *Journal of Sound and Vibration*, **46**, 395–417, 1976.

10.3 H.H. Jenkins, J.E. Stephenson, G.A. Clayton, G.W. Morland, and D. Lyon. The effect of track and vehicle parameters on wheel/rail vertical dynamic forces. *REJ*, 2–16, January 1974.

10.4 S.G. Newton and R.A. Clark. An investigation into the dynamic effects on the track of wheelflats on railway vehicles. *Journal of Mechanical Engineering Science*, **21**, 287–297, 1979.

10.5 R.A. Clark, P.A. Dean, J.A. Elkins, and S.G. Newton. An investigation into the dynamic effects of railway vehicles running on corrugated rails. *Journal of Mechanical Engineering Science*, **24**, 65–76, 1982.

10.6 J.C.O. Nielson and A. Igeland. Vertical dynamic interaction between train and track – influence of wheel and track imperfections. *Journal of Sound and Vibration*, **187**, 825–839, 1995.

10.7 A. Johansson and J.C.O. Nielsen. Out-of-round railway wheels – wheel-rail contact forces and track response derived from field tests and numerical simulations. *Proceedings of the Institution of Mechanical Engineers, Part F: Journal of Rail and Rapid Transit*, **217F**, 135–146, 2003.

10.8 C. Andersson and T. Dahlberg. Wheel/rail impacts at a railway turnout crossing. *Proceedings of the Institution of Mechanical Engineers, Part F: Journal of Rail and Rapid Transit*, **212F, 123–134, 1998.**

10.9 T.X. Wu and D.J. Thompson. Theoretical investigation of wheel/rail non-linear interaction due to roughness excitation. *Vehicle System Dynamics*, **34**, 261–282, 2000.

10.10 T.X. Wu and D.J. Thompson. A hybrid model for the noise generation due to railway wheel flats. *Journal of Sound and Vibration*, **251**, 115–139, 2002.

10.11 T.X. Wu and D.J. Thompson. On the impact noise generation due to a wheel passing over rail joints. *Journal of Sound and Vibration*, **267**, 485–496, 2003.

10.12 D.J. Thompson and T.X. Wu. The effects of non-linearities at the wheel/rail interface on the generation of rolling noise. *Proceedings of Internoise 2001*, The Hague, The Netherlands, 129–134, 2001.

10.13 T.X. Wu and D.J. Thompson. On the parametric excitation of the wheel/track system. *Journal of Sound and Vibration*, **278**, 725–747, 2004.

10.14 D.J. Thompson, T.X. Wu and T.D. Armstrong. Wheel/rail rolling noise – the effects of non-linearities in the contact zone. *Proceedings of 10th International Congress on Sound and Vibration*, Stockholm, 7–10 July, 2003.

10.15 T.X. Wu and D.J. Thompson. A model for impact forces and noise generation due to wheel and rail discontinuities. *Proceedings of the 8th International Congress on Sound and Vibration*, Hong Kong, China, 2905–2912, 2001.

10.16 T.X. Wu and D.J. Thompson. Wheel/rail non-linear interaction with coupling between vertical and lateral directions. *Vehicle System Dynamics*, **41**, 27–50, 2004.

10.17 T.D. Armstrong. Use of a scale model rig to investigate non-linear wheel/rail interaction. PhD thesis, University of Southampton, 2004.

CHAPTER 11

Bridge Noise

11.1 INTRODUCTION

Bridges carrying the railway line are a necessary part of the infrastructure, whether they allow the track to pass over roads, waterways, other railway tracks or natural features such as valleys or gorges. When a train runs across a bridge, it usually produces more noise than when it runs on plain track at grade. This increase in noise varies considerably from one bridge to another but it can typically be 10 dB or more. Bridges, therefore, often form 'hot spots' in noise maps and it is important that they are given due consideration when formulating Action Plans against noise.

There are two main reasons for this amplification of the noise when a train crosses a bridge. The first is that vibrations are transmitted from the rails into the bridge structure which then also radiates noise. Due to the large surface area of the bridge, it acts as a 'sounding board'; the noise radiated can therefore be considerable even though the vibration amplitude is attenuated by the track fastening system. The second reason, as will be seen, is that, depending on the method of rail fastening used on a particular bridge, the noise radiated by the vibration of the rail itself may be considerably greater than for track at grade.

Bridges vary greatly in design and construction. When the early railways were built in the first half of the nineteenth century, bridges and viaducts were mostly constructed of timber or masonry. Iron was also used, an early example being a four arch bridge on the Stockton and Darlington Railway in 1825. After a cast-iron bridge over the River Dee at Chester collapsed under the weight of a train in 1847, research into stronger wrought iron led to its use in many bridges in the second half of the century. Steel was first used towards the end of the nineteenth century, for example in the Forth Bridge in Scotland. The first concrete railway viaducts were constructed at the start of the twentieth century and reinforced concrete is now a widely used material.

Three examples of railway bridges are shown in Figures 11.1–11.3, the noise from which will be considered later in this chapter. The first, in Figure 11.1, is a steel bridge 20.8 m long with the track supported on a steel deck suspended between longitudinal lattice girders. Here, the limited headroom below the bridge is critical. A much larger steel bridge is shown in Figure 11.2, where a steel span of 150 m is used to cross a shipping channel. A concrete viaduct section is used on either side of the main span. The bridge in Figure 11.3 has a reinforced concrete deck supported on longitudinal steel bearers. In the background a transition to an all-concrete structure can be seen, which requires more closely spaced piers.

FIGURE 11-1 Steel bridge at Gavignot in France. Photo: SNCF

A number of attempts have been made to quantify the noise from different bridge types on the basis of measured data. Kurzweil [11.1], for example, divided bridges into 11 categories. Figure 11.4 (based on data from Ungar and Wittig [11.2]) shows a wider collection of measured data of bridge noise divided into seven categories.

FIGURE 11-2 Årsta bridge, Stockholm. Photo: Pandrol

FIGURE 11-3 A concrete/steel composite bridge on the Docklands Light Railway, London. Photo: Olly Bewes

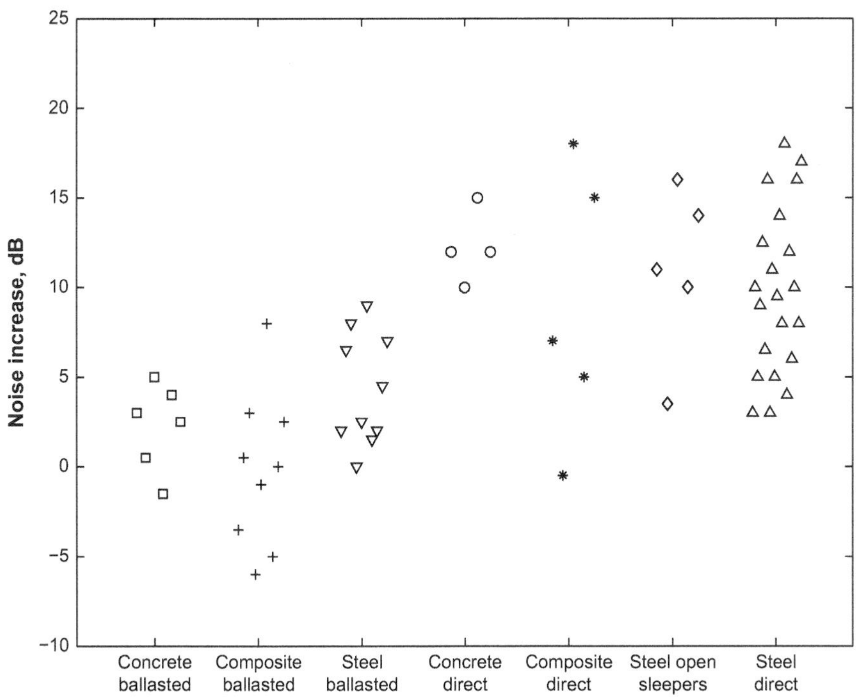

FIGURE 11-4 Increase in noise level as a result of various types of bridge and track fastening compared with plain ballasted track (adapted from Ungar and Wittig [11.2])

These results are shown as the increases in A-weighted noise level relative to the same train on plain track, and contain values between −6 and +20 dB.

It is clear from this that steel bridges are generally noisier than concrete ones, and bridges with direct rail fastenings are mostly noisier than those with ballasted track. Nevertheless, there are several specific counterexamples, for example a steel bridge with direct fasteners at +3 dB and a concrete bridge with ballasted track at +5 dB.

Clearly, many old bridges remain in use, which would be expensive to replace, including many made of steel or wrought iron. But why aren't all new bridges constructed of concrete and using ballasted track? There are many non-acoustical factors that are important as well.

For example, bridges with large spans, such as the one in Figure 11.2, are more easily constructed from steel. Moveable bridges, such as lifting or swing bridges, are inevitably made of steel and without ballast. Steel bridges can also provide a smaller construction height under the rails, which is important for bridges in urban areas, such as that shown in Figure 11.1.

Ballasted track adds considerable weight which has to be carried by the bridge structure. Moreover, direct fastening systems allow the alignment of the rails to be defined and maintained more precisely than for ballasted track, which may reduce potential problems with gauging. On the other hand the alignment of ballasted track can be adjusted more easily.

Although some overall trends can be seen from the results in Figure 11.4, it should be clear that bridges, even within the same category, contain large variations. These may be due to wide differences in bridge design not adequately covered by the categories used. Such design parameters will be discussed in Sections 11.2 to 11.4. However, there are a number of other limitations of the use of measured data in this way which mean that such measured results should be treated with caution:

Barrier effects: bridge parapets may shield the noise, particularly from the wheel/rail region, although the lower parts of the bridge itself are usually not shielded. If not effectively isolated from the bridge, the parapets could also contribute to the radiated noise. The shielding effects can be predicted using the methods applicable for other sorts of barriers (see Section 7.5.2).

Bridge length: if the bridge is short compared with the measurement distance, the measured level may be affected by adjacent plain track. To prevent this, the measurement position should not be further away than about half the bridge length.

Frequency content: the amplification due to the bridge is frequency dependent. This means that the difference in A-weighted level will depend on train speed and roughness spectrum.

To illustrate the frequency dependence, Figure 11.5 shows examples of measured amplifications in one-third octave bands. These are for the steel bridge shown in Figure 11.1 on which the rails are mounted on wooden sleepers laid directly onto the bridge deck [11.3]. In this case there is considerable amplification in the mid-frequencies as well as around 40 Hz. The difference in A-weighted level in this case was 12–14 dB.

Large amplifications at low frequencies, as seen here, are common. At larger distances, and especially inside buildings, sound at these low frequencies will be transmitted more effectively than at higher frequencies. The increase in noise due to the bridge could then be considerably greater than indicated by the differences in A-weighted levels measured near to the track.

In Sections 11.2 to 11.4 a model for the generation of bridge noise is introduced. Based on the parameters identified in this model some guidelines will be given in

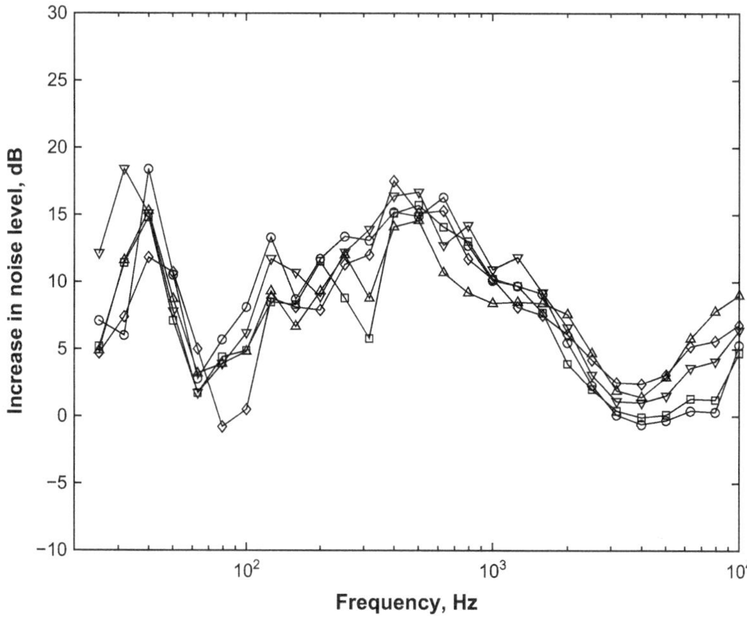

FIGURE 11-5 Difference in noise level measured on the steel bridge at Gavignot with the rails mounted on wooden sleepers directly onto the bridge deck compared with ballasted track for five different train types, 50–80 km/h [11.3]

Section 11.5 for how to control bridge noise. Some case studies will be described briefly in Section 11.6.

11.2 THE EXCITATION OF BRIDGE NOISE

Some drawbacks of empirical data have already been seen, but perhaps the greatest weakness of a purely empirical approach is that it cannot be used for studying design changes unless these are implemented in practice. Following the general approach adopted in this book, therefore, a model is introduced that allows the various parameters that influence the noise to be assessed. In fact, it is an extension of the rolling noise model of earlier chapters.

The way in which a train on a bridge generates noise is shown schematically in Figure 11.6. This forms the basis for calculation models developed in [11.4, 11.5, 11.6]. The interaction between the wheel and rail can be calculated using similar models to those introduced in previous chapters. To determine the power injected into the bridge, a model of the interaction between the rail and the bridge is required. Finally, the vibration of the bridge components and their sound radiation can be calculated using an energy balance method such as Statistical Energy Analysis, see [11.7]. An alternative approach is presented in [11.8] but this is only valid for frequencies where (propagating) waves are present in the rail (i.e. above at least 250 Hz).

In the following sections, the model of [11.4, 11.5, 11.6] is introduced briefly to illustrate the way in which a bridge radiates noise, and to identify which parameters have an influence on the noise. This model was developed originally in [11.4] for steel

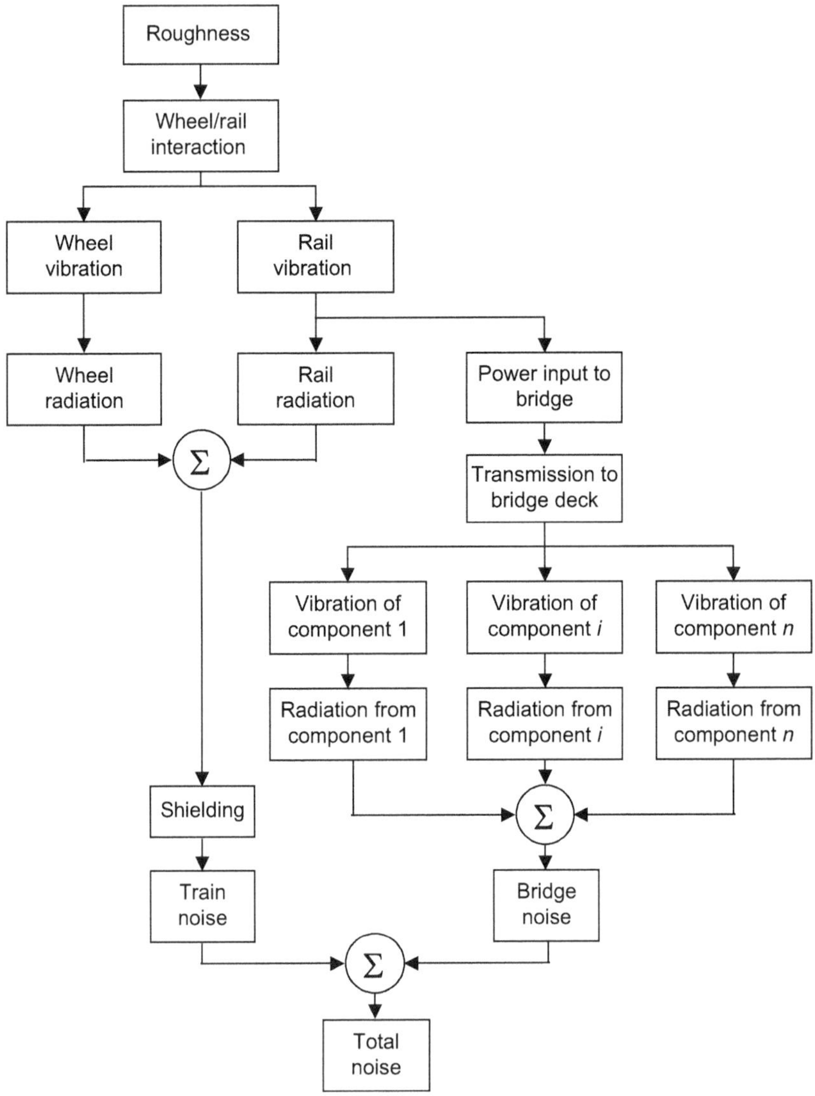

FIGURE 11-6 Block diagram for model of bridge noise

ballastless bridges, but has also been applied to concrete bridges and to bridges with ballasted track. As bridges vary considerably in details of their design, only a broad outline of the modelling technique is presented, illustrated with some examples.

11.2.1 Wheel/rail interaction

In the same way as for rolling noise, the main source of excitation of the bridge is the surface roughness at the wheel/rail interface. The combination of wheel and rail roughnesses causes a relative displacement between wheel and rail at the contact patch, according to their mobilities. As has been seen in Chapter 5, for at least the frequency region 100–1000 Hz the rail mobility is larger than those of both the wheel and contact zone so that the velocity of the rail at the wheel/rail contact point is approximately equal to the roughness velocity, $v_r \approx i\omega r$, see equation (5.17). This remains the case for the mobility of a rail on a bridge. In this frequency region, therefore, changes to the rail mobility will have no direct effect on bridge noise.

The coupling with vibration in the lateral direction can usually be ignored, on the basis that the vertical vibration contributes most to the energy transfer into the bridge, although in some situations, particularly on curves, the lateral direction should be included [11.9].

The same principles for controlling roughness can be applied on bridges as on plain track. In particular, rail grinding is a standard practice for removing excessive roughness, see Section 7.2. The greater importance of low frequencies for bridge noise means that longer wavelengths of roughness are also important.

As well as roughness excitation, the presence of rail joints can lead to impact excitation on the bridge (see Chapter 10). Bridges usually require expansion joints in the rails and it is normally more convenient to locate these on the bridge itself. Rail joints, of course, are also essential for moveable bridges (swing or lifting bridges).

At low frequencies two effects are more important on bridges than on plain track. First, parametric excitation is found at the sleeper passing frequency, due to the spatial variations in the track stiffness between locations over a fastener and between fasteners (see Section 5.7.5). Second, the resonance associated with the wheel mass bouncing on the track stiffness at around 50–100 Hz, discussed in Section 5.2, is stronger for direct fasteners than for ballasted track due to lower damping in the track. When these two frequencies coincide, large amplitudes of vibration are transmitted to the bridge and radiated as noise, especially if these also coincide with bridge or panel resonances. Continuous support of the rails can be used to eliminate the sleeper passing effect.

11.2.2 Rolling noise on bridges

The models described in Chapters 3–6 can be used to predict the rolling noise from trains on bridges, with appropriate parameters used to represent the track. The main difference that must be taken into account is that the track decay rates will, in general, be lower on bridges with direct fasteners than on ballasted track. The average vertical rail vibration can therefore be 5–10 dB higher on a bridge with such a fastening system than for ballasted track, as shown in Figure 11.7. This means that the component of noise radiated by the rail itself will also be higher on such a bridge than on ballasted track. The wheel component of noise should not be affected. On the other hand, for direct fastenings the sleeper component will be eliminated and is replaced by the bridge noise.

Overall, the rolling noise will be increased compared with plain track in the frequency range 400–2000 Hz (see also Figure 11.5). It is important to remember, therefore, that the increase in noise during passage over a bridge is not just due to the radiation from the bridge but also due to the increase in rail noise. The relative importance of these two effects should be established before attempting noise mitigation measures, as different techniques are required to deal with them.

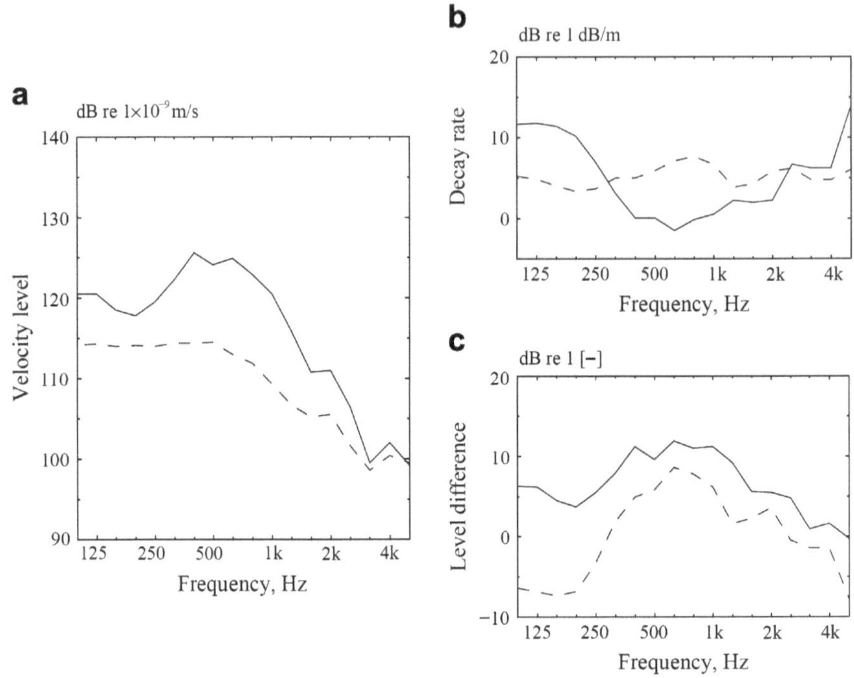

FIGURE 11-7 (a) Measured vertical vibration levels: —, on a bridge with direct fasteners; – – –, on normal track with wooden sleepers. (b) Measured track decay rates (shown as $10\log_{10}\Delta$): —, on bridge; – – –, on plain track. (c) Comparison of differences in vibration levels and decay rates: —, difference in rail vibration level; – – –, difference in $10\log_{10}\Delta$, [11.4]

11.3 POWER INPUT TO THE BRIDGE

11.3.1 Coupling of the track to the bridge

For direct (ballastless) fastening of rails onto bridges, one or more resilient elements are installed between the rail and the bridge. These provide for a certain amount of isolation, so that the vibration level on the bridge is generally less than that on the rail. For ballasted track, the ballast itself also acts as a resilient element, introducing a similar effect.

In general, for effective vibration isolation using such a resilient layer, it is important that the ratios of mobilities (rail: resilient layer: bridge) should be in the proportions small: large: small (i.e. a large 'impedance mismatch' [11.10]). In other words, if the resilient layer is too stiff, or the bridge too flexible, the system will not be well isolated.

The bridge and the rail are not coupled at a single point but over the whole length of the rail. To illustrate the effect this has on the coupling between the track and a bridge, it is useful to consider a generic model based on two beams, representing the rail and the bridge, as shown in Figure 11.8. For simplicity these will be considered here to be infinite Euler–Bernoulli beams. Although discrete fasteners are

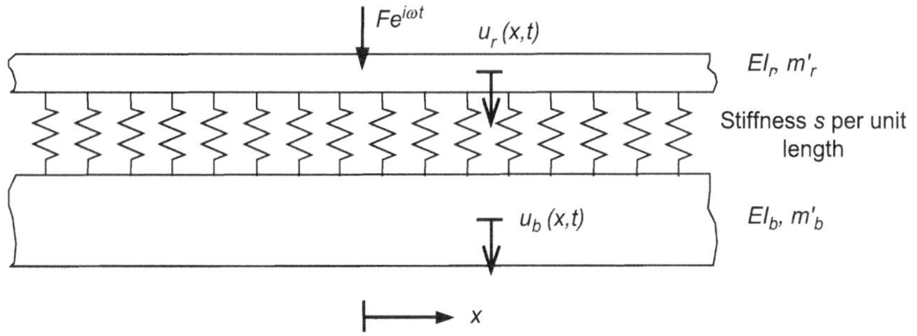

FIGURE 11-8 Two infinite beams coupled by a resilient layer

mostly used, the model will be based on a continuous connection through an elastic layer. This is reasonable, as discussed for track modelling in Chapter 3, particularly for frequencies below 1 kHz.

The two beams have bending stiffness EI_r and EI_b and mass per unit length m'_r and m'_b, respectively. They are coupled by a continuous elastic layer of stiffness s per unit length. As in Chapter 3, damping can be introduced into the beams and the resilient layer by making E and s complex. The bending vibration amplitudes of the beams are denoted $u_r(x, t)$ and $u_b(x, t)$, where x is the coordinate along the beam. The equations of free motion are

$$EI_r\frac{\partial^4 u_r}{\partial x^4} + s(u_r - u_b) + m'_r\frac{\partial^2 u_r}{\partial t^2} = 0 \tag{11.1}$$

$$EI_b\frac{\partial^4 u_b}{\partial x^4} + s(u_b - u_r) + m'_b\frac{\partial^2 u_b}{\partial t^2} = 0 \tag{11.2}$$

Assuming free wave motion at frequency ω and wavenumber k, similar to equation (3.2), the wavenumber satisfies

$$\left(\begin{bmatrix} EI_r & 0 \\ 0 & EI_b \end{bmatrix}k^4 + \begin{bmatrix} s - \omega^2 m'_r & -s \\ -s & s - \omega^2 m'_b \end{bmatrix}\right)\begin{Bmatrix} U_r \\ U_b \end{Bmatrix} = \begin{Bmatrix} 0 \\ 0 \end{Bmatrix} \tag{11.3}$$

where U_r and U_b are the complex amplitudes of the rail and bridge vibration in a particular wave. Solving this eigenvalue problem yields two solutions for k^4, giving wavenumbers $\pm k_1$, $\pm ik_1$, $\pm k_2$ and $\pm ik_2$. k_1 and k_2 are taken here to be the solutions in the fourth quadrant of the complex plane (with positive real part and negative imaginary part). For each of the two eigenvalues there is a corresponding eigenvector $[U_r\ U_b]^{\mathrm{T}}$ defining the ratio between the amplitudes of the two beams in this wave. The same eigenvector applies to all four roots for a given eigenvalue. Since the eigenvector has an arbitrary magnitude it can be written as $[1\ \Lambda_i]^{\mathrm{T}}$ where $\Lambda_i = U_b/U_r$.

At low frequencies there is a single propagating wave in each direction, the other six waves being evanescent. At higher frequencies there are two propagating waves in each direction, and at high enough frequency the wavenumbers k_i tend to those of the free waves in each beam. The cut-on frequency of the second pair of waves can be found by setting $k = 0$ in equation (11.3), giving

$$\omega_0 = \sqrt{\frac{s}{m'_r} + \frac{s}{m'_b}} \tag{11.4}$$

which is known as the decoupling frequency [11.4, 11.11]. For a point force $Fe^{i\omega t}$ applied to the upper beam at $x = 0$, four coupled waves exist in the beams on the left and four on the right of the forcing point. Using symmetry, the response can be written as

$$u_r(x) = A_1 e^{-k_1|x|} + A_2 e^{-ik_1|x|} + A_3 e^{-k_2|x|} + A_4 e^{-ik_2|x|} \tag{11.5}$$

$$u_b(x) = A_1 \Lambda_1 e^{-k_1|x|} + A_2 \Lambda_1 e^{-ik_1|x|} + A_3 \Lambda_2 e^{-k_2|x|} + A_4 \Lambda_2 e^{-ik_2|x|} \tag{11.6}$$

where A_i are the wave amplitudes. To find these amplitudes, four boundary conditions are required at $x = 0$. These are that the slope of each beam is zero and the difference in shear force is equal to the external force (0 for the lower beam). These can be expressed as a matrix equation

$$\begin{bmatrix} -k_1 & -ik_1 & -k_2 & -ik_2 \\ -k_1\Lambda_1 & -ik_1\Lambda_1 & -k_2\Lambda_2 & -ik_2\Lambda_2 \\ -k_1^3 & ik_1^3 & -k_2^3 & ik_2^3 \\ -k_1^3\Lambda_1 & ik_1^3\Lambda_1 & -k_2^3\Lambda_2 & ik_2^3\Lambda_2 \end{bmatrix} \begin{Bmatrix} A_1 \\ A_2 \\ A_3 \\ A_4 \end{Bmatrix} = \begin{Bmatrix} 0 \\ 0 \\ F/2EI_r \\ 0 \end{Bmatrix} \tag{11.7}$$

which can be solved to find the amplitudes A_i. Hence, setting $x = 0$ in equation (11.5), and multiplying by $i\omega$, the point mobility at the rail can be obtained, $Y_r = i\omega u_r(0)/F$.

Example results for the point mobility are shown in Figure 11.9 based on the parameters listed in Table 11.1. At low frequency the two beams are strongly coupled together by the resilient layer and form a composite beam with bending stiffness $EI_r + EI_b$ and mass per unit length $m'_r + m'_b$. Above about 30 Hz the mobility is stiffness controlled and gradually rises to a peak at the decoupling frequency, 300 Hz. Above this frequency the mobility tends to that of the free rail. The transfer mobility $i\omega u_b(0)/F$ to a point on the lower beam directly below the force is also shown. This has a similar response to the rail at low frequencies but is progressively isolated from it at high frequency.

Bewes *et al.* [11.6] presented a model derived from this in which the beams were based on Timoshenko beam theory. A model using two finite beams was considered, with simply supported boundary conditions at the ends. A case with two infinite Timoshenko beams was also considered for comparison. Figure 11.10 shows the real part of the point mobility of the rail for a bridge of length 16 m with the excitation at 4 m from one end. Other parameters are as in Table 11.1. At low frequencies a series of resonances of the composite beam (bridge plus rail) are found, the first being at 10 Hz for these parameters. At higher frequencies the infinite beam model provides a reasonably good asymptote to the result. Comparison of these results for infinite beams with those of Figure 11.9 shows that shear deformation is important for the bridge above about 200 Hz for these parameters.

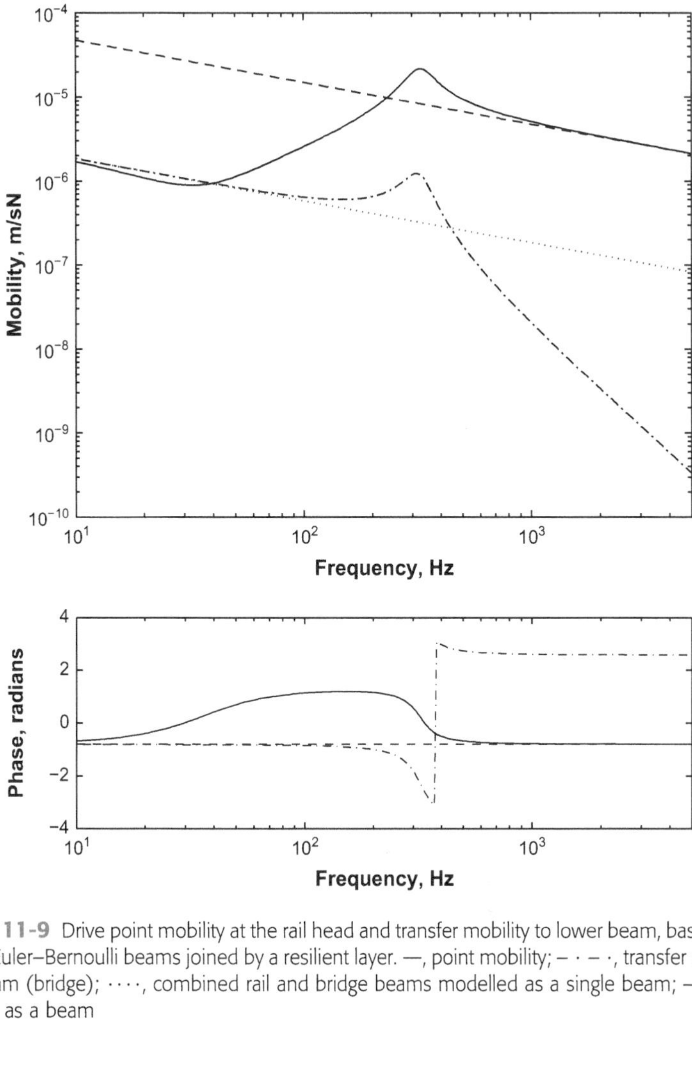

FIGURE 11-9 Drive point mobility at the rail head and transfer mobility to lower beam, based on two coupled Euler–Bernoulli beams joined by a resilient layer. —, point mobility; – · – ·, transfer mobility to lower beam (bridge); · · · ·, combined rail and bridge beams modelled as a single beam; – · – ·, rail modelled as a beam

The power input to the rail by a harmonic force of amplitude F is given by

$$W_{in,r} = \frac{1}{2}\mathrm{Re}\{F^* v_r\} = \frac{1}{2}\mathrm{Re}\{Y_r\}|F|^2 \tag{11.8}$$

where v_r is the amplitude of the rail velocity, Y_r is the rail mobility when coupled to the bridge and * denotes the complex conjugate. To find the power transmitted to the bridge beam, the force per unit length acting on the bridge through the elastic layer is required, $F_b'(x) = s(u_r(x) - u_b(x))$. Similar to equation (11.8), the power per unit length is given by $\{1/2\}$ $\mathrm{Re}(F'^*_b v_b)$ where $v_b(x)$ is the velocity of the bridge beam. Hence, the total power transmitted to the bridge can be calculated using

TABLE 11-1 PARAMETERS USED FOR SIMPLE BRIDGE MODEL

		Rail beam		Bridge beam
Mass per unit length, kg/m	m'_r	54	m'_b	500
Bending stiffness, MN/m^2	EI_r	5	EI_b	2000
Damping loss factor	η_r	0.02	η_b	0.1
Stiffness per unit length of resilient layer, MN/m^2	s	200		
Damping loss factor of resilient layer	η	0.25		

$$W_{in,b} = \frac{1}{2}\text{Re}\int_{-L_1}^{L_2} \{s(u_r(x) - u_b(x))\}^* i\omega u_b(x)dx \tag{11.9}$$

where the bridge is located between $-L_1 \leq x \leq L_2$. This power is shown in Figure 11.11 for the model of two finite Timoshenko beams [11.6]. For comparison, the result for two infinite Timoshenko beams is again shown. As with the response directly beneath the forcing point, the power input to the bridge reduces considerably above the decoupling frequency.

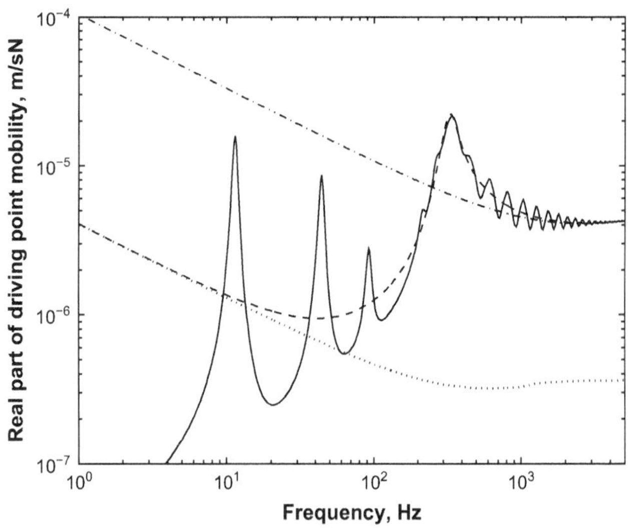

FIGURE 11-10 Real part of the driving point mobility at the rail head. —, two finite Timoshenko beams coupled by a resilient layer; – –, two infinite Timoshenko beams coupled by a resilient layer; ····, combined rail and bridge beams modelled as Timoshenko beams; – · – ·, rail modelled as a Timoshenko beam [11.6]

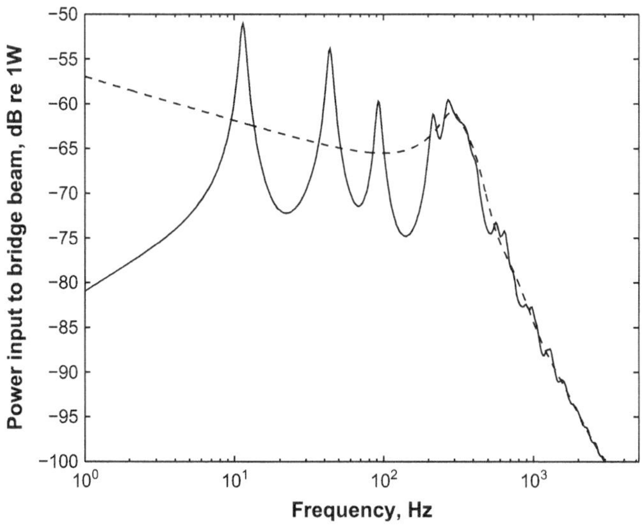

FIGURE 11-11 Power input to the bridge beam for a force on the rail of amplitude 1 N. —, two finite Timoshenko beams coupled by a resilient layer; – –, two infinite Timoshenko beams coupled by a resilient layer [11.6]

11.3.2 Approximate calculation of the excitation of the bridge

Above the decoupling frequency, given by equation (11.4), the rail and bridge vibration increasingly resemble their free vibration. Although the connection is over the whole length of the bridge, the propagating waves in the rail have a wavelength that is much shorter than those in the bridge. It is found that a free wave in the rail would exert no net force on the bridge, due to the cancellation that occurs over a wavelength. This is a similar phenomenon to the acoustic short-circuiting that occurs for sound radiation from bending waves below their critical frequency. Only the region of the rail around the forcing point transmits a net power to the bridge, due to the presence of the near field waves in the rail.

Following [11.4, 11.11], an approximation for the power transmitted to the bridge can be derived. Above the decoupling frequency, provided that the wave-number of free waves in the rail, k_r, is much greater than that for the bridge, k_b, the power input to the bridge $W_{in,b}$ can be written as

$$W_{in,b} \approx \frac{1}{2}\text{Re}(Y_b)\left(\frac{K_{eq}|v_r(0)|}{\omega}\right)^2 \tag{11.10}$$

where Y_b is the point mobility of the bridge and K_{eq} is the equivalent stiffness of the fastener system at this point. The rail velocity $v_r(0)$ is the amplitude at the excitation point, $x = 0$. The bracketed term corresponds to the amplitude of the force acting at the base of the fastener system. The equivalent stiffness of the fastener system is given by [11.11]

$$K_{eq} = 2\sqrt{2}\frac{s}{k_r} = 0.45s\lambda_r \tag{11.11}$$

where λ_r is the wavelength in the rail. Figure 11.12 compares the power input to the bridge calculated using this simplified model with the exact result from equation (11.9) for the case of two coupled infinite Euler–Bernoulli beams. The simplified model can be seen to give a good estimate above decoupling frequency, 300 Hz, but at low frequencies it is no longer valid and it overpredicts the input power considerably.

It has thus been seen that, above the decoupling frequency, the power input to the bridge can be calculated using separate models of the track vibration and the bridge mobility, whereas at lower frequencies a coupled model is required.

11.3.3 Bridge mobility

The bridge mobility could be predicted using a finite element model at low frequencies but at higher frequencies the size of such models becomes prohibitive. A simpler model based on an infinite plate or a deep beam can be used instead and provides a reasonable estimate of the frequency-average behaviour, provided that the modal density and modal overlap are high [11.7].

For example, the mobility of an infinite plate of thickness h is given by

$$Y_{plate} = \frac{1}{8h^2}\sqrt{\frac{12(1-\nu^2)}{E\rho}} \tag{11.12}$$

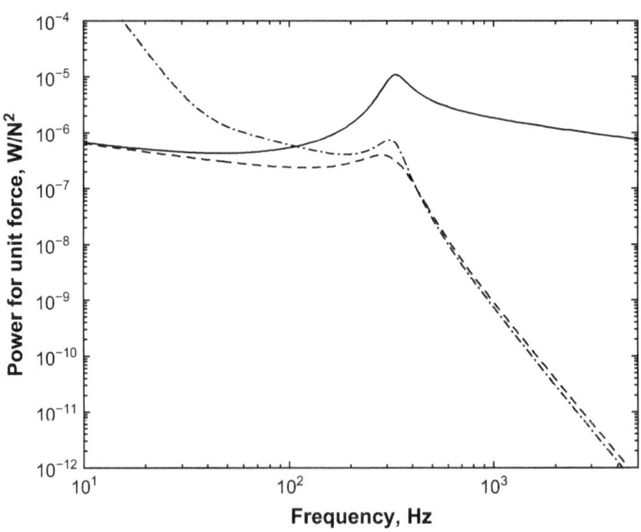

FIGURE 11-12 Power input for a force on the rail of amplitude 1 N calculated for two infinite Euler–Bernoulli beams coupled by a resilient layer. —, power input to the rail; – –, power transmitted to the bridge beam; – · – ·, simplified estimate of power transmitted to the bridge beam

where E is Young's modulus, ρ is the density and ν is Poisson's ratio. The mobility of an infinite Euler–Bernoulli beam is

$$Y_{beam} = \frac{1 - i}{4\omega^{1/2} m'^{3/4} (EI)^{1/4}} \tag{11.13}$$

This can be extended to the case of a Timoshenko beam, see e.g. [11.12]:

$$Y_{beam} = \frac{(1 - i)k}{4\omega m'} \frac{\left(1 - \frac{E}{G^*}\frac{k^4 I^2}{A^2}\right)^{1/2} + i\frac{E}{G^*}\frac{k^2 I}{A}}{\left\{\left(1 - \frac{E}{G^*}\frac{k^4 I^2}{A^2}\right)^{1/2} + i\left(\frac{E}{G^*} + 1\right)\frac{k^2 I}{2A}\right\}^{1/2}} \tag{11.14}$$

where $k = \omega^{1/2}(m'/EI)^{1/4}$ is the wavenumber in the equivalent Euler–Bernoulli beam, $G^* = G\kappa$ is a reduced shear modulus, I is the second moment of area and A is the beam cross-sectional area.

Figure 11.13 shows a comparison of results obtained using an FE model and using models based on a simple plate. These are for a bridge with a concrete deck of length 13.3 m, width 9.5 m and thickness 350 mm supported on four steel I-section beams [11.5]. The FE results are averaged over a number of locations across the bridge span, both above beams and between them. Two results are shown for the simple models, one applying on the plate (based on equation (11.12) above) and the other above the beams (which is a combination of equations (11.12) and (11.14)). It can be seen that,

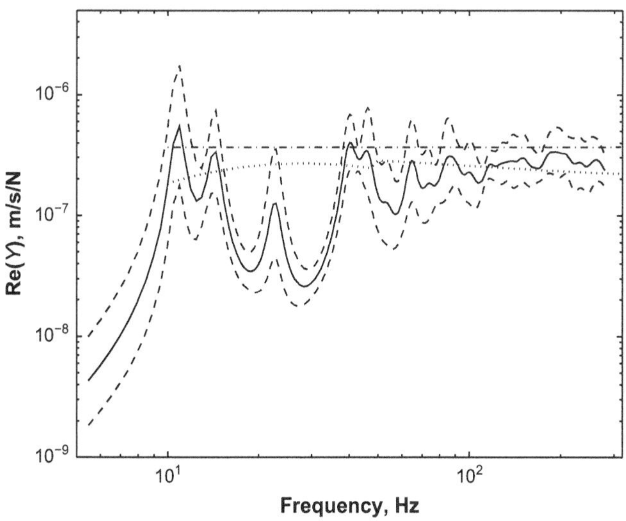

FIGURE 11-13 Comparison of the real part of the input mobility of a concrete/steel composite bridge from a finite element model and from simple models. —, mean of 9 positions from FE model; – – –, mean ± standard deviation from FE model; – · – ·, from infinite plate representing deck; · · · ·, from simplified model of deck plus supporting beam [11.5]

although there are some clear resonances below about 50 Hz, at higher frequencies the simple models give a good approximation.

Where the track is supported above a vertical web, as in Figure 11.14, this acts as a stiff beam and equation (11.14) can be used. However, at high frequencies compressional effects in the height of the beam occur, leading to an increase in the mobility. For an I-section beam (assumed symmetrical) expressions for the mobility are given in [11.6, 11.9]. Above a transition frequency (which is around 1 kHz for typical steel I-beams) the mobility is given by [11.6]

$$Y = \left\{ \frac{1}{\dfrac{\omega(1-\nu^2)}{4Eh_w} + \dfrac{1}{\rho h_w c_R}} + 8\sqrt{\frac{E\rho h_f^4}{12(1-\nu^2)}} \right\}^{-1} \tag{11.15}$$

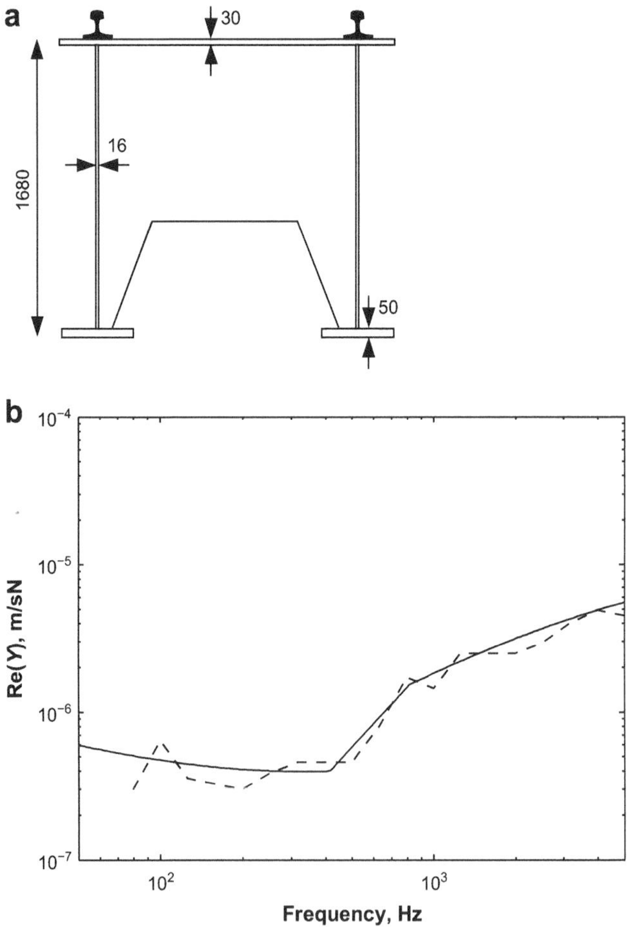

FIGURE 11-14 Comparison of results from simple deep beam model and measured bridge mobility with rail removed [11.4]. – – –, measured; —, simple model

where h_w is the thickness of the web, h_f is the thickness of the flange and c_R is the Rayleigh wavespeed (approximately $0.93(G/\rho)^{1/2}$ for a Poisson's ratio of 0.3 [11.12]). The first term in the above expression is the impedance of an edge-excited semi-infinite plate representing the web and the second is the impedance of the flange plate.

The transition frequency can be found from the first resonance frequency of a rod, representing the web, with point masses at either end, representing the flanges. This can be determined by finding the first root (k_L) of [11.9]

$$\frac{2\mu k_L}{\mu^2 k_L^2 - 1} = \tan(k_L L_w) \tag{11.16}$$

in which $k_L = \omega/c'_L$ is the wavenumber of longitudinal waves in the web, $c'_L = \sqrt{E/\rho(1-\nu^2)}$ is the longitudinal wavespeed in a plate, L_w is the height of the web and $\mu = h_f L_f / h_w$ is the ratio of the end mass to the mass per unit height of the web. Below this transitional frequency the mobility is assumed to be proportional to f^2.

Figure 11.14 shows results from this model which are compared with measurements of mobility on a bridge with the rail removed, from [11.4]. This is modelled as an I-section beam with $L_w = 1.68$ m, $h_w = 0.016$ m, $L_f = 0.5$ m, $h_f = 0.03$ m and $\kappa = 0.6$. Excellent agreement is found with the measurements. Below about 400 Hz the bridge behaves as a Timoshenko beam, equation (11.14). Between about 400 and 800 Hz the mobility increases in proportion to f^2 and above this the deep beam model of equation (11.15) applies.

To minimize the input mobility of a bridge, it is important that the rail is placed directly above stiffening girders. If the distance between the rail and such girders is more than about a quarter of the bending wavelength in the deck plate, they will have no effect on the mobility. Figure 11.15 illustrates this by comparing the mobility of

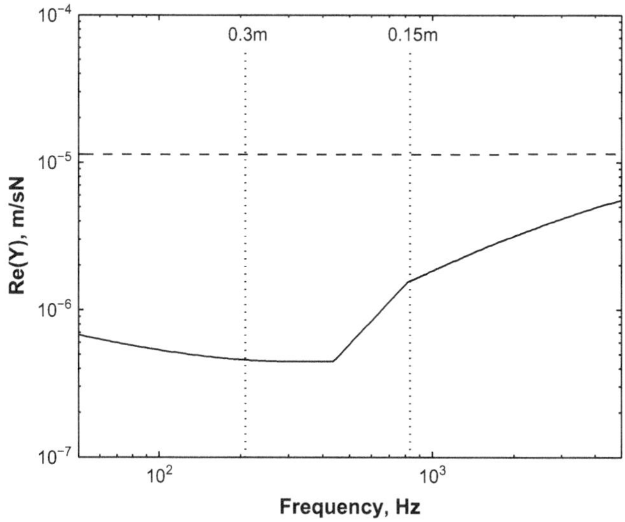

FIGURE 11-15 Real part of bridge mobility. —, deep beam (as in Figure 11.14); – – –, deck plate (30 mm steel); · · · ·, frequencies at which $\lambda/4$ in the deck plate equals 0.3 m or 0.15 m

the I-beam model from Figure 11.14 with the mobility of an infinite plate representing the deck. At low frequencies the wavelength in the plate is long and the beams will dominate the mobility. However, at high frequency the wavelength in the plate reduces and, once about a quarter of a wavelength becomes less than the distance to the beam, the mobility will be dominated by that of the plate. Thus, for a forcing point that is 0.3 m away from the beam, the mobility will rise to that of the plate at around 200 Hz; at 0.15 m from the beam it will rise to that of the plate at around 800 Hz. In each case this leads to an increase in the mobility by a factor of 10 and consequently a 10 dB increase in power transmitted to the bridge. In practice, there will be a gradual transition between the plate and beam mobilities over a broad frequency range.

11.3.4 Ballasted track: spring with internal resonances

Some examples of measured vibration level difference between the rail and the bridge, from [11.13], are given in Figure 11.16. In general, it can be seen that the difference increases towards higher frequencies. It can also be seen that the isolation afforded by ballasted track is greater than that found for direct fastening systems, especially on steel structures.

At low frequencies ballast can be taken to act as a spring, but the wavelength of compressional waves within the ballast layer is such that internal resonances begin to occur at around 200–400 Hz. This results in an increase in the effective stiffness for higher frequencies. Ballast can be treated approximately as an elastic medium.

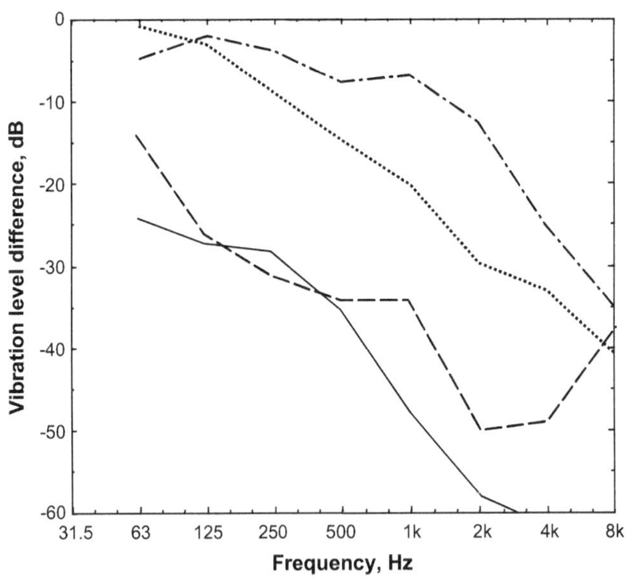

FIGURE 11-16 Examples of vibration level difference between rail and bridge measured on various bridges with different rail fastener systems. —, concrete bridge with ballast; – – –, concrete bridge with direct fastening; ····, steel bridge with ballast; – · – ·, identical steel bridge with direct fastening [11.13]

Assuming an elastic modulus E, density ρ, cross-sectional area (plan view) A and depth h, the point and transfer stiffnesses are [11.14]

$$K_{11} = \frac{EAk}{\tan(kh)}; \; K_{12} = \frac{EAk}{\sin(kh)} \tag{11.17}$$

where $k = \omega\sqrt{\rho/E}$ is the wavenumber of compressional waves in the layer. At low frequencies these stiffnesses tend to $K_{11} = K_{12} = EA/h$ but at high frequencies they show standing wave behaviour. This is illustrated in Figure 11.17 for typical parameters, $E = 1.2 \times 10^8$ N/m^2, $\rho = 2000$ kg/m^3, $A = 0.25$ m^2 and $h = 0.3$ m. The damping loss factor is set to 0.1, making E and hence k complex in equation (11.17). The point stiffness shows a resemblance to the measured data in Figure 3.55. The transfer stiffness drops at high frequencies due to attenuation of waves propagating through the ballast; for higher loss factors this drop becomes greater.

From these results it can be seen that, although ballast acts as a resilient layer between the track and the bridge, its stiffness is not particularly low and increases towards higher frequencies. Hence, the higher 'isolation' afforded by a ballast layer may be due to the addition of mass and damping to the bridge structure, rather than the resilience afforded by the ballast.

Noise measurements were reported in [11.15] on identical steel bridges with ballast and with direct fastening. It was found that the bridge with ballasted track was

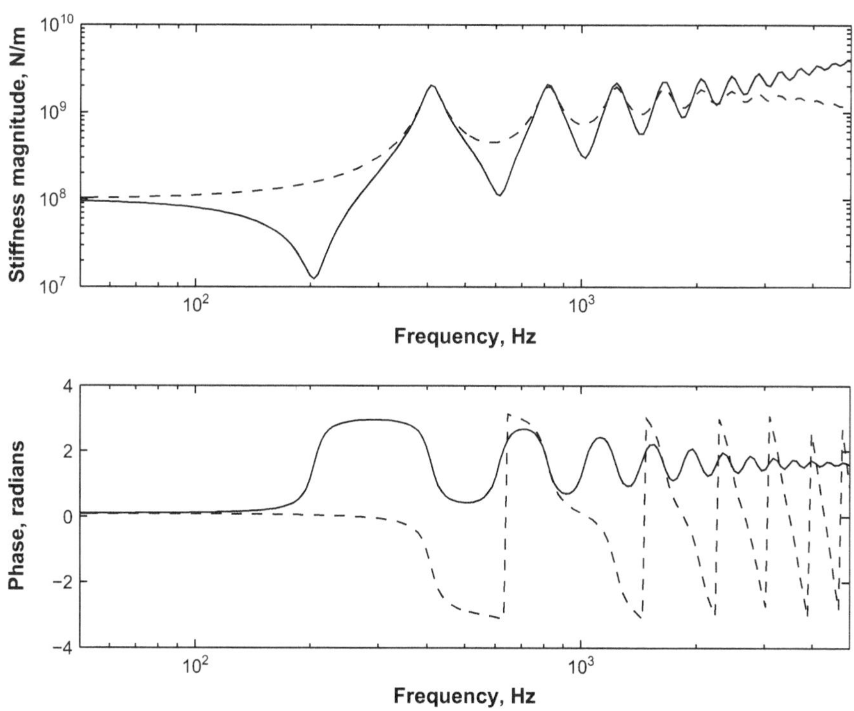

FIGURE 11-17 Calculated stiffness of a layer of ballast of depth 0.3 m. —, point stiffness K_{11}; – – –, transfer stiffness, K_{12}

13 dB(A) quieter, see also [11.16]. However, care should be taken in interpreting this result, as the track on these bridges was mounted on a thin deck plate supported between outer girders. The bridge mobility would consequently be quite high and the ballast would reduce this by increasing its mass and damping. In [11.17] the steel deck of a similar bridge was covered with a layer of sand. This reduced the vibration of the deck plate above 125 Hz, with a reduction of more than 10 dB between 250 and 630 Hz. The noise level under the bridge was reduced by approximately 10 dB(A).

11.4 VIBRATION TRANSMISSION AND RADIATION OF SOUND

11.4.1 Transmission of vibration through the bridge structure

Once vibrational power has been injected into the bridge, this is transmitted into neighbouring components, where it is dissipated by damping or radiated as sound. Bridges are large and complex structures, which have many modes of vibration in the frequency range of interest, especially steel bridges. Finite element models are therefore usually only relevant for low frequency analysis. As frequency increases, and finer detail is required in an FE model, models rapidly become too large and an alternative approach is required.

An alternative numerical approach that can be used to study more complex bridge structures is the Waveguide Finite Element Method [11.18]. Where the cross-section of the structure in the y–z plane is constant along the x-direction, it is sufficient to mesh the cross-section while assuming wave solutions in the x-direction. This method can be used to calculate the bridge mobility or to investigate the coupling between the rail and the bridge directly.

Another approach is to use a power balance method. This allows the transmission of vibration and radiation of sound to be modelled without the need to know much of the detail of the components of the bridge. The actual modes of vibration are not considered; instead an average response is calculated on the basis that many modes are present within a frequency band.

Statistical Energy Analysis (SEA) [11.7] is one such method that has been used successfully to model the response and radiation of railway bridges. For the Gavignot bridge, which is a relatively short single track steel bridge, an FE model was used up to 200 Hz and an SEA model based on 259 subsystems was used for frequencies above 200 Hz [11.3]. In [11.19, 11.20] a similar combination of FE and SEA models was used to calculate the noise radiation by concrete bridges.

In an SEA model the structure is divided into subsystems. The power flow between two subsystems is assumed to be proportional to the difference in their average modal energy [11.7]. The constant of proportionality is usually written in terms of a coupling loss factor, by analogy with the damping loss factor used to describe dissipation. The power balance between the subsystems is then established and solved to find the energies of each subsystem.

For a single subsystem, such as a plate, the power balance equation can be written

$$W_{in} = W_{diss} + W_{rad} \approx W_{diss} \tag{11.18}$$

where W_{diss} is the dissipated power and W_{rad} is the radiated sound power. The radiated sound power is usually much smaller than the dissipated power and can be neglected in the above equation. The dissipated power can be written as

$$W_{diss} = \omega\eta E = \omega\eta M\langle\overline{v^2}\rangle \tag{11.19}$$

where E is the vibrational energy of the subsystem, η is the damping loss factor, M is the mass of the subsystem and $\langle\overline{v^2}\rangle$ is the spatially averaged mean-square velocity. Equations (11.18) and (11.19) can be used to determine the average vibration of the system

$$\langle\overline{v^2}\rangle = \frac{W_{in}}{\omega\eta M} \tag{11.20}$$

from which the radiated power can subsequently be found (see Section 11.4.2 below).

If a second plate with no external excitation is coupled to the first, according to the SEA assumptions the net power transmitted to the second plate can be written as

$$W_{1\rightarrow 2} = \omega\eta_{12}E_1 - \omega\eta_{21}E_2 \approx \omega\eta_{12}E_1 \tag{11.21}$$

where E_i is the vibrational energy of each subsystem and η_{ij} are the coupling loss factors. The approximation in equation (11.21) relies on the assumption of 'weak coupling' between subsystems. This transmitted power can be equated with the power dissipated in the second subsystem:

$$W_{1\rightarrow 2} = W_{diss,2} = \omega\eta_2 E_2 \tag{11.22}$$

to give

$$E_2 = \frac{\eta_{12}}{\eta_2}E_1 \tag{11.23}$$

For more complex SEA networks a matrix equation is established for the power balance and solved to find the subsystem energies.

Here, a simpler model will be presented which avoids the need to determine the coupling loss factors [11.4, 11.21]. If the two subsystems are 'strongly coupled', their average *modal* energies will be equal. Since the number of modes in a frequency band may differ in the two subsystems, their energies are in the ratio

$$E_2 = \frac{n_2}{n_1}E_1 \tag{11.24}$$

where n_i is the modal density of subsystem i. The modal density of a plate (the average number of natural frequencies in a frequency band of width 1 Hz) is given by

$$n(f) = \frac{\sqrt{3}S}{hc'_L} \tag{11.25}$$

where S is the surface area, h is the thickness and $c_L' = \sqrt{E/\rho(1-\nu^2)}$ is the longitudinal wavespeed in the plate. Hence for two plates of the same material

$$E_2 = \frac{S_2 h_1}{S_1 h_2} E_1 \tag{11.26}$$

Thus the mean-square velocities of the two plates are in the ratio

$$\frac{\langle \overline{v_2^2} \rangle}{\langle \overline{v_1^2} \rangle} = \frac{S_1 h_1 E_2}{S_2 h_2 E_1} = \frac{h_1^2}{h_2^2} \tag{11.27}$$

and thinner plates will have a higher vibration level than thicker ones.

Extending this model to an assembly of an arbitrary number of plates, the spatially averaged mean-square velocity on plate n, $\langle \overline{v_n^2} \rangle$ is given in terms of the acoustic power input to the system W_{in} by

$$\langle \overline{v_n^2} \rangle = \frac{W_{in}}{\omega \eta \rho_s h_n^2 \sum_j (S_j/h_j)} \tag{11.28}$$

Compared with a full SEA model, this approach will underestimate the response of subsystems close to the excitation point and overestimate the response of those further away. However, if all components have fairly similar radiation ratios the overall radiated sound power is not much affected by this simplification. Moreover, it is not critical in this model how much of the bridge structure is included, provided that all components have roughly similar properties. For example, in representing a two-track bridge it would be possible to include both tracks in the model or just one; in the latter case the vibration would be 3 dB higher but the radiating area would be only half as large so the sound power would be the same.

The above approach is not well suited to situations where plate thicknesses differ greatly, for example for a thick concrete deck supported by beams formed of thin steel plates [11.9].

From equations (11.20) and (11.28) it can be seen that the vibration of the bridge depends strongly on the damping loss factor. Damping loss factors measured on steel bridges are in the region between about 0.01 and 0.1, as shown in Figure 11.18. The loss factors for concrete structures are generally expected to be a little higher than this, although measured values are not available and are difficult to obtain.

11.4.2 Sound radiation

The sound power radiated by a vibrating plate can be determined using

$$W_{rad} = \rho_0 c_0 \sigma S \langle \overline{v^2} \rangle \tag{11.29}$$

where ρ_0 is the density of air, c_0 is the speed of sound, S is the surface area of the plate and σ is its radiation ratio (see Chapter 6). In general $\sigma \ll 1$ at low frequencies, and $\sigma \approx 1$ at high frequencies. For plates, there is also a region in the mid-frequencies

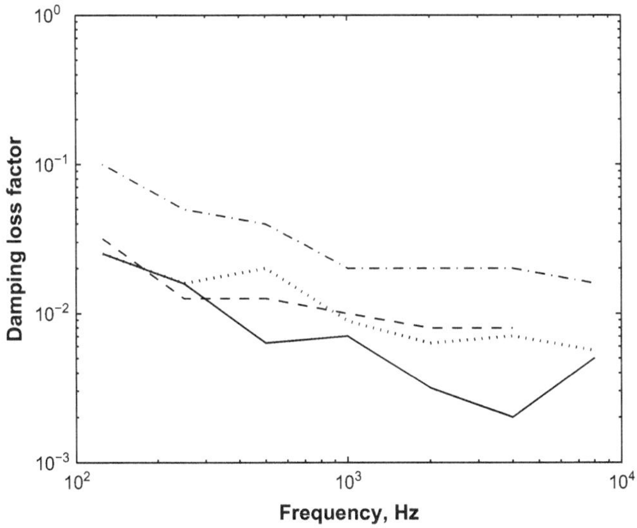

FIGURE 11-18 Examples of damping loss factors measured on steel bridges. —, box girder 54 m long; – – –, swing bridge, riveted girder 50 m long; · · · ·, swing bridge, open girder 43 m long; – · – ·, lifting bridge, open girder 11 m long [11.22, 11.23]

where 'acoustic short-circuiting' occurs (when bending waves have a shorter wavelength than that in air), which has the effect of lowering σ.

The radiation ratio of a plate can be estimated using asymptotic formulae such as those derived by Maidanik [11.24] for a simply supported baffled plate of dimensions $a \times b$, see also [11.25]. These will be summarized here. At very low frequencies:

$$\sigma = \frac{4S}{c_0^2} f^2, \qquad \text{for} \quad f < f_{1,1} \tag{11.30}$$

where $S = ab$ is the surface area and $f_{1,1}$ is the fundamental natural frequency of the plate, given by

$$f_{1,1} = \frac{\pi}{2}\left(\frac{1}{a^2} + \frac{1}{b^2}\right)\sqrt{\frac{D}{\rho h}} \tag{11.31}$$

where $D = Eh^3/12(1 - \nu^2)$ is the plate bending stiffness. In the so-called corner mode region the radiation ratio is given approximately by

$$\sigma = \frac{4\pi^2}{c_0^2 S} \frac{D}{\rho h}, \qquad \text{for} \quad f_{1,1} < f < f_e \tag{11.32}$$

where the upper limit of this region, f_e, is given by

$$f_e = \frac{3c_0}{P} \tag{11.33}$$

in which $P = 2(a + b)$ is the perimeter length. As frequency increases, the 'edge mode' region is reached in which the radiation ratio is given approximately by

$$\sigma = \frac{Pc_0}{4\pi^2 S f_c} \times \frac{(1-\alpha^2)\ln\left(\frac{1+\alpha}{1-\alpha}\right) + 2\alpha}{(1-\alpha^2)^{3/2}}, \qquad \text{for} \quad f_e < f < f_c \tag{11.34}$$

where $\alpha = \sqrt{f/f_c}$. The critical frequency f_c is given by

$$f_c = \frac{c_0^2}{2\pi}\sqrt{\frac{\rho h}{D}} \tag{11.35}$$

Around the critical frequency the radiation ratio is limited to a maximum of

$$\sigma \leq 0.45\sqrt{\frac{Pf_c}{c_0}} \tag{11.36}$$

while above f_c it is given by

$$\sigma = \frac{1}{\sqrt{1 - f_c/f}} \tag{11.37}$$

The radiation ratios for steel plates of various thicknesses, estimated using these formulae, are shown in Figure 11.19. The degree of acoustic short-circuiting, below f_c, can be seen to increase as the plate dimensions increase or as the plate becomes thinner.

For a structure consisting of a number of plates, if σ_n is the radiation efficiency of plate n, the total radiated sound power, W_{rad}, is given by

$$W_{rad} = \rho_0 c_0 \sum_n \sigma_n S_n \langle v_n^2 \rangle \tag{11.38}$$

The above formulae are based on a model of a plate set in an infinite baffle. For an unbaffled plate the low frequency radiation is reduced further [11.26].

It is possible to reduce greatly the radiation efficiency of a plate by introducing perforation [11.27]. This increases the potential for acoustic short-circuiting between the front and back of the plate. No applications are known in practice for bridges although it has been considered in research projects. Nevertheless, girders constructed of open lattices of beams can be expected to radiate less noise than solid girders.

The directivity with which sound is radiated from a bridge may also be important, so that different sound pressures can be experienced at different locations around the bridge. However, it seems that a simple omnidirectional model can often be used with reasonable success to estimate the sound pressure level at a particular position.

11.4.3 Results

Some results of using the calculation model described above are given in Figure 11.20, taken from [11.4]. These results show octave band spectra of

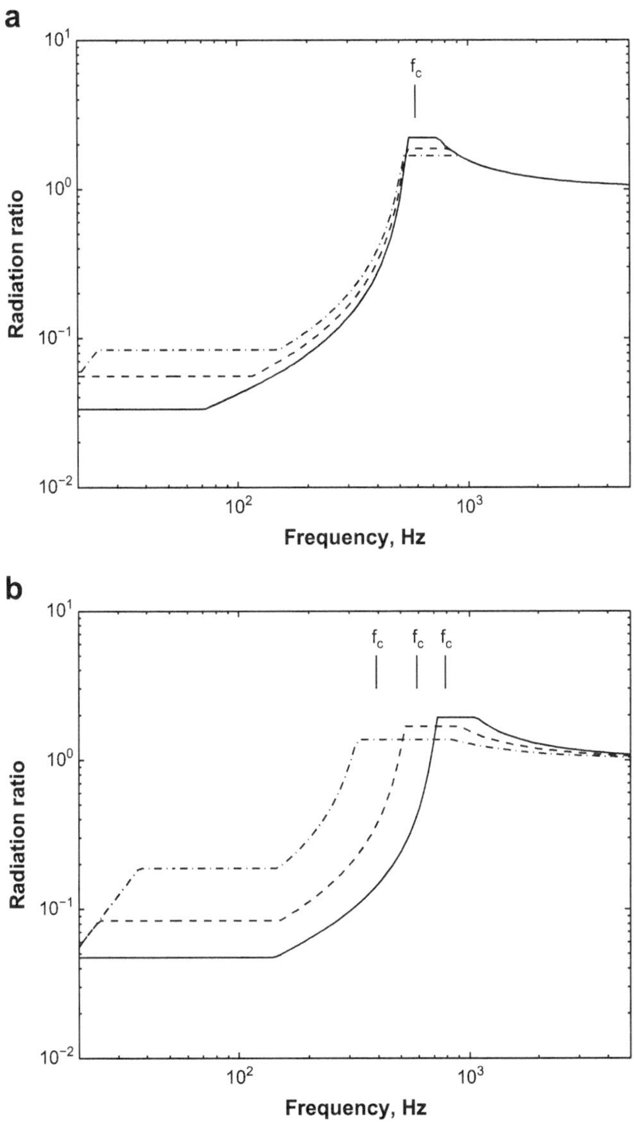

FIGURE 11-19 Radiation ratios of steel plates. (a) 20 mm thick for different dimensions: —, 5 × 2 m; – – –, 3 × 2 m; – · – ·, 2 × 2 m. (b) 2 × 2 m for different thicknesses: 20 mm thick: —, 15 mm; – – –, 20 mm; – · – ·, 30 mm. The critical frequency, f_c, is indicated in each case

A-weighted sound at 7.5 m from the nearest track (in one case 25 m). It can be seen that the agreement with measured data is mostly reasonably good. In each case bridge noise is dominant at low frequencies, whereas above about 1 kHz the rolling noise (wheel and rail) is dominant.

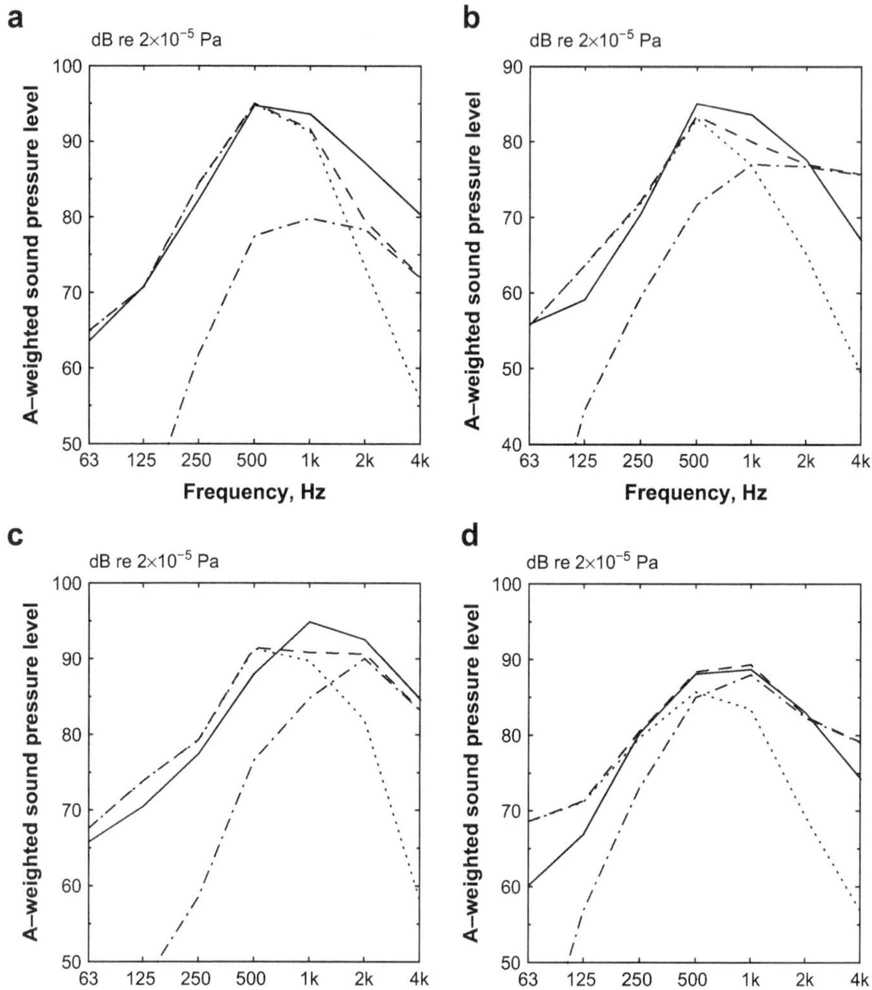

FIGURE 11-20 Measured and predicted noise at 7.5 m from various steel bridges. —, measured noise; ····, predicted bridge noise; – · – ·, predicted rolling noise; – – –, total predicted noise. (a) Swing bridge, open girder 43 m long, (b) swing bridge, open girder 50 m long with soft rail fasteners (noise at 25 m), (c) bow-string girder bridge 250 m long, (d) box girder 54 m long with embedded rails [11.4]

Particular differences can be noted between these results. The bridge in Figure 11.20(b) was fitted with resilient rail fasteners which have reduced the noise from the bridge (a direct comparison with the other results is not possible as the measurement distances differ). The bridge in Figure 11.20(d) had embedded rails, also giving reduced bridge noise but a clear increase in the noise from the rails can be seen (embedded rails are discussed in Section 7.4.5).

Figure 11.21 shows results from [11.6]. These were obtained on the concrete/steel composite bridge shown in Figure 11.3. This had a 0.4 m thick concrete deck supported on four steel I-beams 1 m high. The vibration of the deck is predicted reliably over much of the frequency range. The noise is also predicted well for frequencies between about 50 and 500 Hz. At higher frequencies the rolling noise, not included in the prediction, is responsible for the additional measured noise.

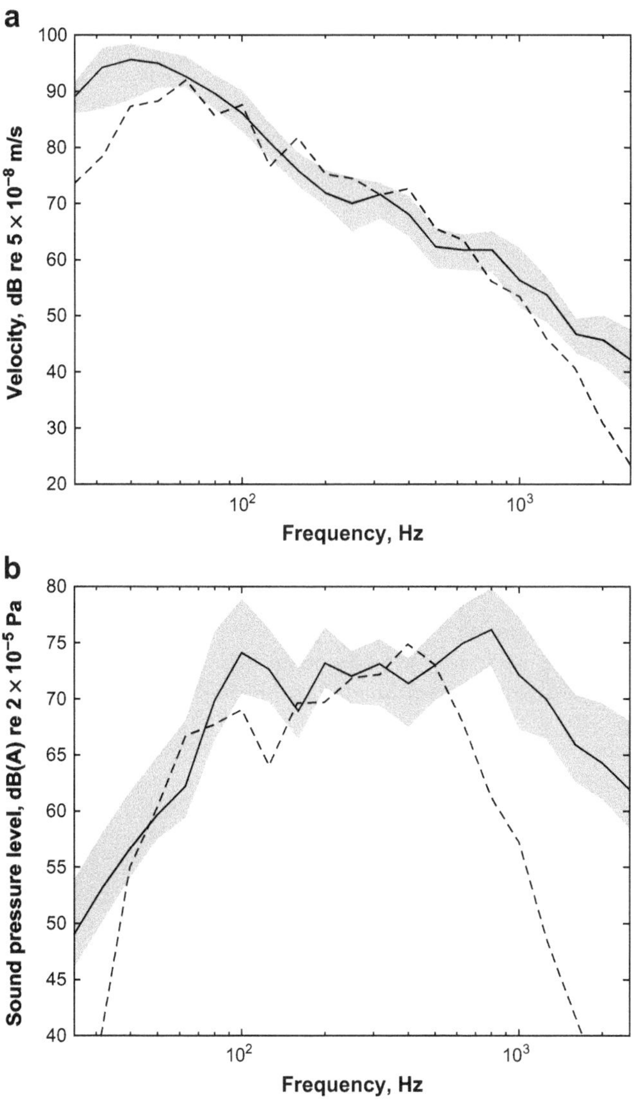

FIGURE 11-21 Comparison of measurements and predictions for concrete/steel composite bridge of Figure 11.3. (a) Average vibration on the bridge deck, (b) sound pressure level at 5 m below the viaduct. —, measured average; – –, predicted; shaded region shows range of levels measured [11.6]

11.5 REDUCING BRIDGE NOISE

Based on the model outlined above, it is possible to identify the parameters that have an effect on the noise from a train on a bridge. The main parameters that can be influenced will be discussed further here, to give some guidelines on how to control bridge noise. This will be illustrated by practical examples. Clearly, some of the techniques discussed will not be applicable to existing bridges but only to new designs.

11.5.1 Fastener stiffness

At first sight it should always be beneficial to reduce the stiffness of the resilient track supports in order to increase the vibration isolation between rail and bridge. However, in practice this does not necessarily produce as large an effect as might be expected.

For example, some results were given in [11.28] in which a range of fasteners were used with stiffnesses varying by up to a factor 20. They were applied on a steel twin-girder bridge with the track mounted on wooden sleepers directly across the girders. The effect on noise spectra was found to be mostly less than 5 dB. The main reason appears to be that, in this application, the wooden sleepers also provided considerable resilience in series with the fasteners. In the case of the stiffer fasteners, this will be greater than the resilience of the fasteners themselves. In [11.29] the transfer stiffness of a wooden sleeper was measured as 265 MN/m.

Another limitation of applying softer rail fasteners is that the track decay rates may be reduced, see Chapter 7. This in turn increases the average vibration of the rail and hence its noise radiation. Application of soft baseplates on a 250 m long steel bow-string girder bridge at Weesp in the Netherlands illustrates this effect. From the reduction in stiffness, a noise reduction of 8 dB was expected above the decoupling frequency, but the actual noise reduction was much smaller, largely due to the increase in rail vibration [11.30].

Although the application of soft baseplates is beneficial for reducing the bridge noise, they should clearly be implemented with care. Nevertheless, in many cases the application of soft rail fasteners has been successful in producing some reduction in noise. In a trial on the bridge at Gavignot a reduction of 3–4 dB(A) was achieved using two alternative rail fasteners in place of wooden sleepers (see Section 11.6.1).

On the Årsta bridge, Figure 11.2, resilient fasteners with a stiffness of 19.5 MN/m were applied in place of wooden sleepers. An overall noise reduction of 3 dB(A) was achieved, with reductions of over 10 dB below 250 Hz [11.31]. These results are shown in Figure 11.22. The predictions indicated that the bridge noise was reduced by about 15 dB(A) while the rolling noise was increased by about 3 dB(A). Clearly, to achieve any further reductions it would be necessary to treat the rolling noise by damping the rail or introducing shielding.

In another application in Switzerland [11.32], resilient baseplates were installed on a 57 m long bridge. This was initially fitted with wooden sleepers with no rail pad. Two types of resilient baseplate were fitted, both with a stiffness of about 20 MN/m. In addition, hanging steel sleepers between the wooden sleepers (used to form a walkway) were also replaced. A reduction of about 10 dB was obtained in the frequency range 80 to 400 Hz, where bridge noise is dominant, while the A-weighted level was reduced by 2–4 dB.

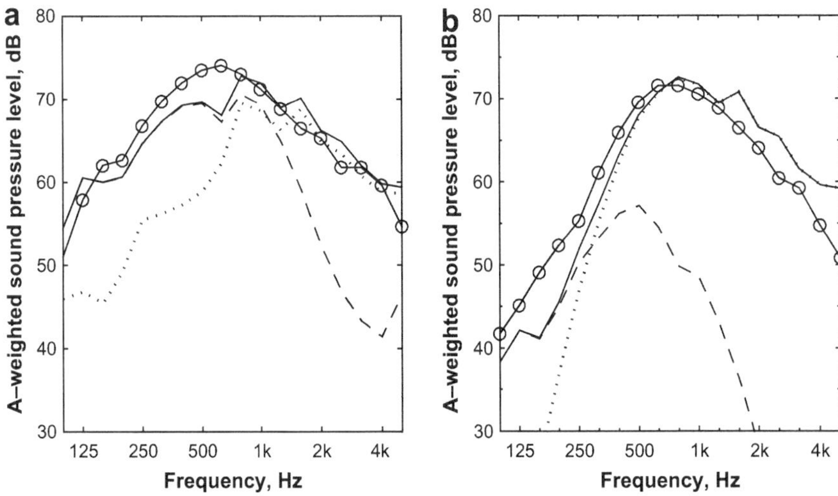

FIGURE 11-22 A-weighted noise levels at about 50 m from Årsta bridge, Stockholm, (a) before installation of resilient rail fasteners, (b) after installation [11.31]. o—o, measured noise; —, total predicted noise; – – –, predicted bridge noise; · · · ·, predicted rolling noise

11.5.2 Rail damping

By applying a damping treatment to the rail, it is possible to reduce the average rail vibration by increasing the attenuation with distance along the rail. The main effect of this is to reduce the component of rolling noise on the bridge, which as has been seen is usually greater with direct fasteners than on ballasted track. Methods of adding damping to the rail are discussed in Chapter 7. It is unclear whether increasing the rail damping will also affect the excitation of bridge noise. According to the model described in Section 11.3 it will not, since only the rail in the vicinity of the forcing point transmits power to the bridge and this is unaffected by rail damping.

The application of a rail damper to the track on the Gavignot bridge is described in [11.3]. The dampers were similar to those shown in Figure 7.31(b). The noise

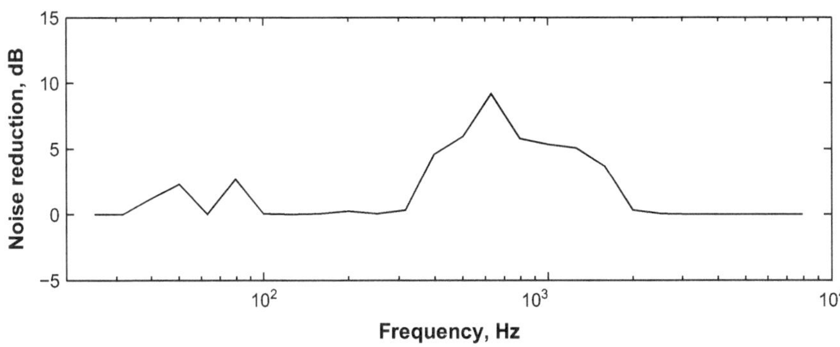

FIGURE 11-23 Noise reduction due to use of rail absorbers on Gavignot bridge [11.3]

reduction achieved is plotted in Figure 11.23. Over the region 400–1600 Hz a considerable reduction in noise was obtained due to a reduction in the noise radiated directly from the rail. This is a region where a large amplification of noise occurred on the bridge, as seen in Figure 11.5, and the A-weighted level was reduced by 3–4 dB. Below 400 Hz, however, the main noise source is probably due to radiation from the bridge and this was not affected by the rail damper. Above 2 kHz the noise radiation will be dominated by the wheels and was similarly unaffected.

11.5.3 Ballast mats

For ballasted track, it is possible to introduce ballast mats under the ballast in order to increase the vibration isolation. Such mats are often used for track in tunnels – see Chapter 13. A typical stiffness per unit area is of the order of 30 MN/m^3, which gives a stiffness per sleeper end of around 20 MN/m. Improvements in vibration isolation of typically 15 dB for 100–400 Hz have been found in tunnels. Measurements of vibration isolation on bridges are much less common, and in any case the mats used on bridges are often stiffer than for tunnels.

As with the ballast itself, internal resonances can occur at relatively low frequencies within a ballast mat. Unfortunately, most measurements of the effects of such mats are limited to frequencies below 400 Hz, and it is not clear whether the improvements found here are also always found at higher frequencies.

Measured results from Japan quoted in [11.1] include an 8 dB(A) reduction for use of a ballast mat on a steel bridge deck. In [11.33] a reduction of 7 dB(A) under a concrete viaduct is reported, with the improvement said to be particularly conspicuous between 250 and 1000 Hz. Some spectral results are also given in [11.34] showing improvements of 10 dB or more in much of the frequency range.

11.5.4 Damping of bridge structure

It has been seen in Section 11.4.1 that the noise can be reduced if the damping of the bridge structure can be increased. There are several examples in the literature of the application of constrained layer damping treatments to steel box girder bridges.

In [11.35] two small steel box girder bridges of height 1 m were treated with different constrained layer damping treatments. Noise reductions of 13 and 18 dB(A), respectively, at 25 m were achieved, which is remarkably successful. Because relatively thick constraining plates were used, and the whole structure was treated, the increase in weight was approximately 25%. The initial damping of these bridges seems to have been very low. The initial loss factor was estimated in [11.35] as 0.0015. This was increased after the treatment to around 0.04, which is comparable to the results in Figure 11.18. It seems, therefore, that application on other bridges would not be so successful, even if the large added weight was allowed.

Results are given in [11.2] for a twin track bridge which had a 9 m wide concrete deck supported by a steel box girder. A damping treatment was applied to the steel girder, resulting in significant reductions in its vibration (10–15 dB), but only small differences in sound level (about 1 to 2 dB(A)). It is clear that here, in contrast to the previous example, rolling noise and/or noise from the concrete deck contributed more significantly to the measured sound level than the girders which were treated.

Tests carried on the bridge at Gavignot [11.3] included the application of tuned absorber dampers on the deck of the bridge. These are shown in Figure 11.24.

FIGURE 11-24 Tuned absorbers applied on steel bridge deck. Photo: SNCF

This achieved a 4 to 6 dB reduction between 30 and 50 Hz but had no effect on the A-weighted level.

11.5.5 Plate thickness

As seen in Section 11.4.1, thicker plates tend to have a lower vibration amplitude. However, as seen in Figure 11.19, the effects of acoustic short-circuiting are smaller for thicker plates, and these two effects tend to cancel each other out in much of the frequency range. It is instructive to consider the ratio W_{rad}/W_{in}. For a single plate, this is given by

$$\frac{W_{rad}}{W_{in}} = \frac{\rho_0 c_0 \sigma}{\omega \eta \rho h} \tag{11.39}$$

By averaging this ratio over the frequency range 250–1000 Hz, an indication of the effect of plate thickness is found [11.4]. This is shown in Figure 11.25 for a typical plate of dimensions 2 × 2 m and damping loss factor 0.01. It can be seen that the sound radiation is very insensitive to plate thickness in the range between about 15 and 40 mm, which is typical of the thicknesses found in steel bridge structures. Only quite extreme changes of thickness could lead to a significant noise reduction. It can therefore be concluded that it is more effective to minimize the input power, in ways discussed above. The same applies to the addition of stiffeners to the plates of a bridge.

11.5.6 Closed structures

Figure 11.26 compares two bridge designs. On the left is an open girder consisting of a deck plate and two webs, while on the right a box girder is shown which is closed by a bottom plate and also by plates at either end. For a closed structure, such as this

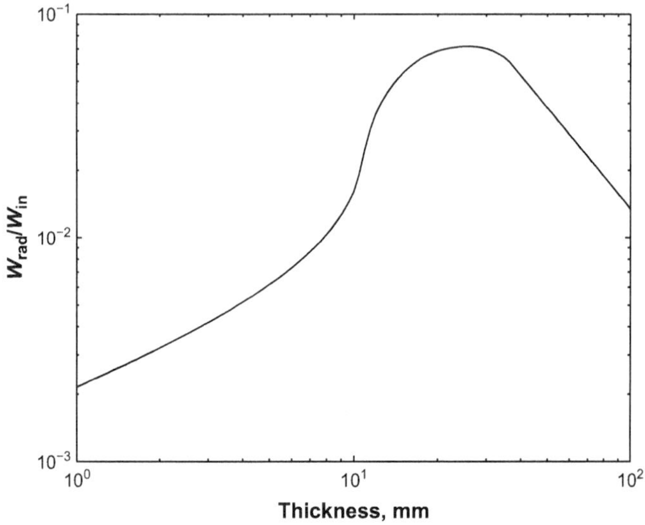

FIGURE 11-25 Ratio of radiated sound power to vibrational power input to a plate (averaged over 250–1000 Hz)

box girder, the sound radiation from the inner surfaces is contained within the structure. This has the effect of reducing the effective radiating area by up to a factor 2, and hence the radiated sound power by 3 dB. The structure must be completely closed, however, as otherwise the noise from the inner surfaces could escape through any gaps.

11.5.7 Barriers and enclosures

Barriers can be used on a bridge to shield rolling noise in the usual way. An absorbent barrier was applied on the Gavignot bridge which gave a 4–5 dB reduction.

Bridge noise itself, however, is radiated from a much larger area, which makes it much more difficult to attenuate by the use of barriers. There is often limited

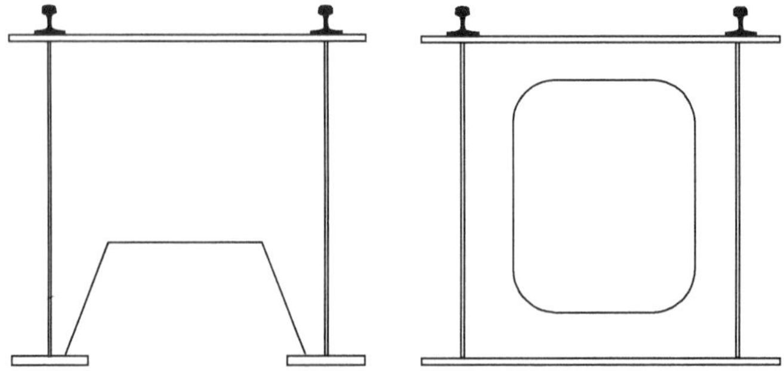

FIGURE 11-26 Typical open and closed girder designs

clearance under the bridge which places severe limitations on the ability to shield radiation from the underside.

Nevertheless, on various different types of bridge on the Shinkansen lines in Japan, extensive tests were carried out into the effectiveness of various barriers, see [11.1]. These results are summarized in Table 11.2. The effectiveness of simple sidewalls on bridges (a), (d) and (e) indicates that noise from the top of the bridge (including wheel/rail noise) was dominant in these cases. It was found that large reductions could be achieved by partial or total enclosure of the bridge. The largest attenuation found was 27 dB for a total enclosure.

These undercovers are also described in [11.33]. They consisted of 12 mm thick plasterboard, to which 0.3 mm lead plates were glued. The reduction in the noise directly under a bridge was about 8–9 dB(A).

The bridge carrying the Channel Tunnel Rail Link over the East Coast Main Line in London was constructed with a complete enclosure above rail level to minimize noise at nearby sensitive locations [11.36]. The single span bridge of 70 m was of a lightweight construction to allow it to be lifted into place. The track was also fitted with a floating slab on the bridge to minimize transmission to the bridge. A similar full enclosure has also been used at some locations on the West Rail viaducts in Hong Kong, see Section 11.6.3 below.

11.6 CASE STUDIES

In this final section three specific examples are discussed where various countermeasures have been considered in an integrated solution.

11.6.1 SNCF/RFF Gavignot bridge

The RFF funded research project on the Gavignot bridge [11.3], see Figure 11.1, has been mentioned several times in this chapter. This was used as a testbed for a number of modifications. The initial increase in noise on this bridge relative to plain track at grade was shown in Figure 11.5 and corresponded to 12–14 dB(A).

The results of various noise reduction measures are summarized in Table 11.3. In this case the main source was found to be the rail, especially once the resilient fasteners had been installed. Absorbent screens were installed inside the main girders and gave a reduction of 4–5 dB. Similar reductions were achieved using either rail absorbers, see Figure 11.23, or two-stage resilient baseplates. However, the combination of resilient fasteners and rail absorbers was particularly effective. This gave a greater reduction than the two measures in isolation, as one reduced the bridge noise and the other the rail noise. Tuned absorbers applied to the bridge deck, shown in Figure 11.24, only had an effect at low frequencies.

11.6.2 Nieuwe Vaart 'Silent Bridge'

Research in the Netherlands by the 'Silent Bridges' consortium led to the development of a quiet steel bridge with integral embedded rails [11.37, 11.38]. As a first step the calculation model was run for a large number of notional bridge girder geometries. The parameters varied were deck and bottom plate thickness (assumed

TABLE 11-2 SUMMARY OF NOISE REDUCTION MEASURES APPLIED TO VARIOUS BRIDGES ON SHINKANSEN [11.1]

Bridge type	Treatment	Noise reduction, dB
(a) Wooden sleepers on steel plate girders beneath the track	1. Side barrier (reflective) and footboard below sleepers	7
	2. As 1 plus damping of girders and footway	9
	3. As 2 plus undercover and absorptive barriers	19
(b) Wooden sleepers on bearers with lattice girders at the sides	1. Pads under sleepers	0
	2. As 1 plus undercover and 2 m high absorptive sidewalls	11
	3. As 2, sidewalls 4 m high	17
	4. As 2, sidewalls 6 m high	17
	5. Complete enclosure	27
(c) Ballasted track on steel deck plate between steel plate girders	1. Ballast mat, 25 mm thick	8
	2. As 1 plus barrier on side girder	13
	3. As 2 plus damping of side girders and footway	13
	4. As 3 plus undercover, side barrier extending above track	16
(d) Concrete bridge with direct fasteners	1. Sidewall 1.9 m high, reflective	7
	2. Overhanging 2 m high barrier	10
	3. As 2 plus absorptive undercover	11
	4. As 3 plus ballast layer on slab	11
(e) Concrete bridge with ballasted track	1. Sidewall 1.9 m high, reflective	7
	2. As 1 plus ballast mat	10
	3. Absorptive sidewall 2.4 m high plus ballast mat	11
	4. As 3 plus 1.5 m absorptive overhang	12

the same for simplicity), web thickness and total height. Structures were only included if they had sufficient mechanical strength to act as a railway bridge.

Figure 11.27 shows the results plotted against relative section mass. Compared with the reference design it was found to be possible to obtain considerable reductions in noise but only by allowing an increase in the mass, and in some cases also the height.

Next, scale models (1:4) were constructed of three bridge girders – the reference girder and two low noise variants, see Figure 11.28. These variants were predicted to reduce the bridge noise by around 5 to 7 dB. One of the most important aspects was to reduce the input mobility which meant increasing the thickness of the web directly beneath the rail. Measured results on these scale models confirmed the results of the calculation model.

TABLE 11-3 SUMMARY OF RESULTS FROM GAVIGNOT BRIDGE

Solution	Measured noise reduction, dB
Absorbent screen	4–5
Rail absorbers	3–4
Resilient baseplates	3–4
Absorbent screen plus rail absorbers	7–8
Resilient baseplates plus rail absorbers	10–11
Absorbers on bridge deck	0*

* Reduction of 4–6 dB in low frequency range 30–50 Hz.

Based on these principles a new single track bridge consisting of two 12.5 m spans was constructed and installed on the line between Groningen and Leeuwarden in the Netherlands at a location called Nieuwe Vaart. The bridge cross-section also included integral channels for embedded rails and integral derailment guards. Measurements showed it to be *quieter* than the adjacent ballasted track on timber sleepers by about 2 dB(A). Spectral results are shown in Figure 11.29(a). Compared with another bridge in the Netherlands, similar to the reference design, it was about 10 dB(A) quieter, see Figure 11.29(b).

11.6.3 Hong Kong West Rail viaducts

The KCRC West Rail line in Hong Kong, completed in 2003, included 13.4 km of elevated double track viaduct structures [11.39]. The environmental noise restrictions placed on the development were considered the most stringent ever, with a limit of 64 dB(A) L_{max} at 25 m from the line during the passage of a nine-car train travelling at 130 km/h. This required more than 20 dB of mitigation measures compared with a typical viaduct [11.40]. Floating slab track together with very soft

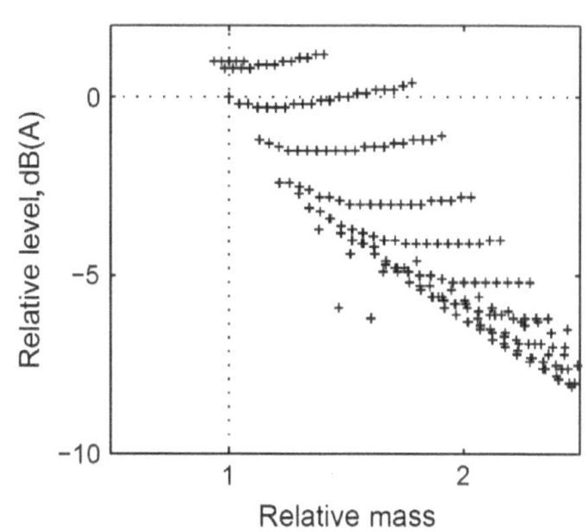

FIGURE 11-27 Results on radiated sound of varying bridge design parameters plotted against increase in section mass [11.37]

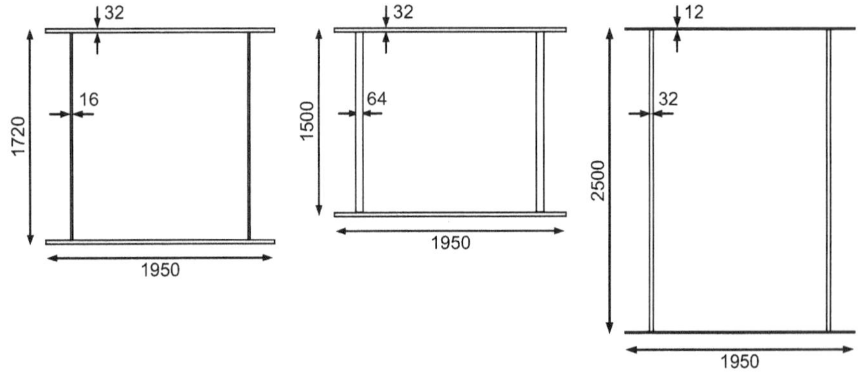

FIGURE 11-28 Bridge girders tested as scale models (1:4): reference (left) and two quiet variants [11.37]. Dimensions indicated (mm) are for full-scale equivalent

rail fasteners was employed throughout. Rolling noise was reduced by the use of vehicle skirts, sound-absorbing plenums formed under the walkways, and integral parapets acting as noise barriers, see Figure 11.30. At sensitive locations a full enclosure above track level was applied.

Since large noise reductions had been achieved from the upper part of the bridge, it was important that structure-radiated noise from the lower part of the bridge was also controlled. The use of floating slab track assisted in this but the bridge structure itself had to be designed carefully. Extensive calculations were carried out leading to optimization of the concrete box girder design, see Figure 11.30 [11.39]. In the optimized design the mobility was kept low by positioning the webs of the box girder directly under the rails. A thinner concrete section could therefore be used. This

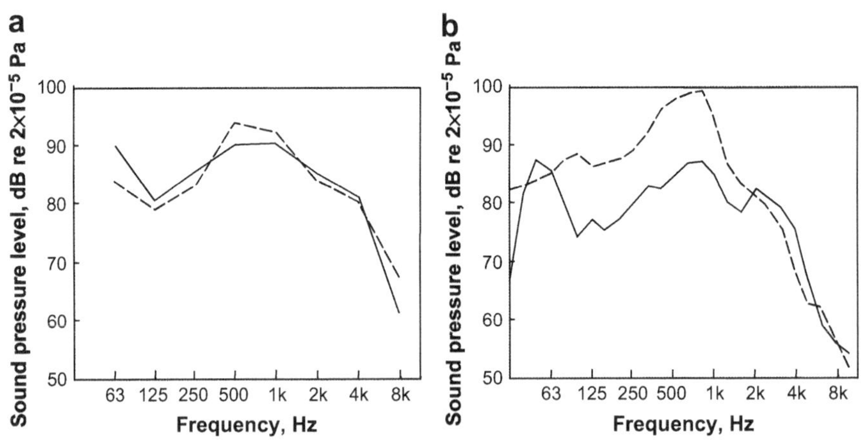

FIGURE 11-29 Measured sound pressure of 'Silent Bridge' for the Nieuwe Vaart (—) [11.38]. (a) Comparison in octave bands with adjacent ballasted track (– – –), (b) comparison in one-third octave bands with another steel bridge (– – –)

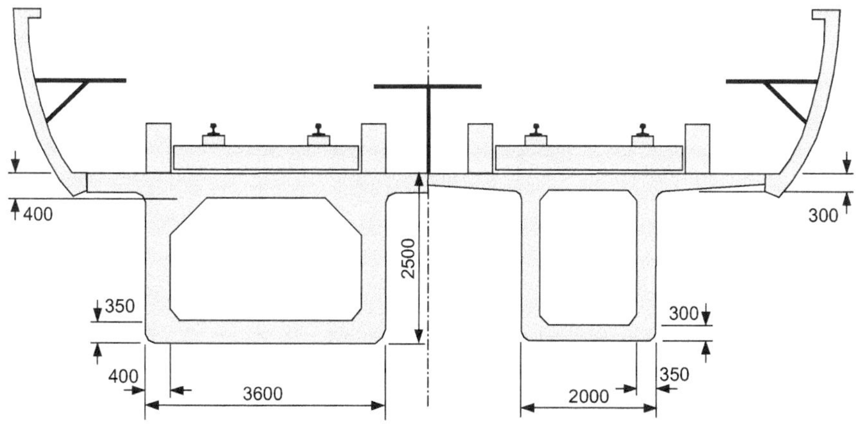

FIGURE 11-30 Initial (left) and improved (right) designs for West Rail viaducts including parapet with integral noise barrier and noise absorbing plenums under walkway (not to scale) adapted from [11.39]

allowed a 30% weight saving compared with the initial design, while achieving the target noise levels.

REFERENCES

11.1 L.G. Kurzweil. Prediction and control of noise from railway bridges and tracked transit elevated structures. *Journal of Sound and Vibration*, **51**, 419–439, 1977.

11.2 E.E. Ungar and L.E. Wittig. Wayside noise of elevated rail transit structures: analysis of published data and supplementary measurements. US Department of Transportation Report no. UMTA-MA-06-0099-80-6, December 1980.

11.3 F. Poisson and F. Margiocchi. The use of dynamic dampers on the rail to reduce the noise of steel railway bridges. *Journal of Sound and Vibration*, **293**, 944–952, 2006.

11.4 M.H.A. Janssens and D.J. Thompson. A calculation model for noise from steel railway bridges. *Journal of Sound and Vibration*, **193**, 295–305, 1996.

11.5 M.F. Harrison, D.J. Thompson, and C.J.C. Jones. The calculation of noise from railway viaducts and bridges. *Proceedings of the Institution of Mechanical Engineers, Part F (Journal of Rail and Rapid Transit)*, **214F**, 125–134, 2000.

11.6 O.G. Bewes, D.J. Thompson, C.J.C. Jones and A. Wang. Calculation of noise from railway bridges and viaducts: experimental validation of a rapid calculation model. *Journal of Sound and Vibration*, **293**, 933–943, 2006.

11.7 R.H. Lyon and R.G. DeJong. *Theory and Application of Statistical Energy Analysis* 2nd edition. Butterworth-Heinemann, Boston, 1995.

11.8 P.J. Remington and L.E. Wittig. Prediction of the effectiveness of noise control treatments in urban rail elevated structures. *Journal of the Acoustical Society of America*, **78**, 2017–2033, 1985.

11.9 O.G. Bewes. The calculation of noise from railway bridges and viaducts. EngD thesis, University of Southampton, 2005.

11.10 F.J. Fahy. Fundamentals of noise and vibration control, Chapter 5 in *Fundamentals of Noise and Vibration*. In: F.J. Fahy and J.G. Walker (eds.). Spon Press, 1998.

11.11 R.J. Pinnington. Vibrational power transmission from a finite source beam to infinite receiver beam via a continuous complaint mount. *Journal of Sound and Vibration*, **137**, 117–129, 1990.

11.12 L. Cremer, M. Heckl and E.E. Ungar. *Structure-borne Sound*, 2nd edition. Springer 1988.

11.13 J.C. Tukker and J.W. Verheij. Noise emission of steel railway bridges. Calculation model for noise reduction measures (in Dutch). TNO report TPD-HAG-RPT-89-0065, November 1989.

11.14 P. Gardonio and M.J. Brennan. Mobility and impedance methods in structural dynamics, Chapter 9 in *Advanced Applications of Acoustics, Noise and Vibration*. In: F. Fahy and J. Walker (eds.). E&FN Spon, London, 2004.

11.15 C. Stüber. Air- and structure-borne noise of railways. *Journal of Sound and Vibration*, **43**, 281–289, 1975.

11.16 Office for Research and Experiment (ORE) of the International Union of Railways: Question D105, Noise Abatement on Bridges: 1. Report RP1, Noise development in steel railway bridges, 1966.

11.17 C. Stüber. Geräuschentwicklung beim Befahren stählemer Eisenbahnbrücken und Abwehrmassnahmen. *Nr. VDI-Berichte* 69, 1963.

11.18 D.C. Herron, A. Wang, D. Rhodes, C.J.C. Jones, D.J. Thompson, and C.M. Nilsson. Modelling railway bridge noise and vibration: application of advanced finite element methods. *Proceedings of World Congress on Rail Research*, Seoul, Korea, May 2008.

11.19 R. van Haaren and A. Koopman. Prediction of noise radiation by concrete railway bridges. *Proceedings of Internoise 99*, Ft Lauderdale, Florida, USA, 1807–1810, 1999.

11.20 R. van Haaren and S. van Lier. Prediction of noise radiation by concrete and composite railway bridges. *Proceedings of Transport Noise 2000*, St Petersburg, Russia, 2000.

11.21 M.G. Dittrich. Application of a fast statistical computational method for assessing the sound transfer of plate-like structures. *Proceedings of International Symposium Prediction of the Noise Emitted by Vibrating Structures*, Senlis, France, March 1991. Special issue of *Revue Française de Mécanique*, 391–402, 1991.

11.22 D.J. Thompson and M.H.A. Janssens. Noise from steel railway bridges – development of the theoretical model, 1991. TNO report TPD-HAG-RPT-92-0077, Delft, 1992.

11.23 M.H.A. Janssens and D.J. Thompson. Rekenmodel voor geluid van stalen spoorbruggen, ontwikkelingen 1992–1993 (Calculation model for noise from steel railway bridges, developments 1992–1993). TNO report TPD-HAG-RPT-93-0151, Delft, 1993.

11.24 G. Maidanik. Response of ribbed panels to reverberant acoustic fields. *Journal of the Acoustical Society of America*, **34**, 809–826, 1962.

11.25 L.L. Beranek and I. Vér. *Noise and Vibration Control Engineering* 2nd edition. John Wiley, 2005.

11.26 A. Putra and D.J. Thompson. The radiation efficiency of point-excited rectangular baffled and unbaffled plates. *Proceedings of Institute of Acoustics*, Vol. 28, Pt 1, Spring Conference, Southampton, 27–28 March 2006, 465–474.

11.27 A. Putra and D.J. Thompson. Sound radiation from a perforated unbaffled plate. *International Congress on Acoustics*, Madrid, 3–7 September 2007.

11.28 J.T. Nelson. Steel elevated structure noise reduction with resilient rail fasteners at the NYCTA. *Proceedings Inter Noise*, **90**, 395–400, Gothenburg, 1990.

11.29 D.J. Thompson and J.W. Verheij. The dynamic behaviour of rail fasteners at high frequencies. *Applied Acoustics*, **52**, 1–17, 1997.

11.30 M.H.A. Janssens and M.G. Dittrich. Noise from steel bridges: comparative measurements of different rail fasteners on the bow-bridge in Weesp (in Dutch). TNO report TPD-HAG-RPT-91-0153, March 1992.

11.31 A. Wang, O.G. Bewes, S.J. Cox, and C.J.C. Jones. Measurement and modelling of noise from the Arsta bridge in Stockholm. *Proceedings of 9th International Workshop on Railway Noise*, Feldafing, Germany, September 2007.

11.32 K.P. Köstli, C.J.C. Jones, and D.J. Thompson. Experimental and theoretical analysis of railway bridge noise reduction using resilient rail fasteners in Burgdorf, Switzerland. *Proceedings of 9th International Workshop on Railway Noise*, Feldafing, Germany, September 2007.

11.33 Y. Ban and T. Miyamoto. Noise control of high-speed railways. *Journal of Sound and Vibration*, **43**, 273–280, 1975.

11.34 R. Wettschureck and G. Hauck. Geräusche und Erschutterungen aus dem Schienenverkehr Chapter 16 in *Taschenbuch der Technischen Akustik*. In: M. Heckl and H.A. Muller (eds.), 2nd edition. Springer-Verlag, Berlin, 1994.

11.35 J.J. Hanel and T. Seeger. Schalldämpfungsgrossversuch an zwei stählemen Eisenbahn-Hohlkastenbrücken (Full scale tests of sound damping on two steel railway box girder bridges). *Der Stahlbau* 47 (12) 353–361, 1978.

11.36 R. Greer, T. Allett and C. Manning. Channel Tunnel Rail Link Section 2, a review of innovative noise and vibration mitigation. *Proceedings of 8th International Workshop on Railway Noise*, Buxton, UK, 533–544, September 2004.

11.37 M.H.A. Janssens, D.J. Thompson, and J.W. Verheij. Application of a calculation model for low noise design of steel railway bridges. *Proceedings of Internoise 97*, Budapest, 1621–1624, 1997.

11.38 J. Bos. Dutch group cuts steel bridge noise. *International Railway Journal* 15–19, September 1997.

11.39 J.H. Cooper and M.F. Harrison. Development of an alternative design for the West Rail viaducts. *Proceedings of Institution of Civil Engineers, Transport*, **153**, 87–95, 2002.

11.40 A.R. Crockett and J.R. Pyke. Viaduct design for minimization of direct and structure-radiated train noise. *Journal of Sound and Vibration*, **231**, 883–897, 2000.

CHAPTER 12

Low Frequency Ground Vibration*

12.1 DIFFERENT TYPES OF RAILWAY-INDUCED VIBRATION

As well as airborne noise, trains induce vibration that is transmitted through the ground, which can also be a source of disturbance to nearby residents. The issue of ground vibration is increasing in importance and includes a number of different, but related, phenomena. These can be divided into the following categories, mainly according to the different types of railway situation:

(1) Heavy axle-load freight traffic, travelling at relatively low speeds, causes high amplitude vibration at the track. This excites waves in the ground that propagate along the ground surface. This type of vibration is especially associated with soft soil conditions, where it is found that significant levels of vibration may be propagated up to distances of the order of 100 m from the track. It often has significant components at very low frequency (below 10 Hz) and induces vibration of nearby buildings, which 'rock' or 'bounce' on the stiffness of their foundations. As well as freight traffic, this type of vibration can be caused by other types of train, notably locomotives or multiple units with high unsprung mass.

(2) High speed passenger trains sometimes travel at speeds in excess of the wave speed of vibration in the ground and embankment. This has been studied by track engineers for some years because of the large displacements that can be caused in the track support structure and in electrification masts, etc. Ground vibration is also produced which propagates away from the track. This may be compared with the bow wave from a ship or, more sensationally, the 'sonic boom' from a supersonic aircraft. Although this phenomenon is comparatively rare, the topic has attracted considerable research attention because of the expansion of the high speed rail network. If such situations arise, high speed railways may thus also cause significant levels of ground vibration at comparatively large distances from the track.

(3) Trains running in tunnels transmit vibration to buildings above and around them. This has higher frequency content than vibration from trains running on surface tracks. This vibration, at the low end of the audible frequency range from about 30 to 250 Hz, may excite bending vibration in the floors and walls of a building which then radiates a rumbling noise directly into the rooms. This noise may be perceived as all the more annoying because the source cannot be seen and no screening remedy is possible. This phenomenon is known as 'ground-borne noise' and will be discussed separately in Chapter 13.

* This chapter has been written by Chris Jones.

This chapter concentrates on low frequency 'feelable' vibration. It thus deals with the first two of the above phenomena and concentrates on vibration from trains running on track at grade. Chapter 13, in discussing ground-borne noise, has particular application to trains in tunnels. Nevertheless, there is considerable overlap between them in terms of the mechanisms of generation, the propagation of vibration and their treatment.

As well as feelable vibration, ground-borne noise has also increased in importance for surface railways, particularly as the use of noise barriers or secondary glazing has become more commonplace. This is demonstrated by a survey carried out in 2007/8 in Switzerland [12.1, 12.2]. It was found that around 170 km of double track of the Swiss Federal Railways (SBB), or 5% of the Swiss railway network, gave rise to high levels of ground vibration. The trigger level of vibration in this study was defined as 'unacceptable' according to the Swiss government's criteria [12.3]. This affected around 30 000 line-side residents. Interestingly, the two phenomena, low frequency vibration and ground-borne noise, were found to be the main concern for approximately equal lengths of track, yet only five of the 170 km are in tunnels. This case may owe something to the fact that the Swiss network already uses a lot of noise barriers. Thus, where direct noise from trains is treated, vibration is often a concern for line-side residents that comes not far behind.

Other surveys confirm the importance of ground vibration as a significant environmental impact from railways. A field study of the Scottish railway network in 1987 by Woodroof and Griffin [12.4] concluded that 35% of residents within 100 m of the track notice vibrations. A similar conclusion was found in Japan for the Shinkansen lines [12.5]. Fields and Walker [12.6] concluded that most of those who reported experiencing railway-induced building vibration were annoyed by it. They therefore estimate that up to 2% of the UK population could be annoyed by railway vibration.

Assessment criteria for vibration are quite different from those for noise, so in the next section the various standards and measurement quantities that apply will be discussed. The physical modelling of ground vibration will be introduced in Section 12.3 in terms of the types of waves that can propagate in a ground and in Section 12.4 in terms of the interaction between a train and a track on the ground surface. In Section 12.5 the calculation of ground vibration is illustrated by example results from three very different situations. Potential mitigation measures for ground vibration are discussed in Section 12.6.

12.2 ASSESSMENT OF VIBRATION

Whereas the human perception of airborne noise is commonly evaluated by the use of the standardized A-weighting, the evaluation of vibration is more complicated. Here, the topic is only briefly introduced insofar as it applies to the perception of vibration from railways.

There are at least three major sets of standards giving weighting curves and metrics for vibration assessment. The overall principles of assessing vibration for comfort and perception are laid down in ISO 2631 [12.8, 12.9]. However, some details are not specified at this international level and so use is usually made of, for example, the British standards [12.10, 12.11] or the German standards [12.12, 12.13]. All standards are revised from time to time, so it is important to refer to them directly in order to apply them. However, it is useful to consider their general principles here.

12.2.1 Characteristics of vibration and its perception

Simplified spectral base curves for the perception of vibration have been established [12.7] and vibration levels are assessed in terms of how many times they are above this. Figure 12.1 shows base curves presented in BS 6472 (1992)[1] in terms of root-mean-square acceleration level [12.10]. These are near the threshold of perception[2].

By way of example, several measured spectra of railway-induced vibration are shown in Figure 12.2. These were measured at a terraced house in Southampton, in the UK, at 50 m from a track in cutting. The trains were electric multiple unit regional rolling stock travelling at speeds between 110 and 120 km/h in the same direction on one track. A background spectrum of vibration at this site is shown, which was caused by nearby road traffic. It is typical that there are significant components of vibration between 4 and 80 Hz. However, vibration spectra vary a great deal from one site to another, even within small distances, as well as between train types. Therefore no spectrum should be taken as a 'typical' level of vibration from trains generally.

When stating vibration in decibels there is no universal standard reference level, as there is for sound, but reference levels of either 1×10^{-9} m/s or 5×10^{-8} m/s are often used for velocity level, the latter being specified in the German standards. For acceleration level, 1×10^{-6} m/s^2 is commonly used.

The base curve for vibration perception in the spinal axis of the body is shown plotted against the example measurements of vibration in Figure 12.2. Here it is shown

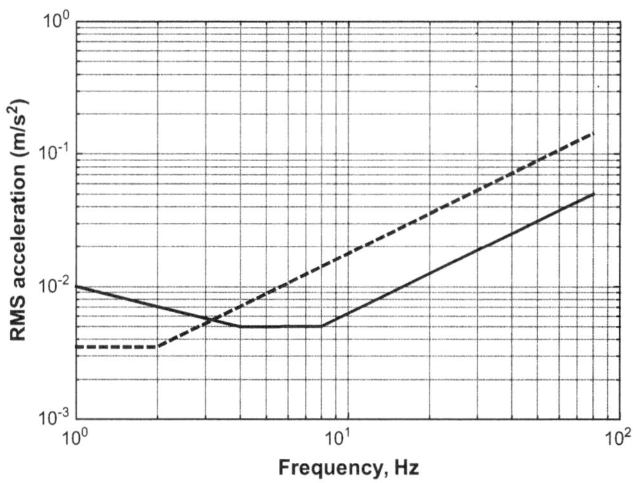

FIGURE 12-1 The vibration acceleration base curves from BS 6472 (1992). —, vertical; – – –, lateral

[1] Since drafting this chapter, BS 6472 has been updated in 2008. This version no longer presents the base curves.

[2] Information about the threshold of perception is stated in the standards in terms of weighted accelerations rather than approximate spectral interpretation made from base curves. The standards state that 50% of the population can perceive a weighted acceleration of 0.01ms^{-2}r.m.s.

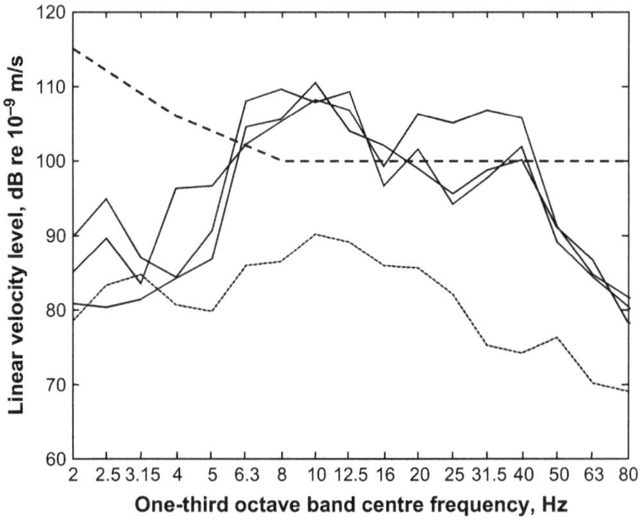

FIGURE 12-2 Measured spectra of vertical vibration in a house 50 m from the track (- - -, background; —, trains; – – –, base curve from BS 6472 (1992))

in terms of the velocity level rather than acceleration as in Figure 12.1. Thus it can be seen that from 8 Hz upwards, vibration velocity level of 100 dB, which corresponds to 0.1 mm/s r.m.s. amplitude in each one-third octave band may be judged approximately as just perceptible. Since the curve is flat, vibration is effectively perceived in this frequency range in terms of its velocity rather than acceleration or displacement.

At the frequencies illustrated in Figure 12.2 the vibration is clearly perceived as 'whole body' vibration which can be felt. As illustrated in Figure 12.2, where vibration from trains is perceived, it is usually only a few dB above the threshold of perception. However, given that annoyance is felt very quickly as the level rises above this threshold [12.4, 12.6, 12.7], it is often appropriate to consider the mitigation of vibration, not in terms of a substantial reduction of vibration level, but in terms of lowering it below the threshold.

12.2.2 Standards for vibration measurement

There is agreement among the standards that, in assessing vibration at a location, both the vertical and lateral components should be measured. The vibration should ideally be measured inside the buildings at the places where it is likely to be perceived. However, it can be problematic to select locations that are not affected by local resonances of the floors and walls of the building. Measurement positions on structural walls are therefore often preferred. If access to the building is not available, measurements on the ground close to the building or on an outside part of the building are acceptable.

Weighting curves are available for assessing vibration, and used in a similar way that the A-weighting is used for noise. The various standards use different weightings to evaluate perception, comfort and motion sickness but only the perception criteria are appropriate for the levels of railway-induced vibration that are normally encountered. Figure 12.3 shows the weighting curves relevant to perception from

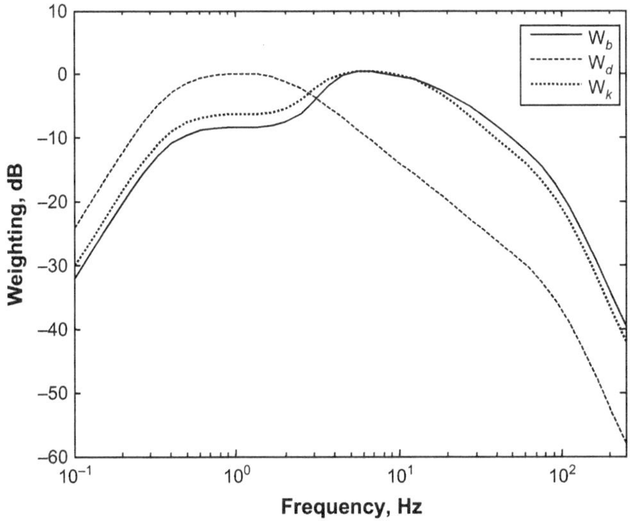

FIGURE 12-3 Weighting curves used for acceleration in ISO 2631-1 and BS 6472/BS 6841

ISO 2631 Part 1 [12.8] and BS 6841 [12.11] (which is used with BS 6472 to evaluate exposure to vibration in buildings [12.10]). The W_d curve is common to both standards and is used to weight the acceleration applying to lateral axes of the body. For the spinal axis (normally vertical), the W_b curve of BS 6841 and W_k of ISO 2631-1 are only slightly different.

BS 6472/6841 can be used to evaluate railway vibration in buildings directly. However, ISO 2631 Part 2 [12.9] is directly applicable to vibration in buildings, rather than Part 1. This makes allowance for the fact that the axis of orientation of a person inside a building is unknown and so uses a single frequency weighting W_m that is applied to the measurements in either the vertical or lateral directions; the more perceptible component is then used in the evaluation. The weighting W_m is shown in Figure 12.4.

The German standard, DIN 4150-2 [12.12], follows a similar procedure but using a filter characteristic called K_b that is specified in the associated standard DIN 45669-1 [12.13]. Unlike the filters in ISO 2631 and the British standards, which apply to acceleration signals, this is used to apply to a measured velocity signal. However, when transformed as it would apply to an acceleration signal, the K_b filter has the same shape as W_m.

12.2.3 Assessment of vibration

To determine the total effect during a day- or night-time period, community noise from trains is assessed using L_{Aeq}. An equivalent is needed for the vibration effects of a number of train passages in a certain period. Different approaches to this are given in the standards. This leads to different metrics that are then used to judge the acceptability of the vibration.

ISO 2631-1 does not offer any rating of perceptible vibration levels except to say that 'adverse comment regarding building vibration in residential situations may

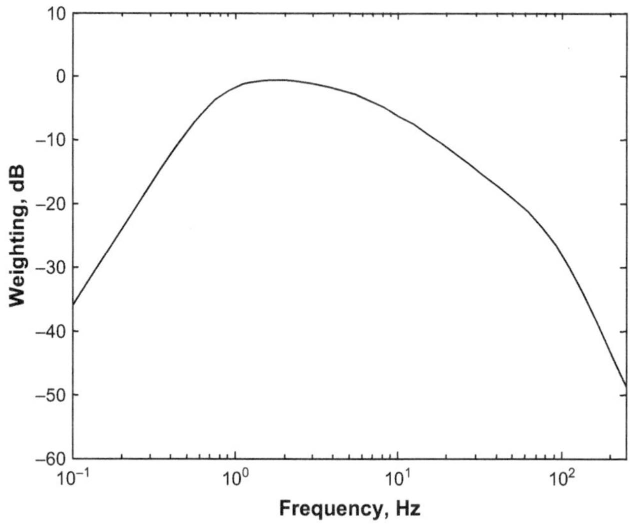

FIGURE 12-4 Weighting curve W_m used for acceleration in ISO 2631-2 and DIN 4150

arise … when magnitudes are only slightly in excess of perception levels'. More guidance is given in the approaches of the British and German standards in the form of Vibration Dose Values and KB values.

(i) Vibration dose values

In BS 6472 (and in ISO 2631-1) a 'vibration dose value' (*VDV*) is defined to quantify intermittent vibration. The *VDV* for a single event is

$$VDV = \left[\int_0^T a^4(t)dt\right]^{0.25} \tag{12.1}$$

which has units m/s$^{1.75}$; $a(t)$ is the frequency weighted (filtered) acceleration as a function of time and T is the duration of the event. The total *VDV* for a number of events is then summed using a fourth power law:

$$VDV_T = [VDV_1^4 + VDV_2^4 + VDV_3^4 + \ldots]^{0.25} \tag{12.2}$$

ISO 2631 Part 1 also contains the *VDV* method but ISO 2631 Part 2, of direct relevance to building vibration, does not.

Once the *VDV* has been calculated, it can be compared with broad criteria for acceptability that are reproduced in Table 12.1. In railway projects it is usual to work to a limit of avoiding the 'low probability of adverse comment'.

For N identical events, from equation (12.2), $VDV_T = VDV\ N^{0.25}$. This implies that to halve the *VDV* for a train service would require a reduction of the number of events by a factor of 16. The magnitude of the vibration of a single event is usually therefore more important than the number of events or the duration. For this reason, at sites with mixed traffic, often a few freight trains, perhaps running at night, are identified as the worst cases and it is these which dominate a *VDV* assessment.

TABLE 12-1 RATINGS FOR RESIDENTIAL BUILDINGS IN VDV ($M/S^{1.75}$) FROM BS 6742

	Low probability of adverse comment	Adverse comment possible	Adverse comment probable
16 hour day	0.2–0.4	0.4–0.8	0.8–1.6
8 hour night	0.13	0.26	0.51

BS 6472 also allows a rough interpretation using root-mean-square acceleration values depending upon their duration although this is may be dropped from future revisions.

(ii) KB value

DIN 4150 [12.12] uses a running root-mean-square vibration velocity measurement (based on a 0.125 second time constant) to form a '*KB* value' KB_{FTr} according to

$$KB_{FTr} = \sqrt{\frac{1}{T_r}\sum_j T_{e,j} KB^2_{FTm,j}} \tag{12.3}$$

where T_r is the evaluation period (day- or night-time) and $T_{e,j}$ is the exposure period of each event, j. $KB^2_{FTm,j}$ is the average of the maximum filtered r.m.s. signal values during each 30 second interval of the whole event. This, or the maximum KB_F value of the event, is then judged against a table of criteria for different types of building and for day- or night-time periods to ascertain what is, or is not, acceptable exposure – there are no terms defining different levels of acceptability.

Since the *KB* value and *VDV* are fundamentally different, as are the acceptance criteria based on them, it is not clear what the comparative or equivalent values of these metrics might be.

12.2.4 Assessment of potential for building damage

People who experience railway-induced vibration often express complaints in terms of concern over possible damage to their property. In fact, the levels of vibration normally encountered from trains are very small when assessed against the criteria for building damage.

Criteria for building damage due to vibration are covered by ISO 4866 [12.14, 12.15] (equivalent to BS 7385), as well as DIN 4150 Part 3 [12.16]. The standards give guidance in terms of peak particle velocities (ppv). Limits are given for cosmetic damage such as cracks in plaster. The dominant frequency corresponding to the ppv also has to be determined. Figure 12.5 shows this baseline for ppv level from the standard. Values that are twice as large as these indicate that minor damage is likely, while for values twice as large again major damage is possible.

It is difficult to compare the metric used for assessment of perception with the ppv values used for building damage assessment. As building damage levels are very rarely reached by rail traffic, an assessment of the vibration with respect to building

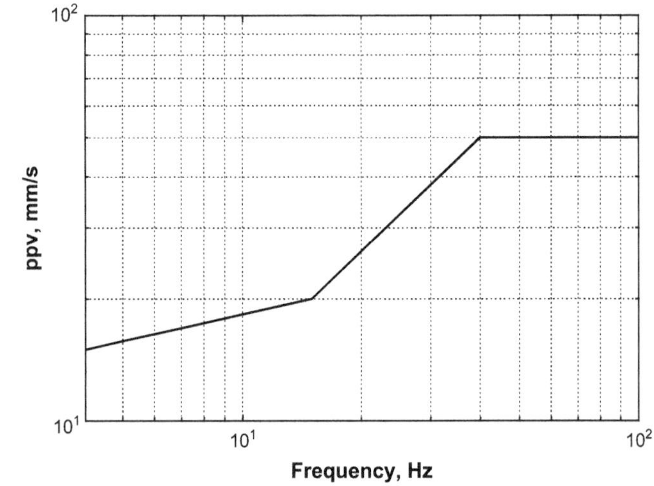

FIGURE 12-5 Criterion for potential cosmetic damage to light-framed buildings from BS 7385 or ISO 4866

damage usually allays the fears of complainants. Regular levels high enough to cause building damage would in fact be intolerable from the point of view of annoyance or disturbance.

Differential settlement of a building may be a factor if damage has actually taken place. However, the standards do not deal with any possible effect of vibration in accelerating such differential settlement.

12.3 SURFACE VIBRATION PROPAGATION

The remainder of this chapter considers the physical modelling of the generation and propagation of ground vibration. This is a complex subject and details of the models used are beyond the scope of this book. The chapter therefore attempts to present a physical understanding of the phenomena on the basis of example calculation results. This section considers vibration in the ground, the effect of the train being deferred to subsequent sections.

12.3.1 Propagation in a homogeneous elastic medium

In infinite solid elastic materials, vibration can propagate by two fundamental mechanisms; shear or dilatation. There are therefore two *fundamental* wave speeds, c_1 and c_2, that are directly related to the material properties of the solid. These are given by

$$c_1 = \sqrt{\frac{\lambda + 2\mu}{\rho}}\,;\, c_2 = \sqrt{\frac{\mu}{\rho}} \tag{12.4}$$

where λ and μ are the Lamé constants:

$$\lambda = \frac{\nu E}{(1+\nu)(1-2\nu)};\mu = \frac{E}{2(1+\nu)} \tag{12.5}$$

E is Young's modulus, ν is Poisson's ratio and ρ is the density of the material.

These two wave speeds correspond to compressional (dilatational) and shear wave motions; μ is equal to the shear modulus, G, while $\lambda + 2\mu$ is equal to the bulk modulus. In surveying terminology they are called the P-wave and S-wave, which stand for 'primary' and 'secondary'. The P-wave is faster than the S-wave, $c_1 > c_2$.

A ground with a free surface is often idealized simply as a half-space of homogeneous elastic material. In a half-space, a free wave can also propagate along the surface, namely the 'Rayleigh' wave. This involves a combination of shear deformation and dilatation due to the free boundary condition at the ground surface. An example of its shape as a function of depth is shown in Figure 12.6. The vertical and horizontal motions are actually $\pi/2$ out of phase with one another, so that particles perform elliptical motion with growing amplitude towards the surface. A snapshot of the particle displacement at an instant of time is shown in Figure 12.7.

The Rayleigh wave is the slowest wave of the half-space, having a speed between 87% and 95% of the shear wave speed (depending on the Poisson's ratio of the material). Being the slowest wave, it is the Rayleigh wave that usually carries the greatest part of the wave energy that is transmitted, particularly to larger distances along the surface, although this also depends on the nature of the load exciting the vibration.

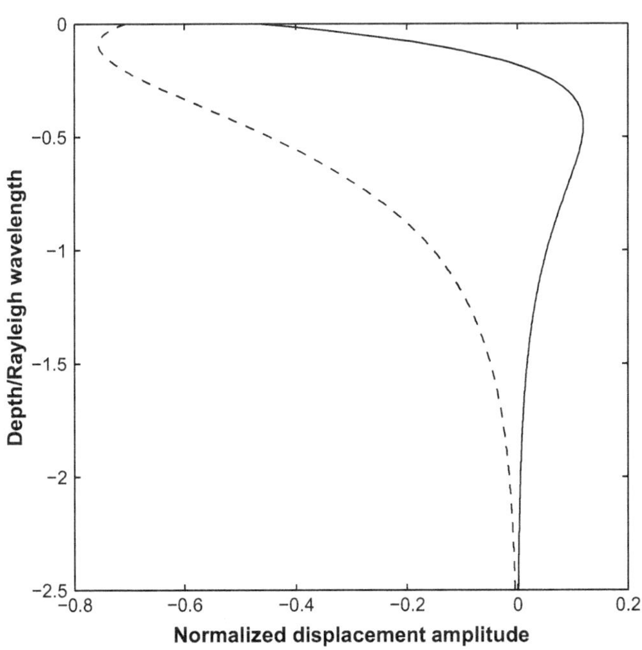

FIGURE 12-6 Example of a Rayleigh wave mode shape for $\nu = 0.3$. —, lateral displacement; – – –, vertical displacement

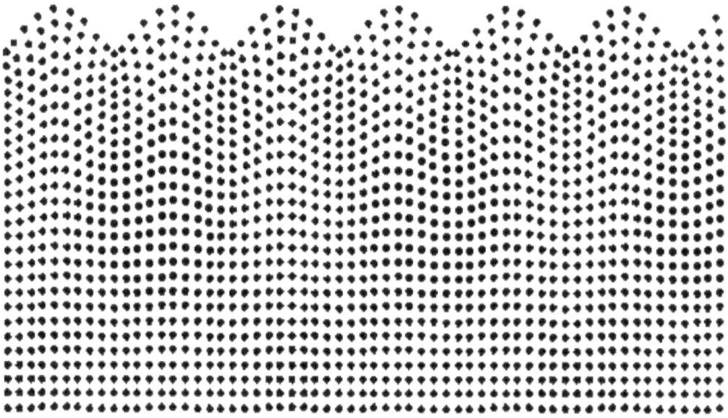

FIGURE 12-7 Instantaneous displacement of particles in Rayleigh wave motion

A soft soil, such as is found near the surface of the ground, may have a Rayleigh wave speed of about 100 m/s, while for a stiffer soil, still near the ground surface, it will be of the order of 300 m/s. Rayleigh waves in these two cases therefore have a wavelength of 20 to 60 m at 5 Hz; at 40 Hz this reduces to about 2.5 to 7.5 m. It can be seen from Figures 12.6 and 12.7 that the displacement in this wave is significant to a depth of somewhat greater than the wavelength on the surface.

12.3.2 Propagation in layered ground

In practice, when compared with such large wavelengths, the ground is not a homogeneous half-space. Any ground is layered on some scale and, typically, grounds have a layer of softer 'weathered' material that is only about 1 to 3 metres deep. Further layers may be present below this, depending on the geology of the site. The layered structure of the ground has important effects on the propagation of surface vibration in the frequency range of interest.

It is impractical to study ground wave propagation by looking only at experimental results because so many measurements would have to be done and because the natural uncertainty in the results would make the 'patterns' hard to see. In order to illustrate how vibration propagates in a layered ground use is therefore made of theoretical models.

Theoretical models of layered ground can be divided into two main categories:

- Analytical models, expressed in terms of wavenumbers in the ground;
- Numerical models, using finite elements or boundary elements.

Two-dimensional models provide rapid calculations that can produce mode shapes in the ground and dispersion characteristics. A method of dynamic flexibility matrices for ground with parallel layers was developed in the 1950s by Thompson [12.17] and Haskell [12.18]. This uses an analytical approach in the wavenumber domain. A useful paper by Kausel and Roësset [12.19] then sets out two-dimensional wavenumber-domain theory in terms of exact dynamic stiffness matrices for layers and a homogeneous half-space. For assembled systems of layers over a half-space, a reverse Fourier transform can be performed on the wavenumber-domain solutions to obtain solutions in the spatial domain.

Similar approaches have been developed to encompass three-dimensional parallel layers of ground [12.20–12.24]. Such an analytical model, coupled with track and vehicle models, is used later in this chapter to demonstrate some physical phenomena of vibration from trains.

For more generalized studies and predictions of real situations, numerical models, such as the finite element (FE) method, are needed. However, it is essential that the wave-propagating behaviour of the ground is adequately modelled; in particular waves should travel outwards through the region of interest as if in an infinite, 'free field', and not be reflected back at artificial boundaries of the model. Conventional finite element models cannot do this efficiently, although infinite elements can be used as non-reflecting boundaries provided that they are placed in the far field.

The boundary element (BE) method [12.25] is much more suited to the task since a medium of infinite extent can be modelled using elements to define only the surfaces or interfaces. Generally, the FE method is used to model the detail of structures such as track, embankment, walls, tunnel or receiver buildings and the BE method to model the ground of infinite extent. The two methods are then fully coupled to build models of the track, its foundation and structures and the propagation path in layered soil. Models may also include reception at the foundations of the building.

Boundary element models have been developed in two dimensions [12.26] and three dimensions [12.27] for studies of railway vibration. Again the two-dimensional models are useful for studying some aspects, because of the efficiency of the calculation. It can be used, for example, to determine relative effects of changes in soil or structure properties or geometry. However, a three-dimensional analysis is required to cover all aspects and to make absolute predictions [12.28]. Unfortunately, three-dimensional FE/BE models for this type of problem require very large computing resources.

For efficient analysis of geometry which is constant in cross-section but 'extruded' into three dimensions along the direction of the track, a number of researchers have developed '2.5-dimensional', 'wavenumber' or similar methods [12.29–12.33]. In these methods a two-dimensional FE/BE model is solved a number of times for different wavenumbers in the axial direction of the extruded geometry. A three-dimensional solution is then recovered by means of a reverse Fourier transform over wavenumber.

These methods have been extended to provide the additional capability of predicting the spectrum of vibration from both the dynamic and quasi-static excitation mechanisms of the moving axles of the train [12.32, 12.33].

12.3.3 Obtaining parameters for theoretical models

To use any of the theoretical models it is necessary to determine the properties of the ground at a site: the layer depths, wave speeds and damping. Use can be made of local seismic survey methods in which the speed of propagation of P- and S-waves is measured directly as a function of depth. In order to ensure that these parameters are appropriate for modelling vibration propagation in the frequency range of interest, a measurement can be made of the transfer mobility to various positions. The ground is excited at a small circular footing using an instrumented hammer and accelerometers are located on the surface of the ground some distance away to measure the response (see also Section 13.5.3). Such measured results can be compared with a relatively simple calculation using an axisymmetric layered ground model [12.19]. Figure 12.8 shows such a comparison, where a variety of loading and response points around a small site have been used at a fixed distance of 15 m to obtain an ensemble of transfer mobility spectra.

At frequencies below 20 Hz the measured results in Figure 12.8 are contaminated by background vibration; very little energy is transmitted into the ground by the hammer used. However, it can be seen clearly that the transfer mobility rises strongly to a peak at about 40 Hz (for this particular site). The frequencies of such features and the differences in levels vary from one site to another but the figure illustrates important features in the frequency range of interest. These are due to the layered structure of the ground.

12.3.4 Modal wave types in a layered ground

In a layered ground, vibration propagates parallel to the surface via a number of wave types or 'modes'. These are often called Rayleigh waves of different order ('R-waves') and Love waves. The Rayleigh waves are also called P-SV waves since they involve coupled components of compressive (P) deformation and vertically polarized shear (S) deformation. Here the name P-SV wave is preferred and the term Rayleigh wave is reserved for the single P-SV wave that exists in a homogeneous half-space. Love waves are decoupled from these and only involve horizontally polarized shear deformation. They are therefore also known as SH waves. Since the vertical forces at the track dominate the excitation of vibration in the ground, the SH waves are not strongly excited; they are not considered further in the present discussion.

To illustrate these wave types, an example notional ground is considered. This ground is used in the rest of the chapter to demonstrate the behaviour of soils. It consists of a layer of soft soil 2 m deep, below which the substratum is assumed to be a half-space of stiffer material. The assumed wave speeds are listed in Table 12.2. In

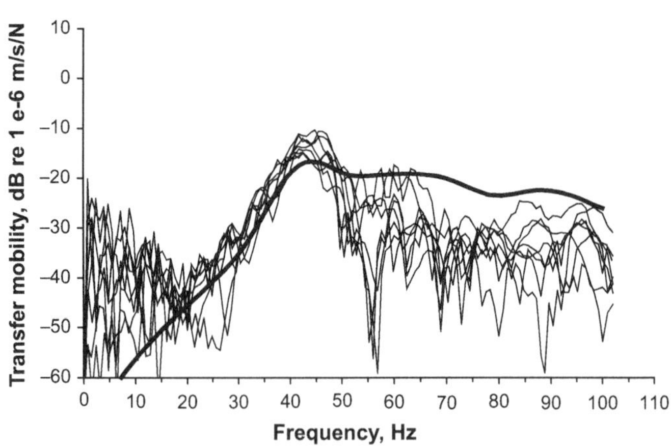

FIGURE 12-8 Measured and modelled transfer mobility across the ground surface from a small circular footing to a vertically orientated accelerometer 15 m away [12.20]

TABLE 12-2 EXAMPLE GROUND PROPERTIES USED IN THIS CHAPTER

	Thickness	P-wave speed	S-wave speed	Density
Upper layer	2 m	360 m/s	118 m/s	1500 kg/m^3
Substratum	infinite	1760 m/s	245 m/s	2000 kg/m^3

results where the damping of the soil material is used it is included in both materials as a loss factor of 0.1.

The wave 'mode' shapes of the P-SV waves that propagate at 40 Hz in this ground are shown in Figure 12.9. At this frequency, the first wave resembles the Rayleigh wave of Figure 12.6, except that the motion is mainly confined to the upper soft layer. The wave speed is similar to the shear wave speed of this upper layer. Figure 12.10 presents a particle motion picture, like Figure 12.7, for this first (lower speed, higher wavenumber) wave type. This shows clearly that the energy is largely transmitted in the top, weathered layer of material that is here only 2 m deep. The

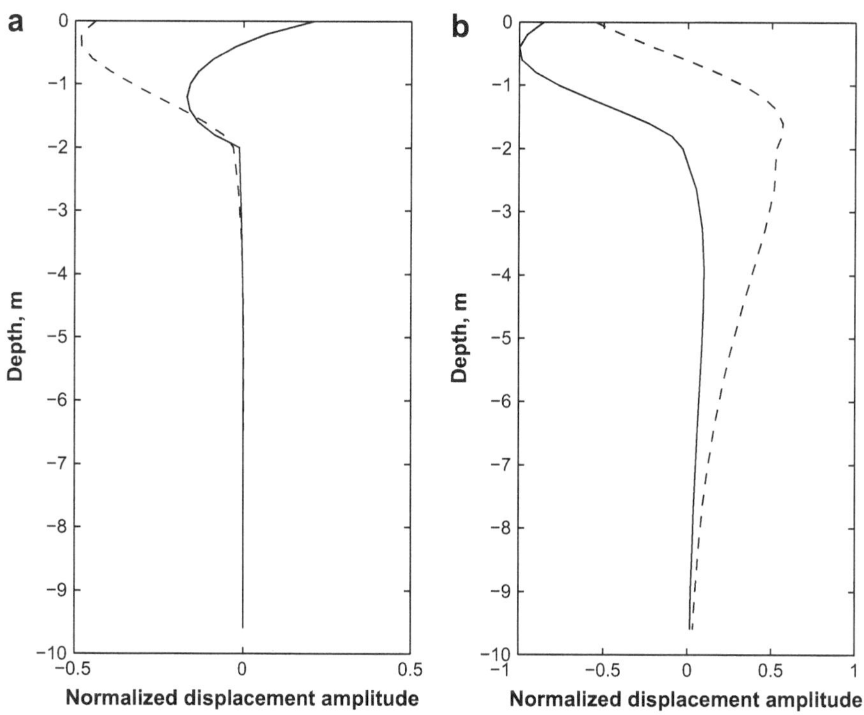

FIGURE 12-9 P-SV modes of the example layered ground structure at 40 Hz. —, lateral displacement; – – –, vertical displacement. The modes have wavenumbers of (a) 2.1 rad/m (118 m/s phase speed) and (b) 1.15 rad/m (218 m/s phase speed)

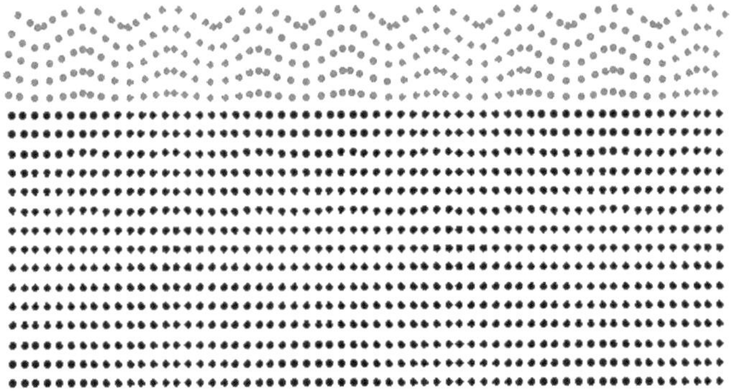

FIGURE 12-10 Instantaneous displacement of particles in the wave motion of the first mode in a layered ground

second wave in Figure 12.9 has a much higher wave speed, close to the shear wave speed of the substratum, and contains considerable motion of this half-space material.

12.3.5 The dispersion diagram

If the wavenumber for each mode ($k = 2\pi/\lambda$, with λ the wavelength) is plotted as a function of frequency, the dispersion curve for the wave type is generated (dispersion curves for waves in track were shown in Chapter 3). Figure 12.11 presents the dispersion diagram for the example soil structure, in which only the propagating P-SV modes are shown. Each line in the diagram represents a wave type associated with a cross-sectional mode of the layered soil. Only propagating waves are shown, although there are also many evanescent waves with high decay rates.

Constructing a line from the origin to a point on a curve, the inverse of the slope of this line is equal to the wave speed (phase velocity) of that wave at a particular frequency, $c = \omega/k$. The inverse slope of the curve itself gives the group velocity of the wave, $c_g = \partial\omega/\partial k$. This is the speed at which energy is transported by the wave type.

For this example set of soil parameters, at very low frequency, only a single propagating mode exists and this has a wave speed close to that of the shear waves in the substratum. At around 15 Hz, a quarter wavelength of the shear wave becomes equal to the depth of the weathered material. Above this frequency the wavenumber rises towards the line representing Rayleigh waves in the upper weathered layer, which it reaches by about 40 Hz. This wave corresponds to motion which involves mostly deformation of this upper layer, as seen in Figure 12.9(a). With the onset of this mode, i.e. propagation predominantly via the upper layer, a rise in the transmitted level of vibration is observed. The peak at 40 Hz in Figure 12.8 corresponds to this effect.

A second mode 'cuts on' at about 23 Hz, which is the wave shown in Figure 12.9(b). Subsequent modes cut on at higher frequencies of 47 and 85 Hz as the second mode, and then the third, are concentrated in the upper layer.

When the ground is excited by a harmonic load [12.34, 12.35], the amplitude of the response can be calculated and plotted against wavenumber and frequency. The

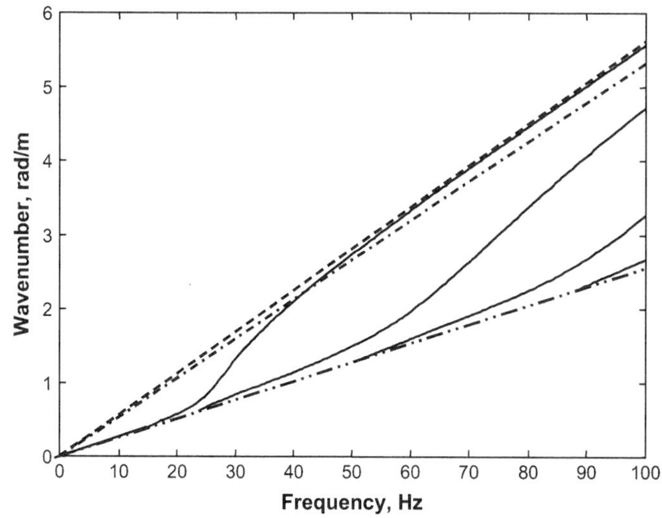

FIGURE 12-11 —, Dispersion diagram for propagating P-SV waves of the example ground; – – –, Rayleigh wave speed of upper layer (110 m/s); - · · - ·, shear wave speed of upper layer; - · · · -, shear wave speed of the half-space [12.34]

results of this for the example ground are shown for the vertical component of the response in Figure 12.12. In order to make the peaks of amplitude distinct over a wide frequency range, the damping of the ground has been set to a low value and the logarithm of the amplitude has been plotted. This reveals the features of the dispersion diagram, showing the relative amplitudes of the different waves excited

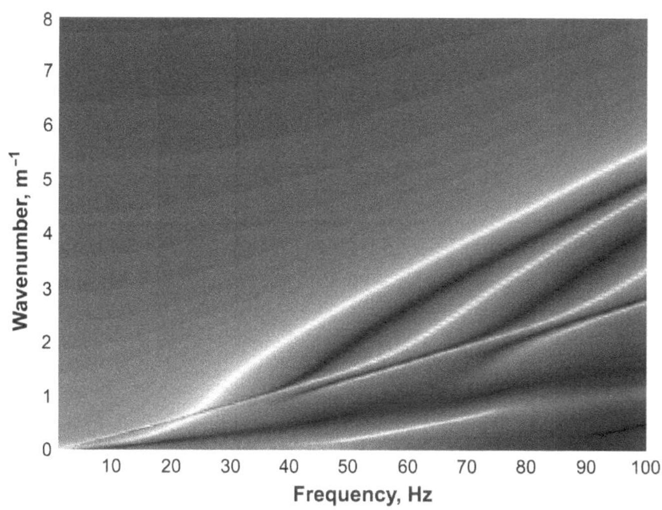

FIGURE 12-12 Vertical amplitude versus wavenumber and frequency calculated for a layered ground loaded vertically over a 3 m by 3 m area (light = maximum) [12.34]

under the oscillating load. A slice through this graph at any frequency represents the amplitudes of propagating waves as a function of wavenumber that would be excited by a harmonic load at that frequency. Since damping is included in the model, the response no longer occurs at discrete wavenumbers but a continuous spectrum of amplitude versus the wavenumber exists with broad peaks at the modal wave propagation wavenumbers of Figure 12.11.

12.3.6 Track and layered ground

Before introducing the interaction of a train with the ground, this section considers the effect on the propagation of vibration of adding the track structure to the ground. For this, an analytical model is used in which an infinite multiple beam model of the track is coupled to a layered ground, as shown in Figure 12.13. The two rails are represented by a single combined beam. For a track with sleepers, the second beam shown consists simply of a layer of mass. Rail pads and ballast are included as layers of stiffness, in the latter case with consistent mass. This model has been used to study the effects of interaction between the track, the ground and the moving loads on the track [12.34, 12.35].

This model can be used first of all to predict the receptance of the track. This is shown in Figure 12.14 for two stiffnesses of ground and compared with the results from a model of the track on a rigid foundation (as in Chapter 3). Here the material parameters for the soils in the stiffer ground are those already used to generate Figures 12.9 to 12.12 and are presented in Table 12.2. For the softer ground the upper layer is replaced by soil with a P-wave speed of 340 m/s and an S-wave speed of 81 m/s.

These results show that a rigid foundation model is good enough for higher frequencies (say above 50 Hz, relevant to rolling noise modelling). However, the

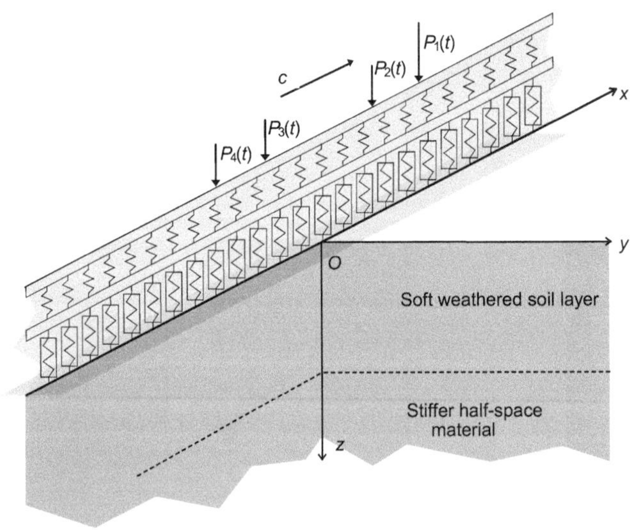

FIGURE 12-13 Analytical model of the ground and track used to study the vehicle, track and ground interaction

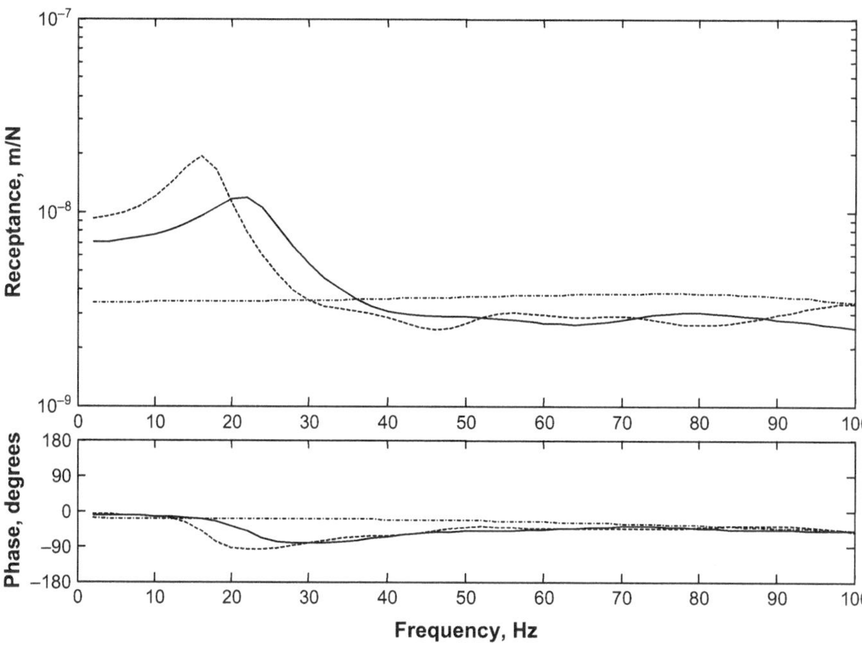

FIGURE 12-14 Track receptance calculated using the coupled track/ground model: - - -, track on softer ground; ——, track on stiffer ground; · – · –, track on rigid foundation [12.34]

ground has a significant influence on the receptance of the track in the frequency range relevant to surface vibration (up to 80 Hz). In particular, the receptance increases at a frequency that is associated with the cut-on of vibration propagation in the layered ground.

Using this combined track/ground model, it is possible to determine the amplitude of vibration excited by an oscillating vertical force on the track as a function of wavenumber in the ground at each frequency. Figure 12.15 presents this for the wavenumber in the direction along the track. This gives the spectrum of the vibration at the ground surface directly beneath the track, i.e. for $y = 0$. Comparing this with the earlier results in Figure 12.12, it can be seen that a new bright line representing propagation at higher wavenumbers has been introduced. Moreover, the amplitude around 20 Hz in the first wavenumber (highest wavenumber, and therefore lowest wave speed) is even stronger than before.

At low frequency the ground wave is modified because the mass (and stiffness) of the track structure adds to that of the ground involved in the wave motion. The wavenumber will be close to that of the unmodified ground wave as long as the track mass is low compared with this ground mass (modal mass of the wave shape). Of course, this wave only propagates along, and close to, the track. At high frequency the wavelengths in the ground become short and less ground material participates in the track wave. For high frequencies, therefore, the track gains a propagating wavenumber separate from that of the ground wave.

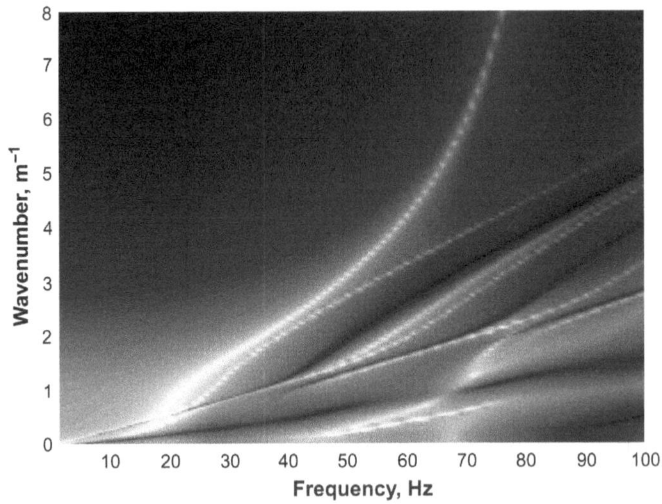

FIGURE 12-15 Vertical amplitude versus wavenumber along the track and frequency calculated for a layered ground loaded via the track (light = maximum) [12.34]

12.4 EXCITATION OF VIBRATION BY A TRAIN

12.4.1 Mechanisms of vibration generation

For a train running on a track on the ground surface a number of mechanisms of vibration generation can be significant. These can be divided into the effects of 'quasi-static' and dynamic loads. In each case the effect of load motion may also be important.

Very close to the track the vibration is dominated by the time-dependent displacement of a fixed point in the ground as the axle loads move past. This is sometimes called the 'quasi-static excitation' mechanism (see Section 5.7.4). However, for conventional train speeds this vibration remains in the near field (about a quarter of a wavelength from the track). Even so, since the wavelengths are long at low frequencies, as noted in Section 12.3.1, buildings close to the track can be affected.

For train speeds that approach or exceed the wave speeds in the ground, the moving axle loads can generate waves that propagate away from the track. In this case it is important that the load speed is included explicitly in the calculation model.

Dynamic forces are generated at the wheel/rail contacts by the combined irregular profile of the track and wheel running surfaces and these can lead to wave propagation in the ground. This is effectively the same mechanism as the excitation of rolling noise (Chapter 5) although longer wavelengths of 'roughness' are involved. On the wheel, at the longer wavelengths, this is represented by out-of-roundness. Additionally, dynamic forces are generated as impacts as the wheels traverse switches and crossings or badly maintained rail joints (see Chapter 10). Uneven track support at sleeper pitch or at longer wavelengths may also give rise to dynamic displacements under the loads of the vehicle (see Section 5.7.5).

TABLE 12-3 **EXAMPLE WAVELENGTHS OF 'ROUGHNESS' (IN M) EXCITING VIBRATION AT DIFFERENT FREQUENCIES FROM TRAINS RUNNING AT DIFFERENT SPEEDS**

	40 km/h	80 km/h	160 km/h	300 km/h
4 Hz	2.8	5.6	11	21
8 Hz	1.4	2.8	5.6	10
16 Hz	0.69	1.4	2.8	5.2
31.5 Hz	0.35	0.70	1.4	2.6
63 Hz	0.18	0.35	0.71	1.3
125 Hz	0.089	0.18	0.36	0.67
250 Hz	0.044	0.089	0.18	0.33

range of 'acoustic' roughness (shaded) — range of track recording car data (hatched)

These dynamic excitation mechanisms are the main mechanism applying to ground-borne noise; as will be seen, they are also often important for feelable vibration. Table 12.3 indicates wavelengths of roughness that excite various frequencies of vibration for different train speeds according to equation (2.1).

It can be seen from Table 12.3 that part of the wavelength range coincides with the acoustic roughness range, particularly for the frequency range applying to ground-borne noise (30–250 Hz). The shorter wavelengths of the acoustic roughness range (below 50 mm) are not significant here. The longer wavelengths, relevant to low frequency vibration, overlap the range applying to 'track top quality'; this is measured using track recording cars and is used to determine whether track maintenance (e.g. tamping) is required.

12.4.2 The critical train speed

The behaviour of the track/ground system when it is excited by a moving constant load can be understood in terms of the propagating waves as displayed in the dispersion diagram. The load speed V can be indicated by a diagonal line with slope $2\pi/V$ passing through the origin.

From the dispersion curves of a layered ground, shown again in Figure 12.16, a moving constant load will excite the propagating waves in the ground if the load speed exceeds the phase speed of the waves. At lower speeds the load-speed line will lie above the dispersion curves and will not intersect them. The minimum speed at which the load speed intersects the dispersion curves is the Rayleigh wave speed of the material in the upper layer. At higher load speeds, the load-speed line moves closer to the frequency axis and will intersect the dispersion curves at lower frequencies and possibly excite multiple waves. Thus at 150 m/s, waves are excited at 28 and 78 Hz.

Turning to the dispersion curves for the track/ground system, shown in Figure 12.15, at relatively low speeds the load-speed line will intersect the dispersion

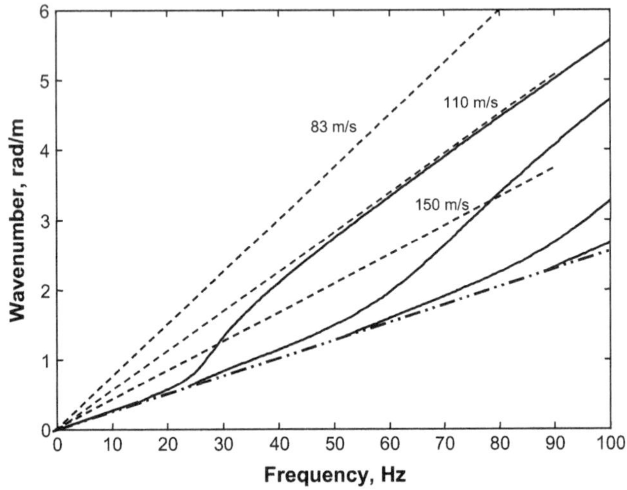

FIGURE 12-16 The dispersion diagram for propagating P-SV waves of the example ground showing various load-speed lines

curve for the track wave, causing a 'resonance' at around 60–70 Hz. This will have only a moderate response due to the influence of damping in the track. As the speed is increased a particularly high response can be expected when the load-speed line intersects the brighter part of the dispersion diagram around 20–40 Hz. In this frequency region the load-speed line will intersect the dispersion curves over a broad frequency region. The corresponding load speed is about 114 m/s.

A more detailed discussion of the critical speed, and the influence of the track parameters on this, is given in references [12.34, 12.35].

Figure 12.17 shows the maximum displacement of the ground at the base of the track as a function of load speed calculated for the two ground stiffnesses and for a heavier track as well as the standard track considered above. For the case shown in Figure 12.15 (stiff ground, lighter track) the displacement has its peak at 114 m/s, corresponding to the intersection noted above. The peak at the critical speed in each case is excited purely by the quasi-static load, not from any roughness of the track which is not included at this stage. Similar results have been measured for trains travelling over sites with soft soils.

In these examples the critical speeds are quite high compared with normal train speeds (100 m/s is 360 km/h). This is the case for most soil conditions. However, it is the ratio of the train speed to ground wave speed that is important. In the example, the Rayleigh wave speed in the upper layer is 110 m/s. In reality, problems occur at sites where ground wave speeds are much lower than this.

12.4.3 Results for a single moving load

Using the theoretical model of Figure 12.13, the wave field for a single constant load moving along the track at a speed below that of any of the waves in the ground is shown in Figure 12.18 [12.24]. The result illustrated is for 83 m/s (300 km/h), not far

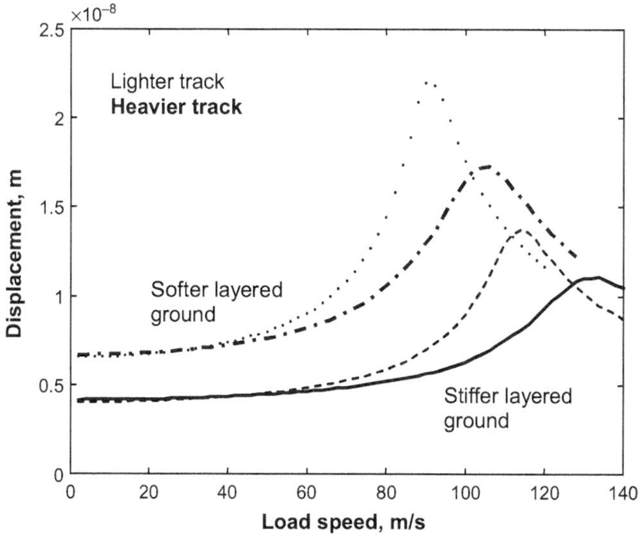

FIGURE 12-17 The maximum deflection under a moving load plotted against load speed for two sets of ground properties and for a light and a heavy track [12.34]

below the minimum ground wave speed of 110 m/s in the example ground (Figure 12.11). This illustrates that this behaviour persists up to velocities that are quite close to the wave speeds in the ground.

The displacement 'dip' under the single load is indicated by the positive (upward) displacement under the track. Although the passage of the quasi-static displacement pattern may be observed close to the track, little effect is observed just a few metres away.

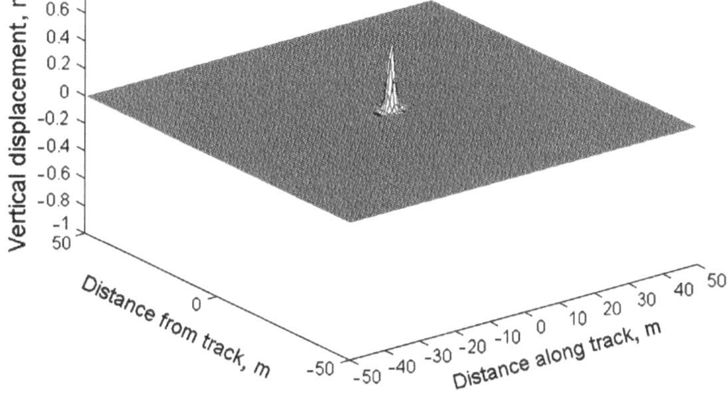

FIGURE 12-18 Displacement pattern in the moving frame of reference for a single non-oscillating axle load on the track moving at 83 m/s, below the wave speeds in the ground [12.24]

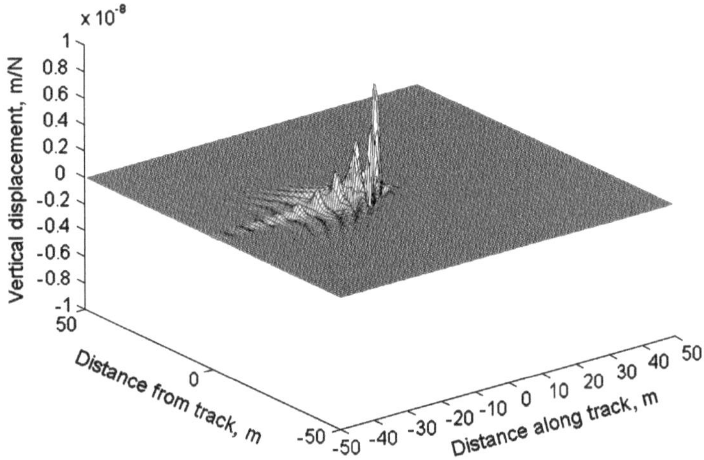

FIGURE 12-19 Displacement pattern in the moving frame of reference for a single non-oscillating axle load on the track moving at 150 m/s, above the critical speed for this track/ground system [12.24]

Figure 12.19 shows the corresponding results when the load travels at a speed that is greater than the critical speed. In this case the load-speed line directly intersects the dispersion curves of propagating waves in the ground. In the field plot, propagating waves may be seen travelling with significant amplitude away from the track. These exhibit the form of a 'bow wave' because the load speed is greater than the speed of waves in the ground.

12.4.4 Excitation by dynamic forces

A stationary harmonic load will, of course, excite all waves that exist at the excitation frequency. The corresponding load line is vertical on the frequency/wavenumber plot. For a moving harmonic load, the excitation is indicated in Figure 12.20 by lines which have slopes corresponding to different speeds but all originating at a non-zero frequency point on the frequency axis, in this case 40 Hz. These lines demonstrate the effect of the speed as the load passes a point, giving rise to a range of frequencies in the ground due to the Doppler effect.

Figure 12.21 shows the response of the ground to a load oscillating at 16 Hz and moving at 40 m/s. The wave speed in the track, which is slower than the wave in the surrounding ground, can be seen to produce a peak in the vibration amplitude running just in front of the load.

Figure 12.22 shows the response to a load oscillating at 40 Hz and moving at 40 m/s. At this frequency the wavelengths can be seen to be much shorter than at 16 Hz and the amplitude under the load is smaller (Figures 12.21, 12.22 and 12.23 are to the same scale), consistent with the lower receptance in Figure 12.14. There is a trail of high amplitude vibration along the track behind the load but the waves propagating out to distances further from the load still have circular wave fronts since the speed of waves in the ground is greater than the speed of the load.

Figure 12.23 shows the results for the 40 Hz load moving at 110 m/s. Here the load is travelling at the speed of the waves in the ground. In this case, further from the track

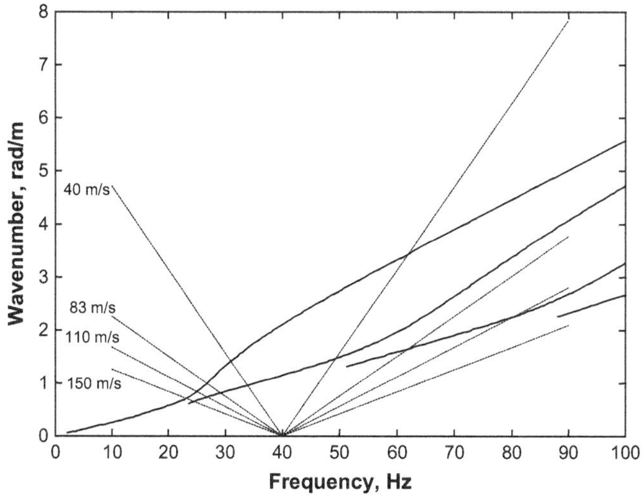

FIGURE 12-20 Excitation by harmonic loads at different speeds showing the Doppler effect that leads to a range of frequencies excited by a single vehicle load frequency

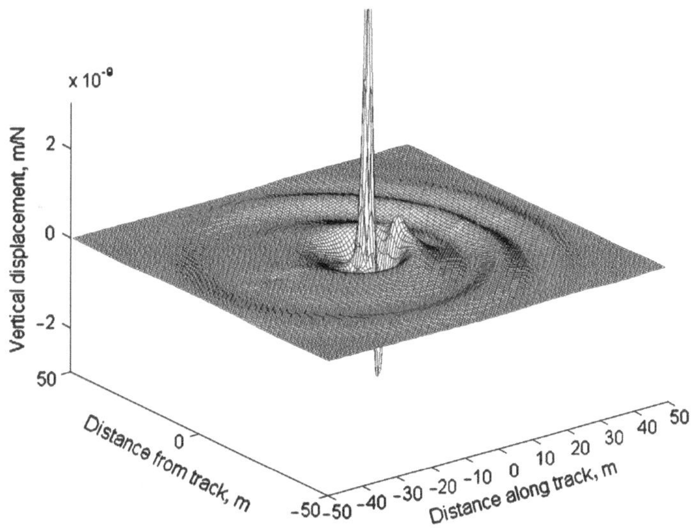

FIGURE 12-21 Response to a 16 Hz load on the track moving at 40 m/s

the waves in the ground do not propagate ahead of the load but are confined to a broad 'Mach cone' behind the load. Strong waves are excited in the track behind the load.

12.5 EXAMPLES OF CALCULATED VIBRATION FROM TRAINS

The model can be extended to include the interaction of a train of vehicles with the track/ground system. The vehicles are represented by multi-body models. Both

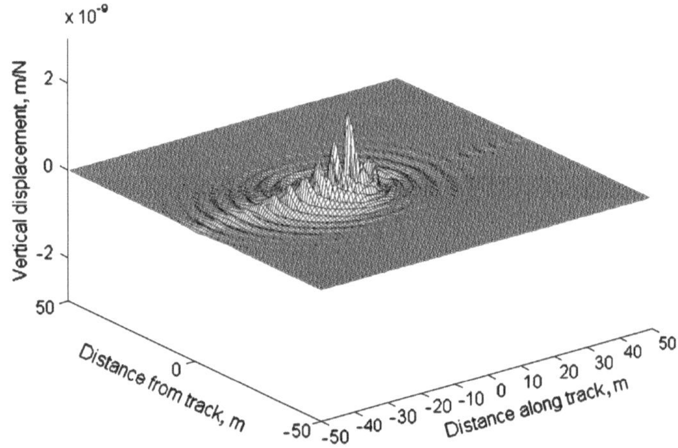

FIGURE 12-22 Response to a 40 Hz load on the track moving at 40 m/s

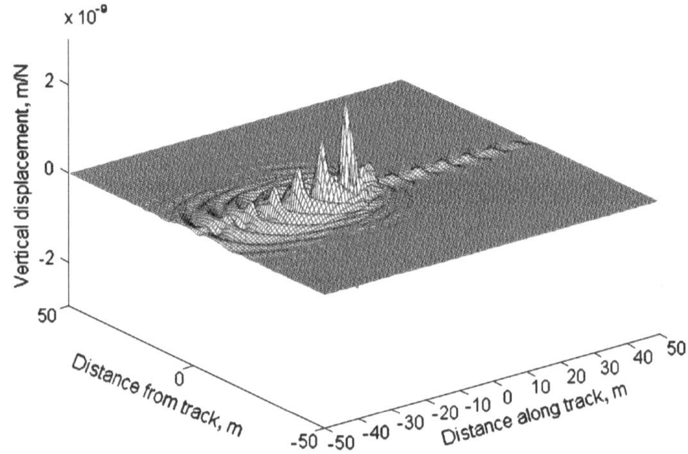

FIGURE 12-23 Response to a 40 Hz load on the track moving at 110 m/s

the quasi-static and dynamic excitation mechanisms can be included [12.36]. The vibration excited by the moving axle loads of the whole train are taken into account and an irregular vertical track profile is introduced to excite dynamic loads at each axle.

Three example cases are examined in the subsections below to illustrate different situations. Comparisons are given with measurements in each case.

12.5.1 Example of train speed exceeding the ground wave speed

In this subsection, the model is applied to the vibrations from the X2000 high speed train at a site called Ledsgård in Sweden [12.37]. At this site very large vibrations were encountered when the trains operated at 200 km/h before the line was

TABLE 12-4 **GROUND PROPERTIES ASSUMED FOR SITE AT LEDSGÅRD, SWEDEN**

	Thickness	P-wave speed	S-wave speed	Density
Upper layer	1.1 m	500 m/s	65 m/s	1500 kg/m^3
Second layer	3 m	500 m/s	32 m/s	1250 kg/m^3
Substratum	infinite	1500 m/s	85 m/s	1470 kg/m^3

opened. The Swedish National Rail Administration (Banverket) carried out an extensive programme of measurements using a test train to investigate the causes.

The ground at this site is modelled here as two layers on a homogeneous half-space using properties identified in reference [12.37]. The properties used are listed in Table 12.4. The track is on an embankment about 1 m high and has monobloc concrete sleepers in ballast. It must be emphasized that the situation at this site is very unusual with the soft conditions of the second layer.

Figure 12.24 shows the dispersion curves of propagating P-SV modes of vibration predicted for the ground at Ledsgård; a load-speed line for the speed of 55.6 m/s (200 km/h) is also shown. It has an intersection with the dispersion curve of the first mode at wavenumber 0.4 rad/m at 4 Hz. The presence of the mass of the track and embankment (not included in the calculation of the dispersion curves) decreases the wavenumber of this intersection slightly. As a result, a propagating wave of about 16 m wavelength is expected to be excited at this speed. Moreover, the intersection between the load-speed line and the dispersion curve for the first wave extends from about 4 to 8 Hz and a further intersection with the second wave occurs above 10 Hz.

Figures 12.25 and 12.26 show the instantaneous displacements of the embankment for the two train speeds, 70 and 200 km/h. At the track, the response to the dynamic wheel/rail forces is small compared with that due to the quasi-static loads. For the low speed case, a quasi-static loading state is indicated with quite large

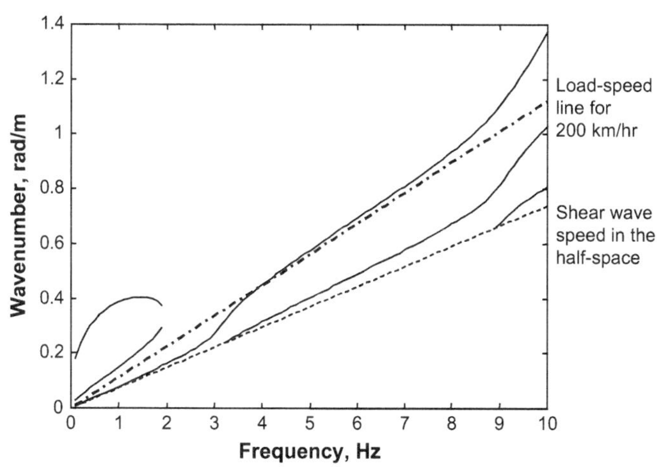

FIGURE 12-24 The dispersion diagram calculated for the site at Ledsgård [12.36]

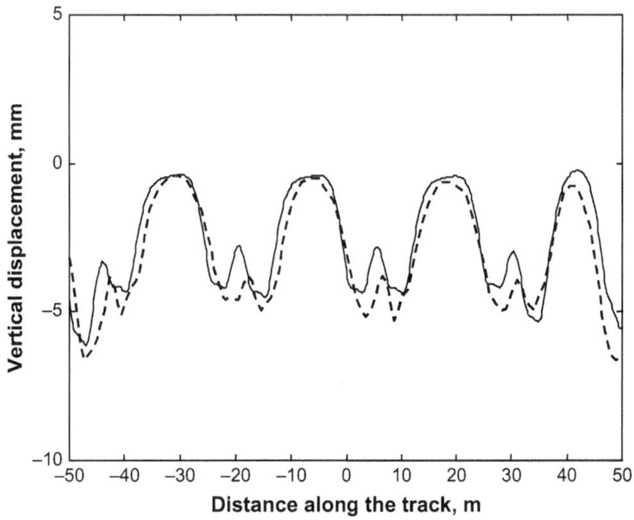

FIGURE 12-25 The displacement of the track at Ledsgård under the X2000 train at 70 km/h [12.36]

amplitudes. However, in the high speed case, since a propagating wave mode in the ground is excited, an oscillating response appears with a much higher amplitude. This propagates along the track from each load and can be seen in Figure 12.26 as an oscillation which continues after the last axle has passed.

The total response generated by the test train has been predicted on the basis of vehicle suspension parameters provided by Banverket and, in the absence of site-specific data, a typical rail profile spectrum measured on a 200 km/h, mixed-traffic

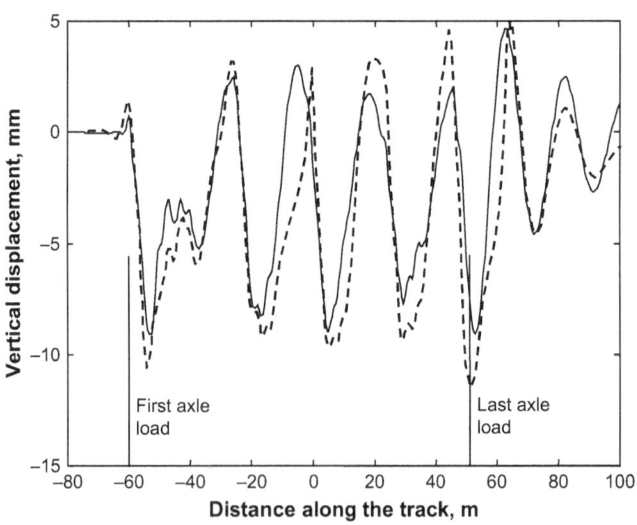

FIGURE 12-26 The displacement of the track at Ledsgård under the X2000 train at 200 km/h [12.36]

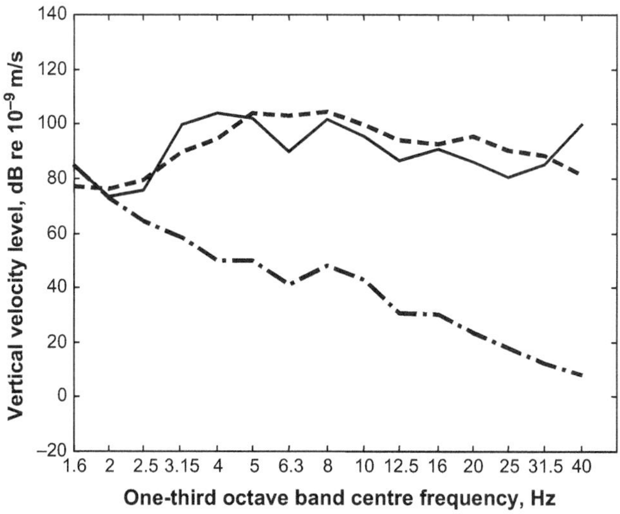

FIGURE 12-27 The spectrum of vibration at 7.5 m from the track at Ledsgård for the X2000 travelling at 70 km/h: —, total prediction (dynamic plus quasi-static); – · – ·, predicted quasi-static component; – – –, measured [12.36]

main line in England has been used. The vertical velocity levels of a measurement point 7.5 m from the track on the ground surface for the two train speeds are shown in Figures 12.27 and 12.28. Figure 12.27 shows that, at 70 km/h, the dynamic components of the wheel/rail forces are dominant over the quasi-static loads, even for very low frequencies. However, in Figure 12.28, for the train at 200 km/h,

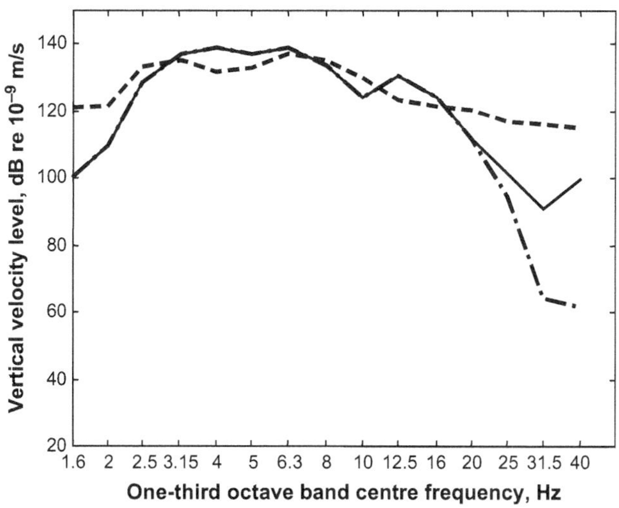

FIGURE 12-28 The spectrum of vibration at 7.5 m from the track at Ledsgård for the X2000 travelling at 200 km/h: —, total prediction (dynamic plus quasi-static); – · – ·, predicted quasi-static component; – – –, measured [12.36]

TABLE 12-5 **GROUND PROPERTIES ASSUMED FOR SITE AT VIA TEDALDA, ITALY**

	Thickness	P-wave speed	S-wave speed	Density
Upper layer	10 m	995 m/s	300 m/s	1800 kg/m^3
Substratum	infinite	1950 m/s	600 m/s	1800 kg/m^3

exceeding the wave speeds in the ground, the response from the quasi-static loads dominates, particularly for the frequency range where the load speed excites the first mode (about 4 to 8 Hz, see Figure 12.24).

12.5.2 Examples of trains travelling at lower speeds for more typical ground parameters

Next, predicted vibrations are compared with measurements for the ETR500 high speed train at a site called Via Tedalda in Italy [12.38]. The average speed of the train passages during the measurement was about 70 to 80 km/h. The vibration has been measured at two points, 13 and 26 m from the track. The assumed ground parameters are listed in Table 12.5. The wave speeds can be seen to be much higher than at Ledsgård and are also somewhat higher than the typical parameters assumed in Table 12.2. The dispersion curves of the ground are shown in Figure 12.29. The first cut-on frequency in the layer is 11.2 Hz.

In the absence of specific parameters, the track structure, other than the embankment, has been assigned parameters typical of a ballasted track with monobloc sleepers. The embankment is 1.5 m high, and its density has been estimated as 1800 kg/m^3.

Figure 12.30 compares predicted and measured vibration (acceleration) spectra at a point 13 m from the track. In the calculations, five ETR500 passenger cars

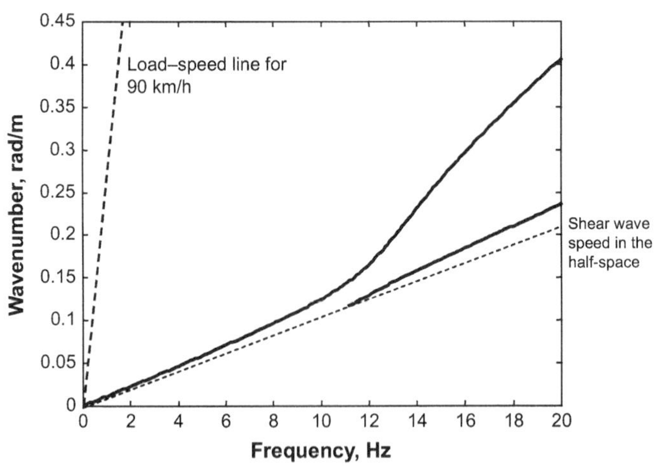

FIGURE 12-29 The dispersion diagram for the Via Tedalda site

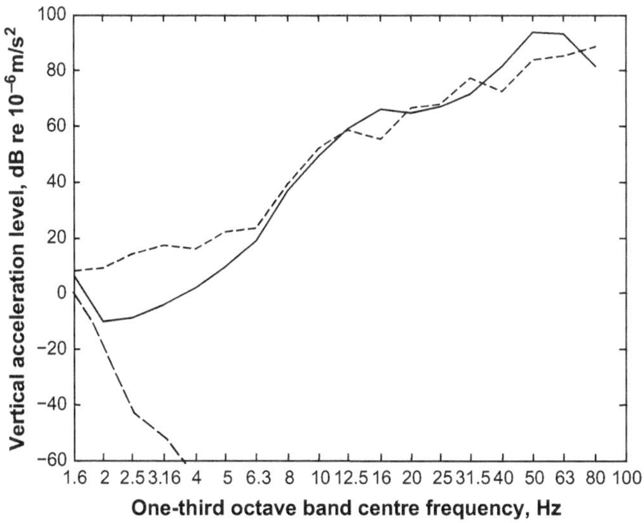

FIGURE 12-30 The spectrum of vibration 13 m from the track for the ETR500 travelling at 90 km/h at Via Tedalda: ——, total prediction (dynamic plus quasi-static); - - - -, measured; — — —, predicted quasi-static component [12.36]

running at 25 m/s (90 km/h) are coupled with the track/ground system and, again in the absence of specific data, the same UK rail profile spectrum has been used. Since the embankment is more extensive at this site than at Ledsgård, better agreement was actually found by modelling its cross-section using finite elements in a finite element/boundary element (FE/BE) scheme [12.24], rather than the analytical model used throughout this chapter. The FE/BE model is described in Chapter 13.

The spectrum shows a rise in vibration level corresponding to the cut-on at about 11 Hz. A close agreement is achieved for frequencies higher than 5 Hz. However, for frequencies of 2 to 5 Hz, the predicted levels are lower than the measured ones. The figure also shows the response due to the quasi-static loads without the dynamic mechanism. At very low frequency, where the wavelength is large, 13 m is close enough to be in the near field. However, it is clear in this case that the quasi-static mechanism of excitation is insignificant for the vibration at 13 m and the dynamic components of the wheel/rail forces dominate the response.

A further comparison is shown for a typical case where heavy axle-load freight wagons produce high levels of ground vibration. Information on the measurement in Nottinghamshire, England, is reported in [12.21]. The average speed of trains during the measurement was about 14 m/s (45 km/h). The ground is modelled as a single layer of 1.8 m depth, overlying a homogeneous half-space, with wave speeds listed in Table 12.6.

The track was ballasted with an embankment of 1.3 m height. In this case, site specific rail profile measurements are available from the time of the vibration measurements and have been used. Figure 12.31 shows the predicted vertical velocity levels on the ground 10 m from the track. For comparison, the range of measured

TABLE 12-6 **GROUND PROPERTIES ASSUMED FOR SITE IN NOTTINGHAMSHIRE, UK**

	Thickness	P-wave speed	S-wave speed	Density
Upper layer	1.8 m	341 m/s	81 m/s	1520 kg/m^3
Substratum	infinite	1700 m/s	216 m/s	2060 kg/m^3

levels from several trains is shown shaded on the figure. The level of response due to the quasi-static loads is less than 40 dB and therefore not shown in the figure.

12.5.3 Summary

In each of the examples presented in this section, at distances of the order of 10 m from the track (and at further distances, as shown in the references) the rise in the spectrum of vibration due to the onset of propagation in the upper layer of soil is clearly seen. It is this that leads to the dominant components of vibration velocity in the spectrum. Thus the perceived vibration is strongly controlled by the conditions of the top layer of soil and the frequency above which the modes propagate in this upper soil layer. In comparison with this upper layer, the track structure (including the embankment) has a significant mass and can therefore also influence the vibration propagation.

At around 10 m from the track and beyond, it is usually the dynamic mechanism of excitation, due to the combined wheel and track irregular profile, that is responsible for the vibration. However, in the unusual case of very soft ground conditions, where the train might travel at speeds comparable with the ground wave

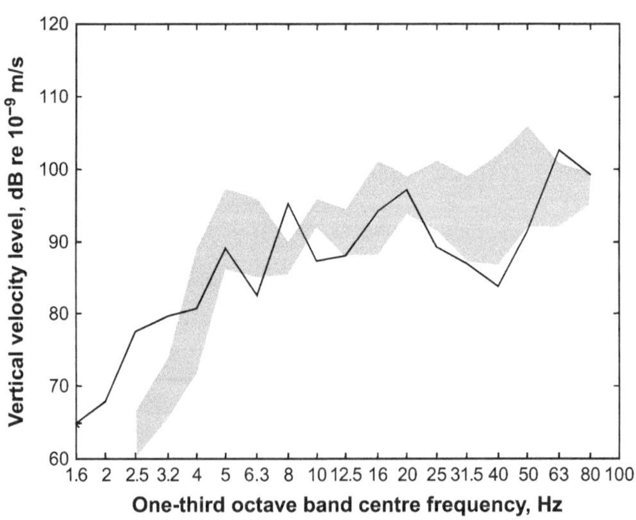

FIGURE 12-31 Measured and calculated vibration 10 m from the track for coal wagons travelling at 45 km/h in Nottinghamshire (shaded range of measurement data) [12.36]

speeds, the excitation of propagating waves via the quasi-static mechanism can become important.

12.6 MITIGATION MEASURES

There are no generally applicable mitigation measures for problems of low frequency vibration from rail traffic. The treatment depends on the dominant mechanism of vibration excitation and on the interaction of the ground and track structure. Given the long wavelengths of vibration in both the track and ground, large-scale civil engineering measures must be adopted. The analysis of mechanisms presented in this chapter gives clues to measures that could be effective but these must be appropriately designed for particular locations.

If dynamic forces are the dominant excitation at the problem frequencies then improving the track alignment by tamping the track (i.e. reducing the long wavelength components of the vertical track profile) should help. However, if the quasi-static excitation is dominant tamping will give no benefit.

12.6.1 Trenches and buried walls

Vibration propagating as modal waves tied to the surface can, in principle, be reflected by a trench or 'in-filled' trench (buried wall). This forms a change in impedance in the propagation medium. These options have, for a long time, been proposed to reduce the transmission of surface-propagating vibration [12.39]. Both measurements and mathematical analysis based on Rayleigh waves in a half-space of homogeneous ground (i.e. not layered) show that a trench will attenuate vibration to about half the amplitude for vibration at wavelengths that are short compared with the depth of the trench. They also show that the benefit is lost beyond a certain distance ('shadow zone'). This is due to the fact that energy is not purely transmitted by surface-mode waves and diffraction can occur around the bottom of barriers.

As has been shown in Figure 12.9, the propagating modes have components of displacement even at large depths within the soil. To achieve high attenuations of vibration in a homogeneous ground at low frequency would require impractical depths of trench. This is indicated in design rules that have been determined using two-dimensional boundary element models. Ahmad and Hussaini conducted a parametric study using a two-dimensional model for open and in-filled trenches close to the source [12.40]. Yang and Hung [12.41] performed a similar study (using an infinite finite element technique) for trenches close to, and therefore protecting, a single property. For practical trenches Yang and Hung showed significant reductions could only be achieved down to about 16 Hz. The relative success of either open or in-filled trenches depends on the geometry and in-fill materials used, as well as the depth.

Having seen, in earlier parts of this chapter, that the relatively thin upper layer of soil is important in determining the nature of surface-wave propagation, it is interesting to examine the effect of soil layering. A trench may be expected to have a greater effect in such a layered ground than in the homogeneous case. Studies that take this into account are few. May and Bolt [12.42] examined this aspect for very low frequencies (1.5 to 6 Hz). Using a two-dimensional model, they showed a case in which, between 3 and 4 Hz, the horizontal component of vibration was increased.

They also showed that in a particular two-trench case, vibration could be amplified because of a resonance of the soil between the trenches. This, at least, shows that there may be 'pitfalls' in the design of trenches and particular designs should be studied carefully.

It may be suspected that the resonance effect of May and Bolt's study might be dependent on the two-dimensional model that was used and 'trapping' of energy between two reflectors can occur more easily than in a three-dimensional case. Hung, Yang and Chang [12.43] extended the analysis of trenches by using wave-number finite and infinite elements but it is not clear how their results compare with the two-dimensional studies. The tools for more advanced studies of trenches exist, e.g. [12.31], but much investigative work has yet to be done.

Some direct testing of trenches alongside railway lines has been carried out; reference [12.1] presents a summary of some results from Switzerland, Germany and Japan. This shows that, at some sites a reasonable depth of trench (3 to 5 m) can produce reductions of the order of 5 dB at frequencies as low as 6 to 8 Hz. However, there are considerable uncertainties in such measured results; without more detail of the sites and validated models, insight cannot be gained as to when and why trenches work or do not. Vibration reduction, of the order of 10 dB, is clearly easier to obtain at higher frequencies (16 Hz upwards is shown). This is, of course, useful for higher frequency vibration problems that are becoming apparent for surface railways, as has been mentioned in the introduction to this chapter.

12.6.2 Wave-impeding blocks ('WIB's)

As the highest components of vibration are controlled by the softer and relatively shallow upper layer of soil that is commonly found, there is some potential in the idea of stiffening the soil to modify the ground layer structure locally. This has been called a 'wave-impeding block' (WIB). These have been proposed for use under or next to railways [12.44–12.47]. The method seeks to stiffen soil or replace it with concrete in order to change the modal propagation regime, i.e. to move the cut-on of propagation in the top layer to higher frequencies. Thus the method uniquely offers the prospect of vibration reduction at very low frequencies, in contrast to barrier methods which are effective above a particular frequency as wavelengths get shorter. Very few practical tests have been conducted, but where experimental results exist these are promising for low frequencies below the cut-on of propagation in the upper soil layer [12.47].

A finite-element/boundary element model for an example analysis of a WIB is shown in Figure 12.32 [12.32]. The WIB extends to 6 m either side of the track centreline. Predictions of the vibration spectrum from the track with this model are shown in Figure 12.33 for distances of 5, 10 and 20 m. The layer depth in this analysis is again 2 m, the upper soil layer is slightly softer than in Table 12.2, having an S-wave speed of 81 m/s and a P-wave speed of 340 m/s. The parameters of the underlying half-space are the same as previously.

It can be seen that the rise in the spectrum due to the cut-on of the first mode in the upper layer of material, between 10 and 25 Hz in the untreated case, is pushed up in frequency by about two one-third octave bands. This results in a substantial reduction in the strongest components of the spectrum between 10 and 40 Hz. At frequencies lower than 10 Hz no consistent reduction is achieved, especially at the larger distances. This example result shows that the method is promising. However,

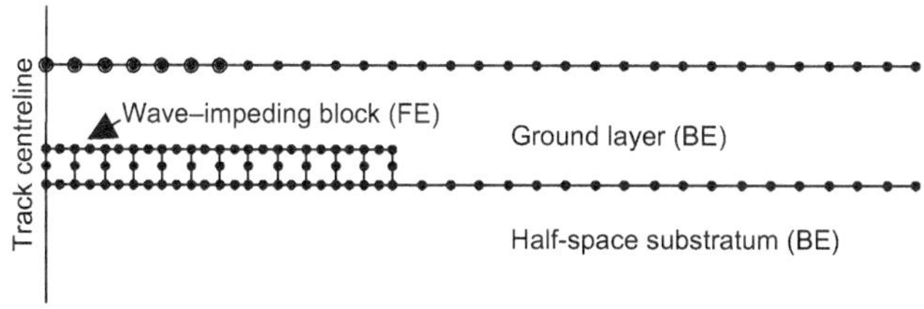

FIGURE 12-32 Model for the WIB in the ground layer (track model attached at circled nodes) [12.32]

more research is required to relate the computational analyses to measurements in practice.

12.6.3 Measures for high speed trains on very soft soils

If the quasi-static axle loads or high speed 'bow wave' are the dominant mechanism then reducing the 'roughness' excitation by track maintenance will give no benefit. Instead, foundation stiffening is appropriate. Since the measurements presented in Figures 12.25 to 12.28 were carried out, the foundations of the embankment at Ledsgård have been stiffened using lime injection techniques and this has alleviated the situation [12.48].

At other sites, where high speed lines pass over areas of very soft soil, the bending stiffness of the embankment has been increased using what is, in effect, a concrete bridge deck supported on piles going through the soft layer into the lower stiffer material. This has been used, for example, at Kungsbaka, in Sweden, close to the site at Ledsgård mentioned above and at Rainham Marshes on the

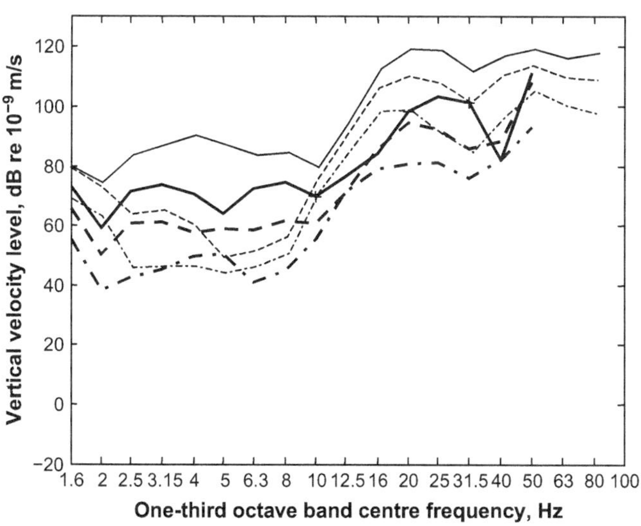

FIGURE 12-33 Vertical velocity levels at 5 m (—), 10 m (– – –) and 20 m (– · –) from the track centreline when a passenger coach runs at 60 m/s: thin lines, without WIB; thick lines, with WIB [12.32]

Channel Tunnel Rail Link in England. However, in each of these cases the main concern has been the stability of the track and associated line-side structures (electrification masts, etc.); there are no building developments next to these railways because of the unsuitable soft nature of the land. Thus, for environmental vibration, the effects are not only unusual but, where they occur, may not affect any line-side buildings.

Another technique that may be appropriate for some very soft soil sites would be to avoid massive embankments by using very light construction materials. This would avoid introducing a slow wave in the embankment structure. To keep the critical speed above the required train speed, the adoption of relatively stiff forms of slab track may also be beneficial.

12.6.4 Vehicle-based measures

As well as measures applied to the track and ground, the vehicle can also have an important influence. This is clear from experience at particular sites where mixed traffic operates but a single train service may give rise the main complaints of vibration.

When the what is now the Central Line of London Underground was opened in 1900, the electric locomotives were totally unsprung and dramatic levels of vibration caused complaints from property owners [12.49]. One company complained that it caused their draftsmen to draw wavy lines. The case was investigated and the locomotives were modified with a suspension, lower overall weight (reduced from 44 tonnes to 31 tonnes) and re-motored. It was concluded that the sprung locomotives produced one-third of the vibration amplitude but multiple unit trains less than one-fifth. By 1903 the locomotives had been scrapped and multiple units were in service.

Many more modern cases of very high vibration levels are associated with freight wagons with friction damped suspensions. When these are used to carry materials such as cement, this may cause the suspensions to seize up and high vibration levels ensue. Again, therefore, effectively unsprung heavy vehicles are known to cause problems. Clearly, the unsprung mass makes a difference to the vibration generated (see Chapter 13 where the effect of the unsprung mass on the dynamic forces generated at the wheel/rail contact is discussed).

Modern passenger rolling stock is often made with low unsprung masses compared with that from prior to the 1990s. This is done in order to reduce track damage. It has been achieved by avoiding electric motors hung directly on the axle. Instead, the motors are hung from the bogie and vibration isolation is included in the transmissions. The scope for reducing the unsprung mass significantly beyond this, for the sake of ground vibration reduction, is probably limited.

The higher unsprung masses and higher suspension frequencies of freight vehicles and locomotives distinguish them as leading to higher levels of vibration at frequencies below 10 Hz. Some bogies exist for freight wagons, designed to reduce track forces, that use a two stage suspension. They have also been observed to cause much less vibration than more conventional freight vehicles at vibration-prone sites.

Different parameters affect the quasi-static and dynamic excitation mechanisms. Reduced unsprung mass leads to reduced dynamic excitation from roughness. However, a reduced axle load has no effect on roughness excitation, see box on page 151, but it directly affects the magnitude of the quasi-static deflections.

A further difference in quasi-statically excited vibration has been observed between bogied and two-axle wagons. The periodic pattern of axle-load deflections at the track caused by long trains of similar wagons causes vibration to be produced with strong harmonic components. Two-axled wagons cause a pattern of deflections with lower frequency, stronger harmonics than bogied (four-axle) wagons. The correspondence between the strength of harmonics in the axle-load deflection pattern in the track and the relative strengths of measured ground vibration has been demonstrated [12.50].

Thus axle load, unsprung mass, suspension frequency, maintenance of friction damped suspensions and axle spacing are all parameters than can affect vibration in different circumstances.

12.6.5 Concluding remarks

This chapter has introduced the issues of environmental vibration from railways but concentrated on the nature of vibration from trains on tracks at grade. This involves surface-propagating modal waves in a layered medium producing 'whole body' vibration in the frequency range from about 4 to 80 Hz.

For mitigation of very low frequency vibration, it is to be expected that remedies must involve the modification of the main dynamic system, namely the ground. Thus mitigation is always going to be very costly. No generic components can be manufactured to reduce vibration, but trenches and WIBs have been shown to hold promise. However, few implementations exist and this is a field for future development as the need to solve the vibration problems increases. Moreover, the success of these measures would depend very much on thorough, site-specific analysis.

Mitigation by vibration-isolating track forms, considered in the next chapter, is rarely appropriate for low frequency vibration as the isolation frequency would be too low to be practical.

As has already been said, higher frequency vibration from at-grade railways is a growing concern. The barrier options discussed in this chapter may be very useful for reducing this higher frequency response. In any case, however, higher frequency vibration problems may have solutions that are common with vibration from railways in tunnel. This is the subject of the next chapter.

REFERENCES

12.1 R. Müller. Mitigation measures for open lines against vibration and ground-borne noise: a Swiss overview. Springer, *Notes on Numerical Fluid Mechanics and Multidisciplinary Design, Noise and Vibration Mitigation for Rail Transportation Systems*, **99**, 2008.

12.2 R. Müller and K. Köstli. *Vibration and vibration induced noise immissions from railways: estimation of retrofit costs for Swiss railways.* SBB report to Swiss federal office of transport, 2008.

12.3 T. Meloni. Verordnung über den Schutz vor Erschütterungen (VSE). Bauwerksdynamik und Erschütterungsmessung, 10, Symposium, EMPA Dübendorf, 2007.

12.4 H.J. Woodroof and M.J. Griffin. A survey of the effect of railway-induced building vibration on the community. ISVR Technical Report no., 160, University of Southampton, 1987.

12.5 S. Yokoshima. A study on factors constituting annoyance due to Shinkansen railway vibration. *Journal of Architecture, Planning and Environmental Engineering*, **526**, 1999.

12.6 J.M. Fields and J.G. Walker. *The effects of railway noise and vibration on the community.* Contract report 77/18, ISVR, University of Southampton, 1977.

12.7 M.J. Griffin. *Handbook of Human Vibration*. Academic Press, London, 1990.
12.8 ISO 2631–1: 1997, Mechanical vibration and shock – Evaluation of human exposure to whole-body vibration – General requirements. International Organization for Standardization.
12.9 ISO 2631–2: 2002, Mechanical vibration and shock – Evaluation of human exposure to whole-body vibration – Part 2: Vibration in buildings (1 Hz to 80 Hz). International Organization for Standardization.
12.10 BS 6472: 1992, Evaluation of human exposure to vibration in buildings (1 Hz to 80 Hz), British Standards Institution.
12.11 BS 6841: 1987, Measurement and evaluation of human exposure to whole-body vibration, mechanical vibration and repeated shock. British Standards Institution.
12.12 DIN 4150–2: 1999, Structural vibration – Part 2: Human exposure to vibration in buildings, (English version). Deutsches Institut für Normung.
12.13 DIN 45669–1: 1995, Mechanical vibration and shock measurement – Part 1: Measuring equipment, (English version). Deutsches Institut für Normung.
12.14 ISO 4866:1990, Mechanical vibration and shock – Vibration of buildings – Guidelines for the measurement of vibrations and evaluation of their effects on buildings. International Organization for Standardization.
12.15 ISO 4866 – Amendment 2:1996, Mechanical vibration and shock – Vibration of buildings – Guidelines for the measurement of vibrations and their effects on buildings International Organization for Standardization.
12.16 DIN 4150–3: 1986, Structural vibration in buildings – Effects on structures. Deutsches Institut für Normung.
12.17 W.T. Thompson. Transmission of elastic waves in plane infinite structures. *Journal of Applied Physics*, **21**, 89–93, 1950.
12.18 N.A. Haskell. The dispersion of surface waves on multilayered media. *Bulletin of the Seismological Society of America*, **43**(1), 17–34, 1953.
12.19 E. Kausel and J.M. Roësset. Stiffness matrices for layered soils. *Bulletin of the Seismological Society of America*, **71**(6), 1743–1761, 1981.
12.20 C.J.C. Jones. Using numerical models to find antivibration measures for railways. *Proceedings of the Institution of Civil Engineers, Transport*, **105**, 43–51, 1994.
12.21 C.J.C. Jones and J. Block. Prediction of ground vibration from freight trains. *Journal of Sound and Vibration*, **193**(1), 205–213, 1996.
12.22 D.V. Jones, D. Le Houdec, and M. Petyt. Ground vibrations due to a rectangular harmonic load. *Journal of Sound and Vibration*, **212**(1), 61–74, 1998.
12.23 X. Sheng, C.J.C. Jones, and M. Petyt. Ground vibration generated by a harmonic load acting on a railway track. *Journal of Sound and Vibration*, **225**(1), 3–28, 1999.
12.24 X. Sheng, C.J.C. Jones, and M. Petyt. Ground vibration generated by a load moving along a railway track. *Journal of Sound and Vibration*, **228**(1), 129–156, 1999.
12.25 J. Dominguez. *Boundary Elements in Dynamics*. Computational Mechanics Publications, Southampton, 1993.
12.26 C.J.C. Jones, D.J. Thompson, and M. Petyt. A model for ground vibration from railway tunnels. *Proceedings of the Institution of Civil Engineers, Transport*, **153**(2), 121–129, 2002.
12.27 L. Andersen. *Wave propagation in infinite structures and media*. PhD thesis, Aalborg University, Denmark, 2002.
12.28 L. Andersen and C.J.C. Jones. Coupled boundary and finite element analysis of vibration from railway tunnels – a comparison of two- and three-dimensional models. *Journal of Sound and Vibration*, **293**(3–5), 611–625, 2006.
12.29 Y.B. Tang and H.H. Hung. A 2.5 D finite/infinite element approach for modelling visco-elastic bodies subjected to moving loads. *International Journal for Numerical Methods in Engineering*, **51**, 1317–1336, 2001.
12.30 P. Jean, C. Guigou, and M. Villot. 2D1/2 BEM model of ground structure interaction. *Journal of Building Acoustics*, **11**(3), 157–173, 2004.

12.31 X. Sheng, C.J.C. Jones, and D.J. Thompson. Modelling ground vibration from railways using wavenumber finite- and boundary element methods. *Proceedings of the Royal Society A*, **461**, 2043–2070, 2005.
12.32 X. Sheng, C.J.C. Jones, and D.J. Thompson. Prediction of ground vibration from trains using the wavenumber finite and boundary element methods. *Journal of Sound and Vibration*, **293**(3–5), 575–586, 2006.
12.33 G. Degrande, D. Clouteau, R. Othman, M. Arnst, H. Chebli, R. Klein, P. Chatterjee, and B. Janssens. A numerical model for ground-borne vibrations from underground railway traffic based on a periodic finite element-boundary element formulation. *Journal of Sound and Vibration*, **293**(3–5), 645–666, 2006.
12.34 X. Sheng, C.J.C. Jones, and D.J. Thompson. A theoretical study on the influence of the track on train-induced ground vibration. *Journal of Sound and Vibration*, **272**, 909–936, 2004.
12.35 C.J.C. Jones, X. Sheng, and M. Petyt. Simulations of ground vibration from a moving harmonic load on a railway track. *Journal of Sound and Vibration*, **231**(3), 739–751, 2000.
12.36 X. Sheng, C.J.C. Jones, and D.J. Thompson. A comparison of a theoretical model for quasi-statically and dynamically induced environmental vibration from trains with measurements. *Journal of Sound and Vibration*, **267**, 621–636, 2003.
12.37 Banverket. *Seminar on High Speed Lines on Soft Ground, Dynamic Soil-Track Interaction and Groundborne Vibration*, produced by the Swedish National Railway Authority, Banverket, Gothenburg, Sweden, 2000.
12.38 C.G. Lai, A. Callerio, E. Faccioli, and A. Martino. Mathematical modelling of railway-induced ground vibrations. *Proceedings of the International Workshop Wave 2000*, 99–110, 2000.
12.39 F.E. Richart, J.R. Hall, and R.D. Woods. *Vibrations of Soils and Foundations*. Prentice Hall International Series in Theoretical and Applied Mechanics, 1970.
12.40 S. Ahmad and T.M. Al-Hussaini. Simplified design for vibration screening by open and in-filled trenches. *Proceedings of the ASCE, Journal of Geotechnical Engineering*, **117**(1), 67–88, 1991.
12.41 Y. Yang and H.H. Hung. A parametric study of wave barriers for reduction of train-induced vibrations. *International Journal for Numerical Methods in Engineering*, **40**(20), 3729–3747, 1997.
12.42 T. May and B.A. Bolt. The effectiveness of trenches in reducing seismic motion. *Earthquake Engineering and Structural Dynamics*, **10**(2), 195–210, 1982.
12.43 H.H. Hung, Y.B. Yang, and D.W. Chang. Wave barriers for the reduction of train-induced vibrations in soils. *Journal of Geotechnical and Geo-environmental Engineering*, **130**(12), 1283–1291, 2004.
12.44 H. Takemiya and A. Fujiwara. Wave propagation/impediment in a stratum and wave impeding block (WIB) measured for SSI response reduction. *Soil Dynamics and Earthquake Engineering*, **13**, 49–61, 1994.
12.45 G. Schmid, N. Chouw, and R. Le. Shielding of structures from soil vibrations. *Proceedings of Soil Dynamics and Earthquake Engineering V*, Computational Mechanics Publications, 1992. 651–662
12.46 A.T. Peplow, C.J.C. Jones, and M. Petyt. Surface vibration propagation over a layered elastic half-space with an inclusion. *Applied Acoustics*, **56**(4), 283–296, 1999.
12.47 H. Takemiya, N. Chouw, and G. Schmid. Wave impeding block (WIB) for response reduction of soil-structure under train induced vibrations. *10th European Conference on Earthquake Engineering*. In: Duma (ed.). Balkema, Rotterdam, 1995.
12.48 A.T. Peplow and A.M. Kaynia. Prediction and validation of traffic vibration reduction due to cement column stabilization. *Soil Dynamics and Earthquake Engineering*, **27**, 793–802, 2007.
12.49 R.L. Vickers. DC Electric Locomotives and Trains in the British Isles. David and Charles. Newton Abbot, 25–26, 1986.
12.50 R.A.J. Ford. The prediction of ground vibrations by railway trains. *Journal of Sound and Vibration*, **116**(3), 585–589, 1987.

CHAPTER 13

Ground-borne Noise*

13.1 INTRODUCTION

In Chapter 12, the issues of vibration from railways were divided into three categories. These are not mutually exclusive but can be treated as separate issues both because of differences in their nature and the mitigation methods that are applicable. Chapter 12 therefore dealt with low frequency surface-propagating vibration from railways at grade.

This chapter deals with higher frequency vibration (approximately 30 to 250 Hz) that leads to noise which is radiated via the vibration response of the walls and floors of a building. The phenomenon of a rumbling noise as trains pass is therefore referred to as 'ground-borne noise', 'vibration-induced noise' or 'structure-borne noise'. It is most associated with trains in tunnels where the direct, airborne noise is effectively screened off.

It has been estimated that 56 000 homes in London are subjected to maximum levels of ground-borne noise during a train pass of over 40 dB(A) and a small number to over 60 dB(A) [13.1]. These are high levels of noise for the low frequency range and considering the nature of the intrusion. This indicates how common ground-borne noise is on metro networks where long lengths of line are in tunnel in densely built-up areas.

As discussed in the introduction to Chapter 12, ground-borne noise can also affect at-grade railways where noise barriers and double glazing are already treating direct airborne noise. However, this chapter is based on examples for the case of tunnels. This, for example, allows it to be shown that there is an important difference between surface-propagating vibration and ground-borne noise. Nevertheless, it must be recognized that, with well-screened surface railways, tramways, railway in cuttings and cut-and-cover tunnels, a blending of the issues presented in Chapter 12 and this chapter will often be the case in practice.

The noise level arising from vibration inside a building is also discussed briefly later in this chapter. However, up to the point at which acoustic radiation occurs in the receiving building, the source and propagation aspects of the problem are those of vibration and this is reflected in the treatment of the topic of ground-borne noise.

Figure 13.1 shows typical vibration spectra measured on the ground floor of a small office building from trains passing 15 m below in a bored tunnel. The background vibration spectrum between the trains is also shown.

In Chapter 12 it was shown, albeit for surface vibration, that quasi-static excitation was limited to about 10 Hz and is only then dominant in some cases. At higher

* This chapter has been written by Chris Jones.

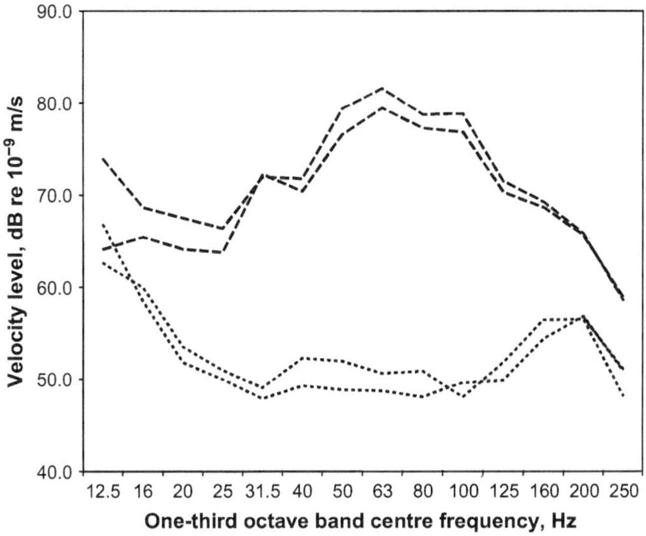

FIGURE 13-1 Typical spectra of vibration from trains in a modern bored tunnel at 15 m depth in clay compared to background vibration. – – –, vibration from trains; - - -, background level

frequencies the mechanism of generation of vibration is, like rolling noise, the uneveness of the wheel and rail surfaces. For the frequency range here, and the higher ground stiffness that exists at depth, it is fairly certain that of these two mechanisms only the dynamically induced vibration will be significant.

The range of roughness wavelengths that are relevant to the ground-borne noise frequency range at various speeds has already been shown in Table 12.3. It extends from the acoustic roughness range, from about 0.05 m, to around 2 m for conventional speed trains.

In this wavelength range the wheel roughness can be influenced by the train's braking mechanism (tread braked or disc braked); longer wavelengths on the wheel are manifested as 'out-of-round' or eccentricity. The range also covers the transition between wavelengths where the rail-head condition is responsible for the 'roughness' and those longer than about 1 m where the 'vertical profile' is controlled by the ballast under the sleepers (i.e. the level maintained by tamping).

There may also be effects of the differential displacement of the rail between the sleeper support positions and mid-sleeper. This can be considered as equivalent roughness, although it is really a parametric excitation (see Section 5.7.5). It excites vibration at the sleeper-passing frequency, which is sometimes significant with hard rail pads but not necessarily with softer rail pads. In addition, vibration may be excited by variations in the sleeper support stiffness.

13.2 ASSESSMENT CRITERIA

Ground-borne noise is perceived as noise rather than as vibration. It is not measured in the same way as rolling noise, at the line-side in the open, because, of course, it only exists inside buildings. The perception is also different from rolling noise because of its frequency range and character.

TABLE 13-1 **TYPICAL ASSESSMENT CRITERIA FOR GROUND-BORNE NOISE FROM [13.2] – US DEPARTMENT OF TRANSPORTATION**

Land use	**Target levels, L_{Amax} (dB re 2 × 10^{-5} Pa)**	
	Frequent events	**Infrequent events**
Residencies and buildings where people normally sleep	35	43
Institutional land use with primarily day-time use (schools, offices, churches)	40	48
Special cases:		
Concert halls, TV studios, recording studios	25	25
Auditoria	30	38
Theatres	35	43

Table 13.1 shows typical assessment criteria that are used for new railway projects. These are stated in the US Department of Transportation guidance on vibration impact assessments, 1995 [13.2].

Note that these are based on the maximum level, L_{Amax}, rather than the long-term averaged level, L_{Aeq}. The A-weighting curve seems to give a good correlation to annoyance [13.3] but, since the whole frequency range of ground-borne noise (30 to 250 Hz) is strongly attenuated by the weighting curve, it would be wrong to compare the noise levels presented in Table 13.1 with those of other types of noise such as airborne railway noise, road traffic noise or aircraft noise, which are usually assessed outside dwellings.

In Table 13.1 'frequent' means more than 70 vibration events per day; most metros or tram systems fall into this category. Most commuter train routes would fall into the 'infrequent' category.

Very similar guidelines to the American ones exist in Germany. Elsewhere, the practice generally is for railway projects to negotiate design aims with the appropriate local authorities or government agencies through the planning or public inquiry process. The targets set rarely differ much in their principles or levels from those of Table 13.1 since they are well known and provide a precedent. Studies show that, in any case, a narrow range of levels is relevant. In [13.4] it was found that, although levels around 32 dB(A) were acceptable, a level of 42 dB(A) gave rise to strong complaints. In the London survey [13.1], already mentioned, exposures above 40 dB(A) were identified. This is not among the more stringent levels for acceptability. The 60 dB(A) exposures found in that survey can be seen, in the light of Table 13.1 and [13.4], to be extreme.

13.3 VIBRATION PROPAGATION FROM A TUNNEL

As with surface vibration propagation, it is useful to visualize the pattern of propagation from a tunnel in order to understand some of the effects that are

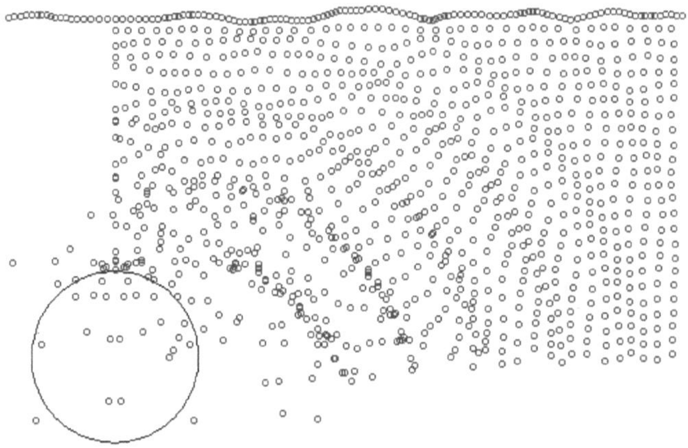

FIGURE 13-2 Vibration field from a circular tunnel at 100 Hz (two-dimensional model)

observed. Figure 13.2 shows a snapshot at one instant in time of the vibration predicted using a two-dimensional boundary element model [13.5]. This shows the deflection of a rectangular grid of points. Close to the tunnel the exaggerated motion is very large and difficult to interpret at this scale, but further away the propagation of waves can be seen more clearly. Unlike surface vibration, 'radiation' of the vibration from some depth within the ground means that 'body waves' propagate rather than surface wave modes. The resulting wave motion involves more shear deformation than compression. Vibration is transmitted to the ground around the whole tunnel ring but most of the vibration is radiated from the tunnel invert (below the track) where the highest vibration response of the ring occurs.

At shallow angles, i.e. to the farther distances on the surface, the propagation adopts a simple, here cylindrical, wave-front shape. Thus, a simple relationship of response level with distance can be expected. However, for the ground surface above the tunnel, the tunnel structure forms a barrier preventing vibration of the invert from radiating upwards. The maximum vibration along the surface therefore often occurs to the side of the tunnel alignment, typically by a distance of the same order as the tunnel depth, as seen in Figure 13.2.

Vibration propagates around the tunnel ring structure rather like surface vibration but this surface is now constrained to a degree depending on the tunnel ring structure. Different tunnel structures therefore propagate vibration of differing amplitudes to the tunnel crown. This is illustrated, again from a two-dimensional finite element/boundary element model [13.5], in Figure 13.3 for a tunnel without lining and one with a continuous concrete tunnel ring. The tunnel structure, then, has a significant influence on the vibration at the surface immediately above the tunnel but not at larger distances away from the tunnel alignment.

This principle is also apparent in the effect of the tunnel depth, illustrated in Figure 13.4 for a 5.6 m diameter tunnel in clay. The responses along the ground surface are plotted as the root of the sum of squared lateral and vertical components. The response converges at large distances, where the radial distance from the tunnel is similar, but above the tunnel the response depends on the distance of the tunnel crown to the surface and the diffraction of vibration around the tunnel.

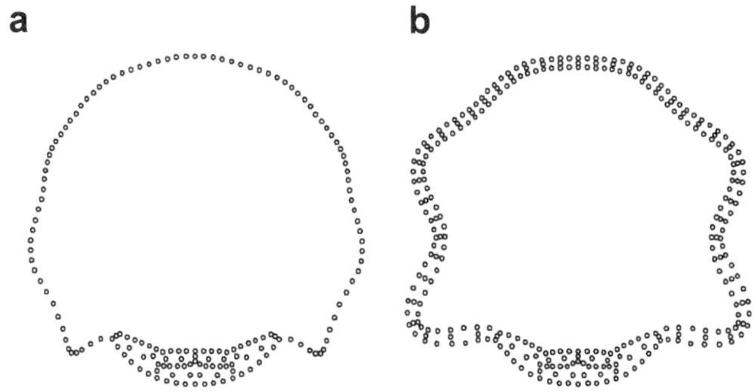

FIGURE 13-3 Vibration pattern at the tunnel ring at 100 Hz for two stiffnesses of tunnel ring structure. (a) Unlined tunnel, (b) tunnel with concrete ring [13.5]

Figure 13.5 presents an example of the propagation from a twin-track cut and cover tunnel at 100 Hz. Here the pattern of vibration involves the resonances of the massive abutment walls in the ground.

The vibration, generated under the tracks, has to diffract around the tunnel (abutment) walls. Close to the walls the decay with distance can therefore differ considerably from one situation to another. At some frequencies, bouncing or rotating resonances of the walls occur. At high frequencies the walls may conduct vibration to the surface so the level is high close to them; at other frequencies the abutment walls shade the ground close to the tunnel from the vibration source at the track.

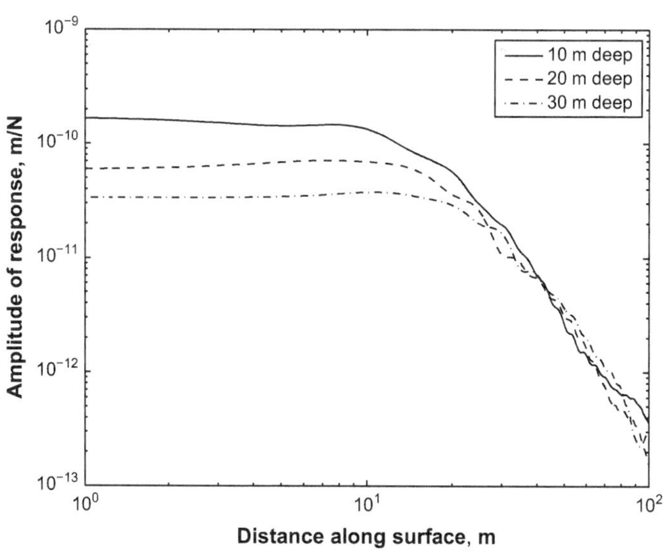

FIGURE 13-4 Results from a two-dimensional model that show the variation of response at 50 Hz from a 5.6 m diameter tunnel in clay at different depths

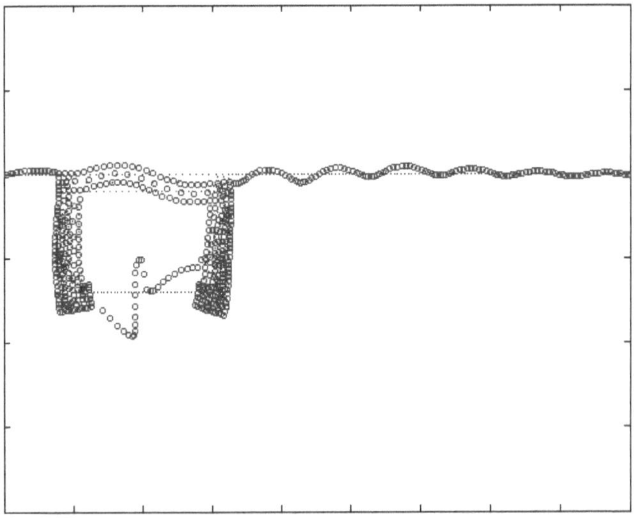

FIGURE 13-5 Propagation from a cut-and-cover tunnel at 100 Hz, modelled using two-dimensional FE/BE model

For either bored or cut-and-cover tunnels, vibration measured at a single point inside the tunnel does not bear a simple relationship to that observed at a distance; the latter is the sum of vibration radiated from all parts of the tunnel section, and all along the tunnel. Near-field waves (reactive field rather than active) influence the area immediately around and inside the tunnel but this vibration does not propagate into the far field.

The vibration directly above a cut-and-cover tunnel is a structural vibration problem, rather than a ground propagation problem. The characteristic wavelengths of vibration and its amplitude are not the same, therefore, as that of the surrounding soil. This can also be seen in Figure 13.5. Often cut-and-cover tunnels are constructed so that only roads are situated on this part of the structure. However, sometimes the tunnel can be part of an integrated development and houses or offices may be built directly on top.

13.4 MODELS FOR GROUND-BORNE NOISE

13.4.1 Vehicle/track interaction model

The model described in Chapter 5 for roughness excitation can also be used for ground-borne noise. In this frequency range it is sufficient to consider the wheel in terms of its unsprung mass, M_W. The track can be represented by a (complex) spring stiffness, K_T, at low frequencies but has a resonance frequency in the range 50 to 100 Hz, above which its mobility resembles an infinite beam. The contact spring between wheel and rail can be neglected at these low frequencies.

As shown in Section 5.2 the velocity response of the rail at frequency ω can be given as

$$v_r = \frac{i\omega r Y_r}{Y_r + Y_w} \tag{13.1}$$

where r is the roughness amplitude, Y_r is the rail mobility and

$$Y_w = \frac{-i}{\omega M_w} \tag{13.2}$$

is the mobility of the wheel. Assuming that the foundation below the track is rigid, the force transmitted to it through the damped spring $\tilde{K}_T = K_T(1 + i\eta)$ is given by

$$F_T = \frac{\tilde{K}_T v_r}{i\omega} = \frac{r\tilde{K}_T Y_r}{Y_r + Y_w} \tag{13.3}$$

The assumption of a rigid foundation is acceptable for a track on a thick tunnel invert but much less satisfactory for track at grade. Above the track resonance frequency, waves propagate along the rail and the above formula is a less realistic approximation to the transmitted force.

To compare two different situations, relative changes in the transmitted force will equate to the corresponding changes in the vibration at a distance. Thus results for two different situations, 1 and 2, can be compared in terms of an insertion loss:

$$IL = -20\log_{10}\left(\left|\frac{v_2}{v_1}\right|\right) = -20\log_{10}\left(\left|\frac{F_{T,2}}{F_{T,1}}\right|\right) \tag{13.4}$$

where v_1 and v_2 are the vibration velocity amplitudes at some fixed receiver location. The insertion loss, IL, is positive if the second situation gives a lower vibration amplitude than the first, which is treated as the reference. Sometimes results are presented in terms of insertion gain, $IG = -IL$.

To explore this further, it is instructive to replace the track by a mass–spring model, with stiffness K_T and mass $M_T = \omega_0^2 K_T$, where ω_0 is the resonance frequency of the track on its foundation given by equation (3.5). This is shown in Figure 13.6(a). It is an acceptable model up to just above the track resonance frequency but at higher frequencies the mobility of a beam reduces less rapidly than this implies. This model leads to an equivalent rail mobility:

$$Y_r = \frac{i\omega}{\tilde{K}_T - \omega^2 M_T} \tag{13.5}$$

from which the transmitted force is given by

$$F_T = \frac{-r\tilde{K}_T\omega^2 M_w}{\tilde{K}_T - \omega^2(M_T + M_w)} \tag{13.6}$$

This transmitted force has a maximum when

$$\omega = \sqrt{\frac{K_T}{M_T + M_w}} \tag{13.7}$$

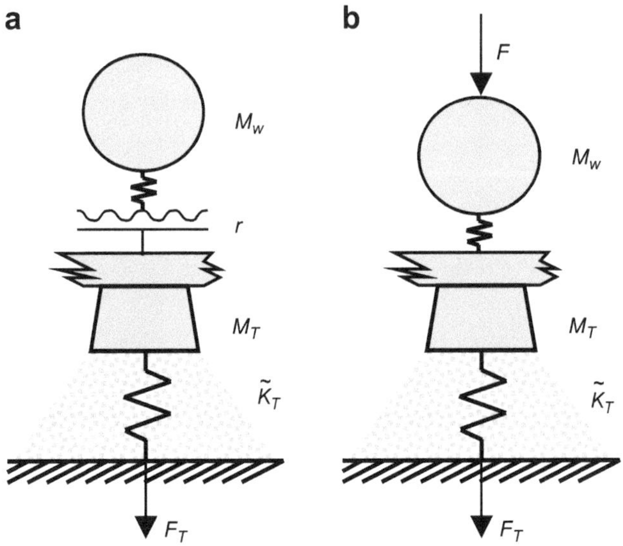

FIGURE 13-6 Mass–spring representations of the wheel/track system. (a) Excited by roughness, *r*, (b) excited by a force, *F*

which can be identified as the coupled wheel/track system resonance as discussed in Section 5.2. This has been modified from equation (5.13) by the inclusion of the track mass, which is important above ω_0.

13.4.2 Equivalent single-degree-of-freedom model

Sometimes in the literature a simple single-degree-of-freedom model is used to represent the excitation of ground-borne noise. This consists of a system in which the vehicle unsprung mass M_w is added to the track mass M_T, all supported by the track stiffness K_T, as shown in Figure 13.6(b). As will be seen, this has some advantages in interpretation, but omits an important aspect of the roughness excitation.

If a harmonic force of amplitude F is applied to such a system, the response amplitude is

$$v_r = \frac{i\omega F}{\tilde{K}_T - \omega^2(M_T + M_w)} \tag{13.8}$$

The resonance frequency of this combined system, where the velocity v_r has a maximum, is the same as the coupled wheel/track resonance frequency identified in equation (13.7). From equation (13.8), the force transmitted to the foundation through the spring $\tilde{K}_T$ is given by

$$F_T = \frac{\tilde{K}_T F}{\tilde{K}_T - \omega^2(M_T + M_w)} \tag{13.9}$$

The ratio $T_F = |F_T/F|$ is known as the force transmissibility and can be used in general to assess the extent of vibration isolation offered by such a single-degree-of-freedom system. The ratio of force transmissibilities in two situations allows the insertion loss to be found:

$$IL = -20\log_{10}\left(\frac{T_{F,2}}{T_{F,1}}\right) = -20\log_{10}\left(\left|\frac{F_{T,2}}{F_{T,1}}\right|\right) \tag{13.10}$$

provided that the force F is the same in the two situations.

Comparing equations (13.6) and (13.9), it can be seen that they are equivalent for a force

$$F = -r\omega^2 M_w \tag{13.11}$$

which is the reaction force required to accelerate the mass over the roughness profile. Hence the insertion loss determined from the equivalent single-degree-of-freedom model is identical to that from the roughness-excited model provided that the wheel unsprung mass is unchanged. Note, however, that the wheel vibration is not correctly predicted in this model.

The force transmissibility of a single-degree-of-freedom system is shown in Figure 13.7. Here the frequency axis is shown in non-dimensional form by dividing by the natural frequency, ω_n. At low frequencies the force is transmitted unattenuated by the spring. At the natural frequency it is amplified, the extent of this amplification depending on the damping. For low values of damping the amplification is large. The force transmissibility only becomes less than unity for frequencies above $\sqrt{2}\omega_n$.

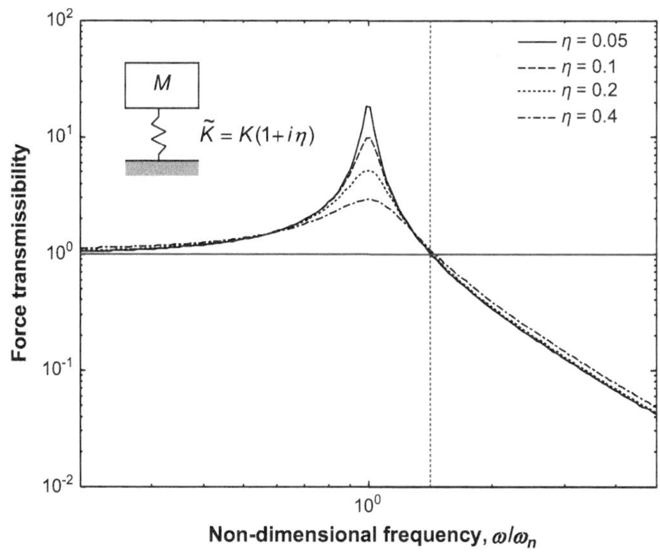

FIGURE 13-7 The force transmissibility of a single-degree-of-freedom mass–spring system with hysteretic damping

For a viscous damping model, as usually presented in vibration text books, the reduction in transmitted force at high frequencies also becomes dependent on the damping rate, with high damping leading to greater transmitted force. However, the hysteretic damping model is more representative of elastomeric materials typically used in track supports. With this damping model the high frequency isolation is independent of damping, as shown in Figure 13.7. In reality a constant damping loss factor implies a weak frequency dependence of the stiffness (see Sections 3.2.7 and 3.8) not included here.

The insertion loss between two situations is illustrated in Figure 13.8 for a change in support stiffness. The parameters used for these are listed in Table 13.2.

Reducing the track support stiffness leads to a reduction in the coupled resonance frequency, here from 75 to 42 Hz. A dip in the insertion loss occurs at the new resonance frequency, above which the insertion loss rises to a peak at the old resonance frequency before levelling off to a constant value. Also shown in the figure are results from the beam model of the track, which are mostly similar except at high frequencies.

Figure 13.9 shows the results of increasing the track mass, here by a factor of 10. Again the simple model gives adequate results, although the beam model shows greater differences than for the results of Figure 13.8. In this example, the large change in track mass has only a small effect on the coupled resonance frequency as it is initially small compared with the unsprung mass of the vehicle.

From these results it is clear that the coupled wheel/track resonance frequency should be made as low as practically possible in order to avoid the increase in vibration from occurring within the frequency range of interest. The damping of this

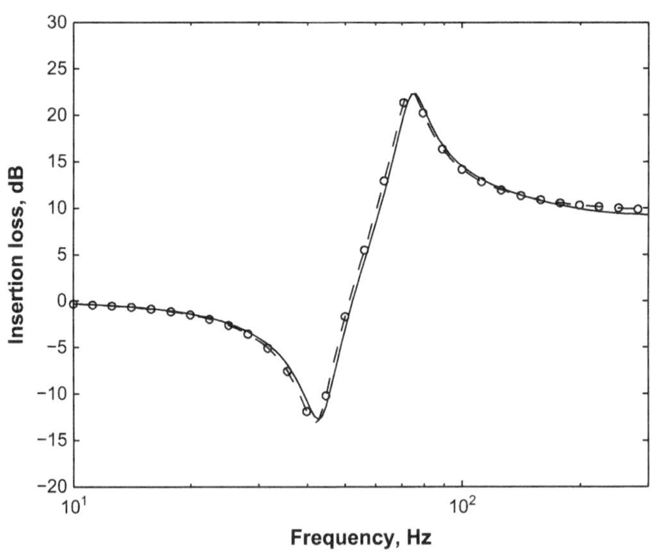

FIGURE 13-8 Insertion loss due to reducing track support stiffness (parameters in Table 13.2). —, track mobility based on beam; – – –, track mobility based on equivalent mass–spring system; o, single-degree-of-freedom model

TABLE 13-2 PARAMETERS USED FOR CALCULATIONS IN FIGURES 13.8–13.10

	Reference case	Variant 1 (Figure 13.8)	Variant 2 (Figure 13.9)	Variant 3 (Figure 13.10)
Rail bending stiffness, EI	6.4 MN/m^2	6.4 MN/m^2	6.4 MN/m^2	6.4 MN/m^2
Track support stiffness, s	100 MN/m^2	25 MN/m^2	100 MN/m^2	100 MN/m^2
Track stiffness, K_T	142 MN/m	50.3 MN/m	142 MN/m	142 MN/m
Track damping loss factor, η	0.2	0.2	0.2	0.2
Track mass per unit length, m_r'	60 kg/m	60 kg/m	600 kg/m	60 kg/m
Track equivalent mass, M_T	85.4 kg	85.4 kg	854 kg	85.4 kg
Wheel unsprung mass, M_w	600 kg	600 kg	600 kg	1200 kg

resonance should also not be too low in order to avoid large amplifications. In Section 13.6, different ways are described in which the coupled wheel/track resonance frequency can be lowered in practice to reduce vibration transmission and therefore lower the level of ground-borne noise.

The effect of changing the vehicle unsprung mass cannot be so adequately represented by the force transmissibility of the single-degree-of-freedom model. Example results are shown in Figure 13.10. The coupled resonance frequency is

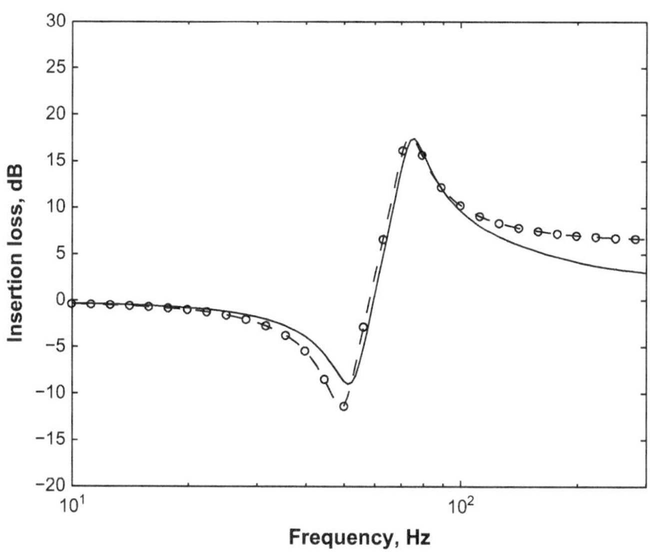

FIGURE 13-9 Insertion loss due to increasing track mass (parameters in Table 13.2). —, track mobility based on beam; – – –, track mobility based on equivalent mass–spring system; ○, single-degree-of-freedom model

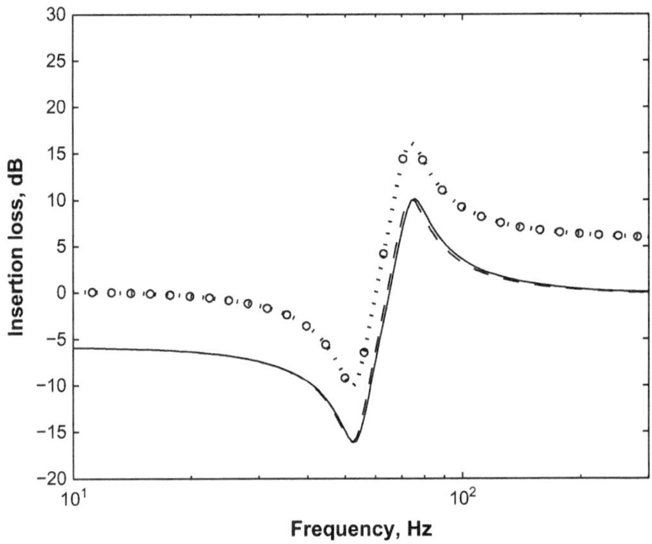

FIGURE 13-10 Insertion loss due to increasing vehicle unsprung mass (parameters in Table 13.2). —, track mobility based on beam; – – –, track mobility based on equivalent mass–spring system; o····o, single-degree-of-freedom model

reduced by an increase in unsprung mass, in the same way as for an increase in track mass. However, this leads to an *increase* in vibration at low frequency, as the force is also modified, equation (13.10). It is therefore seen that *reductions* in unsprung mass are generally to be preferred, especially for lower frequencies but that this may lead to an increase in vibration at the new coupled wheel/track resonance frequency. To prevent this, a large track mass is desirable.

13.4.3 Track-on-half-space model

The single-degree-of-freedom model has too many limitations for practical use. A more realistic, but still relatively simple model is shown in Figure 13.11. This is a simple track/ground model that can be used to determine the effects of changes in vehicle or track parameters. The excitation is modelled as the relative displacement spectrum (roughness) between the unsprung mass of the vehicle and the track. The vehicle can be represented with more of the suspension components than are shown, but for this frequency range the unsprung mass is usually sufficient.

The track is modelled as an infinite layered beam structure; it is sufficient to model both rails as a single beam for this purpose. The track model can be changed to correspond to whatever track form is relevant. It will normally include the tunnel invert as an extra beam at the base of the track. The half-space support represents a realistic frequency-dependent mobility to the track model. This is important, since the energy transmitted to the ground is proportional to the squared force times the real part of the mobility (see Chapter 11).

The use of a simple half-space clearly does not represent the propagation medium from a tunnel to the surface, nor from the tunnel structure other than the invert.

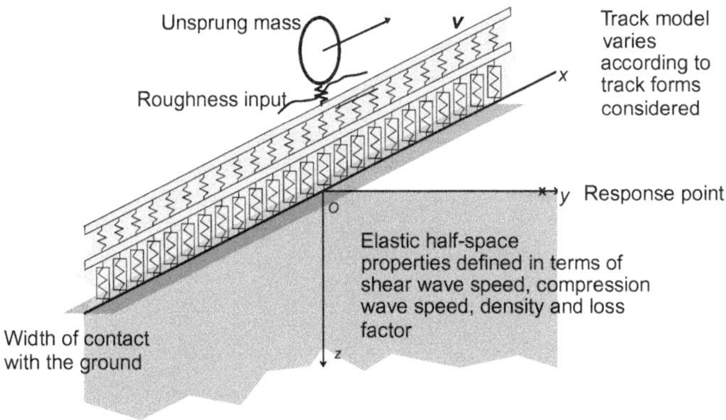

FIGURE 13-11 Schematic view of a model that can be used to predict the changes of vibration at the receiver due to changes in the wheel/rail roughness, track design or unsprung mass of the vehicle

However, the response at some distance from the track does include the effects of the propagation of vibration from all along the track in the far field with some geometric and material damping effects.

The assumption is therefore made that *changes* in the track (or unsprung mass of the vehicle) will result in *changes* in the vibration spectrum at the response point that are approximately independent of other, more specific propagation conditions. This assumption is confirmed by the fact that, when only predicting changes from this model, they are almost independent of the ground material parameters for a wide range of soils. Also, for distances beyond a few metres they are not dependent on the precise distance to the response point.

Of course, in predicting changes from one track to another, it is clear that an insertion loss cannot be stated as an absolute property of the track form; it must always be stated in comparison with another track.

This type of model has been used to predict the vibration mitigation performance of a wide variety of track types for many railway and tramway projects. Some example calculations are given in section 13.6.

13.4.4 Numerical models

To make realistic predictions of the propagation of vibration, the analytical approach described above is too limited. For vibration from surface railways, more extended semi-analytical models can be used including the effects of ground layering (see Chapter 12). However, for vibration from tunnels, cuttings, etc. numerical methods are usually required. These allow the tunnel structures to have arbitrary geometry. They are also useful for situations where the ground is inhomogeneous, for example with non-parallel layers of soil.

Figure 13.12 illustrates one type of model that can be used for predicting the propagation of vibration in this situation. The finite element (FE) method is coupled to a boundary element (BE) method for wave propagation in a linear elastic medium. This is valid for small amplitude wave propagation such as vibration from trains. BE models do this proficiently as only the boundary has to be modelled using

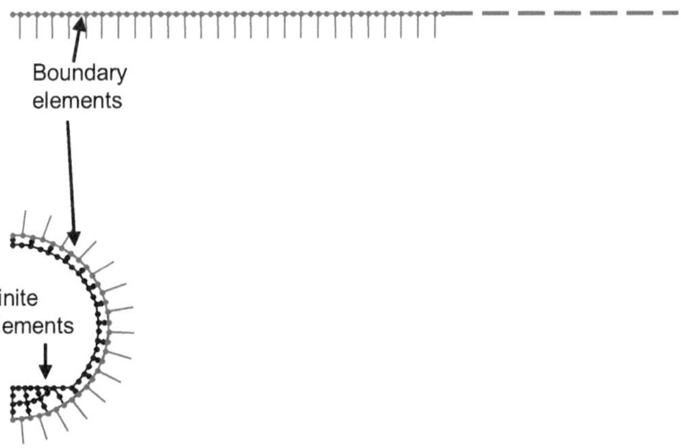

FIGURE 13-12 Coupled finite element model and boundary element model. The boundary elements are illustrated with unit normal vectors pointing into the corresponding medium. A vertical plane of symmetry is assumed

elements; the medium can be infinite. Boundary elements are also used for acoustic radiation problems, see Chapter 6, but in the present case are used for waves in elastic solids.

Some results from a two-dimensional FE/BE model have already been presented to illustrate the nature of propagation from tunnels (Figures 13.2 to 13.5). The model used is described in [13.5]. Full three-dimensional model calculations are possible [13.6] but require extensive computing resources, so in the past two-dimensional models, fast enough to calculate a number of variations of structure, ground lithology, etc. have been preferred.

In recent years a number of '2.5-dimensional' approaches [13.7–13.11] have been developed. These use the 'extruded' geometry of the tunnel. The methods vary but generally the three-dimensional field is obtained by solving a series of two-dimensional problems with different wavenumbers imposed in the axial direction. A Fourier transform is then used to recover the three-dimensional field. This makes the computation efficient enough to carry out on a personal computer within a reasonable time.

These models can be used in conjunction with measurements to determine the effects of changes in the propagation conditions, such as, for example, changes of depth of the tunnel. In addition, coupled with a suitable track and vehicle model, the wavenumber FE/BE method is capable of predicting the spectrum of vibration from trains in the tunnel [13.8]. This can include the excitation from both the moving quasi-static axle loads and the combined track/wheel roughness, although it is the latter that is more important for the ground-borne noise frequency range.

A note of caution should be struck when using theoretical calculations of the propagation of vibration from tunnels. Although the models are advanced and deal with some of the complexity of the situation, it is necessary to make some simplifications in constructing a model. For instance, the ground properties must be taken to be homogeneous and linearly elastic within large regions of the soil. Real soils are, however, complicated and the material properties vary locally along the route as well as close to the surface and at changes of lithology. The type of contact between the tunnel structure and the natural soil (stress relaxation, backfill, etc.) also has an effect. These, and details of structures and ground surface geometry, must be simplified. The result is that there is more uncertainty in using these modelling methods than in making measurements where the results are, at least, bound to be within a representative range of actual vibration levels from railways. Measurements have a variance and therefore additionally lead to a clearer estimate of the uncertainty in predictions; in contrast, theoretical models can give a false impression of precision.

13.5 PREDICTING GROUND-BORNE NOISE FOR ENVIRONMENTAL ASSESSMENTS

13.5.1 Approach

For tunnel sections of new railway projects, ground-borne noise is the most important aspect of the environmental impact on the surroundings that must be evaluated. There are a small number of guides as to how to achieve this (including [13.2] and [13.12]). These break down the prediction into stages corresponding to the 'source', 'propagation' and 'receiver', by analogy to environmental noise prediction. For robust results, methods are based as far as possible on measurements of vibration from existing railways (i.e. infrastructure and rolling stock) that are as close as possible to the situation that is being predicted. Corrections are then introduced to allow for differences from that situation. Interpolation (and in some cases extrapolation) is required to predict vibration at distances other than those that are measured. Corrections are also required for the effects of buildings and finally the noise is estimated from the vibration spectrum.

The following subsections expand on what can be done to fulfil the stages of this process. Because of the variability of the situation and the limited opportunities for relevant measurements, the process is inevitably varied from one project to another. It is not possible, or sensible, to be too prescriptive over methods. Nevertheless, for some cities with large underground railway systems in homogeneous conditions, the railway authority has gathered a large database of measurements made in a systematic way. This is then sufficient to allow estimates to be made for new lines relatively easily and in a relatively standardized manner for that railway.

13.5.2 Vibration spectra

The starting point is to make measurements of vibration from railway vehicles in a situation as close as possible to the one to be predicted. If a city already has underground lines that use similar vehicles with similar track and tunnel construction in the same kind of geology, then the process becomes relatively straightforward and reliable. However, soil lithology differs from one location to the next even in a single city, and tunnel depths also vary. Over the course of development of a network, different contractors build differing tunnel structures. Moreover, new rolling stock is usually bought for new lines making it difficult to rely on measurements of existing rolling stock.

Measurements of vibration have to be made with specialized accelerometers or geophones and some care is required to find the best mounting methods. Experience shows that attachment of a light mounting plate to the ground surface using plaster of Paris gives consistent results, whereas spikes driven into the ground are less reliable.

By measuring at a number of different distances, the decay with distance laws for vibration propagation can be established. Because the wavelengths, damping and 'geometrical spreading' depend on the location, these must be established locally for each railway scheme, usually at more than one location along the alignment.

After having measured spectra for the closest track type and rolling stock, 'correction' spectra can be used to allow for differences between this and the proposed track type and rolling stock. For example, allowance can be made for the influence of the rolling stock unsprung mass and axle density and of the track support stiffness. These corrections can be calculated using the model presented in Section 13.4.3.

13.5.3 Far-field propagation

Where no suitable railway is available with similar ground conditions, it is necessary to establish decay with distance laws for the ground by another means. This requires an artificial vibration source, which can only be a point source rather than the line of sources represented by a train on a track in tunnel.

For a point excitation it is possible to measure at suitable distances with excitation from a drop-hammer device or a hand-held sledgehammer. The measurement scheme is illustrated in Figure 13.13. A small footing attached to the ground helps to ensure that the excitation of the ground is not locally non-linear.

If the response is measured using an accelerometer, the velocity can be obtained by use of an analogue integrator. By recording velocity, rather than acceleration, a lower dynamic range of response signal is obtained making it easier to record or sample. The response is measured on the ground surface at a number of distances from the excitation.

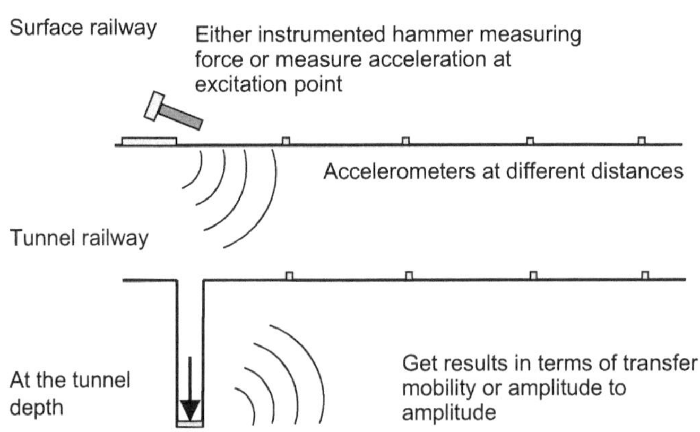

FIGURE 13-13 Measurement scheme for obtaining propagation and decay with distance relationship

Unless a very repeatable drop hammer is used the responses must be normalized, e.g. presented as frequency response functions. In any case the spectrum of excitation will not correspond to that produced by a train through the track and its formation. An instrumented hammer provides a suitable measurement of the excitation force that can be used for normalization. The corresponding frequency response function measurements are in the form of transfer mobility. Alternatively, an accelerometer at the excitation location can be used to give a reference signal, in which case a transmissibility is measured (velocity/velocity).

For measuring transfer functions from a source at tunnel depth, boreholes must be made. Measurements from these are very difficult. Careful thought must be given to anchoring transducers at the bottom of the borehole to measure the vibration of the surrounding soil effectively and not introduce a local resonance of the mass of the transducer on its footing. It is easiest to use the principle of reciprocity and reverse the locations of excitation and response; thus the response is measured down the hole to excitation at a series of positions on the surface. Any measurement using boreholes usually involves the sacrifice of the transducer in the borehole.

Unlike an artificial source, a train is not a point source and some means of adjusting the decay with distance laws appropriately must be found.

The decay with distance *in the far field* is due to two effects: the geometric spreading of the wave and the material damping of the soil. For a particular frequency, this can be expressed as

$$A(d)/A_0 = \underset{\substack{\text{geometric}\\ \text{dispersion}}}{d^{-\alpha}} \times \underset{\substack{\text{damping}\\ \text{losses}}}{e^{-d\beta}} \tag{13.12}$$

where A_0 is the amplitude at the point of excitation and d is the distance. Simple geometric spreading from a point source on the ground surface can be assumed to correspond to $\alpha = \{1/2\}$ (cylindrical wave tied to the surface, e.g. Rayleigh wave), while for a point source at some depth $\alpha = 1$ (spherical wave). The effect of damping, β, has to be determined for the site as a function of frequency.

For the purposes of the attenuation versus distance laws, a train may be modelled practically (though approximately), as a line of incoherent point sources, with a length equal to the (moving) train. Summation of results from expression (13.12) along a finite length for incoherent sources will provide a means of obtaining a suitable decay with distance relationship for a train.

Analysis of measurement data is never simple. Usually, practicality and budgets only allow small numbers of transducer locations to be used. The measured behaviour is varied by local effects, propagation path inhomogeneities (walls, roads, trees, buried pipes) and experimental error. In the light of this variation the process of determining 'corrected' propagation laws for a train source is ill-conditioned, since the attenuations over measurable distances are usually small and there will be unknown systematic deviations from the assumed (indeed, imposed) simple model of equation (13.12).

Figure 13.14 shows some sample results measured in a park next to a railway line. Such locations are useful in that they minimize, but by no means eliminate, the effects of old buildings, foundations, road-beds and trees. Figure 13.14 is a typical imperfect example of surface propagation. Here the decay with distance relationship is found in octave frequency bands. The example shows how much judgement (imagination?) is

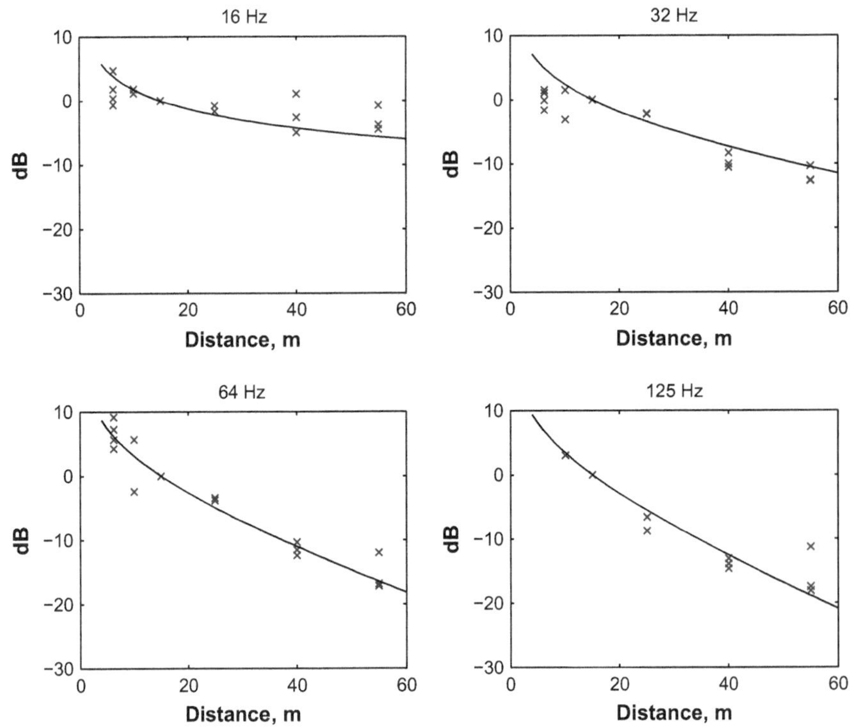

FIGURE 13-14 Example of measured decay with distance from trains; data in octave bands. Vibration levels normalized to reference level at 15 m in each case

involved in extracting the decay with distance, especially at low frequency. Automated curve fitting in these circumstances may produce strange results. The process can be used relatively safely to make predictions of vibration within the measured distance range but extrapolation rapidly becomes very unreliable.

From a borehole it would be difficult to find a measurable response to large distances, adding further to the difficulties.

Sometimes it is possible to use an early or prior tunnel construction to gain access for decay with distance measurements. In that case, care should be taken in using data obtained from, say, the first 20 m from the tunnel alignment. The behaviour described in Section 13.3 and illustrated in Figure 13.4 shows that extrapolation of this near field to, say, 100 m would be grossly misleading. Of course, measurements from boreholes do not exhibit the barrier effects of tunnels.

In the absence of measurements, or where there is only limited opportunity to carry them out, theoretical models can be used to make predictions. This is the case if there is no railway in similar ground conditions with similar structures. Additionally, models can be a very useful complement to measurements.

13.5.4 Near-field propagation

It is clear that propagation in the near field or over the tunnel is not so simple. The situation is complicated by near-field waves, close buildings, cut-and-cover structures, piles and other foundation engineering as well as the effect of the shading from the tunnel already discussed. For trams, transmission through the road pavement will differ from that through the soil, leading to local differences in vibration.

To establish the propagation close to the tunnel in these circumstances, special measurements are required of the transmission from the proposed track alignment into neighbouring buildings. Since the nature of the propagation is more complicated, it can be useful to conduct numerical modelling studies to aid the interpretation of what measurements are possible.

13.5.5 Estimating vibration transmission into buildings

Again, although theoretical modelling is possible, experimental studies are the most direct and reliable way to study the transmission of vibration from the ground into buildings. Measurements of the difference between vibration levels of the ground and those inside a building can be carried out for specific buildings or types of buildings using artificial vibration sources or background vibration due to road traffic.

Because measurement programmes are expensive, most predictions rely on established empirical results. The most well known and comprehensive is a set of tests carried out in Toronto in the 1970s. This, augmented with measurements on various other metropolitan railways in North America, is contained in the US Department of Transportation *Handbook of Urban Rail Noise and Vibration Control*, 1982 [13.12] and its revisions and so has gained extensive use.

For single houses on strip foundations, the empirical data shows the vibration at the base of the building is about 5 dB lower than the vibration level in the ground immediately outside the building. This difference increases to about 15 dB for large masonry buildings on spread foundations.

13.5.6 Effects within buildings

Data from the same source as that for transmission into buildings are available in [13.12] for the effects within buildings. Floors constructed as slabs supported on columns are shown to increase vibration by up to 10 dB. Conversely, going upwards from the base of a building, vibration is generally expected to be attenuated by about 3 dB from one floor to the next.

Alternatively, for specific buildings of special interest, a mathematical model can be used. For this, Statistical Energy Analysis (SEA) is the most appropriate method [13.13], see also Chapter 11.

An alternative calculation approach developed by Hassan [13.14] uses a series of analytical models and dynamic flexibility matrices for propagation in plates and at junctions, etc. to derive expressions for transfer functions of vibration level through the floor levels and supporting columns of tall buildings.

13.5.7 Prediction of noise level from the vibration

Once the vibration spectra of the floor or walls of the building have been determined it is possible to estimate the radiated noise spectrum. If the SEA method is used this can be obtained directly from the method. However, a rapid process applicable generally rather than to specific individual rooms in buildings is usually required.

The 'Kurzweil formula' approach is most often used. Although based in theory it should be thought of also as partly empirical and part of an empirical process such as that outlined in [13.12].

Recalling equation (6.1), the sound power W radiated by vibration of the floor of surface area S is given by

$$W = \rho_0 c_0 \sigma S \left\langle \overline{v^2} \right\rangle \tag{13.13}$$

where $\langle \overline{v^2} \rangle$ is the surface-averaged mean-square velocity, σ is the radiation ratio and $\rho_0 c_0$ is the specific acoustic impedance of air. The average sound pressure in the room can be estimated using Sabine's formula:

$$\left\langle \overline{p^2} \right\rangle = \frac{\rho_0 c_0^2 T_{60} W}{13.81 V} \tag{13.14}$$

where V is the room volume and T_{60} is the reverberation time. This is a measure of the sound absorption of the room and is measured as the time taken for a sound level (from a starting pistol or bursting balloon) to decay by 60 dB [13.15]. It is related to the absorption coefficients of the walls and floors. This model assumes that the sound field in the room is diffuse, a condition that is rarely achieved for the frequency range of ground-borne noise. Nevertheless, it can be used to give a reasonable average result.

Combining these formulae and expressing them in terms of sound pressure level, L_p, in dB re 2×10^{-5} Pa:

$$L_p = L_v + 10 \log_{10}(\sigma) - 10 \log_{10}(H) - 20 + 10 \log_{10}(T_{60}) \tag{13.15}$$

where L_v is the vibration velocity level in dB re 10^{-9} m/s and $H = V/S$ is the height of the room. Now, assuming approximately $\sigma = 1$, $H = 2.8$ m and $T_{60} = 0.5$ s, it is found that

$$L_p \approx L_v - 27\ dB \tag{13.16}$$

Although approximate, this is a convenient equation relating the sound pressure directly to the average velocity on the floor. It should be applied separately to each one-third octave band of a vibration spectrum and the result A-weighted and summed to yield the overall A-weighted level expected in a room.

In this approach, the radiation from walls and ceiling has been neglected and the specific modal behaviour of rooms has been ignored. However, it is an established, appropriate general equation for estimating the sound pressure level from the vibration level prediction. It can be used in the prediction scheme for most buildings without a need for specific information about the interior of the building or other information that is too detailed.

13.5.8 Choice of mitigation options and track structure

Once vibration and noise predictions have been made for a location, there will be a process of predicting impacts at selected receiver locations and choosing the appropriate track forms for different parts of the line in order to bring the L_{Amax}

levels below the agreed targets. Predictions for a single city metro project may involve consideration of a whole range of ground types and tunnel depths.

13.6 MITIGATION MEASURES: TRACK DESIGNS FOR VIBRATION ISOLATION

13.6.1 Introduction

Since the vibration leading to ground-borne noise is of higher frequency than the feelable vibration discussed in Chapter 12, it becomes possible to treat it by vibration isolation, i.e. 'isolating' vibration of the rail from the base of the track and hence from the ground. The same principles apply to the vibration isolation of buildings from the ground. For individual buildings this is also an option that is used in practice, at least for new developments.

The essence of track design for ground-borne noise attenuation is to reduce the coupled wheel/track resonance frequency, as discussed in Section 13.4.2. Figure 13.15 indicates different ways in which this might be achieved in practice. As has been seen, a ballasted track already has a resonance because of the resilience of the ballast. However, in modern tunnels slab track is very often used due to its low cost of maintenance in comparison with the need to keep specialized tamping machinery for underground use. When only rail pads are used to support the rails on

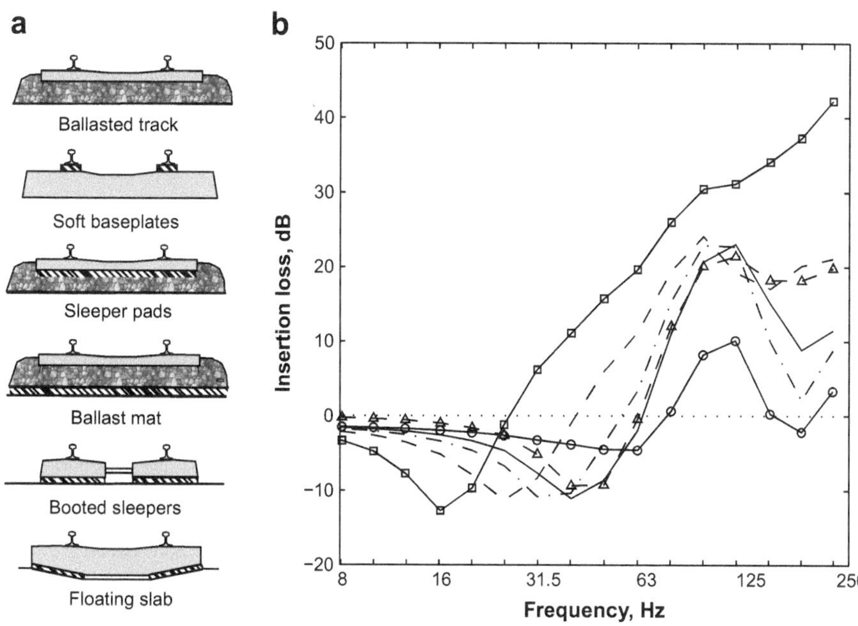

FIGURE 13-15 Different conceptual vibration isolating track designs (left) and insertion losses compared with slab track with rail pads (right). ○—○, ballasted track; △– –△ – · – ·, sleeper soffit pads; —, booted sleepers; – –, ballast mats; □– –□, floating slab track

slab track, it will generally lead to higher levels of ground-borne noise than a ballasted track. Soft rail supports are therefore used to redress the situation. If ballasted track is being used, a further reduction of vibration can be achieved by the use of an elastomeric layer, either between the sleeper and ballast or between the ballast and the tunnel invert. The latter increases the mass above the spring component, thus lowering the coupled resonance further.

For slab track the sleeper masses can be resiliently supported on the slab, or a 'floating slab track' can be used in which the track slab is supported from the tunnel invert by soft supports. The latter has the potential for a very low coupled resonance frequency but is also an expensive option.

Each of these track types is represented pictorially on the left-hand side in Figure 13.15. On the right-hand side, the figure presents calculated insertion losses compared with a slab track with only rail pads supporting the rail. This can be thought of as the basic option for a tunnel where the transmitted vibration does not matter. These insertion losses have been calculated using the track-on-half-space model described in Section 13.4.3. They are meant only to be indicative and are based on typical values for component masses and stiffnesses given in Table 13.3. In practice the parameters for each type of track design can vary considerably, leading to a range of values of insertion loss.

The track designs on the left of Figure 13.15 are arranged from top to bottom roughly in order of increasing potential benefit. Thus it can be seen that the amount of improvement achievable over the basic slab track design increases going from ballasted track down to floating slab track. This is generally also the order of increasing cost. Discussion of the engineering considerations for these different track types is given in the following paragraphs.

These various track forms can also be used for surface railways to reduce ground-borne noise. However, the insertion loss achieved will be less than for a track in tunnel unless a high impedance foundation, for example a concrete raft, is introduced beneath the track.

13.6.2 Soft baseplates

There are many proprietary designs of soft baseplate. Figure 13.16 shows a typical design consisting of a relatively stiff rail pad between the rail and a cast-iron plate, beneath which a thicker soft elastomeric pad is used. Some baseplates, as shown here, have isolated bolts through to the foundation. These must allow for movement under the train load deflection to avoid short-circuiting the resilience of the lower elastomeric pad. To avoid problems with bolts breaking under lateral loads, more advanced baseplates use a clip locking mechanism or a second pair of rail clips to secure the baseplate 'lid', while others use rubber bonding between the components.

The baseplate top is much wider than the rail foot to prevent excessive rail roll and resultant gauge spreading under the lateral forces of the vehicle, particularly during curving. The standard test loadings for rails usually determine the limit to which the vertical stiffness of the baseplate can be lowered.

Baseplates allow the rail support stiffness to be reduced to about 20 MN/m per fastener. They are mostly used on slab track but can be installed on top of sleepers in ballasted track. The same type of baseplates can be used on bridges (on bearers or slab track or straight to a deck) to isolate track vibration from the bridge structure and thus reduce bridge noise (Chapter 11).

TABLE 13-3 **PARAMETERS USED FOR CALCULATIONS IN FIGURE 13.15. FOR THIS MODEL PARAMETERS ARE SPECIFIED PER RAIL OR SLEEPER END AND PER UNIT LENGTH ALONG THE TRACK. THE 'OBSERVATION DISTANCE' IS 25 M AND THE WIDTH OF CONTACT BETWEEN THE TRACK MODEL AND THE GROUND MODEL IS 3.5 M IN ALL CASES. THE WHEEL UNSPRUNG MASS IS 675 KG AND THE CONTACT STIFFNESS IS 1.5×10^9 N/M. THE HALF-SPACE SOIL HAS A YOUNG'S MODULUS OF 372 MN/M^2, DENSITY OF 2000 KG/M^3, A POISSON'S RATIO OF 0.47 AND LOSS FACTOR OF 0.05**

	Reference case, slab track with rail pads	Soft baseplates	Ballasted track	Sleeper pads	Ballast mat	Booted sleeper	Floating slab
Rail bending stiffness (loss factor)	6.4 MN/m^2 (0.001)	6.4 MN/m^2 (0.001)	6.4 MN/m^2 (0.001)	6.4 MN/m^2 (0.001)	6.4 MN/m^2 (0.001)	6.4 MN/m^2 (0.001)	6.4 MN/m^2 (0.001)
Rail mass per unit length	60 kg/m	60 kg/m	60 kg/m	60 kg/m	60 kg/m	60 kg/m	60 kg/m
Rail support stiffness (loss factor)	330 MN/m^2 (0.2)	33 MN/m^2 (0.2)	330 MN/m^2 (0.2)	330 MN/m^2 (0.2)	330 MN/m^2 (0.2)	330 MN/m^2 (0.2)	330 MN/m^2 (0.2)
Sleeper mass	–	–	285 kg/m	285 kg/m	285 kg/m	250 kg/m	
Ballast stiffness (loss factor)	–	–	270 MN/m^2 (0.5)	270 MN/m^2 (0.5)	270 MN/m^2 (0.5)	–	–
Ballast mass per unit length	–	–	600 kg/m	600 kg/m	600 kg/m	–	–
Sleeper pad or ballast mat stiffness (mass neglected) (loss factor)	–	–	–	40 MN/m^2 (0.2)	40 MN/m^2 (0.2)	50 MN/m^2 (0.2)	–
Floating slab bending stiffness (loss factor)	–	–	–	–	–	–	240 MN/m^2 (0.01)
Floating slab mass per unit length	–	–	–	–	–	–	1200 kg/m
Tunnel invert bending stiffness (loss factor)	17.5 MN/m^2 (0.01)	17.5 MN/m^2 (0.01)	17.5 MN/m^2 (0.01)	17.5 MN/m^2 (0.01)	17.5 MN/m^2 (0.01)	17.5 MN/m^2 (0.01)	17.5 MN/m^2 (0.01)
Tunnel invert mass per unit length	1050 kg/m	1050 kg/m	1050 kg/m	1050 kg/m	1050 kg/m	1050 kg/m	1050 kg/m

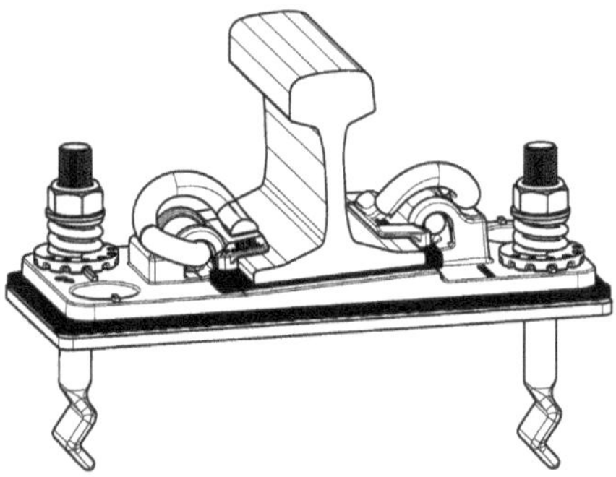

FIGURE 13-16 A conventional baseplate design. Picture courtesy of Pandrol Ltd

An undesirable effect of baseplates is that they lead to an increase in rolling noise. This is because the soft support leads to lower decay rates (see Chapter 3). Moreover, the baseplate top forms an additional radiating area which is only decoupled from the rail at relatively high frequencies, typically of the order of 1 kHz.

Some baseplates address the need to minimize the lateral displacement by using elastomeric elements in shear rather than compression. An example of this is the 'Cologne egg' baseplate.

Figure 13.17 shows an example of a different way of achieving a low vertical stiffness but limiting lateral displacement of the rail head. This allows a lower vertical stiffness than most basepates – under 10 MN/m per fastener. In this case, the fastener assembly allows for vertical adjustment.

13.6.3 Sleeper soffit pads and ballast mats

Sleeper soffit pads ('soffit' means underside) and ballast mats lower the stiffness of the ballast layer and therefore the track resonance frequency. These are illustrated in Figure 13.18. These systems overcome the gauge widening problem, as the rails are mounted with normal rail pads and fasteners onto a standard sleeper. Since the soft component is below the sleeper, the sleeper mass helps to lower the coupled wheel track resonance frequency.

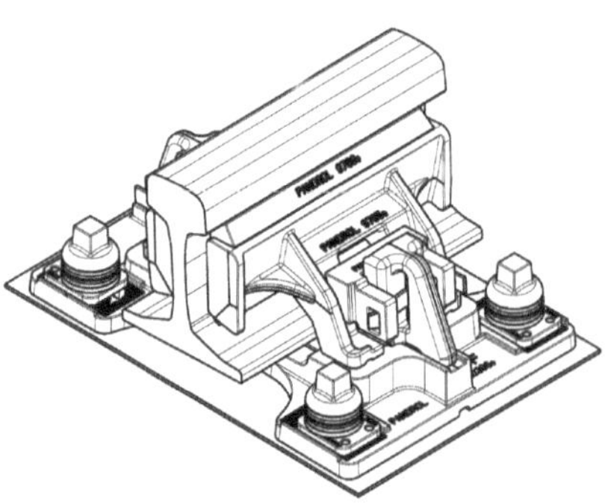

FIGURE 13-17 Fastener with support at the web/head of the rail. Picture courtesy of Pandrol Ltd

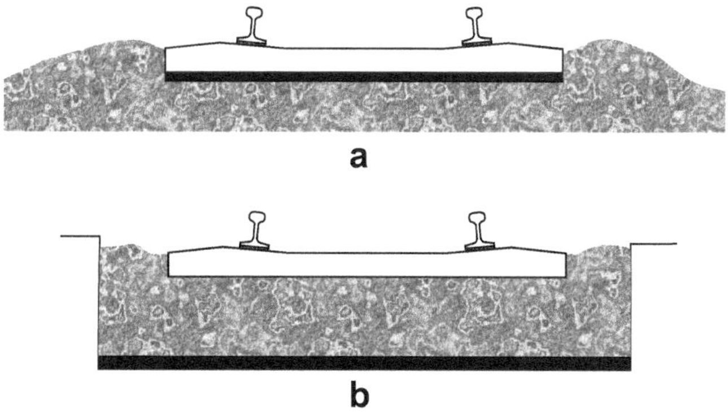

FIGURE 13-18 Illustration of (a) sleeper soffit pad and (b) ballast mat

Sometimes track engineers prefer to retain a ballasted track form rather than use a slab track, especially for short tunnels with normal loading gauge. An advantage of this is that the maintenance does not differ from the standard ballasted track on either side of the tunnel and different practices (often unfamiliar to the local maintenance staff) do not have to be adopted for a short track section. Sleeper soffit pads and ballast mats are attractive options in such circumstances as both allow normal tamping operations.

Sleeper soffit pads have the advantage that they are simple to install during a resleepering operation, since they are delivered already fixed to the bottom of the sleeper.

Ballast mats can be laid on tunnel inverts or a prepared subgrade and have the advantage that the extra mass of the ballast is above the spring in the resonant system. However, if a ballast mat is too soft there is a risk of making the ballast layer unstable under the vibration of passing trains and therefore compromising ride quality and increasing maintenance costs.

13.6.4 Booted sleepers

Figure 13.19 shows the concept of a 'booted sleeper'. There are a number of proprietary booted sleeper systems on the market. These perform in the same way as sleeper soffit pads but the design is integrated with a slab track. The design is usually based around a bi-bloc sleeper design. Again a normal rail pad is used between the rail and the sleeper blocks and a soft pad is used between the sleeper blocks and the slab. This is kept in place and protected by the 'boot'. At installation the track panel (rails with sleepers attached), complete with the boots and integral soft pads, is suspended and levelled. The concrete slab is then cast *in situ* around the boots.

13.6.5 Floating slab track

Floating slab track adds the highest possible mass above the track spring to form a system with a very low resonance frequency. Figure 13.20 shows how a floating slab track is typically designed as part of the tunnel structure. As well as the greater construction cost of the track form itself, great expense can come from any increase

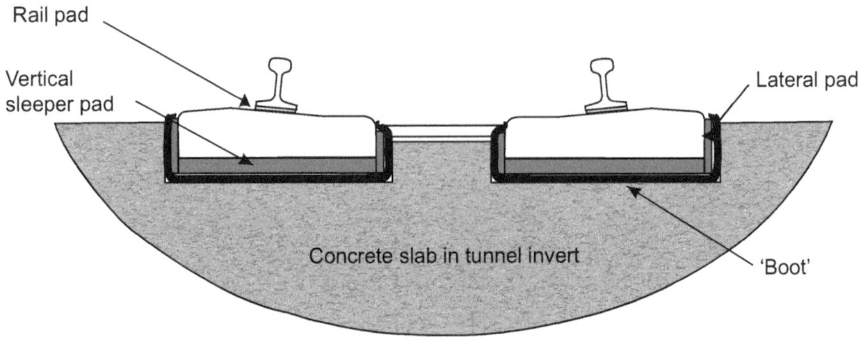

FIGURE 13-19 Arrangement of booted sleepers in slab track

in the diameter of the tunnel that has to be made to accommodate sufficient mass in the floating slab. Thus, the achievable size is often determined by the space in the tunnel.

Limitations also arise because the floating slab usually has to be constructed in finite lengths (from 1.2 m up to about 10 m per block). A soft baseplate is often used as well in order to accommodate safely the differential deflection as the axles move from one block to the next. The rails are often the only bending stiffness connecting the blocks. However, in some designs the differential displacement is limited by introducing additional bonds between the blocks.

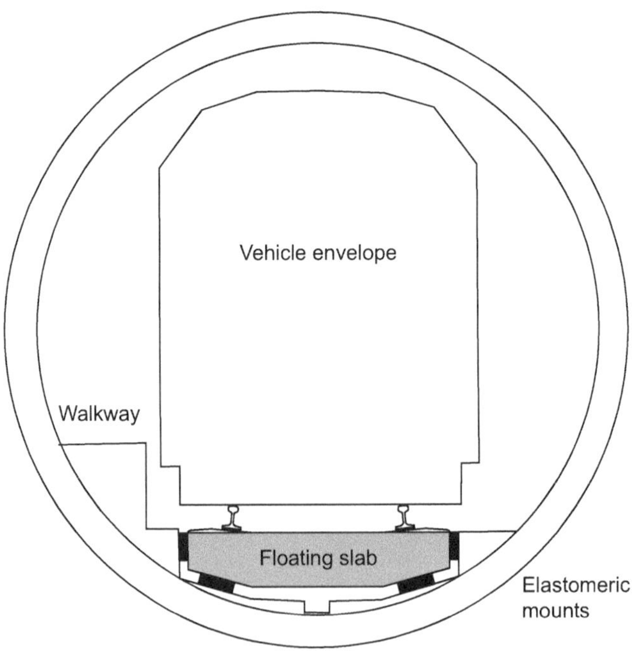

FIGURE 13-20 Floating slab track in a tunnel

Particularly for floating slab tracks with short slab sections, the track has a high mass but a low bending stiffness, in the limit only of that of the rail. Low wave speeds of bending waves along the track structure may therefore be a concern.

There are some designs of continuous floating slab. These have a lower deflection for a given resonance frequency and make maximum use of the tunnel space but have the disadvantage that they are harder to design in such a way that the floating slab mounts can be replaced.

Various materials can be used for the supports. Most are elastomeric but steel coil springs that allow some adjustment can also be used. Some installations use a soft layer of rock wool.

Needless to say, with the need for design specific to the tunnel space available, axle loading and train speeds, and with the maintenance constraints, drainage requirements, etc., floating slab is the most costly option. Nevertheless, there are examples on most modern metro systems that run in tunnels under densely built-up city centres where a high degree of isolation is required.

13.6.6 Transitions and other track design considerations

For every type of track that lowers the overall support stiffness provided by the track, the static deflections under the load of a train are greater than for conventional track. At the connections to standard track, some sort of smooth transition is required in order to avoid impulsive forces that could be very damaging to the subgrade, and cause impulsive ground vibration, discomfort to passengers and rapid fatigue damage to the track components themselves. For tracks with soft baseplates, such transitions are usually achieved by a progressive change of pad stiffness over a suitably long section of track. This may be of the order of tens of metres although it depends on the change to be achieved, the speed of the train and other factors.

For transitions from ballasted to slab track and especially onto floating slab track, short concrete 'bridges' may be used that are allowed to hinge at either end. These still generally need to be accompanied by a transition in the baseplate stiffness.

Switches and crossings cannot generally be mounted on very soft baseplates while their use on floating slab track requires specialized installations [13.16]. This may be a reason for preferring sleeper soffit pads or ballast mat options. In any case, vertical deflections are critical for clearances and alignment at switch rails and noses. For this reason, it may be necessary to design track and tunnel alignments to avoid these near to locations that are sensitive to vibration and ground-borne noise. It is often expected that switches and crossings will give rise to heightened vibration level because of the impulsive forces at joints and the noses. This is not necessarily so for joints that are well made and maintained, but it is wise to reduce their number to as few as possible at locations where noise and vibration are of concern. Clearly, there is a need for noise and vibration to be considered at an early stage of design of new alignments if unfortunate siting is to be avoided.

13.6.7 Other mitigation measures for ground-borne noise

Although ground-borne noise is usually most effectively dealt with by vibration-isolating track forms, other mitigation measures may sometimes be appropriate.

As ground-borne noise is induced by surface roughness, rail grinding, wheel reprofiling and maintenance of rail joints can be effective in reducing the level of excitation. For lower frequencies tamping of ballasted track may improve the vertical alignment and hence reduce the excitation.

As seen in Chapter 12 and Section 13.4 the vehicle unsprung mass has a strong influence on the level of excitation of vibration. Measures to reduce this have been discussed in Section 12.6.4, but should be combined with a high track mass for best effect on ground-borne noise.

It remains the case, however, that vibration-isolating track forms provide the most scope for reducing ground-borne noise.

REFERENCES

13.1 J.W. Edwards. Survey of environmental noise and vibration from London Underground trains. *Proceedings of Internoise '96*, 2029–2032, 1996.

13.2 US Department of Transportation. Transit Noise and Vibration Impact Assessment, April 1995, DOT-T-95–16.

13.3 J.G. Walker and M.F.K. Chan. Human response to structurally radiated noise due to underground railway operations. *Journal of Sound and Vibration*, **193**, 49–63, 1996.

13.4 E.G. Vadillo, J. Herreros, and J.G. Walker. Subjective reaction to structurally radiated sound from underground railways: field results. *Journal of Sound and Vibration*, **193**, 65–74, 1996.

13.5 C.J.C. Jones, D.J. Thompson and M. Petyt. A model for ground vibration from railway tunnels. *Proceedings of the Institution of Civil Engineers, Transportation*, **153**, 121–129, 2002.

13.6 L. Andersen and C.J.C. Jones. Coupled boundary and finite element analysis of vibration from railway tunnels – a comparison of two- and three-dimensional models. *Journal of Sound and Vibration*, **293**, 611–625, 2006.

13.7 X. Sheng, C.J.C. Jones and D.J. Thompson. Modelling ground vibration from tunnels using wavenumber finite and boundary element methods. *Proceedings of the Royal Society A* 461, 2043–2070, 2005.

13.8 X. Sheng, C.J.C. Jones, and D.J. Thompson. Prediction of ground vibration from trains using the wavenumber finite and boundary element methods. *Journal of Sound and Vibration*, **293**, 575–586, 2006.

13.9 G. Degrande, D. Clouteau, R. Othman, M. Arnst, H. Chebli, R. Klein, P. Chatterjee, and B. Janssens. A numerical model for ground-borne vibrations from underground railway traffic based on a periodic finite element-boundary element formulation. *Journal of Sound and Vibration*, **293**, 645–666, 2006.

13.10 Y.B. Yang and H.H. Hung. A 2.5 D finite/infinite element approach for modelling visco-elastic bodies subjected to moving loads. *International Journal for Numerical Methods in Engineering*, **51**, 1317–1336, 2001.

13.11 P. Jean, C. Guigou, and M. Villot. 2D1/2 BEM model of ground structure interaction. *Journal of Building Acoustics*, **11**, 157–173, 2004.

13.12 H.J. Saurenman, J.T. Nelson and G.P. Wilson. *Handbook of Urban Rail Noise and Vibration Control*, Report DOT-TSC-UMTA-81–72. US Department of Transportation, Urban Mass Transportation Administration, 1982.

13.13 R.J.M. Craik. *Sound Transmission through Buildings using Statistical Energy Analysis*. Gower, Aldershot, UK, 1996.

13.14 O.A.B. Hassan. Transmission of structure-borne sound in buildings above railway tunnels. *Journal of Building Acoustics*, **8**, 269–299, 2001.

13.15 D.A. Bies and C.H. Hansen. *Engineering Noise Control, Theory and Practice*, 3rd edition. Spon Press, London, 2003.

13.16 H-G. Wagner and A. Herrmann. Floating slab track above ground for turnouts in tram lines, in Noise and vibration mitigation for rail transportation systems, ed. B. Schulte-Werning *et al. Proceedings of the 9th International Workshop on Railway Noise*, Munich, Germany, Springer, Berlin, 86–93, 2008.

CHAPTER 14

Vehicle Interior Noise*

14.1 INTRODUCTION

From surveys of train passengers it is known that acoustic comfort within trains is considered to be good, yet the acoustic environment is also cited as an important element of comfort: it is thus important to people. While the level of acoustic comfort is generally appreciated, acoustic discomfort is raised as one of the most unfavourable or prominent factors determining levels of discomfort.

As train speeds are raised the levels of interior sound are inevitably increased, requiring additional effort to prevent this from becoming unfavourable. The frequency content of the noise inside a railway vehicle has also changed with the introduction of high speed trains. The widespread use of disc brakes has led to a lower prominence of the rolling noise in the mid and high frequencies while the components at lower frequencies have dramatically increased due to aerodynamic noise. Consequently, the noise spectrum inside a railway modern vehicle is quite similar in shape to that of an automobile.

Two main forms of construction are used for modern railway vehicles. All-steel construction is widely used and comprises thin steel panels (typically 2 mm) spot-welded onto a frame of transverse and longitudinal stiffeners. The alternative, which has also been used extensively in recent decades, is to use aluminium extrusions. As shown schematically in Figure 14.1, these comprise two faceplates, typically 3 mm thick, separated by a lattice of diagonal stiffeners. Although light and stiff, they have quite poor acoustic performance necessitating the use of secondary acoustic treatments such as floating floors.

The calculation approaches used for interior noise have also changed over recent years. At higher frequencies, image source and ray tracing methods were commonly used in the 1990s, along with measurements of panel transmission loss. More recently, statistical energy analysis (SEA) methods have become more widely

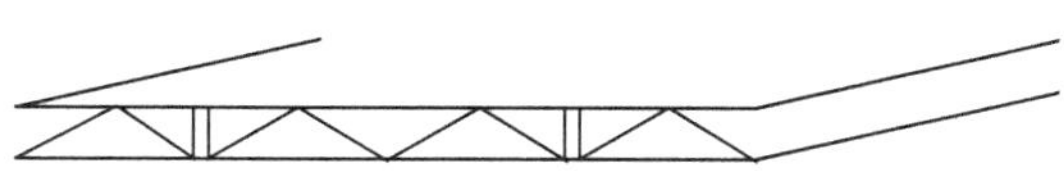

FIGURE 14-1 Schematic view of an extruded aluminium panel

* This chapter has been written in collaboration with Pierre-Etienne Gautier.

adopted. For lower frequencies, modal or semi-modal approaches have been developed alongside conventional finite element analysis.

All the noise sources discussed in previous chapters are also of relevance to interior noise in trains [14.1]. The noise from the wheel/rail region is often the major source. Additionally, on vehicles with underfloor diesel engines, noise from the engine, exhaust, intake, cooling fans and transmission system can all be significant. Noise from the air-conditioning system can also require consideration in modern rolling stock as there is often very limited space in which to package the air-conditioning unit and ducts. Also some care has to be taken in fixing equipment such as compressors under the floor structure of a carriage or on the roof.

Noise is transmitted from each of these sources to the interior by both airborne and structure-borne paths, as shown in Figure 14.2. Structure-borne transmission is often dominant at low frequencies and airborne transmission at higher frequencies.

At high speeds aerodynamic noise becomes important. This consists of noise from the bogie and inter-coach regions transmitted by airborne paths as well as noise

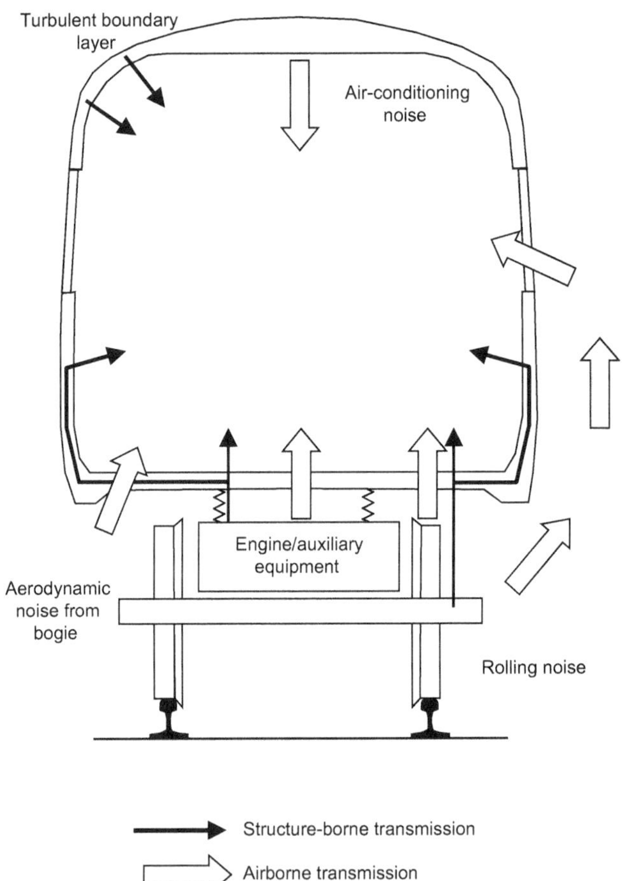

FIGURE 14-2 The main sources of interior noise in a train and their transmission paths into the vehicle

produced by excitation of the carriage structure by the turbulent boundary layer which radiates sound into the carriage interior.

A measurement method for vehicle interior noise is defined in ISO 3381:2005 [14.2]. This requires the use of a measurement site which has a rail roughness spectrum that is compliant with the same limits as in ISO 3095:2005 [14.3].

After an introduction to typical noise levels and measurement quantities in Section 14.2, the sources of interior noise are discussed in more detail in Section 14.3 and factors affecting the transmission paths are described in Section 14.4. Section 14.5 gives an overview of prediction methods and some results are presented in Section 14.6.

14.2 CHARACTERIZING INTERIOR NOISE

14.2.1 Typical noise levels inside rolling stock

In order to illustrate the evolution of sound levels inside rolling stock, some examples of noise spectra in British vehicles are shown in Figure 14.3 (taken from [14.4]). Older vehicles with opening windows, such as the Mk 1 vehicle constructed in 1960, had quite high noise levels in the region 500 to 2000 Hz. This frequency region is important for speech intelligibility and was considerably improved by the

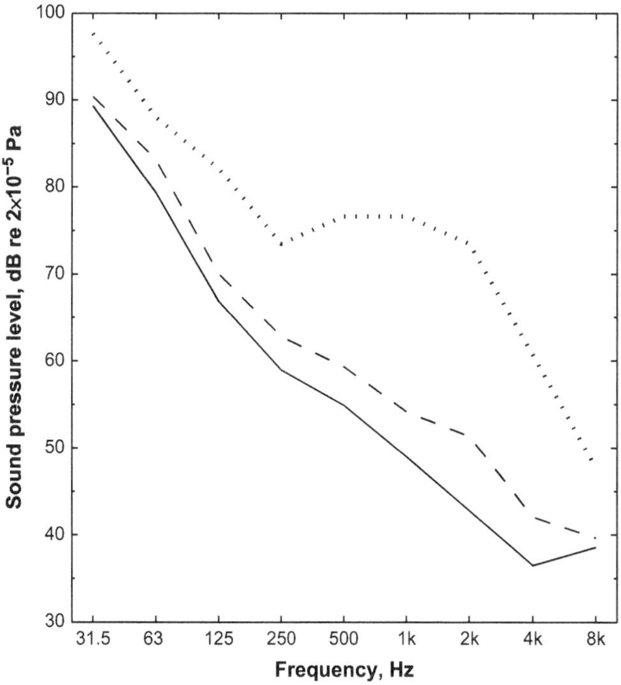

FIGURE 14-3 Examples of octave band spectra measured inside British rolling stock at 145 km/h (from [14.4]). ····, Mk 1 coach, not air-conditioned with cast-iron block brakes (81 dB(A)); – – –, Mk 2d coach, air-conditioned with cast-iron block brakes (63 dB(A)); —, Mk 3 coach, air-conditioned with disc brakes (59 dB(A))

introduction of stock with sealed windows and air-conditioning, such as the Mk 2d coach introduced in 1970. Further improvements in this frequency region were obtained with the introduction of disc brakes in place of cast-iron block brakes, as illustrated by the Mk 3 coach, introduced in 1975. Although modern rolling stock is thus considerably quieter at higher frequencies than older stock, the differences at low frequencies are much more modest. Low frequency noise is therefore particularly dominant in the spectrum of modern rolling stock.

Tables 14.1 and 14.2 list some typical noise levels measured inside TGV trains. Even at high speed, the levels achieved are similar to those for conventional rolling stock. The variation in level occurs due to differences in track quality. The noise levels in the vestibules, i.e. near the doors, are typically 10 dB higher than in the passenger saloons. In terms of overall levels, similar performance at equivalent speeds is achieved for double-deck TGVs (TGV Duplex), see Table 14.2, despite the fact that they use lighter aluminium construction in contrast to the steel construction used in the earlier TGV vehicles.

Examples of one-third octave spectra in the passenger saloons of a TGV Duplex and TGV Atlantique are given in Figure 14.4. Again it can be seen that these spectra have a strong low frequency content. Moreover, in the lower saloon of the TGV Duplex, a significant peak can be observed above 100 Hz, which corresponds to the sleeper-passing frequency. This will be discussed further in Section 14.3.4.

14.2.2 Measurement quantities

Conventionally, the A-weighted sound pressure level has been used to specify acceptable levels inside vehicles. Current specification levels are basically defined in terms of A-weighted levels ranging from about 65 to 68 dB at maximum operating speed. In parallel, a maximum level of 55–60 dB at standstill is often specified to guarantee minimum disturbance from air conditioning and auxiliary systems. However, as seen above, the spectrum of noise inside trains contains considerable energy at low frequency. This low frequency sound energy can be a source of human fatigue, but is not effective in masking speech, for which noise in the range 200 to 6000 Hz is most effective.

Passenger sensitivity to noise inside a train varies considerably from one person to another [14.5]. Clearly, it is desirable that the noise should not interfere with conversation held between neighbours. However, particularly for a modern open saloon-type vehicle, silence would also not be ideal. There should be sufficient background noise so that passengers are not disturbed by others talking further along the vehicle (people talking loudly into mobile phones are a particular source of

TABLE 14-1 **NOISE LEVELS MEASURED AT DIFFERENT SPEEDS IN TGV ATLANTIQUE**

	Noise level (dB(A))	
Speed (km/h)	**Saloon**	**Vestibule**
220	62 to 70	76
270	65 to 70	79
300	67 to 72	80

TABLE 14-2 **NOISE LEVELS MEASURED AT VARIOUS SPEEDS IN DIFFERENT COMPARTMENTS OF TGV DUPLEX**

	Noise level (dB(A))			
Speed (km/h)	**Lower saloon**	**Upper saloon**	**Lower vestibule**	**Upper vestibule**
200	57.5 to 59	58 to 59.5	69	65.5
250	59.5 to 62.5	60 to 63	72.5	66
270	64 to 65	–	–	68
280	62.5 to 66.5	61.5 to 64.5	75	67.5
300	62.5 to 66.5	64 to 67.5	73.5 to 76	68 to 71
330	64 to 68.5	64 to 68.5	77.5	70.5 to 71.5
350	65 to 71	65.5 to 70	75.5 to 79	71 to 75

annoyance). It is therefore probably not desirable to reduce the sound levels in the range 250–2000 Hz any further than those shown in Figure 14.3 for the Mk 3 coach. Reduction at lower frequencies would be desirable but is very difficult to achieve.

In order to define acceptable acoustic environments, various alternative quantities exist that can be used instead of, or as well as, the A-weighted level. These include the B-weighted level, Preferred Speech Interference Level (PSIL), Loudness Level, Alternative Noise Criteria (NCA), Noise Ratings (NR) and Room Criteria (RC) [14.4, 14.6] along with octave or one-third octave spectra. Recent studies have shown that the loudness is better correlated with the sensation of discomfort in vehicles than the L_{Aeq}. Hardy [14.5] shows that PSIL between 50 and 55 dB is optional; below this range passengers tend to find the environment too quiet.

14.2.3 Variation within the interior

The interior sound level varies considerably within a vehicle. Figure 14.5 shows some example measured results [14.7]. This was a British Mk 2 coach dating from the 1960s, although the interior dated from the 1990s. A loudspeaker was placed at one end of the vehicle saloon at a height of about 1 m and the sound pressure was measured along a line down the central gangway at the height of the headrests. Clearly, in a running vehicle other source positions will apply, but these results serve to illustrate the general trends that can be expected.

Results are shown for three one-third octave bands in terms of relative sound pressure level. At low frequencies, strong modal patterns are observed due to the long acoustic wavelength. For a typical interior vehicle width of approximately 2.5 m, half an acoustic wavelength is equal to the width at about 68 Hz, so that below this frequency the carriage will act as a one-dimensional waveguide. Plane waves travelling along its length are reflected to form standing wave patterns. At higher frequencies the sound spreads out in three dimensions. Due to the absorptive properties of the seats, carpets, etc. considerable decay in sound level is observed along the coach. Two glass partial screens on either side of the door at the centre of the coach caused additional attenuation at some frequencies.

To illustrate the variations across the width of the vehicle, measured results are also shown for positions in front of each seat headrest on one side of the vehicle. The

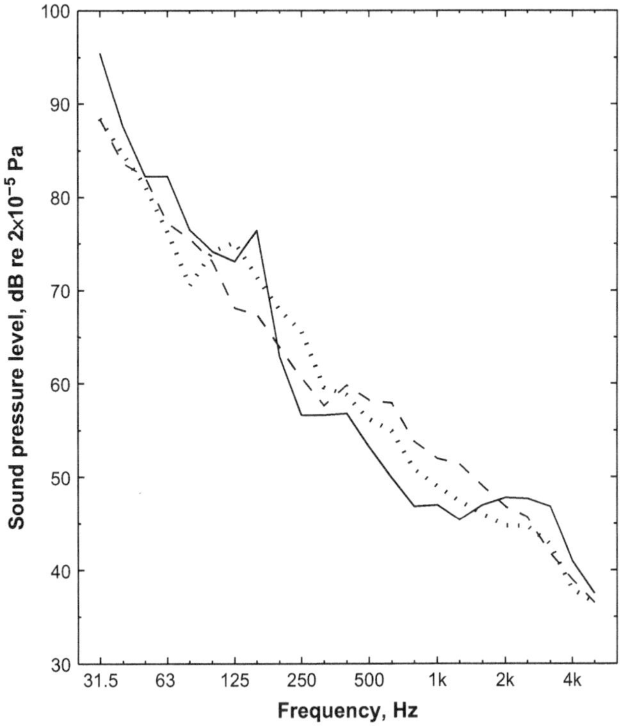

FIGURE 14-4 Examples of one-third octave band spectra measured inside TGV rolling stock at 300 km/h. —, TGV Duplex, lower saloon (67.2 dB(A)); – – –, TGV Duplex, upper saloon (65.5 dB(A)); · · · ·, TGV Atlantique (66.5 dB(A))

seats were arranged in groups of four around tables. At low frequencies these measurements follow the same pattern as the measurements in the gangway, but at higher frequencies considerable differences can be seen between adjacent seated positions. These spatial variations may be significant for passengers in the vehicles; the 500 Hz frequency band, for example, is quite important for speech interference. It can also be expected that differences will occur between left and right ear positions at an individual seat.

14.3 SOURCES OF INTERIOR NOISE

In order to model interior noise, it is necessary to take account of all noise sources that potentially contribute to the noise level. Moreover, it is necessary to bear in mind that for most sources propagation can occur by both airborne and structure-borne paths. The following sources must be taken into account:

- rolling noise;
- excitation phenomena due to the sleeper-passing frequency, and/or defects on wheels;
- aerodynamic noise due to cavities, bogie region, inter-coach gap, etc.;
- turbulent boundary layer noise.

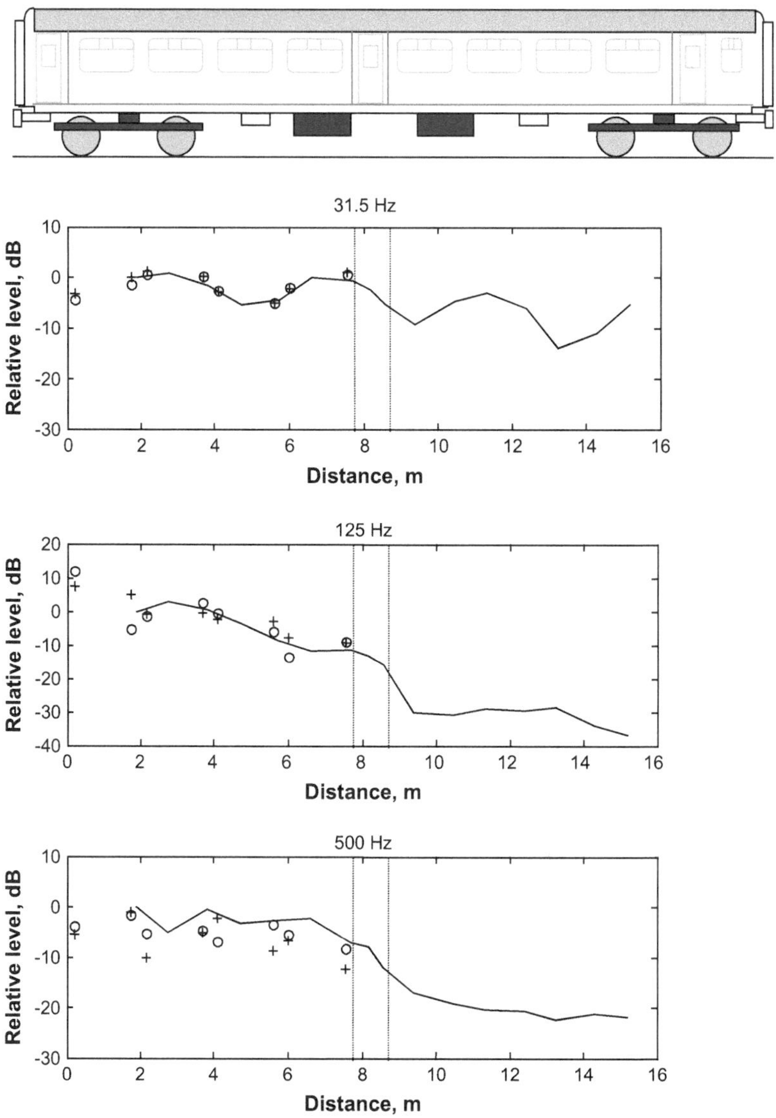

FIGURE 14-5 Relative internal sound levels in selected one-third octave bands in an ex-BR Mk 2 coach due to a sound source located at the left-hand end, 1.05 m above the floor. ——, measured at 1.05 m from floor along central gangway; ○, measured 0.05 m from headrests of aisle seats (height 1.05 m from floor); +, measured 0.05 m from headrests of window seats (height 1.05 m from floor). Positions of partial screens indicated by dotted lines [14.7]

These will be discussed in more detail below.

In addition, modern rolling stock is often formed of electric or diesel multiple units (EMUs or DMUs). In such vehicles electric traction equipment and/or diesel engines may be mounted under the floor (or on the roof) and can contribute significantly to interior noise and vibration. As well as the diesel powerpack, various types of auxiliary equipment such as compressors and air-conditioning units are often located under the floor or on the roof of modern multiple units or trams. Attention should be given to their potential contribution to interior noise.

14.3.1 Rolling noise

Models for rolling noise have been described in Chapters 2–6. These can also be used to estimate the airborne source terms for use in interior noise predictions, accounting for wheel and track characteristics and roughness. Figure 14.6 shows an example for a TGV at 300 km/h in one-third octave bands. This shows a prediction using the TWINS model and the measured acoustic power from the bogie region that forms the input to the vehicle. Good agreement is found from a few hundred hertz and above.

To estimate the structure-borne component of rolling noise a different source input is required. The structure-borne transmission can be estimated by using measurements (or predictions) of vibration at the source, for example at the axlebox, together with measurements or predictions of transfer functions from this position to the interior sound. A similar approach was also used in [14.8] to determine the contribution of structure-borne excitation of a freight wagon bogie and body to the exterior noise (see also section 4.7).

Often, on modern bogies, the anti-yaw damper forms a significant structure-borne path for the rolling noise.

14.3.2 Aerodynamic noise

As explained in Chapter 8, there are a number of aerodynamic noise sources on a high speed train. The ranking of the different sources with respect to their

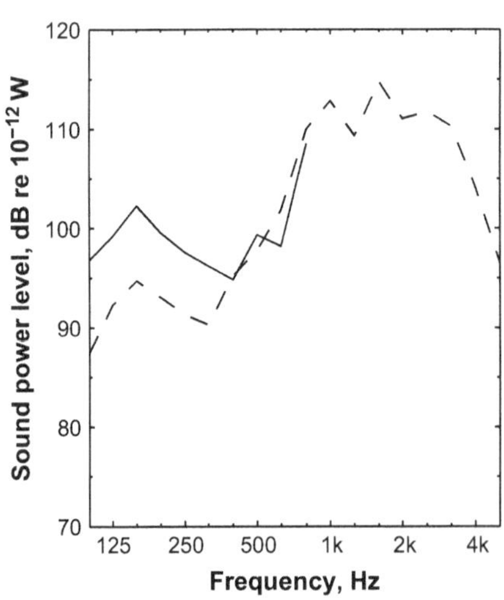

FIGURE 14-6 Comparison between calculated (– – –) and measured (—) acoustic power of rolling noise from a high speed train bogie at 300 km/h

contribution to internal noise may not be the same as for external noise. For the internal noise, aerodynamic noise contributes more significantly and at lower speeds than it does for the external noise.

These sources include the bogie region, inter-coach gap and the turbulent boundary layer.

Aerodynamic noise from the bogie region has up to now only been characterized experimentally. As an example, Figure 14.7 shows an example of a one-third octave spectrum of sound power from the bogie region of a TGV. This was determined from measurements by subtracting the rolling noise derived from a combination of measurements and theoretical analysis. Compared with the rolling noise (see Figure 14.6) the frequency content can be seen to be significant up to around 1000 Hz.

For the turbulent boundary layer (TBL) noise, a Corcos model [14.9] was developed from measurements using an aray of microphones flush mounted on a panel replacing a window, as shown in Figure 14.8 [14.10]. Examples of power spectral densities for various locations on the car body are plotted in Figure 14.9.

Aerodynamic noise from the inter-coach region is particularly important in tunnels, although it is also significant elsewhere. Figure 14.10 shows a measured noise spectrum for a TGV Atlantique. The large peak at 80 Hz is due to the inter-coach gap. Modifications of this region by fitting larger rubber baffles as shown in Figure 14.10(c) eliminated this peak.

14.3.3 Technical equipment

This category includes diesel engines and their related sources, electric motors, air-conditioning equipment, compressors, brakes, etc. Unlike rolling noise and aerodynamic noise, these sources are not directly speed dependent, but work more or less intermittently according to the demand placed on them.

(a) Diesel powerpack systems

In most recent DMUs, the diesel engine, along with some of the intake and exhaust systems, are incorporated into a unit called the 'powerpack system'. The engine itself is an important source of structure-borne as well as airborne sound. Additionally, the intake and exhaust system have to be considered for airborne noise.

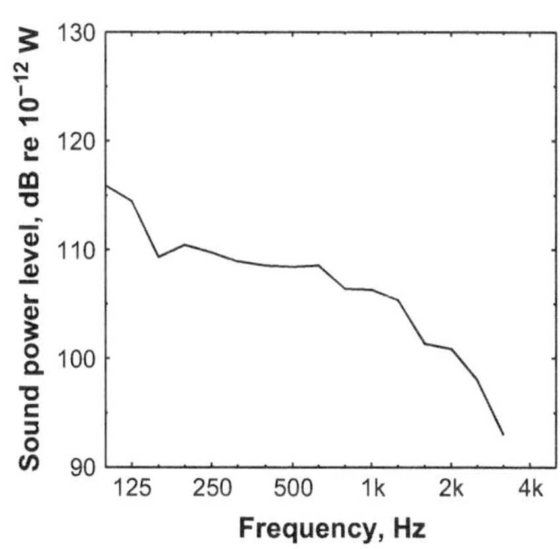

FIGURE 14-7 Example of one-third octave sound power spectrum of aerodynamic noise in the bogie region of a high speed train

FIGURE 14-8 Microphone array flush mounted on a panel replacing a window for the characterization of the boundary layer noise

The mounting of the diesel powerpack system under the floor is particularly important in terms of noise and also vibration level, as noise and vibrations levels are reduced with the new generations of rolling stock. Usually, a double isolation arrangement is used to minimize the structure-borne transmission, in which the engine is resiliently mounted on a frame or raft, which is then resiliently mounted from the underside of the vehicle.

For the airborne sound transmission, the source level can be estimated using an empirical formula for the A-weighted sound pressure level at 1 m from a large turbocharged diesel engine [14.11]:

$$L_{p,A} = 40 \log_{10} \Omega + 50 \log_{10} B + 10 \log_{10} N_{cyl} - 143.5 \quad (14.1)$$

where Ω is the rotational speed (in rev/min), B is the cylinder bore diameter in mm and N_{cyl} is the number of cylinders. A correction of up to 1.5 dB should be added to this for Vee-type engines.

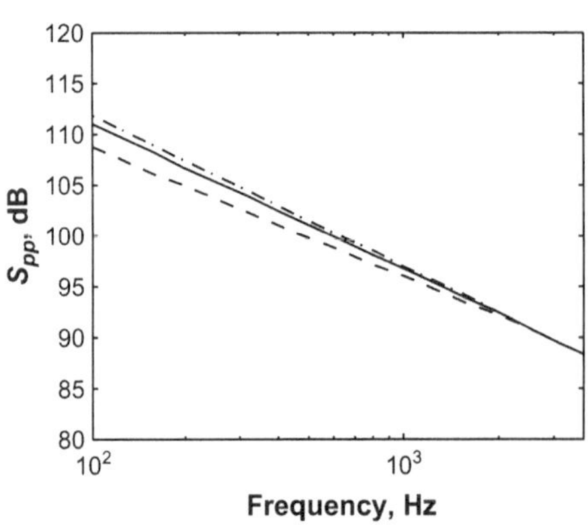

FIGURE 14-9 Turbulent boundary layer noise spectra at different locations of the car body of a high speed train around the centre of the vehicle: —, window; – – –, roof; – · – ·, underfloor regions [14.10]

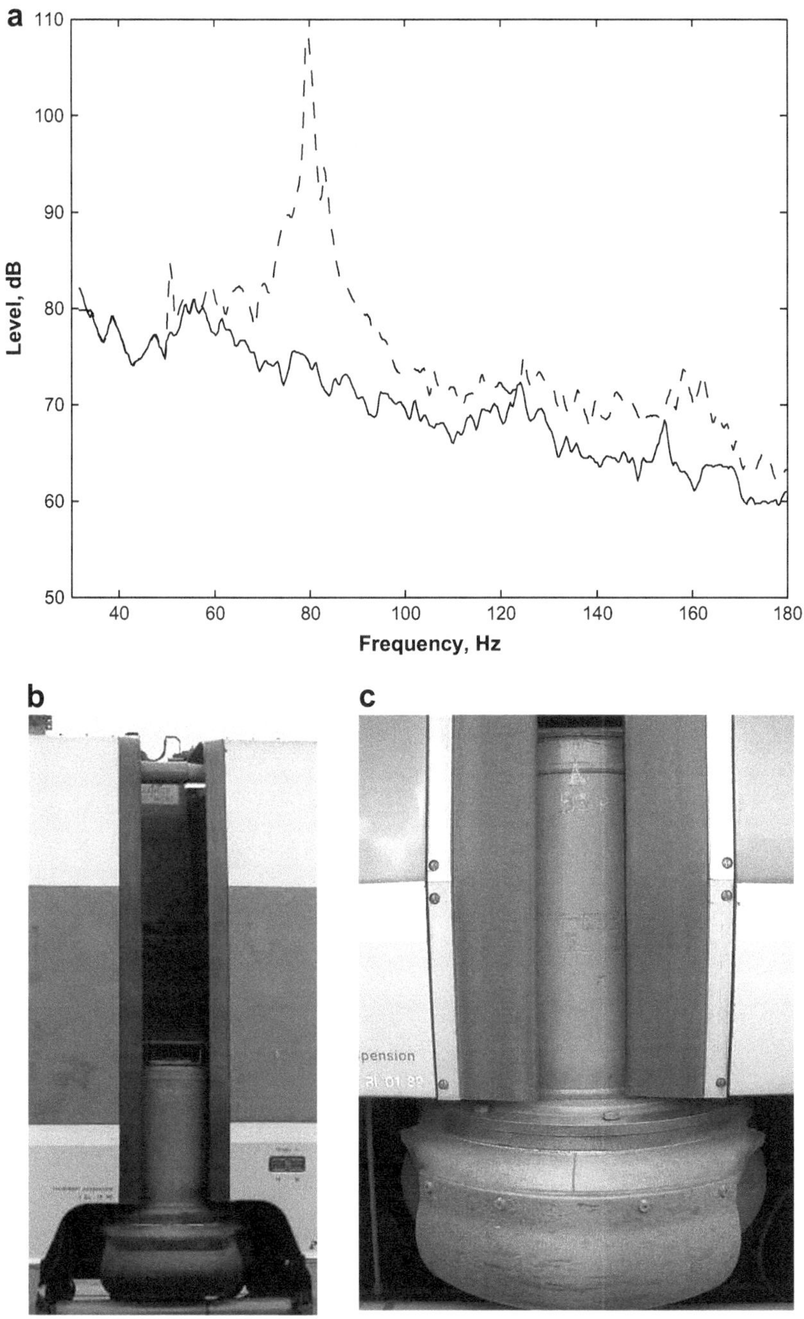

FIGURE 14-10 Results from TGV-A in a single bore tunnel: – – –, with standard inter-coach region, shown on the left; —, with modified region shown on the right

(b) Electric motors and other auxiliary equipment

Electric motors generate noise due to electro-magnetic excitation of the casing as well as tonal noise from air motion within the motor cavity. However, they are usually not a major source of noise inside the train, although their cooling fans are often more prominent. Inverters (and their fans) may be more significant than the motors.

Air-conditioning equipment is another source of fan noise. In this case care is required to prevent it from being transmitted into the vehicle through ventilation ducts.

14.3.4 Other sources to be taken into account

(a) Sleeper passing frequency

Excitation at the sleeper-passing frequency has already been observed in Figure 14.4. In [14.12] it was shown that this effect is compounded when the bogie axle spacing is a multiple of the sleeper spacing. The peak at the sleeper-passing frequency can dominate the A-weighted level in some circumstances. The excitation mechanism is discussed briefly in Section 5.7.5. It is transmitted to the interior by structure-borne paths. In addition, an aeroacoustic excitation at the sleeper-passing frequency is also possible in some situations.

(b) Out-of-round wheels

Out-of-round wheels are considered in [14.13]. Figure 14.11 shows an example of the narrow-band sound spectrum in a TGV Duplex in which the wheels have

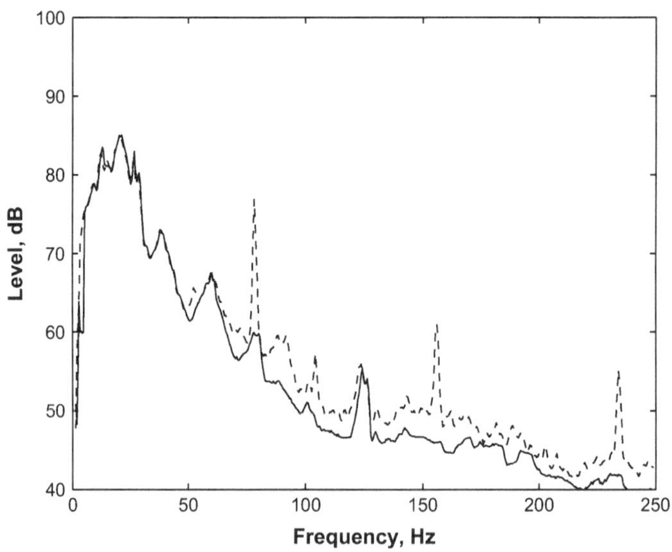

FIGURE 14-11 Influence of wheel reprofiling on interior noise in upper saloon of a TGV Duplex (300 km/h). – – –, before reprofiling; —, after reprofiling

developed a strong out-of-round profile. A strong peak is seen at 78 Hz, which corresponds to a wavelength of about 1 m. This is caused by a wave on the wheel with three wavelengths on the circumference. It is thought that this may be caused by the tooling using during the manufacture of the wheels.

14.4 TRANSMISSION PATHS

As stated in the introduction, sound from most sources can be transmitted to the interior by both airborne and structure-borne paths.

14.4.1 Airborne transmission

Airborne sound transmission into the vehicle occurs due to acoustic excitation of the vehicle floor, walls, windows, doors and roof. These are caused to vibrate and radiate sound to the interior. Additionally leaks and vents are a source of direct airborne transmission.

The acoustic performance of a panel can be measured by placing it between two reverberant rooms and measuring the difference in sound pressure level between the two rooms [14.14]. The sound reduction index (or transmission loss) is the difference between the incident intensity level and the transmitted intensity level, which can be derived from such a measurement after allowing for the size of the panel and the absorption in the receiver room. This can then be used in combination with sound pressure levels on the source side to estimate the sound transmitted to the interior of the vehicle.

A typical sound reduction index of a homogeneous panel is shown schematically in Figure 14.12. Generally, the sound reduction index of panels is dominated by the

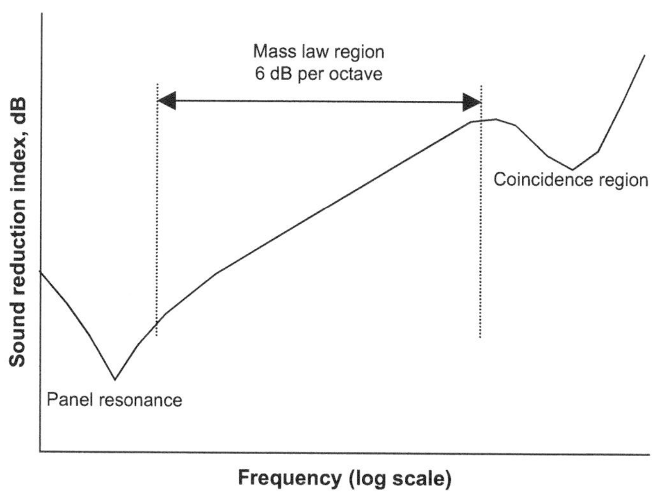

FIGURE 14-12 Typical sound reduction index of a homogeneous panel due to a diffuse incident field

'mass law' behaviour in a wide frequency range. For diffuse field incidence, this is given approximately by

$$R_f = 20\log_{10}(fm'') - 47 \tag{14.2}$$

where f is the frequency and m'' is the mass per unit area of the panel. The mass law behaviour extends from the first resonance of the panel up to just below the critical frequency (see equation (11.35)). In this region the bending stiffness of the panel and its damping have no effect on the sound transmission. At high frequencies, the coincidence region occurs where the wavelengths in the structure and in air are similar. Here a dip in the sound reduction index occurs at the critical frequency; the extent of this dip depends on the damping (see [14.14, 14.15] for more details).

The use of light-weight constructions such as extruded aluminium or corrugated steel leads to a low sound reduction index. This follows from the mass law which states that the sound reduction index reduces by 6 dB for a halving of panel mass. However, such structures tend to have an acoustic performance that is even worse than the mass law would suggest, due to the presence of an extended frequency region over which coincidence effects occur. For example, Figure 14.13 shows measurements of the sound reduction index from a 60 mm thick extruded aluminium floor of a railway vehicle with a 3 mm wall thickness taken from [14.16] (similar results are also found in [14.17]). Also shown is the 'field incidence' mass law estimated for a homogeneous panel of similar mass, from equation (14.2). Clearly, the extruded panel exhibits a much lower sound reduction index than this. It can be brought closer to the mass law behaviour by the use a suspended inner floor and by adding a damping treatment to the extruded section. However, the added mass of such treatments tends to negate the advantage of extruded aluminium as a light form of construction.

14.4.2 Structure-borne transmission

As well as the airborne path, considerable sound is transmitted to the vehicle interior through structural paths, particularly at low frequency. This originates from

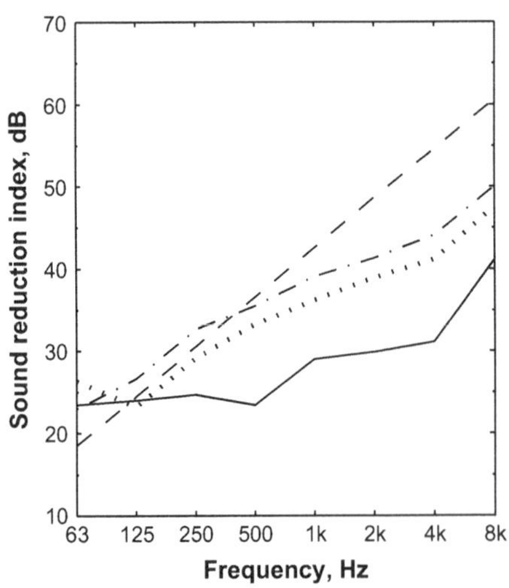

FIGURE 14-13 Octave band sound reduction index of extruded aluminium floor. —, measured on bare floor panel; · · · · ·, measured for bare floor panel plus 12 mm suspended wooden deck; – · – ·, measured for damped floor panel plus 12 mm suspended wooden deck; – – –, field incidence mass law for 30 kg/m^2 (data from [14.16])

the wheel/rail region as well as from underfloor diesel engines where these are present. Structure-borne engine noise can be reduced significantly in many cases by applying good mounting practice [14.1]. The mount stiffness must be chosen taking into account the frequency characteristics of the engine, as an incorrect choice of stiffness can lead to amplification rather than attenuation of transmitted vibration. Care should also be taken to avoid noise transmission along flanking paths such as pipes and hoses. As stated above, double isolation systems are commonly used to mount the engine on the vehicle.

Rolling noise is transmitted from the bogie into the vehicle via the primary and secondary suspensions. In addition, the anti-yaw dampers form a significant path. On articulated trains such as the TGV the arrangement is more complex, but the yaw dampers are still the main path for the transmission of vibrations.

14.5 PREDICTION OF INTERIOR NOISE

14.5.1 Low frequency modelling

Deterministic methods such as finite elements (FEM) may be applied at low frequencies to predict the vehicle interior noise. An example is found in [14.18]. In practice, these are usually limited to frequencies below about 100 Hz due to the large models required for higher frequencies. Due to the regular geometry an analytical model of the interior may also be used to construct the interior acoustic field on the basis of simple room modes [14.19].

14.5.2 Higher frequency modelling

At high frequencies the number of structural and acoustic modes becomes prohibitive for such approaches. The preferred analysis method for frequencies above about 250 Hz is therefore Statistical Energy Analysis (SEA). This can be used in both predictive mode [14.16, 14.20] and experimental mode [14.21]. However, in predictive mode it is not straightforward to define the coupling loss factors between the various subsystems, especially where use is made of aluminium extrusions [14.17, 14.22, 14.23] or other inhomogeneous constructions [14.24]. This is an area of continuing research. Moreover, as SEA is a statistical method, it provides an average result and cannot account easily for the spatial variations in sound field such as seen in Figure 14.4.

Ray tracing methods, as used in architectural acoustics, are also used to predict the distribution of sound within a vehicle interior at higher frequencies.

14.6 MODEL ASSESSMENT AND RESULTS

Finally, in this section some example results are presented on the TGV to illustrate the use of the SEA modelling technique in particular. Figure 14.14 shows the overall A-weighted noise level at various positions within a TGV Atlantique. Quite good agreement can be seen between the predictions and measurements.

The contributions of airborne and structure-borne paths to the total noise spectrum are shown in Figure 14.15 for one microphone position at 300 km/h. The

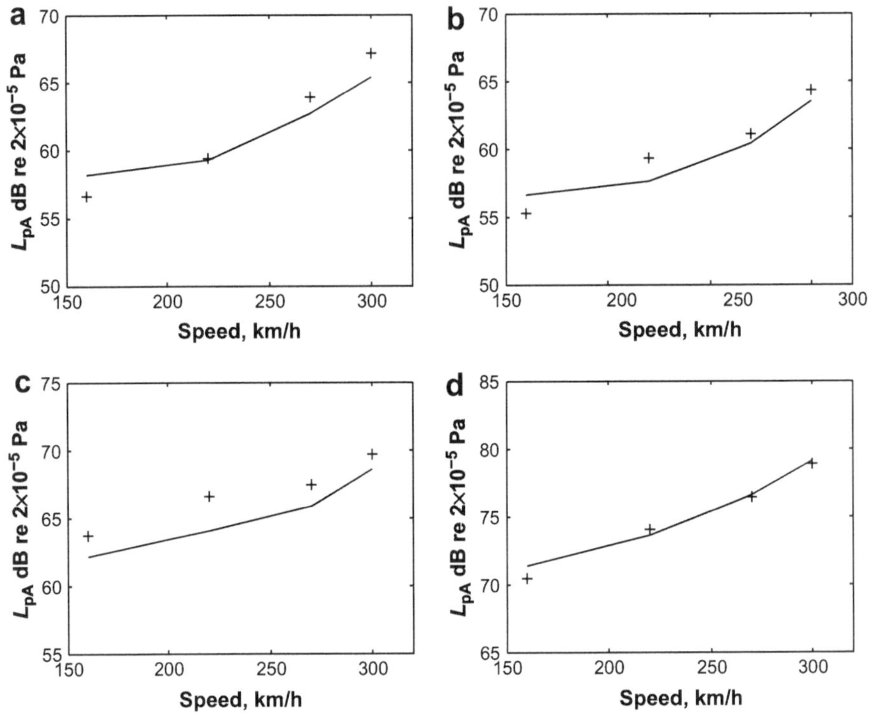

FIGURE 14-14 A-weighted sound pressure level at various positions within a TGV. Comparison of measurements (+) and predictions using SEA (—). (a) M1 (end of saloon), (b) M3 (centre of saloon), M6 (end of saloon), M10 (vestibule)

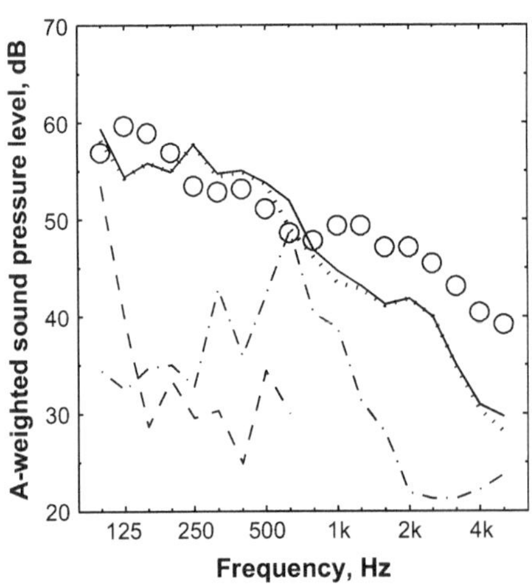

FIGURE 14-15 Comparison of airborne and structure-borne contributions to interior noise in TGV at position M1 at 300 km/h. ○, total measured; —, total predicted; ·····, predicted airborne; – – –, predicted structure-borne; – · – ·, background noise

TABLE 14-3 **RELATIVE CONTRIBUTIONS TO TOTAL SOUND LEVEL INSIDE A TGV**

Speed (km/h)		Rolling noise	Aerodynamic (bogie and inter-coach gap)	Turbulent boundary layer
160	Free field	88%	10%	2%
220	Single track tunnel	50%	49%	1%
	Twin track tunnel	43%	56%	1%
	Free field	49%	46%	5%
270	Single track tunnel	31%	68%	1%
	Twin track tunnel	25%	74%	1%
	Free field	32%	63%	5%
300	Twin track tunnel	21%	75%	4%
	Free field	24%	60%	16%

predicted noise follows the measured spectrum quite well, although it is under-predicted at high frequencies. In this case the airborne component dominates while the structure-borne noise has negligible contribution above 100 Hz. The arrangement of the articulated vehicles means that structure-borne sound is transmitted to one vehicle ('carrier') more than to the other ('carried'). This may be part of the reason why the noise level at position M6 in Figure 14.14 is higher than at M1.

Table 14.3 shows the relative contributions of various sources to the total sound level inside a TGV determined using this model. This shows that rolling noise is clearly dominant at conventional speeds but, as speed increases, the contributions from aerodynamic noise and turbulent boundary layer excitation become more significant. The noise level increases in tunnels due to an increase in the transmission of rolling noise and aerodynamic noise, the inter-coach gap being particularly important. Consequently, the turbulent boundary layer excitation is less relevant as these other sources increase in level.

14.7 CONCLUDING REMARKS

Vehicle interior noise is a complex problem to which this chapter has only given a brief introduction. All the sources described in the rest of the book apply to interior noise, so that noise control techniques described elsewhere apply equally to interior noise. Other sources are also present, such as traction and air-conditioning equipment.

Additionally, it is necessary to take account of the filtering effect of the vehicle structure on the sound transmitted through both airborne and structure-borne paths. As a result of this filtering, the frequency spectrum of interior noise is strongly dominated by low frequencies. This is a consequence of the fact that the effects of both the isolation due to suspension elements and the sound reduction due to airborne transmission through panels increase towards higher frequencies [14.14, 14.15]. Moreover, absorptive materials tend to be effective only at high frequencies;

they are often very ineffective at low frequencies where they are too thin compared with the acoustic wavelength. Similarly damping treatments often tend to be more effective at high frequencies, where bending wavelengths are shorter.

Apart from this frequency-dependence of the noise spectrum, acoustic treatments can in principle be applied to achieve whatever noise level is desired. However, this inevitably implies a penalty in terms of their added weight, their space requirements and their cost implications. Especially as lighter vehicles are sought to improve energy efficiency, it will be increasingly challenging to maintain acoustic comfort levels.

Moreover, another aspect that requires attention is sound quality. The use of simple A-weighted sound pressure levels can be particularly misleading in determining the suitability of an interior noise spectrum and other measures should be used such as PSIL or Loudness Level [14.5].

Finally, it is worth noting that active noise control is especially relevant for low frequency noise control of enclosed spaces, e.g. [14.25]. It has been applied successfully for tonal propeller noise in aircraft and has also been demonstrated in cars for both engine noise and low frequency road noise. An experimental application has been made for the driver's cab of a train but it was found that the region that could be controlled, around the driver's head, was rather limited in size [14.26]. Apart from the driver's cab, the interior of a train is quite large (compared with a car) and the noise is broad-band (unlike propeller noise). Consequently, an active control system would involve many transducers (microphones, loudspeakers or structural actuators) which would probably rule out its widespread adoption on cost grounds.

REFERENCES

14.1 A.E.J. Hardy and R.R.K. Jones. Control of the noise environment for passengers in railway vehicles. *Proceedings of the Institution of Mechanical Engineers*, **203F**, 79–85, 1989.

14.2 EN ISO 3381, Railway applications – Acoustics – Measurement of noise inside railbound vehicles. International Standards Organization ISO, 3381: 2005.

14.3 EN ISO 3095, Railway applications – Acoustics – Measurement of noise emitted by railbound vehicles. International Standards Organization ISO, 3095: 2005.

14.4 P.W. Eade and A.E.J. Hardy. Railway vehicle internal noise. *Journal of Sound and Vibration*, **51**, 403–415, 1977.

14.5 A.E.J. Hardy. Railway passengers and noise. *Proceedings of the Institution of Mechanical Engineers*, **213F**, 173–180, 1999.

14.6 A.E.J. Hardy. Measurement and assessment of noise within passenger trains. *Journal of Sound and Vibration*, **231**, 819–829, 2000.

14.7 D.J. Thompson and C.J.C. Jones. Noise and vibration from railway vehicles, Chapter 10 in *Handbook of Railway Vehicle Dynamics*, S. Iwnicki ed. CRC Press, Taylor and Francis, Boca Raton, 279-325. 2006.

14.8 F. de Beer and J.W. Verheij. Experimental determination of pass-by noise contributions from the bogies and superstructure of a freight wagon. *Journal of Sound and Vibration*, **231**, 639–652, 2000.

14.9 M. Corcos. The resolution of turbulent pressure at the wall of a boundary layer. *Journal of Sound and Vibration*, **6**, 59–70, 1967.

14.10 German–French Cooperation Deufrako Annex K2, 1999, Noise sources from high speed guided transport, final report.

14.11 D.J. Thompson and J. Dixon. Vehicle noise, Chapter 6 in *Advanced Applications in Acoustics, Noise and Vibration*, eds. F. Fahy and J. Walker. Spon Press, London, 2004.

14.12 J. Färm. Interior structure-borne sound caused by the sleeper-passing frequency. *Journal of Sound and Vibration*, **231**, 831–837, 2000.

14.13 A. Johansson. Out-of-round railway wheels – assessment of wheel tread irregularities in train traffic. *Journal of Sound and Vibration*, **293**, 795–806, 2006.
14.14 F. Fahy and J. Walker (eds). *Fundamentals of Noise and Vibration*. E&FN Spon, London, 1998.
14.15 D.A. Bies and C.H. Hansen. *Engineering Noise Control, Theory and Practice*, 3rd edition. Spon Press, London, 2003.
14.16 N.J. Shaw. The prediction of railway vehicle internal noise using statistical energy analysis techniques. MSc dissertation, Heriot–Watt University, 1990.
14.17 T. Kohrs. Structural acoustic investigation of orthotropic plates, Diploma Thesis, TU Berlin, 2002.
14.18 A. Bracciali and C. Pellegrini. FEM analysis of the internal acoustics of a railway vehicle and its improvements. *Proceedings WCRR 97*, Florence, November 1997.
14.19 F. Létourneux, S. Guerrand, and F. Poisson. Low-frequency acoustic transmission of high-speed trains: simplified vibroacoustic model. *Journal of Sound and Vibration*, **231**, 847–851, 2000.
14.20 B. Stegeman. *Development and validation of a vibroacoustic model of a metro rail car using Statistical Energy Analysis (SEA). MSc dissertation*, Chalmers University, Gothenburg, Sweden, 2002.
14.21 K. de Meester, L. Hermans, K. Wyckaert and N. Cuny. Experimental SEA on a highspeed train carriage. *Proceedings ISMA21*, Leuven, Belgium, 151–161, 1996.
14.22 P. Geissler and D. Neumann. SEA modelling for extruded profiles for railway passenger coaches. *Euro-Noise 98*, München, Germany, 189–194, 1998.
14.23 G. Xie, D.J. Thompson, and C.J.C. Jones. A modelling approach for the vibroacoustic behaviour of aluminium extrusions used in railway vehicles. *Journal of Sound and Vibration*, **293**, 921–932, 2006.
14.24 D. Backström. Analysis of the sound transmission loss of train partitions. MSc dissertation, KTH Stockholm, 2001.
14.25 S.J. Elliott and P.A. Nelson. Active noise control. In: F. Fahy andJ. Walker (eds), *Advanced Applications in Acoustics, Noise and Vibration*, Spon Press, London, 2004.
14.25 T. Loizeau and F. Poisson, Acoustic active control inside a locomotive cabin. Proc. World Congress on Railway Research, Montreal, Canada, 2006.

APPENDIX **A**

Measurement of Train Pass-by Noise

A.1 MEASUREMENT QUANTITIES

Considering a notional time history of the noise during a train pass-by, shown in Figure A1, several different single number quantities are used to define the noise level.

Maximum level

This is the maximum A-weighted sound pressure level (with the averaging set to 'fast') over the pass-by, L_{Amax}.

Commonly, the term 'maximum level' is also used to refer to the average level of the plateau region during a pass-by. This is more useful than the actual maximum level, which can be influenced by a single noisy wheel, but is less well defined.

SEL (sound exposure level)

The SEL is formed from the integral of the squared pressure over the whole pass-by (including the rising and falling parts), normalized to 1 second:

$$\mathrm{SEL} = 10\log_{10}\left\{ \int_{t_1}^{t_2} \frac{p^2(t)}{p_{\mathrm{ref}}^2} dt \right\} \tag{A1}$$

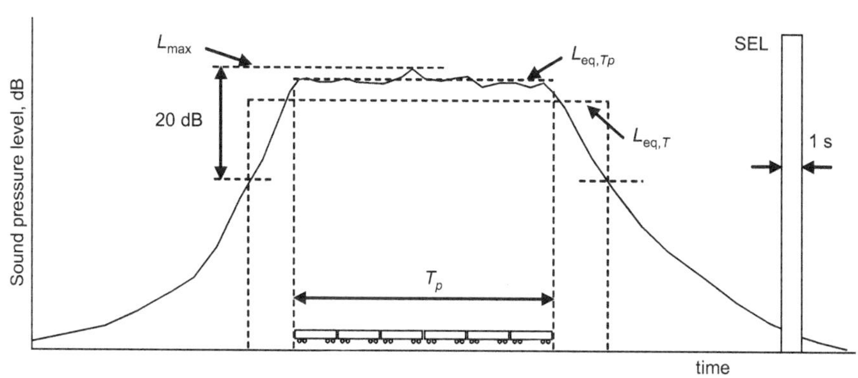

FIGURE A1 Notional time history of train pass-by noise indicating various quantities where $p(t)$ is the pressure at time t, p_{ref} is the reference pressure and the times t_1 and t_2 are chosen to include the whole pass-by, or more practically they are usually defined as the points at which the level is 10 or 20 dB below the maximum level, as shown in Figure A1. As the passage time is usually longer than 1 s this will give a level that is higher than the maximum level. The SEL can readily be used as input to a calculation of long-term noise exposure based on equivalent sound levels, L_{Aeq}.

Short-term equivalent levels

The short-term equivalent levels, $L_{\text{Aeq},T}$ are defined in a similar way to long-term L_{eq} values. These have the form

$$L_{\text{Aeq},T} = 10\log_{10}\left\{\frac{1}{T}\int_{t_1}^{t_2}\frac{p^2(t)}{p_{\text{ref}}^2}dt\right\} \tag{A2}$$

where $T = t_2 - t_1$. The duration T may be chosen to represent the length of a vehicle or of the whole train (T_p, the length of the train over buffers, divided by its speed) or it may be the time between points at which the level is 10 or 20 dB below the maximum level.

For a train of similar vehicles, $L_{\text{Aeq},Tp}$, the equivalent pass-by level, is a more formalized means of measuring the 'average' level during the pass-by. However, it does not take account of the rising and falling parts of the measurement and so it does not include the whole energy of the train pass-by. This makes it less suitable for calculating long-term L_{eq} values. On the other hand, the equivalent level over the -10 or -20 dB points takes account of the whole pass-by but presents a level that is lower than the $L_{\text{Aeq},Tp}$ and which is sensitive to the duration included in the average.

TEL (transit exposure level)

The TEL is formed from the same integral as the SEL, i.e. over the whole pass-by, but is normalized by the passage time T_p rather than the measurement time:

$$\text{TEL} = 10\log_{10}\left\{\frac{1}{T_p}\int_{t_1}^{t_2}\frac{p^2(t)}{p_{\text{ref}}^2}dt\right\} \tag{A3}$$

Since $T_p < t_2 - t_1$, again the TEL may be greater than the maximum level. The TEL results measured for various high speed trains [A1] were found to be between 0.5 and 1.5 dB higher than the corresponding $L_{\text{Aeq},Tp}$ results.

A.2 MEASUREMENT PROCEDURES

ISO 3095-1975

Since 1975, ISO 3095 provided a standard method of measuring train pass-by noise [A2]. This was based on a measurement location at 25 m from the centre of the track. The measurement quantity was the maximum A-weighted sound pressure level (with the averaging set to 'fast') over the pass-by, L_{Amax}.

The principal difficulties with ISO 3095-1975 were that the track condition was not defined other than that it should be in good condition. Large variations can be found for a given vehicle depending on the test location.

ISO 3095:2005

A modified standard was under development for a considerable time and was finally approved in 2005, ISO 3095:2005 [A3]. This includes a limit on the rail roughness and definition of the track in terms of decay rates. These are intended to ensure that the measurement is influenced as little as possible by the track, and mainly determined by the properties of the vehicle (both its roughness and its vibro-acoustic properties). The noise level is defined in terms of the TEL for the whole train or the $L_{\mathrm{Aeq},Tp}$ for parts of the train.

TSI High Speed Trains 2002

The Technical Specifications for Interoperability (TSI) were produced for the European Union following Directives and include requirements for noise levels. These define their own measurement procedure and also set limit values for trains. The TSI for High Speed Trains [A4] came into effect in 2002. It includes a limit curve for roughness which is more strict than that in ISO 3095:2005. The track was defined in terms of its type of rails, sleepers and pad stiffness. The noise level is defined in terms of the TEL for the whole train or the $L_{\mathrm{Aeq},Tp}$ for parts of the train, as in ISO 3095:2005. However, the TSI-HS is undergoing revision and these requirements are likely to change to bring them into line with the TSI for Conventional Rail.

TSI Conventional Rail 2004

The TSI for railway vehicles intended to operate at conventional speeds (up to 190 km/h) was introduced subsequent to that for high speed trains [A5]. Consequently, the measurement procedure defined was updated. The track specification, referred to as ATSI, includes a different rail roughness specification and the track was defined in terms of its decay rate rather than the pad stiffness. The measurement quantity is now $L_{\mathrm{Aeq},Tp}$ for the whole train as well as individual vehicles or parts of the train.

A.3 THE NOEMIE PROJECT

The European NOEMIE test campaign provided trackside noise emission values of several high speed trains at speeds up to 320 km/h including comparisons of measurements on the same train made in different countries [A1]. Operational tracks were chosen in five European countries, intended to satisfy the conditions for the TSIs and were ground to ensure low roughness. The roughness levels of these and several other tracks considered are shown in Figure A2 along with the limit values of ISO 3095:2005, TSI-HS and ATSI (TSI-CR).

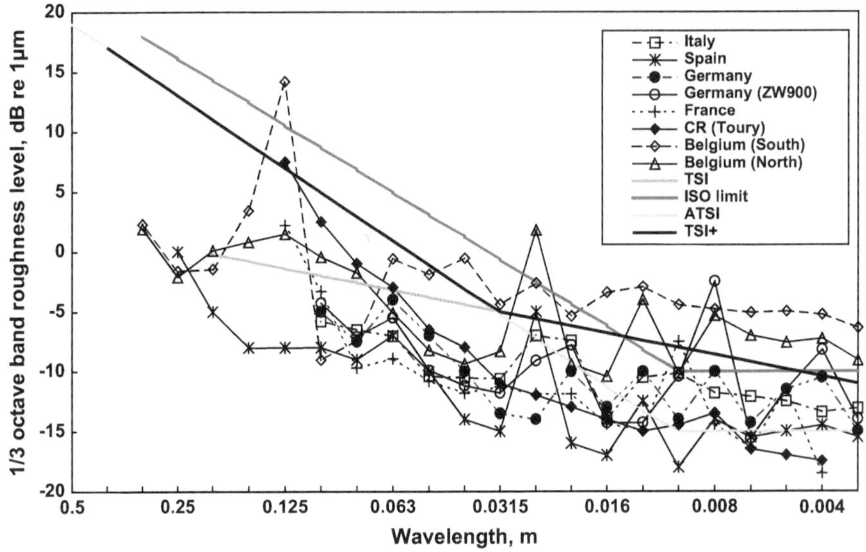

FIGURE A2 Roughness measured at test sites in NOEMIE along with various limit values (from [A1])

Finally, an alternative limit was proposed referred to as TSI^+ which corresponds to the original TSI limit at short wavelengths and the ATSI limit at long wavelengths, which was intended to be more readily achievable. Most of the tracks can be seen to satisfy this limit in most frequency bands, although the effect of grinding can be seen in some cases at a wavelength of 25 mm. In contrast, the TSI rail roughness limit was met in some cases, but the ATSI limit appeared to be too demanding for the existing operational rail grinding techniques [A1].

In addition, a revised limit for track decay rates was proposed in the TSI^+ specifications intended to reduce further the potential influence of the track in the measurements [A1].

REFERENCES

[A1] P. Fodiman and M. Staiger. Improvement of the noise technical specifications for interoperability: the input of the Noemie project. *Journal of Sound and Vibration*, **293**, 475–484, 2006.

[A2] International standard 'Acoustics – Measurements of noise emitted by railbound vehicles', ISO 3095–1975.

[A3] International standard 'Railway applications – Acoustics – Measurements of noise emitted by railbound vehicles', ISO 3095:2005.

[A4] Commission decision of 30 May 2002 concerning the technical specification for interoperability relating to the rolling stock subsystem of the trans-European high-speed rail system referred to in Article 6(1) of Directive 96/48/EC, reference 2002/735/EC. *Official Journal of the European Communities*, **12**, September 2002.

[A5] Directive 2001/16/EC of the European parliament and of the Council of 19 March 2001 on the interoperability of the trans-European conventional rail system. *Official Journal of the European Communities*, **2**, April 2001.

APPENDIX **B**

Short Glossary of Railway Terminology

At-grade	track which is neither on an **embankment** or in a **cutting**.
Articulated train	a train in which bogies or wheelsets are shared between adjacent vehicles.
Axle	shaft connecting two wheels on either side of the vehicle. The wheels are forced to rotate at the same speed. Vehicles with **independent wheels** have 'stub axles' that do not connect the two wheels on either side of the vehicle.
Axle load	the vertical load applied to the track by a single **axle**.
Ballast	layer of coarse stones supporting the **sleepers**.
Ballast mat	a layer of resilient material placed beneath the **ballast** to give additional vibration isolation.
Baseplate	a **track** component designed to hold the **rail** in place, usually with resilience to provide improved vibration isolation.
Bibloc sleeper	a **sleeper** constructed from two concrete blocks connected by a metal bar.
Block brakes	(see tread brakes)
Bogie	frame holding two **wheelsets**, which in turn supports the vehicle body. A bogie vehicle is one which has two (or more) bogies (as distinct from a **two-axle vehicle**).
Bored tunnel	a tunnel which has been constructed by drilling usually with a tunnel boring machine (TBM).
Carriage	a passenger-carrying vehicle.
Clips	sprung or bolted elements holding the **rail** to the **sleeper** or **track** structure.
Crossing	a place in the track where two tracks cross each other, usually at a shallow angle.
Conicity	the cone angle on the wheel running surfaces used to ensure stable running. This is variously 1:20, 1:30 or 1:40.

Cut-and-cover tunnel a shallow tunnel which has been constructed by digging a **cutting** and then covering it over after construction.

Cutting a place where the track is lower than the surrounding land.

Derailment accident occurring when a train leaves the rails.

Disc brakes braking system on trains in which callipers act on a brake disc rather than brake blocks acting on the **wheel** itself. Brake discs may be mounted on the **axle** or directly on the **web** of the **wheel**.

DMU Diesel Multiple Unit: a **multiple unit** train set powered by diesel engines.

Embankment a place where the track is raised up above the surrounding land on earthworks.

EMU Electric Multiple Unit: a **multiple unit** train set powered by electric traction.

Fishplates plates held on either side of the rail **web** used to connect two lengths of **rail** together.

Fixed formation a train which always runs with the same combination of carriages, typically driven by **power cars** at either end.

Flange the horizontal parts at the top and bottom of an I beam. The part of the wheel **tread** that projects below the **rail head** and prevents **derailment** (see Figure 4.1).

Foot the flat part of the **rail** at its base.

Gauge distance between inner face of each rail. Standard gauge is 1435 mm.

Gauge face the inner surface of the **rail head**.

Grinding a process for removing a thin layer of metal from the top of the **rail head** in order to remove roughness and/or to restore the correct profile. Special grinding trains are used for this.

Inclination of rails the angle of the **rails** to the vertical used together with **conicity** on the wheels to ensure stable running. This is variously 1:20, 1:30 or 1:40.

Independent wheels situation in which the wheels on either side of the vehicle can rotate independently.

Jakobs bogie a form of **bogie** which is shared between adjacent vehicles in an **articulated** train.

Joint (or **rail joint**) a connection between two lengths of rail, often held together by an arrangement of bolts and **fishplates**.

Light rail an urban transit or tram system using light railway vehicles and usually lighter track construction.

Loading gauge standard maximum envelope within which the vehicle must remain. This varies between railway authorities (see also **structure gauge**).

Locomotive a powered vehicle used to draw or propel a train of **carriages** or **wagons** (as opposed to a **multiple unit**)

Monobloc sleeper a **sleeper** constructed from a single piece, usually concrete.

Multiple Unit a self-powered train set with a driving compartment at each end. It comprises several carriages in a fixed formation, some or all of which are powered.

Pantograph device on the top of the train used to collect electric current from the overhead electrification wires.

Plain track track which is not **switches** or **crossings**.

Points see **switch**.

Power car a **locomotive** that is part of a permanently coupled fixed formation train.

Powered axle one which is driven by a motor

Primary suspension arrangement of springs and dampers between the **wheelsets** and the **bogie** (or vehicle body).

Rail the longitudinal steel beams on which the train runs.

Rail head the bulbous part at the top of the **rail**.

Rail pad elastomeric mat between the **rail** and the **sleeper** (or **baseplate**).

Running surface that part of the **wheel** that is in contact with the **rail** and vice versa.

Secondary suspension arrangement of springs and dampers between the **bogie** and the vehicle body.

Slab track a form of **track** construction in which the **rails** are held on a continuous concrete layer (the slab), which may include sleepers. There is usually no ballast.

Sleeper a transverse beam under the **rails** (see Figure 3.1) used to maintain track **gauge** and to distribute loads from the **wheels**. These may be wooden, concrete or steel.

Sleeper soffit pad a resilient layer placed directly under the **sleepers**.

Structure gauge standard minimum envelope beyond which all lineside structures (such as bridges, tunnels and platforms) must remain. This varies between railway authorities

	and may vary from one line to another (see also **loading gauge**).
Subgrade	the ground beneath the **ballast** and any sub-ballast layer.
Suspension	arrangement of springs and dampers supporting the vehicle and isolating it from the **track** (see **primary** and **secondary suspensions**).
Switch (or **points**)	an arrangement in the track to allow trains to follow alternative routes.
Tamping	track maintenance in which the vertical profile of the **track** is restored after a period of traffic. It is usually carried out by machines by squeezing the ballast together under the sleepers.
Track	this consists of two **rails** held by **clips** onto **sleepers** (or **slab**) and where present the **ballast** beneath it.
Trailer axle	(or **trailer vehicle**) one which is not powered.
Tread	part of the **wheel** which runs on the **track**.
Tread brakes	braking system based on brake blocks acting on the wheel **tread**.
Tunnel crown	the arch at the top of the tunnel or structure.
Tunnel invert	the inverted arch at the bottom of the tunnel or structure below the track. It is usually thickened by concrete to provide a level surface.
Two-axle vehicle	a vehicle, usually for freight, that has no **bogies** but is supported directly by two **wheelsets.**
Tyre	the region of a **wheel** near the running surface (see Figure 4.1). Originally this part was shrunk onto the inner part of the wheel, but most wheels are now constructed of a single piece.
Wagon	a freight-carrying vehicle.
Web	the thin part of a beam, such as the middle part of the **rail**, the vertical part of an I beam or the inner annular region of a **wheel** (see Figure 4.1)
Wheel load	the vertical load applied to the track by a single **wheel**.
Wheelset	two **wheels** connected by an **axle**.

List of Symbols

A	Beam cross-sectional area
A_n	Wave amplitude
B	Analysis bandwidth (Section 5.4)
B	Engine bore diameter (Section 14)
C	Viscous damping (per unit length) (Section 3.2)
C	Viscous damping (Section 4.3)
C_{eq}	Equivalent damper (Section 5.2)
C_1, C_2, C_3	Coefficients in solution for Timoshenko beam (Section 3.4)
C_{11}, C_{22}, C_{23}, C_{33}	Kalker creep coefficients (Section 5.3)
C^*_{pq}	Frequency-dependent creep coefficients (Section 5.3)
C_H	Constant of non-linear Hertzian contact stiffness (Section 10.2)
D	Directivity factor (Section 6)
D	Plate bending stiffness (Section 11.4)
D_n	Modal constant (Section 5.2)
E	Young's modulus
E'	Plane strain elastic modulus (Section 5.3)
E_i	Energy of subsystem i (Chapter 11)
F	Force amplitude
F_0	Normal force
F_T	Force transmitted to foundation (Chapter 13)
G	Shear modulus
G	Transfer function from force to velocity (Section 9.5)
H	Contact filter (Section 5.4)
H_1, H_2	Transfer functions from velocity or force to force (Section 9.5)
H_0, H_1	Hankel functions (Section 6)
I	Second moment of area of beam
$[I]$	Identity matrix
I	Sound intensity (Section 6)
Im	Imaginary part
J_n	Bessel function
K	Stiffness
K_b	Weighting for vibration (Chapter 12)
K_H	Hertzian contact spring stiffness (Section 5.2)
K_T	Static stiffness of track (Section 3.2)
L	Length of sleeper (Section 3.7), beam (Section 4.4)

L	Length of finite line source (Section 6.6)
L_{Aeq}	Equivalent A-weighted sound level
L_p	Sound pressure level
L_W	Sound power level
L_1, L_2	Bridge lengths either side of excitation point (Section 11.3)
M	Mass
M	Mach number (Chapter 8)
N	Speed exponent (Section 2.2)
N	Fresnel number (Section 7.5)
N	Normal force (Section 9.4)
N_{cyl}	Number of cylinders (Section 14)
P	Perimeter length of rail (Section 6.4) or plate (Section 11.4)
R	Curve radius (Chapter 9)
Re	Real part
Re	Reynolds number (Chapter 8)
Res	Residue in complex integration
R_0	Combination of wheel and rail radii of curvature (Section 5.3)
R_f	Field incidence sound reduction index (Chapter 14)
R_r	Rail radius (=0)
R_{rT}	Rail radius (transverse profile)
R_w	Wheel radius
R_{wT}	Wheel radius (transverse profile)
S	Surface area
S_v	Spectral density of wheel velocity (Section 4)
St	Strouhal number (Chapter 8)
T	Analysis time
T_{60}	Reverberation time
T_F	Force transmissibility (Chapter 13)
T_{mn}	Term in equations for rotating wheel (Section 4.5)
U	Vibration displacement amplitude
U	Vibration displacement in wavenumber domain
U	Flow velocity (Chapter 8)
V	Train speed
W	Sound power
W'	Sound power per unit length
W_b, W_d, W_k, W_m	Weighting functions for vibration (Chapter 12)
W_{diss}	Dissipated power (Chapter 11)
W_{in}	Input power (Chapter 11)
W_{rad}	Radiated power (Chapter 11)
Y	Mobility (velocity/force)
a	Semi-axis length of contact patch in rolling direction
a_i	Coefficients of rational polynomial model for track (Section 10.2)
b	Semi-axis length of contact patch in transverse direction
b	Bogie wheelbase (Chapter 9)

b_j	Coefficients of rational polynomial model for track (Section 10.2)
c	Size of contact patch $= (ab)^{1/2}$
c_0	Speed of sound in air
c_g	Group velocity
c_L	Longitudinal wave speed
c'_L	Longitudinal wave speed in a plate
c_R	Rayleigh wave speed
d	Distance between sleepers (Chapter 3)
d	Distance between two monopoles (Section 6.1)
d	Horizontal distance between source and receiver (Section 6.6)
d	Wheel flat depth (Section 10.3)
d	Rail joint dip (Section 10.4)
$d_{\pm}$	Term in equations for rotating wheel (Section 4.5)
e	Euler's number (base of natural logarithm)
f	Frequency
f_c	Critical frequency (Section 6)
f_r	Ring frequency (Section 4)
f_2	Dynamic component of lateral force (Section 9)
g	Aspect ratio of contact patch ($=a/b$)
h	Height
h	Plate thickness (Section 6)
h	Rail joint step height (Section 10.4)
h	Thickness of resilient layer (Section 11.3)
i	Imaginary unit ($\sqrt{-1}$)
k	Wavenumber
k_B	Wavenumber in free beam
k_e	Wavenumber of evanescent wave
k_p	Wavenumber of propagating wave
l	Flow width (Chapter 8)
l_0	Wheel flat length (Section 10.3)
m	Number of nodal circles in wheel mode
m'	Mass per unit length
m''	Mass per unit area
m_n	Modal mass of nth normal mode (Chapter 4)
n	Number of nodal diameters in wheel mode (Chapter 4)
n	Index in modal summation
n	Speed exponent (Chapter 8)
n	Modal density (Chapter 11)
p	Sound pressure
p	Normal pressure in contact zone (Chapter 5)
r	Radial coordinate (in cylindrical polar coordinates, Chapter 4; in spherical polar coordinates, Chapter 6)
r	Roughness amplitude
s	Stiffness per unit length
t	Time

u	Vibration displacement
v	Vibration velocity
w	Numerical distance (Section 6.6)
w	Rail joint gap width (Section 10.4)
x	Coordinate in direction along the track
y	Coordinate perpendicular to the track
z	Coordinate vertically downwards
z_n	Normal specific acoustic impedance (Section 6.6)
z'_n	Non-dimensional normal specific acoustic impedance (Section 6.6)
Γ_i	Normalized creep (Section 5.3)
Δ	Track decay rate
Δ	Relative frequency bandwidth (Section 6.6)
Δf	Modal frequency bandwidth (Chapter 4)
$\Delta\omega$	Modal frequency bandwidth (Section 5.2)
Λ_i	Ratio of bridge to rail displacement amplitude in wave i (Section 11.3)
Ψ	Ratio of rotation to displacement amplitude for beam (Section 3.4)
$\underline{\Psi}_{mn}$	Term in equations for rotating wheel (Section 4.5)
Ω	Rotational speed of wheel
Ω	Rotational speed of engine (Chapter 14)
α	Receptance
α	Correlation parameter in contact filter (Section 5.4)
α	Adhesion coefficient (Section 9)
α	Rail joint dip angle (Section 10.4)
α	Ratio of wavenumber to wavenumber at critical frequency (Section 11.4)
α	Geometric spreading factor (Chapter 13)
β	Wavenumber of static beam deflection (Section 5.7)
β	Non-dimensional factor (Section 9.4)
β	Exponent representing damping losses (Chapter 13)
γ	Propagation coefficient (Section 3.5)
γ_1, γ_2	Creepages (Section 5.3)
δ	Kronecker's delta
δ	Reduction factor for creepages (Section 5.3)
δ	Path length difference (Section 7.5)
∂_t	Differential operator, d/dt (Section 10.2)
ε	Increment in wavenumber pole (Section 3.2, Section 4.4)
ε	Increment in frequency (Section 5.2)
ε_{jk}	Term in equations for rotating wheel (Section 4.5)
ζ_n	Damping ratio of nth normal mode (Section 4)
η	Damping loss factor
η_{ij}	Coupling loss factor (Chapter 11)
θ	Angular coordinate (in cylindrical polar coordinates, Section 4)

θ	Parameter used in describing contact (Section 5.3)
θ	Angular coordinate (in spherical polar coordinates, Section 6)
κ	Timoshenko shear coefficient
λ	Wavelength
λ	Amount that friction coefficient falls at large sliding velocity (Chapter 9)
λ	First Lamé constant (Chapter 12)
μ	Coefficient of friction (Chapter 5, Chapter 9)
μ	Viscosity (Chapter 8)
μ	Second Lamé constant (Chapter 12)
μ_d	Dynamic coefficient of friction
μ_s	Static coefficient of friction
ν	Poisson's ratio
ξ	Coordinate for moving excitation (Section 3.2)
ξ	Non-dimensional parameters used to determine contact patch deflection (Section 5.3)
ρ	Density (of structure)
ρ_0	Density of air
σ	Radiation ratio (Chapter 6, Chapter 11)
σ_1, σ_2	Non-dimensional parameters used to determine contact patch dimensions (Section 5.3)
σ_e	Flow resistivity (Section 6.6)
τ	Ratio of sleeper to rail displacement (Section 3.3)
ϕ	Rotation of Timoshenko beam cross-section (Section 3.4)
ϕ	Azimuthal angular coordinate (in spherical polar coordinates, Section 6)
χ	Ratio of transverse to normal contact stiffness (Section 5.3)
ψ	Angular coordinate (in spherical polar coordinates, Section 6)
ψ	Yaw angle of wheelset (Section 9.5)
ψ_n	Modal amplitude of nth normal mode (Section 4)
ω	Circular frequency
ω_0	Cut-on frequency of beam on elastic foundation (Section 3.2)
ω_n	Natural frequency of nth normal mode (Section 4)
ω_T	Cut-on frequency of higher order wave in Timoshenko beam (Section 3.4)
ω_3	Spin creepage (Chapter 5)

Subscripts

b	Ballast (Chapter 3)
b	Bridge (Chapter 11)
c	Contact (spring)
f	Bridge girder flange (Chapter 11)
p	Rail pad
r	Rail

s	Sleeper
w	Wheel
w	Bridge girder web (Chapter 11)
1,2,3	Directions *x*, *y* and *z*

Abbreviations

BR	British Railways, national railway company until 1996
BEM	Boundary Element Method
BS	British Standards Organization
CAA	Computational Aeroacoustics
CEN	European Standards Organization
CFD	Computational Fluid Dynamics
DI	Directivity Index
DIN	German Standards Organization
DNS	Direct Numerical Simulation (CFD method)
DB	German Railways
DMU	Diesel Multiple Unit
DPRS	Discrete point reacting springs (contact model, Chapter 5)
EMU	Electric Multiple Unit
ERRI	European Rail Research Institute (former research arm of UIC)
ETR500	Italian high speed train
FEM	Finite Element Method
FRF	Frequency Response Function (receptance, mobility or accelerance)
HPF	High Positive Friction (friction modifier)
HST	High Speed Train
ICE	Inter City Express (German high speed train)
ICES-STV	Research project in the Netherlands (Quieter Train Traffic)
IL	Insertion Loss
ISO	International Standards Organization
JR	Japanese Railways
K-blocks	Composite brake blocks
KB	German standard weighting for vibration
LCF	Low Coefficient of Friction (friction modifier)
LES	Large Eddy Simulation (CFD method)
LL-blocks	Composite brake blocks that have same friction characteristics as cast-iron blocks
LVDT	Linear Variable Differential Transformer (displacement transducer)
MONA	French research project on roughness in the 1990s (Chapter 7)
NR	Noise Rating
NS	Netherlands Railways
OFWHAT	Optimized Freight Wheels and Track (research project)

ORE	Organization for Research and Experiment of UIC (later became ERRI)
ppv	Peak particle velocity
PSIL	Preferred Speech Interference Level
PSE	Paris-Sud Est (TGV route)
RANS	Reynolds-Averaged Navier–Stokes (CFD method)
RC	Room Criteria
r.m.s.	Root mean square
RFF	French infrastructure authority
RONA	French research project on wheels in the 1990s (Chapter 7)
RTRI	Railway Technical Research Institute (Japan)
SBB	Swiss Federal Railways
SEL	Sound Exposure Level
SNCF	French National Railways
TBL	Turbulent Boundary Layer
TEL	Transit Exposure Level (Appendix A)
TGV	Train à Grande Vitesse (French high speed train)
TSI	Technical Specifications for Interoperability
TWINS	Track-Wheel Interaction Noise Software
VDV	Vibration Dose Value (see Chapter 12)
VONA	French research project on track in the 1990s (Chapter 7)
UIC	International Union of Railways

Index

Railway Noise and Vibration

Tables in **bold**
Figures in *italic*
Boxes shown with 'B' eg, 234B

MIX
Papier | Fördert
gute Waldnutzung
FSC® C083411

Kolibri 360 GmbH
Ferdinand-Jühlke-Straße 7
99095 Erfurt, Deutschland
produktsicherheit@kolibri360.de